MW00592629

COLLEGE PHYSICS

Preliminary Edition

COLLEGE PHYSICS

DOUG DAVIS

EASTERN ILLINOIS UNIVERSITY

Saunders College Publishing

Saunders Golden Sunburst
Series

Copyright © 1994 by Saunders College Publishing

All rights reserved. No part of this publication may be reproduced or transmitted in any form or by any means, electronically or mechanically, including photocopy, recording, or any information storage and retrieval system, without permission in writing from the publisher.

Requests for permission to make copies of any part of the work should be mailed to: Permissions Department, Harcourt Brace College Publishers, 8th Floor, Orlando, Florida 32887.

Text typeface: Caledonia
Compositor: York Graphic Services, Inc.
Acquisition Editor: John Vondeling
Developmental Editor: Lloyd Black/York Production Services
Managing Editor: Carol Field
Project Editor: Sally Kusch/York Production Services
Copy Editor: Irene Nunes
Manager of Art and Design: Carol Bleistine
Art Director: Christine Schueler
Text Designer: Tracy Baldwin
Layout Artist: Gene Harris
Text Artwork: Rolin Graphics Inc.
Cover Designer: Lawrence R. Didona
Manager of Photos and Permissions: Dena Digilio-Betz
Marketing Manager: Margie Waldron
Director of EDP: Tim Frelick
Production Manager: Charlene C. Squibb
Cover credit: SkyBridge, British Columbia, Canada/BC Transit Photo

ISBN: 0-03-007488-6
Printed in the United States of America
Library of Congress Catalog Card
Number: 94-26395

To

AMY and EMILY,
the joys of my life

and

to
JUDY
who *is* my life

T his book is intended for a one-year course in introductory physics usually taken by students in the life sciences—including pre-medicine, pre-optometry, pre-veterinary, and associated health fields—as well as other fields of science and technology or by anyone wanting a full and complete exposure to physics. The mathematical background assumed throughout the book includes algebra and trigonometry; no calculus is used.

My main purpose in writing *College Physics* was to provide students with a clear, easy-to-understand explanation of the ideas of physics, which includes almost everything around us. Careful presentation of concepts is a central theme of this text. If students fail to understand the basic ideas of physics, then the development of problem-solving skills becomes merely a mechanical exercise. The physicist strives to understand concepts and processes, and students need to know that the exploration of physics is not memorizing lists of equations.

This book presents the standard topics usually covered in a two-semester algebra-trigonometry-based introductory physics course:

- Mechanics (Chapters 1–11)
- Thermodynamics (Chapters 12–14)
- Vibrations and Waves (Chapters 15–16)
- Electricity and Magnetism (Chapters 17–22)
- Optics (Chapters 23–26)
- Modern Physics (Chapters 27–31)

Features

To provide students the utmost in clarity and understanding, a unique blend of pedagogical features has been created. It is my intention that this book be useful, effective, and efficient for the student who is learning introductory physics. Of particular note are the following features:

Physics At Work. Each chapter begins with a mini-essay and accompanying photograph that describes a real-world application of physics ideas and concepts. Students immediately get to see "physics at work" in the world they live in. They learn that physics is much more than some arcane processes carried out in a sterile laboratory. The principles of physics can be found everywhere around us.

Key Issues. The main ideas or concepts of a chapter are listed at the beginning of each chapter. This list of ideas provides a concise preview of what will be discussed in the chapter. With these main concepts to work from, students can better fit together all the details that follow.

Concept Overview. Each major section within a chapter is preceded by an overview of the main points discussed in that section. Students often become so immersed in the details of an explanation or derivation that they lose sight of the important concepts. By telling students what these main ideas are *before* they read a section, they will not have to worry about missing an important point. Also, these key concept "signposts" provide a convenient tool for review.

Problem-Solving Tips. Many of the worked examples are preceded by "Problem-Solving Tips" that give students a heads-up concerning the way to approach a problem or the things to look for.

Worked Examples. Numerous worked examples help students make the connections between the "big ideas" of physics, their application, and the techniques of problem solving. Each example contains a statement of the problem, a discussion of the reasoning involved, and the worked-out solution.

Practice Exercises. Numerous worked examples are immediately followed by the statement of a slightly altered problem that will give students a chance to apply what they have just learned. The answer to each practice exercise is given directly following the exercise statement so students can receive immediate feedback.

Marginal Notes. Throughout the text relevant applications, important ideas, or interesting anecdotes about history's great physicists are presented in marginal notes. The notes are identified by one of two icons. A *satellite dish* icon

denotes a technology-based application. A *tree* icon identifies a life science-based application.

Illustrations. The range of physics is enormous—from the subatomic to the cosmic. Yet, much of physics deals with phenomena that we humans can readily see with our eyes. Thus the study of physics has become an increasingly visual endeavor. This text strives to show students the beauty and fascination of physics through nearly 400 color photographs and 1,000 line drawings. The line drawings follow a color convention (for example, force vectors are always blue and velocity vectors are always red) that helps students keep concepts such as force and velocity in their correct context no matter what the topic area. See the Pedagogical Use of Color chart on the endpaper for more details.

Important Equations. Important equations are highlighted to remind students of their importance and to make it easy for students to find them later as they review.

Units. The international system of units (SI) is consistently used throughout. British engineering units (common units) are used only occasionally and for comparison, as in discussing the relationship between kilowatts and horsepower. Care has been taken to adhere to accepted conventions for the use of significant figures.

Summary. The main ideas of each chapter are summarized at the end of the chapter. This summary is intended to remind students of what they have learned in the chapter and to pull together the important ideas. Of course, it is an excellent tool for review.

Discussion Questions. Each chapter contains about 12 to 15 discussion questions that require careful thought and explanation. Most of these questions address concepts and are useful for classroom discussion, small group discussion, or individual reflection.

Problems. A very large number of numerical problems complete each chapter. These problems are graded in difficulty and arranged by section. Easy problems are designated by no symbol, intermediate problems by the ○ symbol, and challenging problems by the ● symbol. A final set of General Problems not tied to a single specific section close each chapter. Multiple data set problems (signified by the ❖ icon) may be solved by hand but also lend themselves nicely to solution by spreadsheet or computer programming.

As you can see from the brief table of contents, this textbook is structured to begin with mechanics and continue naturally with thermodynamics. It continues with vibrations and waves, electricity and magnetism, optics, and ends with modern physics—special relativity and an introduction to quantum physics. Each section is fairly independent, and the order of the sections may be rearranged fairly easily. For example, optics could easily follow directly after vibrations and waves if an instructor is willing to omit or explain later the electromagnetic nature of light. The chapter on special relativity could certainly follow directly after the section on mechanics. Additionally, "starred sections," marked with an asterisk (°), may be omitted without any loss of continuity. Topics within sections may also be rearranged to suit an instructor's preference.

Chapter 6 covers work, power, and energy, and Chapter 7 covers momentum and collisions; i.e., energy is covered before momentum. These topics could be covered in the reverse order, although the discussion in Chapter 7 distinguishing between elastic and inelastic collisions then should be covered after the remaining parts of Chapter 7 and all of Chapter 6 are covered. I have tried to write the book so that each instructor can customize to a degree the order in which he or she presents the material.

Ancillaries

Available to all adopters of *College Physics* are the following ancillaries:

Instructor's Manual with Complete Solutions. This manual contains answers and worked-out solutions to all questions and problems in the text, chapter outlines, and a videodisc barcode guide. Free to adopters.

Study Guide. This guide enhances students' learning with selected solutions, extensive problem-solving strategies, chapter summaries, and a videodisc/videotape demonstration guide and question set. For sale to students.

Overhead Transparency Acetates. This set contains 150 color overhead transparency acetates of conceptually based artwork from the textbook. Free to adopters.

Physics Problems Set for Interactive Physics™ II. This set contains 50 examples and problems taken from the text and offered in the Interactive Physics format. Versions are available for Macintosh® and IBM/Windows®. Free to adopters.

Discovery Exercises for Interactive Physics. This workbook encourages the use of the Physics Problems Set for Interactive Physics by offering directed exercises and tips for preparing simulations. Versions will be offered without disks, with Macintosh disks, or with IBM/Windows disks. For sale to students.

Saunders Physics Videodisc, Version 2. Over 500 still images from Saunders physics titles plus 70 physics demonstrations by J. C. Sprott. The videodisc is accompanied by a barcode manual, LectureActive software for both IBM/Windows and Macintosh, and the LectureActive User's Guide. Free to qualified adopters.

LectureActive™ Interactive Videodisc & Lecture Presentation Software. This distinctive software (available in IBM/Windows and Macintosh formats) helps instructors customize lectures by giving them quick, efficient access to the video clip and still frame data on the Saunders Physics Videodisc, Version 2. All video clip and still frame data from the videodisc are entered and listed on the software, so it is easy to customize lectures swiftly and simply by calling up the desired footage or images. Free to qualified adopters.

Physics Demonstration Videotape. This two-hour videocassette contains 70 physics demonstrations by J. C. Sprott and is accompanied by an Instructor's Manual. Free to adopters.

NOVA Videotapes. Adopters can choose from a selection of NOVA videotapes. Free to qualified adopters; see your sales representative for details.

Printed Test Bank. This bank contains approximately 1,700 conceptual questions and problems in multiple choice and other formats. Free to adopters.

ExaMaster+™ Computerized Test Bank for IBM and Macintosh. All the questions from the printed test bank are presented in a computerized format that allows instructors to edit, add questions, and print assorted versions of the same test. Also includes ExamRecord gradebook software. Free to adopters.

RequesTest™. Instructors without access to a personal computer may contact Saunders Software Support Department at (800)447-9457 to request tests prepared from the computerized test bank. The test will be mailed or faxed to the instructor within 48 hours.

For information about these ancillaries contact your local Saunders sales representative or call Harcourt Brace College Publishers at (800)237-2665 or (708)647-8822. In Canada call (416)255-4491.

Acknowledgments

Where do I begin? From my own reading of Prefaces to other introductory physics texts, I long ago realized that it was standard to thank one's wife and children for their patience and support. Little did I realize how much patience and support, endurance, and frustration are required of (or extracted from) the spouse and children of a textbook author. Others may be "golf widows" or "football orphans," but my wife Judy has been a "computer widow" and my daughters Amy and Emily have been "textbook orphans" while I have hidden myself away for countless hours and more years than I care to admit, working on this text. I do appreciate their endurance, their support, and their love. I hope they like the result.

Books and learning are special to me. Special thanks go to my parents, Bruce and Lillian Davis, for reading many, many books to me—especially about Robin Hood, King Arthur, and *White Prince* (a circus horse)—and for their love and encouragement throughout the years. Professors Henry Unruh, Gerald Loper, and Henry Pronko at Wichita State University and Professor Steven Moszkowski at UCLA have had special impact on me; I greatly appreciate and gratefully acknowledge their help.

Whom should I thank next? I appreciate the support and encouragement of the Thursday evening Men's Bible Study of Charleston Alliance Church, who have heard more about this textbook than any of them probably ever wanted to know. While this textbook carries my name on its cover (and that makes me feel very good indeed), I also realize that it is the product of a great deal of work by a great many people. John Vondeling, Lloyd Black, Sally Kusch, Christine Schueler, Charlene Squibb, and Marjorie Waldron at Saunders College Publishing, Kirsten Kauffman at York Production Services, and Jeff Holtmeier (my first editor at Harcourt Brace) are the people I have talked and labored with on this project over these many years. This project bears their imprint as well as mine. Thanks go to Lloyd Black, Lori Eby, and Dena Digilio-Betz of Saunders for their photo research and for the use of a number of Lloyd's photographs taken expressly for this book. The artists at Rolin Graphics did a good job in rendering the line drawings. Thanks, too, go to Irene Nunes for her very close reading and editing of the final manuscript and her many, many contributions.

I wish to thank the reviewers of this text for their careful attention and many good suggestions at all stages of this project.

Zaven Altounian, *McGill University*
Ronald Baltzer, *Glendale Community College*
Paul A. Bender, *Washington State University*
Carl Bromberg, *Michigan State University*
Victor Cook, *University of Washington*
George Dixon, *Oklahoma State University*
Miles J. Dresser, *Washington State University*
Robert J. Endorf, *University of Cincinnati*
Nathan Folland, *Kansas State University*
James Garner, *Bradley University*
Tom Hudson, *University of Houston, University Park*
Mario Iona, *University of Denver* (emeritus)
Kirby W. Kemper, *Florida State University*

Anthony W. Key, *University of Toronto*
Peter Landry, *McGill University*
David Markowitz, *University of Connecticut, Storrs*
Laurence C. McIntyre, *University of Arizona*
Ed Oberhoffer, *University of North Carolina, Charlotte*
Richard E. Pontinen, *Hamline University*
James E. Purcell, *Georgia State University*
Michael Ram, *S.U.N.Y. at Buffalo*
Marilyn V. Rands, *Lawrence Technological University*
Jay D. Strieb, *Villanova University*
Michael A. Simon, *Housatonic Community College*
James E. Stewart, *Western Washington University*

I expressly want to thank Mario Iona and David Markowitz for their painstakingly thorough reviews that helped shape the final draft manuscript. Others who labored in support of the book include accuracy reviewers Diane Costa and Matthew Bierer, both of the University of Alaska–Fairbanks, and Charles E. Miller, Jr., of Eastern Illinois University. Hilda N. Mireles of the University of Texas, Pan American, prepared the Instructor's Manual with Complete Solutions as well as the Test Bank. I appreciate their efforts to ensure quality and accuracy in this book.

Doug Davis

Charleston, Illinois

Welcome to physics! "Physics Is Fun!" has become a byword or proverb of many physics professors. And yet sometimes students still shudder a little when they hear the word *physics* or realize that "a full year of physics is required" in their major. How can this be?

Physics is the most basic or fundamental of the sciences. The concepts, principles, ideas, and techniques you will learn in this course will be helpful in your further endeavors—whether in the life sciences, geology, technology, nursing, or just for general interest. This course should *help* you proceed and succeed in your chosen career. Please do *not* view it as an impediment or an obstacle.

Please read the preceding Preface, which describes the particular features of this book. Being aware of these features will help you use them more effectively. The textbook for any course should be a primary source for *learning* the material—not a last-minute reference book.

How should you study for this physics course? Of course, effective study habits vary from individual to individual. But there are some basic ideas that should prove helpful. As a general rule, for most of your classes you should expect to spend at least two hours of reading, studying, and problem solving for every one hour in class. Perhaps, a bit more time may be necessary for physics. A typical physics class meets three hours a week for lecture and two hours a week for laboratory. You should therefore expect to spend at least eight to ten hours per week outside of class working on physics.

Read the material in the textbook *before* listening to the lectures. Ask questions of your instructor during or after class. Go back and review the material after hearing it explained in the lecture. Reading the material several times will be more effective than trying to absorb everything during a single pass through the chapter. Think of this process as a quick preview, actual reading, and one or more quick reviews. Use the Key Issues, Concept Overviews, marginal notes, highlighted equations, and chapter Summary to help you. These items will provide systematic guidance through the chapter.

Physics is like scuba diving. You cannot learn it by watching someone else do it. You have to get wet—and messy—doing it yourself! That means struggling with the homework problems. "I understand the material but I just can't work the problems" is an oxymoron. I enjoy watching Jacques Cousteau, but I am no scuba diver. I enjoy reading about Amelia Earhart, but I am no pilot. The objective of this text—and of your college physics course—is to teach you "real physics" and not just to tell you a little bit "about physics."

Only as you work physics problems do you really understand all the implications of the ideas and the interrelationships between them. "Working problems" includes struggling with approaches that do not work. Knowing how *not* to approach a problem in the future is important. Reviewing the worked examples in the text is an important part of the process. Take note of the Problem-Solving Tips that precede some of the examples. These tips can help you select the correct approach. Take time to study the reasoning part of each example's solution. Mere memorization of a particular solution's steps will not help you solve more involved problems. There is no substitute for working through the solution on your own or with a small group. Use the Practice Exercises that follow many of the worked examples to see if you understand the process and reasoning. Working these can be confidence boosters.

How should you begin to solve a physics problem? First, make *sure* you understand what the problem is asking. Read the problem carefully. Do not start looking for equations. Read the problem again. Rewrite it in your own words if that will make its meaning clearer. Take time to make a clear diagram. Make the diagram large and label everything that is known on it. Then label what is being asked for. Think about the diagram. What does it show? What physics ideas or principles come into play in the diagram? Use this visual or graphic approach to begin the problem-solving process.

Only now you are ready to think about equations. What equations summarize the concepts or principles dealt with by the problem and your diagram? Write these down. Re-write them with the given values from the problem and with the unknown(s) that you seek. Can you solve for the unknowns? Two unknowns will require two equations. Break complex problems into a series of smaller problems.

Once you have a numerical solution, ask if it is reasonable. Is it what you might expect? Is it believable? If the acceleration of an automobile is calculated to be $1\,000$ m/s^2, something is wrong. If the current to a kitchen toaster is calculated to be 923 A, an error was made. If the focal length of the eyepiece to a microscope comes out to be 1.23 km, you *know* that cannot be. Units are important; never use numbers alone without proper units. By keeping track of the units as you multiply, divide, or take the square root of the associated numbers, you will be more likely to spot errors. Answers that are far out of the realm of possibility should cause alarm bells to ring. Developing a sensibility about the reasonableness of an answer is an attribute of a good problem solver.

Though you need to gain confidence working problems on your own, it is a good idea at times to work collaboratively with some classmates. Particularly challenging problems may require a group effort to obtain the solutions. Sharing of personal problem-solving techniques can help make all members of the group better problem solvers. In some courses teaching assistants are available to help with the homework problems. The Study Guide and the use of software such as Interactive Physics™ are also good sources of help as you tackle the problems.

Physics laboratory experiments are an integral part of most college physics courses. These labs help to build the connections between theory and practice, between ideas and the real world, between equations and actual situations. As with all of physics, concentrate on *understanding*. Always pause and ask "What have I done?" or "What does this mean?" or "What ideas and principles have I used or demonstrated?" Never be satisfied with "We got the right answer!"

I wish you an enjoyable year studying physics, and I hope this textbook makes that study easier, and, yes, even exciting. Again, welcome to physics!

CONTENTS OVERVIEW

CONTENTS

Knowledge and appreciation of the principles of physics are necessary for successful space travel. Here the space shuttle launches the *Magellan* spacecraft on its way to map the surface of Venus. The laws of motion are used to launch the shuttle into Earth orbit, move *Magellan* out of the cargo bay while the craft is in a state of weightlessness, and send it on the correct orbital path to a rendezvous with Venus. While in orbit around Venus, the spacecraft *Magellan* uses a special radar to obtain high-resolution images of the cloud-shrouded surface. While looking at this picture, can you list some of the physics principles at work?

The Magellan spacecraft, as it was launched from a space shuttle in 1989 (NASA)

Introduction

When you can measure what you are speaking about and express it in numbers, you know something about it; but when you cannot express it in numbers, your knowledge is of a meager and unsatisfactory kind.

—Lord Kelvin

Key Issues

- Physics, once known as natural philosophy, seeks to understand and explain the world around us. Starting with mechanics, we will try to explain the motion of objects. Next, we will try to understand electricity and magnetism, which greatly influence our lives—from lightning bolts to ordinary room lights, television sets and computers, batteries and starters for our cars, and electric motors around our homes. In optics, we will learn how cameras work, why some of us wear eyeglasses, and how images are projected on a movie screen. In modern physics, we will investigate the unusual behavior of things we have never observed before—from objects traveling near the speed of light to single atoms. Physics is everywhere.

- Carefully defined units are required to pass information and measurements from one group of people to another. The International System (SI) of units is used by scientists worldwide.

- Every measurement includes some uncertainty, so that only some of the figures of a measurement or calculation are significant. Also, calculators and computers may give a result with 14 digits when only two or three are meaningful.

- Scientific notation—using powers of ten—lets us express both very small and very large numbers in a convenient manner with less risk of error.

In one sentence, physics is understanding the world around us. Why are the brakes on a Mack truck so heavy? How do bats detect bugs? How do dolphins detect fish? How can the space shuttle stay in orbit with its rocket engines turned off? How can a communications satellite appear to remain fixed and unmoving as it relays television signals? Why will an orchestra sound different in different concert halls? How can you make your house cool on a hot August day? Why is the sky blue? How do you see colors on a color television set? With some strings of Christmas tree lights, the entire string goes out if one bulb burns out; with other strings this is not the case—why the difference? What happens as an electronic flash unit charges? Why can an electric motor overheat if it stalls? Why do some of us need glasses? How can the shiny rainbow patterns on a compact disc produce such beautiful music? These are all questions for physics.

1.1 What Is Physics?

CONCEPT OVERVIEW

- Physics, once known as natural philosophy, deals with matter and energy and with the structure of matter—in short, with everything in the physical world around us.

- Physical theories are accepted or rejected as their predictions agree or fail to agree with measurements of real events.

Physics deals with matter and energy, with the structure of matter, and with the interaction of matter and energy. In the Universities of Scotland, physics was known as natural philosophy until only a few years ago. The *key* idea in physics is *understanding*. A cursory view of physics may leave the impression that the most important part of physics is numbers or equations, but that is wrong! Physics depends heavily upon measurements and experiments. Measuring and experimenting are how we obtain information about the world, and these procedures require numbers. The ideas we discuss in physics are often best handled through equations, but the equations are just a form of shorthand, a summary of the ideas. The equations must never become a substitute for the ideas!

Physics always deals with the real world—with things that can be measured. Observations—preferably careful measurements—are made of how something behaves. Then a theory may be created to explain that behavior and predict the value of a measurement, but a theory must always be validated by further experimentation. A theory is validated by how well it *predicts* the outcome of an experiment. A theory may be elegant and may sound reasonable. It may be proposed by very prominent scientists. But if its predictions disagree with the results of experiments—with the measurements that are taken—then the theory must be modified. Physics is an experimental science.

Galileo may well be considered the first physicist—or the first modern scientist. In the fourth century B.C., the Greek philosopher Aristotle had stated that heavy bodies always fall faster than light ones. This statement seemed reasonable, for bricks almost always fall faster than feathers on Earth, but it was accepted just because Aristotle had stated it. It was seldom questioned—until Galileo in the 16th century. Galileo experimented both in his mind with "thought experiments" and in the real world with real objects. Legend has it

Galileo Galilei (1564–1642) established the foundations of mechanics—indeed, of modern science. *(Tintoretto, National Maritime Museum)*

Figure 1.1
Legend has it that Galileo dropped objects of various sizes and masses from the Leaning Tower of Pisa.
(David Morrison)

that he dropped objects of various sizes and masses from the bell tower of Pisa (which was already leaning, even in Galileo's time), shown in Figure 1.1.

In the 17th century, Johannes Kepler expressed his three laws of planetary motion, which were based on a lifetime of observations of the planets by Tycho Brahe. Kepler's laws made predictions that were verified by data from astronomical measurements. Isaac Newton's Law of Universal Gravitation was accepted because it agreed with Kepler's laws.

Because measurements are so important in physics, we need to discuss how they are made and what they mean.

1.2 Units and Standards

CONCEPT OVERVIEW

■ The standards and units used in measurements are important for the easy and ready exchange of data as well as commerce.

For measuring almost anything, we must give a number—how many?—and a unit—of what?—to explain the measurement. Both the number and the unit are important. In some everyday situations, the unit may be so obvious that it is unstated. You might go into a donut shop and say, "I'll take one dozen." In discussing the acceleration of a sports sedan, you might say, "I took it from zero to 60 in seven and a half seconds." On a hot August day, you might complain, "It's over a hundred today." In each case the meaning of the statement is clear; it really cannot be misunderstood. In physics, however, things often can be misunderstood. So it is very important to include the unit that tells what you are talking about—"I'll take one dozen *donuts*" or "from zero to 60 *miles per*

Isaac Newton (1642–1727) influenced mathematics and physics as few, if any, others have. He developed calculus, formulated the laws of motion and the law of universal gravitation, invented the reflecting telescope, and made additional great advances in optics. *(National Portrait Gallery, London)*

Figure 1.2
Measuring the length of your car provides an example of an operational definition.

With the rise of the great civilizations of Egypt and Mesopotamia a widely accepted system of units became necessary for levying taxes, managing commerce, and settling land disputes.

hour" or "it's over a hundred *degrees Fahrenheit* today." Units aren't extras in physics; they are essential.

We shall be able to study physics with a surprisingly small number of basic units. Length, mass, and time will be needed first. What are these, anyway? Or, more important, how can we measure length, mass, and time? Physics is based upon *operational definitions,* which define quantities by specifying the procedures used to measure them. This means if something is physics, we can define an operation or rule or procedure to measure it. If we cannot measure it, it is not physics. How do you measure love, truth, beauty, anger, or fear? We have no operational definitions to measure them. They are certainly real; they exist. They are important, but they are not physics.

What is length? We might say it is the distance between two points, but then we have to define distance. That is a circular definition that explains very little. How would you measure the length of, say, your car? The easiest way is to take a steel tape measure, hold one end at the front bumper, and read the length of the car by looking at the numbers on the tape that are beside the rear bumper. An equivalent way is to lay a ruler on the road with one end even with the front bumper. Then (carefully) mark the position of the ruler and repeatedly move it end over end until you come to the back bumper. Note how many times the ruler can be placed end to end between the front and rear bumpers. That is the length of your car. Both these procedures are sketched in Figure 1.2.

In either case we define the length in terms of some previously agreed upon *units*—the units marked on the tape or those marked on the ruler. How can we decide which units to use or how can we call one unit better than another? Almost any unit can be used—as long as we all agree to it. Suppose, for instance, you decide to measure the car length in *cubits.* A cubit is the length of your forearm from your elbow to the end of your fingers. How many cubits would you measure for the length? How many cubits would your mechanic measure for the same length? The problem is obvious: because everybody uses a different length for her or his cubit, everyone measures a different number of cubits for the length of your car. We need a *standard.*

1.3 Units of Length

CONCEPT OVERVIEW

- The basic unit of length in the SI is the meter.

- As with all other basic SI units, the meter can be used in smaller or larger units that are multiples of ten of a meter: 1000 m = 1 km; 1 m = 100 cm; 1 m = 1000 mm.

While it is fun to look back at the standards that have been used throughout history, as sketched in Figure 1.3, let us concentrate on standards used today. Tape measures and rulers are useful because we know they are going to give the same readings whether they were manufactured in Toronto or Tucumcari. This is because they are made to specifications that ultimately compare them to a length standard agreed upon by the National Institute of Standards and Technology (formerly the National Bureau of Standards) in Gaithersburg, Md., (near Washington, DC) and Boulder, CO, or a similar organization in some

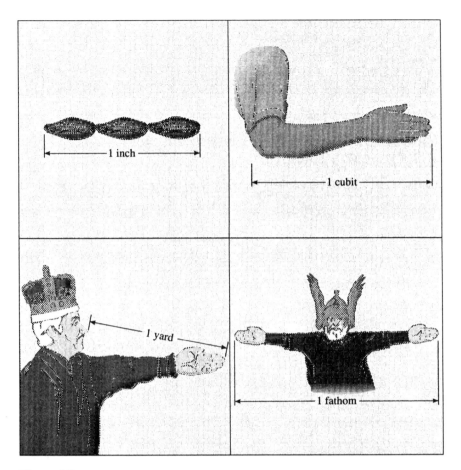

Figure 1.3
Throughout history, an interesting variety has been used in establishing standards for measurements. Standards are important for waves and commerce.

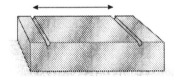

Figure 1.4
A working length standard may be a carefully machined block or bar of metal or a bar of metal with two lines carefully etched upon it.

other country, such as the *Bureau International des Poids et Mesures* (the International Bureau of Weights and Measures) in Sèvres (near Paris), France. A working length standard may be a carefully machined block or bar of metal or a bar of metal with two lines carefully etched upon it. An example is shown in Figure 1.4. We shall discuss a more fundamental length standard momentarily.

Now that we know how to make a length standard, what size shall we make it and what shall we call it? Lengths in inches, feet, yards, and miles, all units of what is called the British system of measurement even though it is no longer officially used in Great Britain, are still in common use in the United States. Essentially all the rest of the world uses the metric system (more properly known as the International System, *Système International,* or SI) with lengths based upon a standard *meter* (m). Undoubtedly the United States will eventually change to SI, too, but change often comes slowly. It seems, therefore, that in the U.S. both systems will be in use alongside each other for the immediate

The growing importance of today's global economy has caused many manufacturing companies in the United States to convert to metric-based designs.

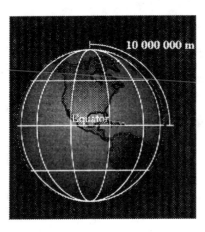

Figure 1.5
Originally, the meter was intended to
be 1/10 000 000 of the distance from
Earth's equator to the North Pole.

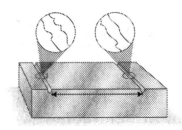

Figure 1.6
The thickness of and irregularities in
the lines etched in the standard
meter present a problem.

More precise measuring standards are
needed in this age of microcircuits
and nanotechnology.

future. So you may need to handle both. However, this book will use SI units,
for they are the common language of science—of medicine, engineering, technology, geology, biology, chemistry, and physics.

The meter was originally intended to be one ten-millionth of the distance
from Earth's equator to its pole, as sketched in Figure 1.5. The standard meter,
though, was the length of a platinum-iridium bar kept at Sèvres, France. However, the exactness of this—or any other—physical, metal-bar standard is limited by the coarseness of the lines scratched to define its ends. Even the temperature of the bar affects the accuracy. See Figure 1.6 for a sketch of a
microscopic view of a line scratched on the metal. How do we measure the
standard length? From the center of the scratch? Where is that? From the
inside—or outside—edge of the scratch? How do we determine that?

In 1960 a new standard of length was adopted that avoids this difficulty. An
excited krypton-86 atom emits orange light of a definite wavelength. This light
is accurately reproducible; its wavelength is the same in any laboratory. The
meter was redefined as 1 650 763.73 wavelengths of the light emitted by
krypton-86. In 1983 an even newer standard was adopted to define the meter; a
meter is now defined as the *distance traveled by light in a vacuum during a
time of 1/299 792 458 second(s)*. (This effectively defines the speed of light in
vacuum as *exactly* 299 792 458 m/s.) Length standards, such as those in Figure
1.7, are ultimately compared to this new standard.

In SI units, all other length units are related to the meter by powers of ten;
these multiples are given in Table 1.1. These prefixes are used for all SI units,
not just for length. That is, a nanometer (nm) is 10^{-9} m and a nanosecond (ns)
is 10^{-9} s. A millimeter (mm) is 10^{-3} m and a milliliter (ml) is 10^{-3} liter (l).

Of course, the British system and the SI can both be used to measure the
same object, so there must be some relationship between units in the two

TABLE 1.1 Multiples and submultiples of the meter	
Kilometer, km	$1\ \text{km} = 10^3\ \text{m} = 1000\ \text{m}$
Meter, m	
Decimeter, dm	$1\ \text{dm} = 10^{-1}\ \text{m} = 0.1\ \text{m}$
Centimeter, cm	$1\ \text{cm} = 10^{-2}\ \text{m} = 0.01\ \text{m}$
Millimeter, mm	$1\ \text{mm} = 10^{-3}\ \text{m} = 0.001\ \text{m}$
Micrometer, μm	$1\ \mu\text{m} = 10^{-6}\ \text{m} = 0.000\ 001\ \text{m}$
Nanometer, nm	$1\ \text{nm} = 10^{-9}\ \text{m} = 0.000\ 000\ 001\ \text{m}$
Picometer, pm	$1\ \text{pm} = 10^{-12}\ \text{m} = 0.000\ 000\ 000\ 001\ \text{m}$
Femtometer, fm	$1\ \text{fm} = 10^{-15}\ \text{m} = 0.000\ 000\ 000\ 000\ 001\ \text{m}$

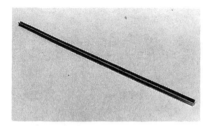

Figure 1.7
Length standards like the Standard Meter Bar, constructed of a platinum-iridium alloy, are ultimately compared to the standard based on the distance light travels in a vacuum in a given time. *(National Institute of Standards and Technology, U.S. Department of Commerce)*

systems. The inch is *defined* to be *exactly* 2.54 cm. That definition is enough to produce the British–SI relationships given in Table 1.2.

To illustrate how to convert from one unit to another, let us determine the number of feet in, say, 7.4 yd. The answer is so easily arrived at that you may readily calculate it in your head (22 ft). Do not do that; write out the full details of the solution this first time. **Doing so allows this easy problem to be used as a prototype for the more difficult problems that will surely follow.** Start with a length, $L = 7.4$ yd. If this number is multiplied by unity, it is not changed. If we wanted to, for instance, we could multiply it by 4/4 to write it as

$$L = 7.4\ \text{yd} \times \frac{4}{4} = \frac{29.6}{4}\ \text{yd}$$

The quantity (29.6/4) yd is not very useful, but it is the same thing as the 7.4 yd we started with. Now let us do the same sort of thing but in a useful manner. Because 3 ft is the same thing as 1 yd, we can multiply our length by 3 ft/1 yd without changing that length. That is, we can write

$$L = 7.4\ \text{yd} \times \frac{3\ \text{ft}}{1\ \text{yd}} = 22\ \text{ft}$$

Notice that multiplying by the fraction 3 ft/1 yd, which equals unity, provides yards on the bottom to cancel the yards on the top, leaving feet, which are the units we seek for our answer.

TABLE 1.2 Relationships between common British and SI units	
1 in. = 2.54 cm[a]	1 cm = 0.3937 in.
1 ft = 0.3048 m	1 m = 3.281 ft
1 yd = 0.9144 m	1 m = 1.094 yd
1 mi = 1.609 km	1 km = 0.6214 mi

[a]One inch is *defined* as *exactly* 2.54 cm. That means 1 ft = 0.3048 m and 1 yd = 0.9144 m, exactly; the other conversions have been rounded to four significant figures.

Problem-Solving Tip: Units Conversion

- For *all* conversions, treat the units as ordinary factors that can be multiplied and divided. Begin with the initial quantity in its original units. Write the conversion factor (for example, 365.25 days = 1 year) as a fraction that is equal to unity, (365.25 days/1 year) or (1 year/365.25 days). Choose that form of the conversion factor that will allow the initial units to be canceled by the fraction and then multiply. If the new unit is smaller than the initial one, you should expect to obtain a numerical value larger than the original.

Surveying instruments using lasers permit much greater precision in measuring distances than was previously available. (© *David R. Frazier Photolibrary*)

If the new unit is larger than the initial one, you should expect to obtain a numerical value smaller than the original. This conversion process may have to be carried out more than once.

Example 1.1

How many inches are there in 1 m?

Reasoning

We know there are 2.540 cm in 1 in., but are we to multiply something by 2.540 or divide something by 2.540? Remembering when to multiply and when to divide can become awkward as more and more possibilities come along. However, multiplying by a conversion factor that is equal to unity will take care of this decision automatically. We are starting with meters, and so a conversion factor with meters on the bottom to cancel the initial unit of meters is called for. Eventually, we want a result with inches on the top.

Solution

Start with a length of 1 m and multiply by conversion factors that are equal to unity. First, convert the initial quantity, 1 m, to centimeters by writing

$$1 \text{ m} = 1 \text{ m} \left(\frac{100 \text{ cm}}{1 \text{ m}} \right) = 100 \text{ cm}$$

Continue the conversion of 1 m to inches. Multiply by the factor for converting from centimeters to inches written as a fraction equal to unity, $\left(\frac{1 \text{ in.}}{2.540 \text{ cm}} \right)$:

$$1 \text{ m} = 1 \text{ m} \left(\frac{100 \text{ cm}}{1 \text{ m}} \right) \left(\frac{1 \text{ in.}}{2.540 \text{ cm}} \right) = 39.37 \text{ in.}$$

Notice, again, that the units have taken care of any questions of whether to multiply or to divide. The centimeters on the bottom cancel the centimeters on the top; only inches remain, and so the result is

$$1 \text{ m} = 39.37 \text{ in.}$$

Example 1.2

How many kilometers are there in 1 mile (mi)?

Reasoning

As before, we shall multiply by conversion factors that are equal to unity and have units on top and bottom to cancel the initial unit of miles and eventually provide the unit of kilometers.

Solution

Using only the conversion factor for the relationship 1 in. = 2.54 cm, we can carry out this conversion by changing 1 mile to inches, then inches to centimeters, and then centimeters to kilometers:

$$1 \text{ mi} = 1 \text{ mi}\left(\frac{5280 \text{ ft}}{1 \text{ mi}}\right)\left(\frac{12 \text{ in.}}{1 \text{ ft}}\right)\left(\frac{2.54 \text{ cm}}{1 \text{ in.}}\right)\left(\frac{1 \text{ m}}{100 \text{ cm}}\right)\left(\frac{1 \text{ km}}{1000 \text{ m}}\right)$$

$$1 \text{ mi} = 1.61 \text{ km}$$

Practice Exercise   How many *feet* are there in a kilometer?

Answer   3281 ft = 1 km

1.4 Units of Time

CONCEPT OVERVIEW

- The basic unit of time in the SI is the second.

- Units of time smaller than the second use SI prefixes—for example, 1 millisecond = 1 ms = 10^{-3} s; 1 microsecond = 1 μs = 10^{-6} s; 1 nanosecond = 1 ns = 10^{-9} s.

- Units of time larger than the second use common units, such as minutes, hours, days, and years.

What is time? The best answer to that question is a description of how to measure it. Time is what a clock measures. Anything that repeats over and over the same way can be used to measure time. A cup with a small hole that allows water to drip out regularly could be used as a water clock. A pendulum—whether a chandelier in a cathedral, a child on a playground swing, or a more common one in a grandfather clock—marks the passage of time. Examples are shown in Figure 1.8.

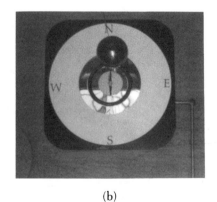

(a)   (b)   (c)

Figure 1.8

(a) An example of "Galileo's lamp." Galileo supposedly observed a lamp at the end of a long, thin rod that was gently moving back and forth. Measuring this motion with his pulse (the most accurate timepiece he had), Galileo determined that the period was independent of the size of the amplitude. What happens if the length of the swinging lamp, which we call a pendulum, is lengthened or shortened? *(Comstock, Inc.)* (b) A Foucault pendulum proves that the Earth rotates on its axis in 24 hours. *(© Lloyd W. Black)* (c) A stroboscopic photograph of a clock pendulum, showing its motion through a period. *(© Robert Mathena/Fundamental Photographs, NY)*

The desire to make clocks that are ever more reliable and accurate was a response to the demands of commerce. Now scientists construct ultra-accurate devices for measuring time so that they can, for example, test Einstein's theory of general relativity.

Again, we must ask about a standard. Because astronomers make amazingly precise measurements, until recently the time standard was based upon astronomical measurements of the time required for Earth to complete one orbit around our Sun, as sketched in Figure 1.9. With the astronomers' accuracy, it was found that there is some variation in even that! We now use a time standard based upon the vibration of a cesium-133 atom. One **second,** the basic unit of time both in the SI and in the British system, is *defined* as the time required for 9 192 631 770 vibrations of a cesium-133 atom. An atomic clock is shown in Figure 1.10.

The other units of time—minute, hour, day, year—are also the same in the SI and the British system. All the metric prefixes we discussed earlier concerning length can be used with time as well (a millisecond is 10^{-3} s and a nanosecond is 10^{-9} s). However, because it is easier to think in terms of minutes, hours, days, or years rather than kiloseconds, megaseconds, or gigaseconds, such SI units of time are rarely used.

Figure 1.9
Until fairly recently, a standard year—and thus a standard second—was defined in terms of Earth's orbital motion around our Sun.

Figure 1.10
NIST-7, the world's most accurate clock, is the seventh generation of atomic clocks at the National Institute of Standards and Technology. It was unveiled in April 1993 and initially kept time to an accuracy of one second in one million years. After becoming fully operational in early 1994, it attained an accuracy of one second in three million years. *(National Institute of Standards and Technology, U.S. Department of Commerce)*

Example 1.3 _____

How many seconds are there in 1 year?

Reasoning

As before, we shall multiply by conversion factors that are equal to unity and have units on top and bottom to cancel the initial unit of year and eventually provide the unit of seconds.

Solution

Start with what you know, 1 year, and continue to multiply this by fractions that equal unity (like 365.25 days/year) until the fractions finally provide the unit needed (seconds):

$$1 \text{ year} = 1 \text{ year}\left(\frac{365.25 \text{ days}}{1 \text{ year}}\right)\left(\frac{24 \text{ h}}{1 \text{ day}}\right)\left(\frac{60 \text{ min}}{1 \text{ h}}\right)\left(\frac{60 \text{ s}}{1 \text{ min}}\right)$$

$$1 \text{ year} = 31\,557\,600 \text{ s} = 3.1558 \times 10^7 \text{ s}$$

1.5 Units of Mass

CONCEPT OVERVIEW

■ The basic unit of mass in the SI is the kilogram.

■ Mass describes how difficult it is to accelerate an object or to stop it once it is moving.

Mass is a description of how much matter—or material or "stuff"—an object is made of. Mass describes how difficult it is to accelerate an object or to stop it once it is moving—its "inertia." A shot put has more mass than a styrofoam cup. A Mack truck has more mass than a Miata roadster. As with other basic quantities, to move from a verbal, qualitative description to a specific, quantitative definition of mass, we must define a unit and a standard.

In the SI, the basic unit of mass is the **kilogram** (kg) and the standard is a cylinder of platinum-iridium alloy kept at Sèvres by the *Bureau International des Poids et Mesures*. The National Institute of Standards and Technology in the United States and similar organizations in other countries have secondary standards that are copied from this primary standard with great precision. A standard kilogram is shown in Figure 1.11. Mass is the only basic unit for which we do not have an atomic standard.

All the metric prefixes we discussed earlier concerning length can be used with mass as well. In particular, we often have occasion to talk about a *gram* (g). There are 1000 grams in one kilogram. That is, $1000 \text{ g} = 1 \text{ kg}$. A milligram (mg) is 10^{-3} g.

Figure 1.11
The standard kilogram.

Example 1.4 _____

How many milligrams are there in 1 kg?

Reasoning

The pattern we have established continues; we may multiply by conversion factors that are equal to unity and have units on top and bottom to cancel the initial unit of kilogram and eventually to provide the unit of milligrams.

Solution

$$1 \text{ kg} = 1 \text{ kg}\left(\frac{1000 \text{ g}}{1 \text{ kg}}\right)\left(\frac{1000 \text{ mg}}{1 \text{ g}}\right)$$

$$1 \text{ kg} = 1\,000\,000 \text{ mg} = 10^6 \text{ mg}$$

However, the elegance of the SI is that these conversions can be made in our heads just by moving the decimal point. To convert kilograms to grams, the decimal point is moved three places to the right:

$$1 \text{ kg} = 1000 \text{ g}$$

To convert any mass given in grams to milligrams requires moving the decimal point three places to the right:

$$1000 \text{ g} = 1\,000\,000 \text{ mg} = 10^6 \text{ mg}$$

Mass is related to weight. In physics we shall find an important distinction between mass and weight; they are related, but they are not the same thing. Masses are often compared by using a balance. This procedure compares the weight of an unknown mass with the weight of a known mass, but mass and weight are proportional (at Earth's surface) so that equal weights commonly mean equal masses. In the United States, we commonly express the weight of objects in pounds (lb), and it is commonly stated that 1 kg is the same as 2.2 lb. That means that an object whose *mass* is 1 kg has a *weight* of 2.2 lb. The difference between mass and weight will be explained in Chapter 4, where force, mass, and acceleration will concern us.

1.6 Significant Figures

CONCEPT OVERVIEW

- Every measurement, by its very nature, has some uncertainty in it.

- The uncertainty in a measurement determines how many significant figures should be used to represent the measurement or the result of a calculation.

There is always some uncertainty in any measurement. Figure 1.12 shows a meter stick marked in tenths of meters (decimeters). Figure 1.13 shows a

Figure 1.12
This meter stick is marked in tenths of a meter, decimeters (1 dm = 0.1 m).

Figure 1.13
This meter stick is marked in hundredths of a meter, centimeters (1 cm = 0.01 m). Why is this stick better than the one shown in Figure 1.12 for measuring the dimensions of objects such as books?

meter stick marked in hundredths of meters (centimeters). How well can you measure with this latter stick? The centimeter markings clearly tell you that the length of a particular box is between, say, 24 cm and 25 cm, but you must estimate any additional information. With some care and practice, you might estimate to tenths of divisions and determine that the length is 24.7 cm; there is some doubt as to the accuracy of the 7 in that measurement.

Again with care and practice, you might use this meter stick to measure the width of the box to be 16.4 cm. There is some uncertainty in the 4 in that measurement, too. The area of that side of the box is found by multiplying the length by the width: $A = LW = (24.7 \text{ cm})(16.4 \text{ cm}) = 405.08 \text{ cm}^2$. But can this meter stick really measure areas accurately to hundredths of square centimeters? Look again at the uncertainty in the last digit of each measurement we made.

When the length was measured as $L = 24.7$ cm, there was some question about the 7. The length is greater than 24.65 cm and less than 24.75 cm. Likewise, the width, $W = 16.4$ cm, is greater than 16.35 cm and less than 16.45 cm. If we take the smaller values, we have for the minimum value of A

$$A_{min} = LW = (24.65 \text{ cm})(16.35 \text{ cm}) = 403.028 \text{ cm}^2$$

If we take the larger values, we have for the maximum value of A

$$A_{max} = LW = (24.75 \text{ cm})(16.45 \text{ cm}) = 407.138 \text{ cm}^2$$

The uncertainties in our initial measurements mean that our result is really something between 403.028 cm² and 407.138 cm². Therefore it makes little sense to claim that our result is exactly 405.08 cm², measured to a hundredth of a square centimeter. How many of our figures are trustworthy, or significant?

We measured the length *to three significant figures* when we wrote 24.7 cm; the italicized words are just another way of saying that there is uncertainty in the third digit, the 7. We measured the width of 16.4 cm *to three significant figures,* and so there is uncertainty in the third digit, the 4. When we multiply those together, the result is accurate only to three significant figures; the area should be written as 405 cm². There is always some uncertainty in the last significant figure, but notice that our answer of 405 cm² falls between the minimum of 403.028 cm² and the maximum of 407.138 cm² obtainable from our initial measurements.

We have measured the length and width of the box and know the area of one side, each quantity to three significant figures. Now measure its thickness very precisely with the micrometer of Figure 1.14, which can measure to a thousandth of a centimeter, 0.001 cm. Suppose the thickness is 2.345 cm. What is the volume of the box? Volume is area A multiplied by thickness T:

$$V = AT = (405 \text{ cm}^2)(2.345 \text{ cm}) = 949.725 \text{ cm}^3$$

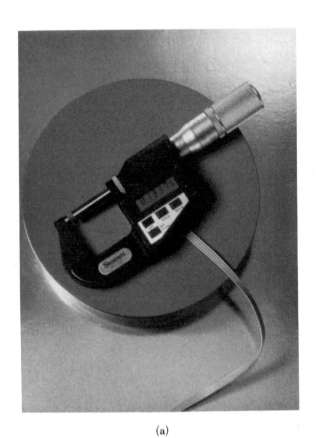

(a)

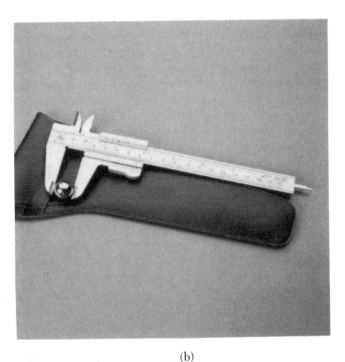

(b)

Figure 1.14
Two devices for measuring small distances accurately. (a) This digital micrometer can measure distances to an accuracy of 0.001 mm. *(Courtesy of the L.S. Starrett Company)* (b) These calipers can measure distances to an accuracy of 0.05 mm. *(Courtesy of PASCO Scientific)*

As before, however, we must ask if we can really claim to know the volume as accurately as that. We were not certain if the area was 403 cm^2 or 405 cm^2 or 407 cm^2, and so we certainly do not know the volume down to a thousandth of a cubic centimeter. The area 405 cm^2 is known to three significant figures. The thickness 2.345 cm is known more accurately, to *four* significant figures. The less-well-known factor provides the limit on our knowledge. *When multiplying (or dividing), the result can be accurately known to the number of significant figures of the least-well-known factor.* In our case, that means $V = 950$ cm^3.

We must also be careful of significant figures when numbers are added or subtracted. Suppose a tower is to be constructed in three sections, each built by a different contractor. Parts of the specifications sheet are lost, and each contractor measures the length of his or her section as he or she chooses. The results are

$$L_1 = 61.3 \text{ m}$$
$$L_2 = 33.193 \text{ m}$$
$$L_3 = 8.34 \text{ m}$$

What is the total height of the tower? Of course, we expect the total height to be the sum of these three numbers:

$$H = L_1 + L_2 + L_3 = 102.833 \text{ m}$$

However, do we really know the height that well? Length L_1 is given as 61.3 m, and that means there is some question about the 3. If we know that length to within only a tenth of a meter—it could be 61.35 m or 61.25 m—then we cannot tell the total length to an accuracy of a thousandth of a meter! The best we can honestly say is

$$H = 102.8 \text{ m}$$

When adding (or subtracting), the result can be accurately known only to the position of the least-well-known number. Both L_1 and L_3 have been measured to three significant figures, but L_1 is known only to within a tenth of a meter; L_3, to within a hundredth. Therefore L_1 is the limiting number here. The *position* of the least-well-known number (tenths of meters in L_1) determines the accuracy of the answer; the *number* of significant figures is not critical for addition and subtraction.

Notice that most hand-held calculators and most computers do not worry about significant figures. Both calculators and computers are wondrous tools, but they are just that—tools for you to use and control. A calculator's answer with a long string of numbers in it may look impressive, but it is probably misleading.

If we measured a length to be 2.5 cm to within a thousandth of a centimeter, we can write that as 2.500 cm. The trailing zeros are significant figures. They indicate that the accuracy is much greater than another measurement made only to a tenth of a centimeter, which we would indicate by 2.5 cm. The two lengths may be the same, but the former has been measured a hundred times more accurately than the latter.

We shall see more of significant figures in the next section.

Compare the portion of a meter stick (*top*) with that of the adjacent yardstick (*bottom*). Which one will give a more precise measurement?

1.7 Scientific Notation

CONCEPT OVERVIEW

■ Numbers in physics are conveniently written in the form of a number (between 1 and 9.999...) multiplied by 10 raised to a power. An example is 1.234×10^5. This is known as scientific notation.

Numbers in physics range from the astronomically large to the microscopically small, as indicated in Figures 1.15 and 1.16. The distance from Earth to our Sun is about 93 000 000 mi; or 147 000 000 km. The radius of a hydrogen atom is about 0.000 000 000 053 m. It soon becomes tedious to write such numbers. A misplaced zero drastically changes an answer! Because of this, powers-of-ten notation, or scientific notation, is often used.

The number of times something is multiplied by itself can be shown by a superscript, called an exponent, as we have already done in writing cm^2 and cm^3. We shall do this with the number 10, so that

$$10 = 10^1$$
$$1000 = 10 \times 10 \times 10 = 10^3$$
$$1\,000\,000 = 10^6$$

Division will be indicated by a negative superscript:

$$0.1 = \frac{1}{10} = 10^{-1}$$

$$0.001 = \frac{1}{10 \times 10 \times 10} = \frac{1}{10^3} = 10^{-3}$$

$$0.000\,001 = \frac{1}{1\,000\,000} = \frac{1}{10^6} = 10^{-6}$$

The number 10 to the zeroth power is just unity ($10^0 = 1$). Notice that multiplication of two powers of ten means adding the exponents, as in the following example.

Figure 1.15
Earth is about 147 000 000 km from our Sun.

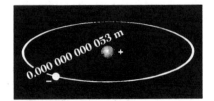

Figure 1.16
In a hydrogen atom, a single electron "orbits" a proton at a radius of about 0.000 000 000 053 m.

Example 1.5 ⎯⎯⎯⎯⎯⎯⎯⎯⎯⎯⎯⎯⎯⎯⎯⎯

Using scientific notation, multiply 100 and 10 000 to obtain 1 000 000.

Reasoning and Solution

First, express 100 and 10 000 as powers of ten:

$$100 = 10^2 \quad \text{and} \quad 10\,000 = 10^4$$
$$100 \times 10\,000 = (10^2)(10^4)$$
$$= [(10)(10)][(10)(10)(10)(10)] = 10^{2+4} = 10^6$$

Of course, this addition of exponents is also true for negative exponents.

Example 1.6 ⎯⎯⎯⎯⎯⎯⎯⎯⎯⎯⎯⎯⎯⎯⎯⎯

Multiply 0.01 and 1000 using powers of ten.

Reasoning and Solution

First, express these numbers as powers of ten:

$$0.01 = 10^{-2} \quad \text{and} \quad 1000 = 10^3$$
$$0.01 \times 1000 = 10^{-2} \times 10^3$$
$$= \left[\frac{1}{(10)(10)} \right][(10)(10)(10)] = 10^{-2+3} = 10^1 = 10$$

We can express any number as a number between 1 and 9.999 . . . multiplied by some power of ten. The distance from Earth to our Sun, in this notation, is $147\,000\,000$ km $= 1.47 \times 100\,000\,000$ km $= 1.47 \times 10^8$ km, and the radius of a hydrogen atom is $0.000\,000\,000\,053$ m $= 5.3 \times 0.000\,000\,000\,01$ m $= 5.3 \times 10^{-11}$ m. In this notation, then, the power of ten is the most important thing to look at when you want to know which of two numbers is larger; for instance, 3.47×10^8 is less than 1.23×10^9.

How can we add numbers like 2.235×10^5 and 3.759×10^6? A moment's reflection will show that the answer is *not* 5.994×10^{11}. That answer is clearly wrong, but what is right? We must expand this notation a bit and rewrite the numbers so that they both have the *same* power of ten:

$$
\begin{array}{rl}
2.235 \times 10^5 = & 2.235 \times 10^5 \\
+\ 3.759 \times 10^6 = & \underline{37.59 \times 10^5} \\
& 39.825 \times 10^5 = 3.9825 \times 10^6
\end{array}
$$

However, by our consideration of significant figures, when we add 2.235 and 37.59, the correct answer is 39.82 since the 9 in 37.59 is uncertain. Therefore, our answer must be 3.982×10^6. Both 2.235×10^5 and 3.759×10^6 are known to four significant figures, but there is a difference in the uncertainty (the *accuracy*) of the two. This can be seen in a further example.

Example 1.7

Using scientific notation and being careful about significant figures, determine the sum of 3.567×10^5 added to 4.298×10^3.

Reasoning

Both numbers are known to four significant figures, but one is known much more accurately than the other where addition is concerned.

Solution

We can see this difference in accuracy when we write both numbers raised to the *same* power of ten before we add them:

$$
\begin{array}{r}
3.567 \times 10^5 = 356.7 \times 10^3 \\
+\ 4.298 \times 10^3 = \underline{4.298 \times 10^3} \\
360.998 \times 10^3 = 361.0 \times 10^3 = 3.610 \times 10^5
\end{array}
$$

Remember that the 7 is uncertain in 3.567; that is just the nature of measurements. This number may really be 356.74×10^3, or it may be 356.68×10^3. Because of this uncertainty, it would be misleading to write the sum as 360.998×10^3. The accuracy of our initial numbers determines that the final result ought to be written 3.610×10^5.

Practice Exercise   What is 3.567×10^5 added to 5.432×10^3?

Answer 3.621×10^5

Determining the correct number of significant figures in multiplication and division is even easier. To see this, multiply 4.567×10^6 by 7.89×10^{-3}. As we write this out, the idea to remember is that the *order* of multiplication does not matter; 2×3 is the same as 3×2. We shall use this fact to let us multiply the decimal-number parts together and multiply the powers of ten together separately. We can write our multiplication as $(4.567 \times 10^6) \times (7.89 \times 10^{-3}) = 4.567 \times 7.89 \times 10^6 \times 10^{-3} = (4.567 \times 7.89) \times (10^6 \times 10^{-3}) = 36.03363 \times 10^3$. As always, though, we must be careful with the significant figures. One of our factors is known only to three significant figures, and so the answer is valid only to three significant figures. So the answer must be written as $36.0 \times 10^3 = 3.60 \times 10^4$.

SUMMARY

Physics is understanding the world around us. Numbers and equations are useful tools to help understand and explain ideas, but the ideas are most important.

Physics is an experimental science. Theories are accepted or rejected as they agree or disagree with measurements.

Measurements require two pieces of information—a number telling how many and a unit telling what is being measured. Units can always be changed by multiplying by a fraction that is equal to unity, as, for example,

$$\frac{2.54 \text{ cm}}{1 \text{ in.}} \quad \text{or} \quad \frac{1 \text{ km}}{1000 \text{ m}} \quad \text{or} \quad \frac{1 \text{ day}}{24 \text{ h}} \quad \text{or} \quad \frac{60 \text{ s}}{1 \text{ min}}$$

To ensure that people in different places (or at different times) know what other people mean by any measurement, standard units are required. The SI provides these standard units and a convenient method of relating small units to larger units by multiples of ten. In the SI, the **meter** is the basic unit of length, the **second** is the basic unit of time, and the **kilogram** is the basic unit of mass.

When multiplying (or dividing), the result can be accurately known only to the *number* of significant figures of the factor with the smaller number of significant figures.

When adding (or subtracting), the result can be accurately known only to the *position* of the least-well-known number.

The wide range of numbers used in physics is often conveniently expressed as a number between 1 and 9.999 . . . multiplied by ten raised to some power, such as 1.619×10^{-14} or 1.50×10^{12}. This is known as scientific notation.

QUESTIONS

1. Long ago, in Great Britain, the inch was defined as the length of three barley grains placed end to end. In what ways was this a good way to define a standard? In what ways was this a poor way to define a standard?
2. Why are units required in making measurements?
3. In what ways is it important to have units that other people also use?
4. Suppose you can communicate with people in a distant, isolated country by means of radio signals only—no television or transportation. Compose a verbal message that enables them to know how long 1 m is so that they may compare it with their unit of length. Compose a message to tell them how long 1 s is. Compose a message telling them what 1 kg is so that they can compare it with their unit of mass.
5. Why do you think the United States uses a system of units that is different from the system (SI) used by most of the rest of the world?
6. Do you feel the United States should adopt SI units? If so, explain why. If not, explain why not.
7. In what ways would the adoption of SI units by the United States be beneficial? In what ways would it be detrimental?

PROBLEMS

1.1 What Is Physics?

1.2 Units and Standards

1.3 Units of Length

1.1 How many square centimeters are there in 1 m^2?

1.2 How many cubic centimeters are there in 1 m^3?

1.3 How many square millimeters are there in 1 m^2?

1.4 A room is 11 ft by 19 ft. How many square yards of carpeting is necessary to carpet that room?

1.5 A tabletop measures 4 ft by 5 ft. What is the area of the tabletop in square meters?

1.6 A metric unit of area is the hectare, which is $10\,000 \text{ m}^2$. The common British unit of area is the acre, which is 4840 yd^2. How many acres are there in one hectare?

1.7 The wavelength of red light from a helium-neon laser is 632.8 nm. How many of these wavelengths will fit in 1 cm?

1.4 Units of Time

1.8   How many minutes are there in a 30-day month?

1.9   If you sleep 8 h a day, how many seconds will you sleep in a year?

1.10   The speed limit on Canada's major highways is 100 km/h. What is that in miles per hour?

1.11   The speed limit on an exit ramp is 35 mi/h. What is that in kilometers per hour?

1.12   Express 65 mi/h in kilometers per hour.

1.13   In the laboratory, we usually measure speed in meters per second. Convert 100 km/h to meters per second.

1.5 Units of Mass

1.14   How many grams are there in 1.234 kg?

1.15   If an object has a mass of 4.5 kg, what is its mass expressed in grams?

1.16   What is the mass in kilograms of an object having a mass of 15 g?

1.17   A Tylenol caplet contains 350 mg of acetaminophen. Express this mass in grams.

1.18   An aspirin tablet contains 325 mg of acetyl-salicylic acid. Express this mass in grams.

1.6 Significant Figures

1.19   Being careful to give the answer to the appropriate number of significant figures, add the following numbers: 29.34, 35.452, 123.92, 8.934.

1.20   Add the following numbers and write the answer with the appropriate number of significant figures: 8.348, 98.2, 1.0007.

1.21   Add the following numbers, being careful with significant figures: 34.45, 28.334, 15.229, 38.321.

1.22   Two I-beams are to be joined together. One has been measured to be 3.456 m long; the other, 4.3 m. When they are bolted snugly together, what value can you accurately give for the total length?

1.23   The length of a card is measured as 42.3 cm; its width, as 12.764 cm. To the correct number of significant figures, what is the area of the card?

1.24   The edges of a box are measured to be 1.2 m, 0.34 m, and 0.093 m. What is the volume expressed to the correct number of significant figures?

1.7 Scientific Notation

1.25   Express the following numbers in scientific notation: 846 100, 39 432.15, 37 300, 192 300, 3 430.

1.26   Express the following numbers in scientific notation: 234.2, 980.1, 23.98, 180.2, 17.34.

1.27   Express the following numbers in scientific notation: 0.25, 0.005 43, 0.049, 0.000 12, 0.098 7.

1.28   Express the following numbers in scientific notation: 0.0593, 34.231, 10.003, 0.0107, 398.2.

1.29   The diameter of our Moon is 3480 km; express this in scientific notation.

1.30   The diameter of our Sun is 1 390 600 km; express this in scientific notation.

1.31   The radius of the orbit of Mars is 228 000 000 km. Express that distance in scientific notation.

1.32   The characteristic yellow light from sodium has a wavelength of 0.000 000 589 0 m; express this length in scientific notation. Express this length in nanometers.

1.33   Light from a helium-neon laser has a wavelength of 632.8 nm. Express this in meters, using scientific notation.

1.34   Add the following numbers, being careful to keep only the appropriate significant figures in your answer: 9.135×10^2, 3.375×10^4, 1.934×10^3.

1.35   Add the following numbers, being careful to keep only the appropriate significant figures in your answer: 9.135×10^{-3}, 3.375×10^{-4}, 1.934×10^{-5}.

1.36   Multiply the following numbers, being careful to keep only the appropriate significant figures in your answer: $(3.45 \times 10^3) \times (2.0543 \times 10^5)$.

1.37   Multiply the following numbers, being careful to keep only the appropriate significant figures in your answer: $(1.234 \times 10^2) \times (9.87 \times 10^6)$.

1.38   Multiply the following numbers, being careful to keep only the appropriate significant figures in your answer: $(2.39 \times 10^{-2}) \times (1.8324 \times 10^3)$.

1.39   Multiply the following numbers, being careful to keep only the appropriate significant figures in your answer: $(7.214 \times 10^3) \times (4.3 \times 10^{-4})$.

General Problems

○1.40   The distance from our Sun to its nearest
neighboring star is about 10^{17} m. The Milky
Way galaxy is roughly a disk of radius 10^{21} m
and thickness 10^{19} m. Estimate the number
of stars in the Milky Way, assuming the distance between our Sun and its nearest neigh-
bor is typical.

○1.41   A mechanical watch is advertised as not gaining or losing more than 5 s a month. What is
the percent accuracy of this watch? That is,
if it gains (or loses) 5 s in a 30-day month,
how accurate—expressed as a percentage—is
the time it reads?

A moving train on a straight track exhibits all the principles of motion in one dimension. At rest in the station, the train has no velocity and no displacement. As it moves down the track away from the station, its distance from the station at any instant is a measure of its displacement. Because a train on a straight track can move only in the direction the engine is pointed, the velocity (a quantity made up of a speed and a direction and therefore defined as a *vector* quantity) is known once the engine begins to move. Also, the train must accelerate to go from a speed of zero to a speed selected by the engineer.

Such descriptions of motion are interesting for one train or plane, but consider the value of knowing the displacement, velocity, and acceleration of the many vehicles that travel a single track, highway, or airspace during the course of a day. Most large cities are served by commuter rail and subway lines, for instance, some of which have trains arriving at stations every 5 min during rush hour. What would happen if no one paid any attention to the motions of all these trains? How would you regulate train acceleration and velocity to maintain consistent service at all stations on the line?

(© Chris Michaels/FPG International Corporation)

2

Motion in One Dimension

Key Issues

- Kinematics—first developed by Galileo in the 17th century—is the description of motion.

- For any object in motion we must understand the concepts of displacement, velocity, and acceleration. Displacement is defined in terms of position relative to an origin or reference frame. Velocity can then be defined in terms of displacement and time; likewise, acceleration can be defined in terms of velocity and time. These definitions lead to mathematical statements that are shorthand ways of describing these aspects of motion.

- In physics the terms "speed" and "velocity" carry meanings similar to but more precisely defined than those in ordinary conversation. Care must be taken not to confuse the everyday meaning of words with their special uses in physics.

- Graphs are powerful tools for analyzing motion. Graphs help us connect observable motions to the mathematical descriptions of these motions.

We are all familiar with objects in motion—from a sprinter finishing a 100-m dash to a football being kicked through the goal posts. Explaining the *cause* of motion is called *dynamics,* and we shall defer our study of this topic until Chapter 5. At the moment we shall restrict ourselves to merely describing motion. This topic we call *kinematics.* Kinematics and dynamics are two subbranches of the branch of physics called *mechanics.*

2.1 Kinematics: Describing Motion

CONCEPT OVERVIEW

■ Kinematics is the description of motion.

Within kinematics, we shall restrict our present study to *motion in one dimension,* which is also called *straight-line motion.* The easiest example to think of may be a small model railroad train on a straight track, as in Figure 2.1, but the ideas and equations we shall develop are much more general. They also describe a car as it moves down a straight road or a real railroad train as it travels along a straight track, as in Figure 2.2 or in the photograph that opens this

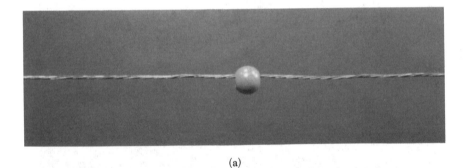

(a)

(b)

Figure 2.1
(a) A bead moving along a wire and (b) a model railroad train on a straight track provide examples of straight line motion. *(Both photos © L. W. Black)*

(a)   (b)

Figure 2.2
The ideas and equations of straight line motion also describe large, extended, real objects like (a) a train traveling along a straight track or (b) an automobile traveling along a straight highway.   *((a) © G. Brimacombe/The Image Bank; (b) © Larry Pierce/The Image Bank)*

chapter. Another example of one-dimensional motion is the motion due to gravity that occurs when we drop an object or throw it straight up, as in Figure 2.3.

Kinematics was first developed and understood by Galileo in the 17th century. He observed swinging chandeliers in the cathedral in Pisa and measured the times of their swings by counting his own pulse. He found that the time required for a pendulum to swing is the same whether the swing is large or small. No one had noticed this detail before. Expanding his studies beyond

Motion experiments by Galileo and others helped spur the creation of more accurate clocks.

Figure 2.3
Either dropping an object—or throwing it straight up—is another example of straight-line motion. This motion is also known as *free fall*.

pendula, Galileo built accurate water clocks to measure the time required for various balls and blocks to roll and slide down inclined planes.

Natural philosophers before Galileo had described and discussed and thought about how the world ought to behave, but Galileo did more: he observed and measured and built. In short, he experimented. That indeed was the beginning of modern science.

2.2 Distance and Displacement

CONCEPT OVERVIEW

- Displacement depends upon initial and final positions. Displacement is a vector quantity; this means that displacement along a straight line may be either positive or negative.

To describe or understand the motion of an object, we must be able to specify its position at various times. Doing so requires a *coordinate system,* or *reference frame.* Figure 2.4 shows a car moving down a straight road. Using s as our symbol for position, we can measure position from the corner traffic light; we shall consider that point to be the origin of our coordinate system and give it the coordinate $s = 0$. Initially the car is at position s_i (the subscript "i" stands for initial). Finally the car is at position s_f ("f" for final). The car's *displacement* from its initial position to its final position is $\Delta s = s_f - s_i$. The symbol Δ is the

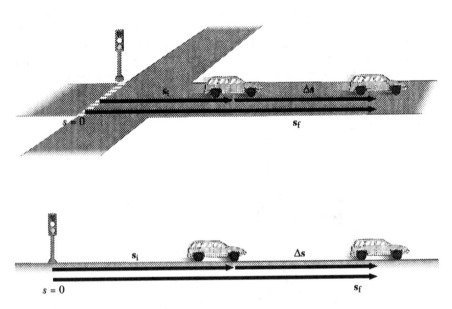

Figure 2.4
Position is measured from some origin ($s = 0$). Displacement Δs is the difference between the final position and the initial position: $\Delta s = s_f - s_i$.

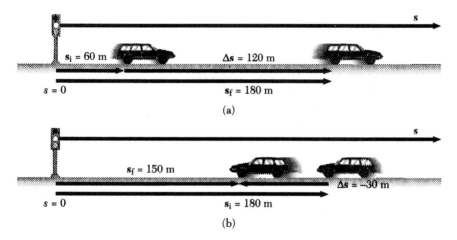

Figure 2.5
(a) A positive displacement means movement to the right. (b) A negative displacement means movement to the left.

Greek letter "delta" and means "the difference in." It always involves a final or later value *minus* an initial or earlier value. The displacement of the car is the "directed distance" it has moved. If it moves to the right, its displacement is positive; if it moves to the left, its displacement is negative.

In Figure 2.4 the displacement Δs is to the right, so if we use numerical values as in Figure 2.5(a), we can say the car's displacement is a positive 120 m: $\Delta s = +120$ m. A positive value for the displacement means it has moved to the right (toward increasing values of s). If the car now turns around, as indicated in Figure 2.5(b), from an initial position of 180 m from the origin to a position that is 150 m from the origin, then its displacement is $\Delta s = s_f - s_i = 150$ m $-$ 180 m $= -30$ m. The negative sign indicates a displacement to the left (toward decreasing values of s).

Many of the quantities we encounter in physics are *vector* quantities, which means direction is required to fully describe them. A vector has *magnitude* and *direction*. The magnitude is a number and a unit such as 2.3 in. or 376 km. In one-dimensional motion, direction can be indicated by the sign on a vector quantity, such as the $+120$ m and -30 m of the displacement in Figure 2.5. Displacement, velocity, and acceleration are all vector quantities, and therefore their signs are quite important. In the next chapter, we shall look at vectors more closely. Some other vector quantities we shall encounter are force, momentum, electric field, and magnetic field. Other quantities we encounter in physics are *scalar* quantities, which have no direction associated with them. Examples of scalars are time, mass, and temperature.

Distance is the scalar companion of the vector quantity displacement. In Figure 2.5(a) the car has traveled a distance of 120 m, and in Figure 2.5(b) it has traveled a distance of 30 m. Distance does not tell whether the car has moved to the right or to the left.

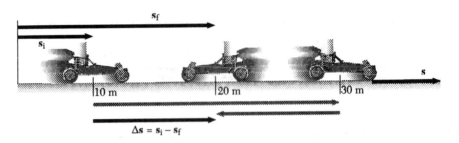

Figure 2.6
Distance and displacement are different things. Here the distance the object moves is 30 m, but its displacement is only 10 m.

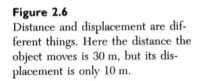

Distance means *total* distance traveled. In Figure 2.6 an object starts at an initial position $s_i = 10$ m, moves to $s = 30$ m, turns around and comes back to a final position of $s_f = 20$ m. The distance traveled is 30 m, for that is the total distance—20 m to the right plus 10 m to the left. The object's displacement, however, is 10 m to the right, for displacement depends only upon the initial and final positions: $\Delta s = s_f - s_i$.

2.3 Speed and Velocity

CONCEPT OVERVIEW

- Average velocity is the displacement of an object divided by the time required for that displacement to take place.

- Instantaneous velocity—usually referred to simply as velocity—is the average velocity in the limit as the time interval involved approaches zero.

In ordinary conversation, the words "speed" and "velocity" are used interchangeably. In physics, however, these words mean different things. Just as distance is the scalar companion of the vector quantity displacement, speed is the scalar companion of the vector quantity velocity. To specify an object's speed, all you need give is how fast the object is moving: the car's speed was 135 km/h. To specify an object's velocity, you must give its speed and the direction of that speed: 135 km/h in a southwesterly direction. Thus if there are two cars, one traveling to the right at 100 km/h and the other traveling to the left at 100 km/h, the two have the same speed but not the same velocity. *Average velocity* is the displacement of an object divided by the time required for that displacement to occur:

$$v_{\text{avg}} = \frac{\Delta s}{\Delta t} = \frac{s_f - s_i}{t_f - t_i} \tag{2.1}$$

where Δs is the displacement and Δt is the time interval. In the coordinate system established in Figure 2.4, a velocity to the right, in the direction of increasing displacement, is positive and a velocity to the left is negative. Also, an upward velocity, in the direction of increasing displacement, is positive, and a downward velocity is negative. *Average speed* is the total distance traveled divided by the total time required. Average speed depends upon the path taken; average velocity does not.

Problem-Solving Tip

■ In working physics problems, listing the "known quantities" and the "needed quantity" is a useful step in deciding which relationships are important or useful.

Example 2.1

Dulles International Airport near Washington, D.C., is about 4400 km from Los Angeles International Airport (Fig. 2.7). If a transcontinental flight from Washington to Los Angeles takes 5.0 h, find the average velocity.

Reasoning

We know the displacement, $\Delta s = 4400$ km, and the time, $\Delta t = 5.0$ h. We want to find the average velocity, v_{avg}, which is defined by Equation 2.1.

Solution

$$v_{avg} = \frac{\Delta s}{\Delta t} = \frac{4400 \text{ km}}{5.0 \text{ h}} = 880 \text{ km/h}$$

The average velocity is 880 km/h west (or approximately west). To be perfectly correct, you would give the average velocity as "880 km/h in the direction from Washington to Los Angeles."

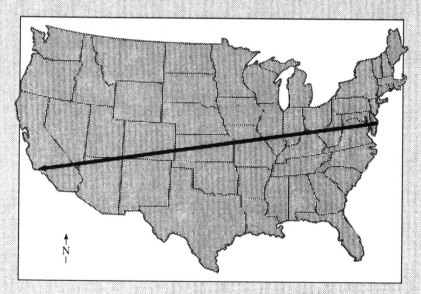

Figure 2.7

A cruise control device in an automobile automatically maintains a constant speed.

A speedometer tells you your instantaneous speed.

In some cases an object may move at a *constant*, or *uniform*, *velocity*. Let us take a car out and drive along a straight, level highway with the cruise control set at 100 km/h. The velocity remains constant, and therefore the average velocity is the only velocity there is in this situation.

In other cases the velocity is not uniform. Example 2.1 looked at a transcontinental airliner flying with an average velocity of 880 km/h to the west. It is clear that the plane could not have traveled the entire 5 h at exactly that velocity. Taxiing on the runway, the plane would be moving at a much lower speed and probably in a different direction. And the speed in an approach pattern would not be 880 km/h (nor would the direction remain the same). Weather conditions along the route may have required changes in velocity (either changes in speed or changes in direction). None of these variations would show up in an average velocity.

Instantaneous velocity is the velocity of an object at a particular instant. *Instantaneous speed* is the speed at a particular instant. Speed is a positive number (or zero) with units; there is no direction for speed. Your car's speedometer indicates instantaneous speed.

Figure 2.8 shows a car undergoing speed trials on a straight track. The table there lists the car's position at 1.00-s intervals. At the end of 10.00 s, the car has traveled 210.4 m; thus its average velocity for those 10.00 s is 21.04 m/s in the direction of the track. But what is it doing at any one particular time? To answer this question, we need to calculate not average velocity but instantaneous velocity.

Time, t (s)	Position, s (m)
0.00	0.0
1.00	2.5
2.00	10.0
3.00	22.5
4.00	40.0
5.00	61.4
6.00	85.6
7.00	112.6
8.00	142.4
9.00	175.0
10.00	210.4

Figure 2.8
A car undergoing speed trials on a straight track provides an example of straight-line motion. (© *Tom Sobolik/Black Star*)

To see how to calculate instantaneous velocity, let us calculate the average velocity over an interval shorter than 10.00 s. As indicated in the expanded table of Figure 2.9, between 2.00 s and 3.00 s, the car travels $\Delta s = 22.5$ m − 10.0 m = 12.5 m, for an average velocity of 12.5 m/s along the racetrack (Fig. 2.9). While also an average velocity just as our 21.04 m/s is, this new value of 12.5 m/s is an average velocity for one particular 1.00-s time interval. And this latter value is closer to what we need as an instantaneous velocity. If more accuracy is required, then we could find the positions at, say, 2.00 s and 2.10 s or 2.50 s and 2.60 s. Figure 2.10 is an expanded data table showing the car's position at 0.10-s intervals between 2.00 s and 3.00 s. At 2.60 s, its position is 16.90 m from the origin; at 2.50 s, 15.62 m. Its displacement during that 0.10-s interval, according to these data, is

$$\Delta s = s_f - s_i = 16.90 \text{ m} - 15.62 \text{ m} = 1.28 \text{ m}$$

So the car's average velocity is

$$v_{avg} = \frac{1.28 \text{ m}}{0.10 \text{ s}} = 12.8 \text{ m/s}$$

along the track. This average velocity during the 0.10-s interval from 2.50 s to 2.60 s is still closer to the instantaneous velocity.

Time, t (s)	Position, s (m)	Change in position, Δs (m)	Average velocity, v (m/s)
0	0.0	—	—
1.00	2.5	2.5	2.5
2.00	10.0	7.5	7.5
3.00	22.5	12.5	12.5
4.00	40.0	17.5	17.5
5.00	61.4	21.4	21.4
6.00	85.6	24.2	24.2
7.00	112.6	27.0	27.0
8.00	142.4	29.8	29.8
9.00	175.0	32.6	32.6
10.00	210.4	35.4	35.4

Figure 2.9
The car's average velocity reflects its position at different times. (© *Tom Sobolik/Black Star*)

Time, t (s)	Position, s (m)	Change in position, Δs (m)	Average velocity, v (m/s)
2.00	10.00	—	—
2.10	11.02	1.02	10.2
2.20	12.10	1.08	10.8
2.30	13.22	1.12	11.2
2.40	14.40	1.18	11.8
2.50	15.62	1.22	12.2
2.60	16.90	1.28	12.8
2.70	18.22	1.32	13.2
2.80	19.60	1.38	13.8
2.90	21.02	1.42	14.2
3.00	22.50	1.48	14.8

Figure 2.10
Average velocity over smaller time imtervals Δt comes closer to instantaneous velocity—the velocity at a particular moment.

We can generalize this procedure to define the *instantaneous velocity v* as the average velocity in the limit of ever-shorter time intervals:

$$v = \lim_{\Delta t \to 0} \frac{\Delta s}{\Delta t} \qquad (2.2)$$

This means that, as Δt gets smaller and smaller, the value of $\Delta s/\Delta t$ gets closer and closer to the instantaneous velocity. In most situations, we shall simply call this the *velocity*. If we mean average velocity, we shall state that explicitly. Likewise, we shall use "speed" to mean instantaneous speed. Like average velocity, instantaneous velocity is a vector quantity and therefore is either positive or negative for one-dimensional motion.

2.4 Acceleration

CONCEPT OVERVIEW

■ Acceleration is the change in velocity divided by the time required for that change to take place.

Look again at the data in Figure 2.9. We have already found that, between $t = 2.00$ s and $t = 3.00$ s, the car's velocity was 12.5 m/s. Between $t = 8.00$ s and $t = 9.00$ s, the car traveled 32.6 m for a velocity of 32.6 m/s. The velocity is changing as time goes by.

As an object's position changes, we find it useful to describe how fast this change occurs; *velocity* is this *time rate of change of position*. When the velocity changes, it is useful to describe how quickly that change occurs. We call this *time rate of change of velocity* the *acceleration*. A rocket may have a very small acceleration during the initial moments of liftoff, and a runner may have a large acceleration even though her velocity is small.

Analogous to velocity and Equation 2.1, we define average acceleration as the change in velocity divided by the time required for that change:

$$a_{avg} = \frac{\Delta v}{\Delta t} = \frac{v_f - v_i}{t_f - t_i} \qquad (2.3)$$

As with average velocity, average acceleration involves only the initial and final motion—details of the motion in between are lost.

If a car changes its velocity from $v_i = 20$ km/h to $v_f = 70$ km/h in $\Delta t = 20$ s, then its average acceleration is

$$a_{avg} = \frac{70 \text{ km/h} - 20 \text{ km/h}}{20 \text{ s}} = \frac{50 \text{ km/h}}{20 \text{ s}} = 2.5 \text{ km/h/s}$$

These units are read "kilometers per hour per second." This means that, on the average, the car's velocity increases by 2.5 km/h during each 1-s interval. In units of meters per second, the initial and final velocities are $v_i = 5.56$ m/s and $v_f = 19.4$ m/s, and so the average acceleration can also be written as

$$a_{avg} = \frac{19.4 \text{ m/s} - 5.56 \text{ m/s}}{20 \text{ s}} = \frac{13.8 \text{ m/s}}{20 \text{ s}} = 0.69 \text{ m/s/s} = 0.69 \text{ m/s}^2$$

These units are read "meters per second per second" or "meters per second squared." In either case, they mean that, on the average, the car's velocity increases by 0.69 m/s during each 1-s interval.

Under ordinary driving conditions, the change in velocity will not be precisely the same during each of those 20 s. The instantaneous acceleration at any particular moment will undoubtedly differ from this average value. As with velocity, average acceleration does not answer the question, how quickly is the velocity changing *at this instant?* Just as we did for velocity, we define *instantaneous acceleration* in the limit of very short time intervals. We use times t_f and t_i that are very close to each other; in other words, we let $\Delta t = t_f - t_i$ approach zero. We can write this instantaneous acceleration as

$$a = \lim_{\Delta t \to 0} \frac{\Delta v}{\Delta t} \qquad (2.4)$$

As with velocity, this definition means that our results will be closer to the instantaneous value as we make the time interval $\Delta t = t_f - t_i$ smaller and smaller. We shall usually call this the acceleration rather than explicitly saying "the instantaneous acceleration."

Figure 2.11 continues the development of our data table for the car on a straight track. During the first few time intervals, the velocity increases by 5.0 m/s every second, which means the car is undergoing an acceleration of 5.0 m/s². During the later time intervals, the acceleration is reduced to 2.8 m/s².

Time, t (s)	Position, s (m)	Change in position, Δs (m)	Average velocity, v (m/s)	Change in velocity, Δv (m/s)	Average acceleration, a (m/s²)
0	0.0	—	—	—	—
1.00	2.5	2.5	2.5	—	—
2.00	10.0	7.5	7.5	5.0	5.0
3.00	22.5	12.5	12.5	5.0	5.0
4.00	40.0	17.5	17.5	5.0	5.0
5.00	61.4	21.4	21.4	3.9	3.9
6.00	85.6	24.2	24.2	2.8	2.8
7.00	112.6	27.0	27.0	2.8	2.8
8.00	142.4	29.8	29.8	2.8	2.8
9.00	175.0	32.6	32.6	2.8	2.8
10.00	210.4	35.4	35.4	2.8	2.8

Figure 2.11
The car's changing position is described by its velocity. Likewise, its changing velocity is described by its acceleration. (© *R. Mackson/FPG International Corporation*)

Like velocity, acceleration is a vector quantity. In one-dimensional motion, we can denote acceleration direction by plus and minus signs. In the data table in Figure 2.11, for instance, note that the acceleration remains positive throughout. That means the velocity is increasing throughout. For each time interval Δt, we find $\Delta v > 0$. That is what a positive acceleration means. However, the acceleration decreases—in other words, the values of Δv, for the same Δt, get smaller. That means there is a decrease in the rate at which the

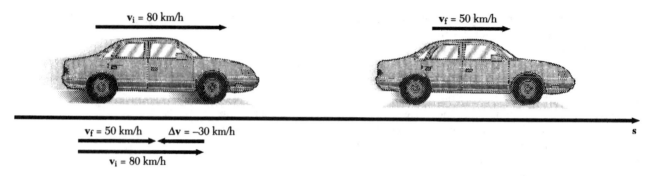

Figure 2.12
A car that slows down as it continues to move in the positive direction (to the right) has a negative acceleration.

velocity is increasing. The distinction between velocity and acceleration is important but sometimes subtle.

What happens as a car slows down, as in Figure 2.12? As shown in the figure, the direction of increasing displacement s is to the right. In this coordinate system, displacements, velocities, and accelerations to the right are positive; those to the left are negative. As long as the car continues to move to the right, its velocity is positive. However, if it slows down from, say, 80 km/h to 50 km/h, the *change* in velocity is negative:

$$\Delta v = v_f - v_i = 50 \text{ km/h} - 80 \text{ km/h} = -30 \text{ km/h}$$

Since acceleration is defined as $\Delta v/\Delta t$, a negative Δv means a negative acceleration. The car is undergoing a negative acceleration as its velocity changes from 80 km/h to 50 km/h. A negative acceleration is sometimes called a deceleration; "negative acceleration" is usually preferable in physics. A negative acceleration means an acceleration in the direction of decreasing (or negative) position—an acceleration to the left in Figure 2.10.

Motion with uniform acceleration is a special case that we shall look at in some detail. Gravity causes objects to fall with a uniform acceleration near Earth's surface. Uniform acceleration is a reasonable approximation to many real-world situations.

Consider an object initially at position s_i at time t_i and moving with velocity v_i. If the object experiences a uniform acceleration a, we would like to know where it will be and how fast it will be going at some later time t. That is, how can we predict (or calculate) position s and velocity v for any later time t if we know initial position s_i, initial velocity v_i, and (constant) acceleration a? This is typically our objective in physics—given initial conditions, predict what happens at some later time.

From the acceleration we can determine the velocity. The velocity will then lead us to the position. With uniform acceleration, we need not distinguish between average and instantaneous values. There is only one acceleration a. From the definition of acceleration, $a = \Delta v/\Delta t$, we can solve for the change in velocity: $\Delta v = a\Delta t$. For convenience, we shall set $t_i = 0$ and let $t = t_f$, so we can write $\Delta v = at$. But $\Delta v = v_f - v_i$. If we use $v = v_f$, then

$$v = v_i + at \qquad (2.5)$$

This relationship answers part of our initial question: we can now determine the object's velocity at time t.

What of its position? The average velocity involves position and time, and so we shall start there. From $v_{avg} = \Delta s/\Delta t$, we can solve for $\Delta s = v_{avg}\Delta t$. Since $\Delta s = s_f - s_i$, we have

$$s = s_i + v_{avg}t$$

where $s = s_f$ and $t = t_f$ as before. Now we must find v_{avg}. For the case of uniform acceleration, the average velocity is just the average of the initial and final velocities:

$$v_{avg} = \frac{v_i + v}{2}$$

We know that $v = v_i + at$, so that

$$v_{avg} = \frac{v_i + v_i + at}{2} = v_i + \frac{at}{2}$$

Now v_{avg} can be used to find the position:

$$s = s_i + \left(v_i + \frac{at}{2}\right)t$$

or

$$s = s_i + v_i t + \frac{at^2}{2} \qquad\qquad (2.6)$$

This equation is very useful. It answers the remaining question about position. If we know initial position s_i, initial velocity v_i, and constant acceleration a, we can predict the position at any future time t, just as Equation 2.5 lets us predict the velocity. Given constant acceleration and initial conditions, how fast is a car going after 20 s? Or how far has a plane traveled in 13 s? These are the kinds of questions we can now answer.

There is an additional question we might ask, though—one that these equations cannot immediately answer. Given constant acceleration and initial conditions, how far does a car travel as it brakes to a stop? Or how far does a train travel as its velocity changes from one value to another? Our present equations depend upon time, and neither question involves time. A new relationship—a new equation—involving only position, velocity, and acceleration is not difficult to derive.

Since we do not need time, solve $v = v_i + at$ for t to find

$$t = \frac{v - v_i}{a}$$

Now substitute this value for t into Equation 2.6:

$$s = s_i + v_i\left(\frac{v - v_i}{a}\right) + \frac{a}{2}\left(\frac{v - v_i}{a}\right)^2$$

Multiplying these terms yields

$$s = s_i + \frac{v_i v - v_i^2}{a} + \frac{v^2 - 2v v_i + v_i^2}{2a}$$

Simplifying gives

$$2a(s - s_i) = v^2 - v_i{}^2$$

Time has been removed, as we needed. Now we can solve for the velocity in terms of v^2:

$$v^2 = v_i{}^2 + 2a(s - s_i) \qquad \textbf{(2.7)}$$

We have now developed three very useful relationships for understanding motion: a velocity-time equation (2.5), a position-time equation (2.6), and a velocity-position equation (2.7). The first two enable us to calculate the velocity or position at any later time. The third allows us to calculate the velocity at some later position. The third equation comes from the first two.

Problem-Solving Tips

- Explicitly stating the "known quantities" and the "needed quantities" is a useful step in deciding which equation(s) to use.
- Diagrams help ensure that origins and directions are used consistently.
- Unit analysis (also known as dimensional analysis) requires that all quantities—both known and unknown—be accompanied by their respective units. Many errors in solving problems can be avoided by performing the algebra or arithmetic on the units to check the accuracy of your solutions.

Example 2.2

An air track glider in a physics experiment has an acceleration of 0.350 m/s². If it starts from rest, how fast is it moving in 3.00 s? How far has it traveled in that time?

Reasoning

We know the acceleration $a = 0.350$ m/s² and the initial velocity $v_i = 0$ ("starting from rest" is just another way of saying "initial velocity is zero"). When asked, "How far has it traveled?" we may safely assume we are being asked, "How far has it traveled *from its initial position?*" and so we can use $s_i = 0$. We know the time $t = 3.00$ s. We want to find the velocity v and position s at that time. We need equations involving time. The first two of our "big three" kinematics equations involve time. They provide velocity and position—just the quantities we need!

Solution

Diagrams are always useful in physics, and usually necessary. Figure 2.13 shows the glider initially at rest at its initial position $s_i = 0$. The diagram shows the glider moving to the right with constant acceleration $a = 0.350$ m/s².

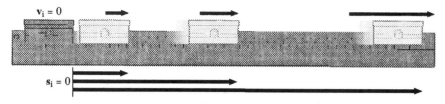

Figure 2.13
Starting from rest, a glider on an air track accelerates to the right.

The velocity at time $t = 3.00$ s is, from Equation 2.5,

$$v = v_i + at$$
$$= 0 + (0.350 \text{ m/s}^2)(3.00 \text{ s}) = 1.05 \text{ m/s}$$

Notice that the units come out as they should. Ending up with the appropriate units is a good check (but not a complete guarantee) that the right equation was chosen and that it was used properly.
The position is given by the position-time equation (2.6):

$$s = s_i + v_i t + \frac{at^2}{2}$$

$$= 0 + 0 + \frac{(0.350 \text{ m/s}^2)(3.00 \text{ s})^2}{2} = 1.58 \text{ m}$$

In 3.00 s, the glider attains a velocity of 1.05 m/s and travels 1.58 m.

Practice Exercise   If the acceleration is 0.150 m/s^2, how fast will the glider be moving in 3.00 s? How far will it have traveled in 3.00 s?
Answer $v = 0.450$ m/s, $s = 0.675$ m

Example 2.3

A car accelerates uniformly from 0 to 50.0 km/h in 8.50 s. How far does it travel during that time?

Reasoning

Figure 2.14 is a sketch of the situation showing what we know and what we need. We know velocities ($v_i = 0$ and $v = 50.0$ km/h) and time ($t = 8.50$ s) and want displacement ($s = ?$). We have equations that relate velocity, acceleration, and time or displacement, acceleration, and time, or even velocity, displacement, and acceleration. Each of our "big three"

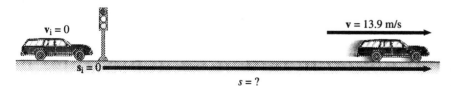

Figure 2.14
How far does the car travel in the 8.50 s required to reach a velocity of 50.0 km/h, or 13.9 m/s?

equations involves acceleration. Therefore we shall first find the value of the acceleration and then use that value in the position-time equation to find how far the car has traveled in 8.50 s.

Solution

First change velocity to meters per second:

$$v = 50.0 \frac{\text{km}}{\text{h}} \left(\frac{1000 \text{ m}}{1 \text{ km}}\right)\left(\frac{1 \text{ h}}{3600 \text{ s}}\right) = 13.9 \text{ m/s}$$

The acceleration is the change in velocity divided by the change in time:

$$a = \frac{13.9 \text{ m/s}}{8.50 \text{ s}} = 1.64 \text{ m/s}^2$$

This means that, during each second, the velocity increases by 1.64 m/s. Now that the acceleration is known, we are free to use the position-time equation (2.6) to find the position:

$$s = s_i + v_i t + \frac{at^2}{2}$$

$$= 0 + 0 + \frac{(1.64 \text{ m/s}^2)(8.50 \text{ s})^2}{2} = 59.2 \text{ m}$$

Practice Exercise   If the car accelerates uniformly from 0 to 50.0 km/h in 7.00 s, how far does it travel during that time?

Answer   48.6 m

What can you do to gain confidence that you have written, say, Equation 2.6

$$s = s_i + v_i t + \frac{at^2}{2}$$

correctly? Is it $v_i t$ or $v_i t^2$? Is it $at/2$ or $at^2/2$? How can we check on our memory? As the tip preceding Example 2.2 reminds you, unit analysis provides a very useful tool that will quickly catch many of the errors you are likely to make. The quantity s on the left side of the equation is a displacement, measured in meters. Likewise, s_i is also a displacement, measured in meters. The old adage "you cannot add apples and oranges" is true and forms the basis of unit analysis. The term $v_i t$ has units of $(\text{m/s})(\text{s}) = \text{m}$, so it can be added to s_i. Had we tried $v_i t^2$, that term would have units of $(\text{m/s})(\text{s})^2 = \text{m} \cdot \text{s}$, which *cannot* be added to s_i in m. A pure number, like the 1/2 in this equation, has no units; you can therefore omit it from the unit analysis. The term $at/2$ has units of $(\text{m/s}^2)(\text{s}) = \text{m/s}$, which *cannot* be added to s_i in m. The term $at^2/2$ has units of $(\text{m/s}^2)(\text{s})^2 =$

meters, which can be added to s_i in m. Dimensional analysis gives us confidence that

$$s = s_i + v_i t + \frac{at^2}{2}$$

is correct. It does not tell us where the factor of 1/2 must go, but it will catch many other common errors.

Example 2.4

Using the acceleration of the previous example, determine how fast the car will be going when it has traveled 100 m. Assume the acceleration remains constant.

Reasoning

We know the acceleration and the displacement, and want to determine the velocity. Velocity appears in the velocity-time equation (2.5),

$$v = v_i + at$$

but using this equation requires knowing the time. We do not immediately know the time required to travel 100 m. We could find the time from the position-time equation (2.6),

$$s = s_i + v_i t + \frac{at^2}{2}$$

since s is known. There is nothing wrong with this approach, but there is a more direct one. We faced this same situation—that we know acceleration and displacement and need velocity—when we derived the velocity-position equation (2.7),

$$v^2 = v_i^2 + 2a(s - s_i)$$

so this is the appropriate equation to use for our situation.

Solution

Using our particular values of $v_i = 0$, $a = 1.64$ m/s², and $s - s_i = 100$ m, we get, from Equation 2.7,

$$v^2 = 0 + 2(1.64 \text{ m/s}^2)(100 \text{ m}) = 328 \text{ m}^2/\text{s}^2$$
$$v = 18.1 \text{ m/s}$$

Of course, this solution carries the correct units; speed is distance divided by time, or units of meters per second.

If you prefer the answer in kilometers per hour, we can readily convert by our usual method:

$$v = \left(18.1 \frac{m}{s}\right)\left(\frac{1 \text{ km}}{1000 \text{ m}}\right)\left(\frac{3600 \text{ s}}{1 \text{ h}}\right) = 65.2 \text{ km/h}$$

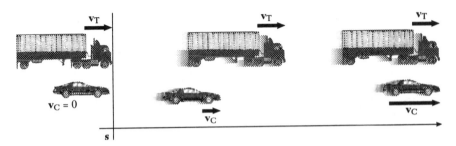

Figure 2.15
A truck traveling at constant velocity, passes a car at $t = 0$. The car, initially at rest, begins to accelerate at $t = 0$. When and where does the car catch up with the truck?

Example 2.5 ───────────────────────────────

Figure 2.15 depicts an interesting and classic situation. A car, capable of a constant acceleration of 2.5 m/s², is stopped at a traffic light. When the light turns green, the car starts from rest with this acceleration. Also as the light turns green, a truck traveling with a constant velocity of 40 km/h passes the car. Clearly the car will eventually travel faster than the truck and will overtake it. Where will the car catch up with the truck?

Reasoning

The car and truck each move with constant acceleration, so the equations we have developed describe their motion. We shall write an equation for the position of the car s_C and another for the position of the truck s_T. When the car catches up with the truck, they are at the same position, so we shall set these two quantities equal: $s_C = s_T$.

The most important step in this solution is a clear diagram. Figure 2.16 is a sketch of the motion just described with the known information

For the truck: $\mathbf{v_T}$ = 40 km/h = 11.11 m/s = constant
$\qquad\qquad\quad \mathbf{a_T} = 0$

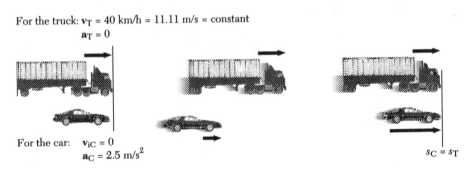

For the car: $\mathbf{v_{iC}} = 0$
$\qquad\qquad \mathbf{a_C}$ = 2.5 m/s² $\qquad\qquad\qquad\qquad\qquad\qquad\qquad s_C = s_T$

Figure 2.16
Explicitly writing the known (or given) information is useful.

indicated. At time $t = 0$, both the car and the truck are at position $s_i = 0$ ($s_{iC} = s_{iT} = 0$). The car starts from rest ($v_{iC} = 0$) and moves with constant acceleration ($a = 2.5$ m/s^2). The truck moves with constant velocity ($v_T = 40$ km/h $= 11$ m/s). At some later time t, they have the same position ($s_C = s_T$); we are looking for that position.

Solution

The position of the car is described by the position-time equation (2.6) with $v_{iC} = 0$ and $a_C = 2.5$ m/s^2:

$$s_C = s_{iC} + v_{iC}t + \frac{a_C t^2}{2}$$

$$= 0 + 0 + \frac{(2.5 \text{ m/s}^2)t^2}{2}$$

The position of the truck is described by the same position-time equation but with $v_{iT} = 11$ m/s and $a_T = 0$:

$$s_T = s_{iT} + v_{iT}t + \frac{a_T t^2}{2}$$

$$= 0 + (11 \text{ m/s})t + 0$$

We must find the time t at which $s_C = s_T$. Thus,

$$\frac{(2.5 \text{ m/s}^2)t^2}{2} = (11 \text{ m/s})t$$

Carrying out the division, moving both terms to the same side, and dividing by m/s^2 give

$$1.25t^2 - (11 \text{ s})t = 0$$
$$t^2 - (8.8 \text{ s})t = 0$$
$$t(t - 8.8 \text{ s}) = 0$$

There are two solutions to this equation:

$$t = 0 \quad \text{and} \quad t = 8.8 \text{ s}$$

The first one, $t = 0$, describes the initial situation: at $t = 0$, the car and truck are together and $s_C = s_T$. The second solution, $t = 8.8$ s, is the interesting one; that is what we are looking for. Now there is just one step more. Where—at what position—does the car catch up with the truck? Putting this value of the time, $t = 8.8$ s, into the car's position-time equation yields

$$s_C = \frac{(2.5 \text{ m/s}^2)(8.8 \text{ s})^2}{2} = 97 \text{ m}$$

Of course, this must also be the position of the truck. You should check this and see for yourself.

The French TGV train *(© Telegraph Colour Library/FPG International Corporation)*

Example 2.6

The French National Railroad holds the world's speed record for passenger trains in regular service. A TGV (*trés grand vitesse*, or very great speed) train traveling at a velocity of 300 km/h requires 1.20 km to come to an emergency stop (Fig. 2.17). Find the braking acceleration for this train, assuming constant acceleration.

Reasoning

We know the velocities and distance involved and want to find the acceleration. Equation 2.7, our velocity-position equation, describes the relationship among velocities, distance, and acceleration, so that equation is our choice.

Solution

First convert 300 km/h to meters per second:

$$v_i = 300 \, \frac{km}{h} \left(\frac{1000 \, m}{1 \, km} \right) \left(\frac{1 \, h}{3600 \, s} \right) = 83.3 \, m/s$$

The initial velocity is 83.3 m/s, and the final velocity is zero. The distance traveled is $s - s_i = 1.20$ km $= 1200$ m.

$$v^2 = v_i^2 + 2a(s - s_i)$$
$$0 = (83.3 \, m/s)^2 + 2a(1200 \, m)$$
$$a = \frac{-(83.3 \, m/s)^2}{2(1200 \, m)} = -2.89 \, m/s^2$$

The acceleration is negative because the train continues to move forward while it slows to a stop.

_____

Practice Exercise   If the TGV requires 1.80 km to come to rest from 300 km/h, what is its braking acceleration?

Answer $-1.93 \, m/s^2$

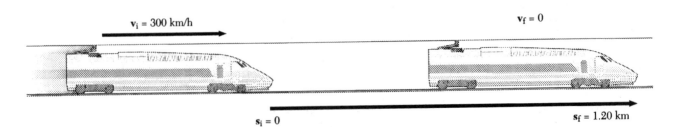

$v_i = 300$ km/h $v_f = 0$

$s_i = 0$ $s_f = 1.20$ km

Figure 2.17
A sketch with known information is an important tool in solving problems.

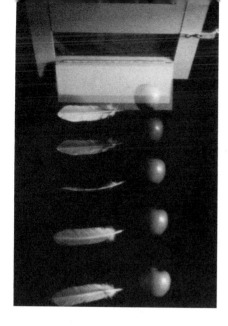

Figure 2.18
In the absence of air resistance (as in a vacuum), all objects fall with the same acceleration (9.80 m/s² downward near Earth's surface). *(© James Sugar/Black Star)*

2.5 Free Fall

CONCEPT OVERVIEW

■ To the degree that we can neglect air resistance, all bodies dropped near Earth's surface fall with the same acceleration, $g = 9.80$ m/s², downward.

How do objects behave when they are dropped? It is widely accepted that all objects fall toward Earth with the same uniform acceleration. But is this really so? Hold a piece of paper in your left hand and a coin in your right hand. Release them at the same time and observe their motion. The coin drops quickly, while the paper glides and flutters slowly to the floor. They do not fall with the same acceleration at all! The ancient Greeks thought of similar motion 20 or more centuries ago and decided that heavier objects fall faster. Indeed, that conclusion would seem reasonable from this simple experiment.

Now let us modify the experiment slightly and do it again. This time, crumple the paper a little before you drop it. Now the paper falls faster than in the first experiment. Crumple it into an even tighter ball or fold it into a tight little bundle. Now the paper and coin fall in very much the same fashion. It is the air and the larger surface area that account for any slowness of the paper; this is known as *air resistance*. If we drop the coin and paper—crumpled or flat—in a vacuum with no air present, they both fall at exactly the same rate. This can be seen with the apparatus shown in Figure 2.18 where an apple and a feather fall side by side in a vacuum chamber. Figure 2.19 shows this in a more dramatic way.

It was Galileo, in the 17th century, who first carefully examined the motion of falling bodies. Galileo found—as we find today—that all objects fall with the same acceleration. That acceleration is 9.80 m/s², directed downward toward Earth. This statement is true only if the air resistance is small enough that we can ignore it; with the tools we now have (that is, with the equations of constant acceleration), we can correctly describe the motion of coins and baseballs but not the motion of feathers and confetti. The statement is also true only if we are near the surface of Earth. We can discuss objects dropped from tall towers or even high-flying jets, but we cannot discuss objects fired far above the atmosphere.

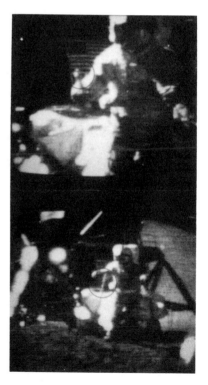

Figure 2.19
The vacuum of the Moon provides a good laboratory for demonstrating that all objects fall with the same acceleration. *(NASA)*

43

TABLE 2.1 Acceleration due to gravity at selected sea-level locations

Location	Latitude (°N)	g (m/s^2)
Quito, Ecuador	0	9.780
Madras, India	13	9.783
Hong Kong	22	9.788
Cairo, Egypt	30	9.793
New York	41	9.803
London	51	9.811
Oslo, Norway	60	9.819
Murmansk, Russia	69	9.831
North Pole	90	9.832

The Earth as photographed from the GOES satellite in geostationary orbit. *(NASA)*

Measurements of Earth's gravitation at various points helped to determine that our planet is a bit more ellipsoidal than spherical.

This uniform acceleration near Earth's surface is the *acceleration due to gravity*. We give it the symbol g. We shall use the value

$$g = 9.80 \text{ m/s}^2 \qquad \textbf{(2.8)}$$

This acceleration is directed downward, toward the center of Earth. It varies slightly with location—it is slightly greater near the poles, slightly less near the equator (because Earth is not a perfect sphere)—and decreases with altitude. Table 2.1 lists a few measured values.

Any object thrown or released near Earth's surface is said to be in free fall if we can neglect the effects of air resistance. Such an object released or thrown directly upward or downward is another example of one-dimensional motion with uniform acceleration.

In a coordinate system in which we designate up as the positive direction, consider a ball thrown upward from the top of a building with an initial velocity of $v_i = 20$ m/s. For the moment, we want to concentrate on the general nature of the motion, so we shall use the approximation $g \approx 10$ m/s^2. Figure 2.20 shows a sketch and a table of the ball's position and velocity. The data in the table come directly from the velocity-time equation (2.5) and the position-time equation (2.6). Look at the data for $t = 3.0$ s:

$$v = v_i + at$$
$$= 20 \text{ m/s} + (-10 \text{ m/s}^2)(3.0 \text{ s}) = -10 \text{ m/s}$$

Note that, because we have chosen up as positive, the acceleration must be negative because it is directed downward, toward the ground. This negative acceleration yields a negative velocity at 3.00 s.

Since we are describing vertical motion, let us replace s by y in Equation 2.6, and then this equation becomes

$$y = y_i + v_i t + \frac{at^2}{2}$$

There is no real difference between this equation and the form of Equation 2.6

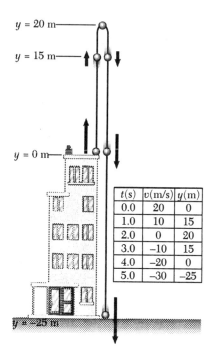

$t(s)$	$v(m/s)$	$y(m)$
0.0	20	0
1.0	10	15
2.0	0	20
3.0	−10	15
4.0	−20	0
5.0	−30	−25

Figure 2.20

A ball thrown straight up from the top of a building provides an example of straight-line motion with constant acceleration. For convenience, the approximation has been used that the acceleration due to gravity is 10 m/s² downward.

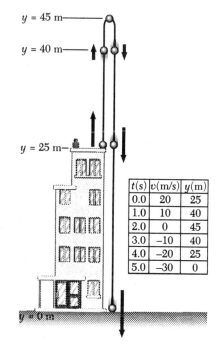

$t(s)$	$v(m/s)$	$y(m)$
0.0	20	25
1.0	10	40
2.0	0	45
3.0	−10	40
4.0	−20	25
5.0	−30	0

Figure 2.21

The initial position need not be zero. This diagram and data table describe exactly the same motion as in Figure 2.20, but this time the origin has been taken as the sidewalk at the base of the building.

we learned earlier. Displacement may be labeled y as easily as it is labeled s. Thus we have

$$y = 0 + (20 \text{ m/s})(3.0 \text{ s}) + \frac{(-10 \text{ m/s}^2)(3.0 \text{ s})^2}{2} = 15 \text{ m}$$

An interesting point about this type of motion is that the velocity is momentarily zero when the thrown object reaches its highest point. In Figure 2.20, the ball moves from $y = 0$ at $t = 0$ to its high point $y = 20$ m at $t = 2.0$ s. On this trip up, the velocity is positive. Once the ball begins falling back to Earth, its velocity becomes negative. At that one instant where the direction of the velocity changes, the speed is zero.

Note that we are measuring distances from the top of the building, so that $y_i = 0$. At time $t = 3.0$ s, the ball's position is $y = 15$ m above the top of the building. Its velocity is $v = -10$ m/s; the negative sign indicates that it is moving downward. The origin need not be the initial position, however. Figure 2.21 shows the same motion as before, but this time the origin is chosen as the sidewalk at the foot of the building. All positions are now measured relative to $y = 0$ at the bottom of the building. Now the motion begins at $y_i = 25$ m. The motion is the same in both cases—as it must be.

Look closely at the sketch and data describing the motion for the first four seconds. There is an interesting symmetry present: for a given position, the speed is the same going up as coming down. If we film this ball being thrown and then show that movie backward, it will look like a ball initially at the ground 25 m below the top of the building being thrown upward with an initial speed of 30 m/s. The motion is reversible; there is no way to tell whether the movie is being shown forward or backward (other than the details of the catch or throw of a person's hand). All the fundamental laws of physics have this property of being time-reversible. If we include air resistance, this symmetry of time-reversal is broken—when air resistance is not negligible, we *can* tell whether the movie is being run forward or backward.

Example 2.7

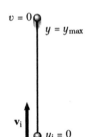

$v = 0$ $y = y_{max}$

v_i

$y_i = 0$

Figure 2.22
How high does the ball go?

Find the maximum height obtained by a ball thrown vertically upward with initial velocity v_i (Fig. 2.22).

Reasoning

We know the initial velocity v_i and shall take as our final velocity the velocity at the moment the ball reaches maximum height. From Figures 2.20 and 2.21, we know this velocity to be zero. We also know the acceleration—it is the acceleration due to gravity. If we choose to measure up as positive, then $a = -g$. We need the position ($y_{max} = $?). Position is related to velocity and acceleration through the velocity-position equation (2.7). So we can solve for the maximum height using that equation.

Solution

For vertical motion, it is again convenient to replace s by y. Equation 2.7 becomes

$$v^2 = v_i^2 + 2a(y - y_i)$$

Replacing a by $-g$, setting $y_i = 0$, $v = 0$, and solving for y, we get

$$y = y_i + \frac{v_i^2 - v^2}{2g} = \frac{v_i^2}{2g}$$

$$y_{max} = \frac{v_i^2}{2g}$$

Example 2.8

A stone is dropped into a well and requires 2.30 s to hit the water. How far is the water's surface from the top of the well?

Reasoning

This is simply asking how far the stone falls in 2.30 s. The position-time equation (2.6) relates what we want to what we have. Because the stone is in free fall, we know that $a = -g$.

Solution

$$y = y_i + v_i t + \frac{(-g)t^2}{2}$$

$$= 0 + 0 + \frac{(-9.80 \text{ m/s}^2)(2.30 \text{ s})^2}{2} = -25.9 \text{ m}$$

As always, the correct units on the answer—that we end up with a distance measured in meters—give us some confidence that the correct equation was used and the correct steps carried out. (Correct units do not guarantee correctness, but incorrect units guarantee that an error has been made.)

The negative sign indicates that, when it hits the water, the stone is 25.9 m below the ground (where we set $y = 0$).

___

Practice Exercise If the stone requires 1.20 s to hit the water, how deep is the well?

Answer 7.06 m

Example 2.9

From atop a building 50 m tall, a ball is thrown upward with an initial velocity of 12 m/s (Fig. 2.23). How long will it take to strike the ground?

Reasoning

The acceleration and positions are known, and we want the time. The position-time equation (2.6) relates acceleration, position, and time.

Solution

Choosing the origin of coordinates to be at the ground, we have $y_i = 50$ m at the top of the building and $y = 0$ when the ball hits the ground. The initial velocity is $v_i = 12$ m/s. Again taking $a = -g$, we have for Equation 2.6

$$y = y_i + v_i t + \frac{(-g)t^2}{2}$$

$$0 = (50 \text{ m}) + (12 \text{ m/s})t + \frac{(-9.8 \text{ m/s}^2)t^2}{2}$$

This is a quadratic equation in t. After rearranging terms, it becomes

$$(4.9 \text{ m/s}^2)t^2 - (12 \text{ m/s})t - 50 \text{ m} = 0$$

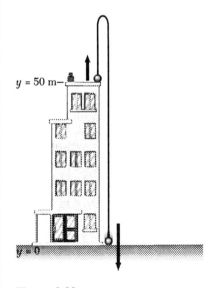

Figure 2.23
If a ball is thrown upward with an initial velocity of 12 m/s, how long will it take to strike the ground?

$y = 50$ m

$y = 0$

Notice that the units are consistent; each term has units of length (m) because the seconds in the denominators are canceled by the seconds in the t factors in the numerators. Being sure the units are consistent, we can drop them to write this equation as

$$4.9t^2 - 12t - 50 = 0$$

which we can solve by the standard quadratic equation:

$$ax^2 + bx + c = 0$$

which has solutions of

$$x = \frac{-b \pm \sqrt{b^2 - 4ac}}{2a}$$

For our particular case, $a = 4.9$, $b = -12$, and $c = -50$:

$$t = \frac{-(-12) \pm \sqrt{(-12)^2 - 4(4.9)(-50)}}{2(4.9)} = \frac{12 \pm 34}{9.8}$$

We have two possible solutions:

$$t = 4.7 \text{ s} \quad \text{and} \quad t = -2.2 \text{ s}$$

Practice Exercise   If the ball is thrown upward with an initial velocity of 18 m/s, how long will it take to strike the ground?

Answer   5.5 s and −1.8 s

How can there be two solutions to a physical problem? After all, the ball does only one thing! It strikes the sidewalk at only one time. If the time is measured from $t = 0$ as the ball leaves the thrower's hand, then a positive time means a time *after* the ball has been thrown. A negative time represents some time *before* it was thrown. Physically, we can ask only what happens *after* the ball is thrown. Both the positive and the negative value are solutions to the quadratic equation. (The negative value means that 2.2 s earlier the ball could have been thrown up from the sidewalk so that at $t = 0$ it passed the top of the building with a velocity of 12 m/s.) However, we are interested in what happens after the ball is thrown, and so we want the solution for positive time. Thus, the ball hits the sidewalk 4.7 s after it is thrown.

2.6 Graphical Interpretation

CONCEPT OVERVIEW

- Velocity is the slope of the line tangent to the curve at every point on a position-time graph.
- Acceleration is the slope of the line tangent to the curve at every point on a velocity-time graph.

Graphs are an important tool of science. With graphs you can immediately see ideas and relationships that would be difficult to visualize working with a data table alone. We can use graphs in describing and understanding motion.

Figure 2.24 is a graph of the position of a car at various times. We shall plot the time along the horizontal axis and any other variable, like position or velocity or acceleration, along the vertical axis. One question that comes to mind is, how fast is this car going? We know that the average velocity is the change in position divided by the change in time:

$$v_{\text{avg}} = \frac{\Delta s}{\Delta t}$$

During the first 5.0 s, the car has traveled 7.5 m. So, during that time, its average velocity is

$$v_{\text{avg}} = \frac{7.5 \text{ m}}{5.0 \text{ s}} = 1.5 \text{ m/s}$$

What about some later time? As you can see from the dotted lines, during the 6-s interval from 2.0 s to 8.0 s, the car traveled 9.0 m. During that time, its average velocity is

$$v_{\text{avg}} = \frac{9.0 \text{ m}}{6.0 \text{ s}} = 1.5 \text{ m/s}$$

In both cases our process has been to find the slope of this graph. The slope of a straight-line graph, you may recall, is the change in the "vertical variable" divided by the change in the "horizontal variable," the *rise* over the *run*.

The curve in Figure 2.24 is a straight line; therefore we can say the position varies *linearly* with time, or—to say it another way—the slope everywhere is the same, which means that the velocity is constant. The average velocity, then, equals the instantaneous velocity.

Figure 2.25 is a graph of velocity versus time. The velocity is a constant 1.5 m/s, and so the graph is a straight, horizontal line. The slope of this line is zero. Acceleration is the change in velocity divided by the change in time, $\Delta v/\Delta t$. That is, acceleration is the *slope* of the line on a graph of velocity versus time. With uniform velocity, the acceleration is zero.

Figure 2.26(a) is more interesting. It describes the position of a body as a function of time. Between 0 s and 5.0 s, the position-time graph is a straight line. For every 1 s we watch, the body moves 15 m. The slope of this part of the line, then, is 15 m/s, and this is the velocity of the body from $t = 0$ s to $t = 5.0$ s. This velocity is shown on Figure 2.26(b) by the horizontal line segment at $v = 15$ m/s running from $t = 0$ s to $t = 5.0$ s.

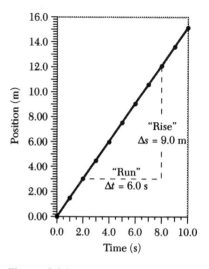

Figure 2.24
The slope of a line is the change in the "vertical variable" divided by the change in the "horizontal variable," the *rise* over the *run*. Velocity is the slope of the line on a position-time graph.

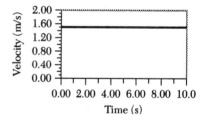

Figure 2.25
A constant-velocity graph has a slope of zero. Constant velocity means zero acceleration.

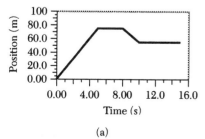

(a)

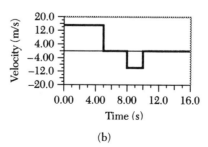

(b)

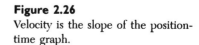

Figure 2.26
Velocity is the slope of the position-time graph.

Between $t = 5.0$ s and $t = 8.0$ s, the body has stopped and remains at $s = 75$ m. In other words, there is no change in position; $\Delta s = 0$. The position-time graph is a horizontal line with a slope of zero. This zero slope means the velocity is zero. That velocity is zero shows up on the velocity-time graph as a horizontal line segment at $v = 0$ from $t = 5.0$ s to $t = 8.0$ s.

From $t = 8.0$ s to $t = 10.0$ s, the position-time graph shows the body to be moving backward—toward smaller values of s. During this time it moves from $s = 75$ m to $s = 55$ m. That is a displacement of $\Delta s = s_f - s_i = 55$ m $- 75$ m $= -20$ m. We write $\Delta s = -20$ m, with the negative sign indicating this decrease. This decrease occurs over a time interval of $\Delta t = 2.0$ s. Or we could look at, say, $t = 8.5$ s and $t = 9.5$ s, use $\Delta t = 1.0$ s, and find $\Delta s = -10$ m from the graph. In either case, the slope is -10 m/s. This slope shows up on the velocity-time graph as a straight line segment at $v = -10$ m/s from $t = 8.0$ s to $t = 10.0$ s. After $t = 10.0$ s, the body remains at rest at $s = 55$ m, the slope of the position-time graph is zero, and the velocity is zero.

Look at Figure 2.27(a), another position-time graph. What can we say about the slope or velocity for this case? A position-time graph need not be composed of only straight lines. The "slope of the line" is now replaced with the "slope of the line tangent to the curve at a particular time." This is illustrated at $t = 3.0$ s and $t = 10.0$ s. Finding "the slope of the line" for such a curve at a particular time by finding the slope of the *tangent line* at that particular time is the same as the limiting process of letting Δt approach zero we used

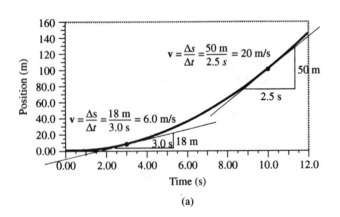

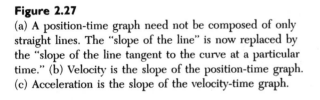

(a)

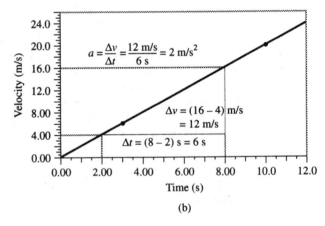

(b)

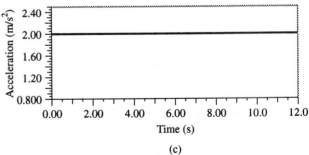

(c)

Figure 2.27
(a) A position-time graph need not be composed of only straight lines. The "slope of the line" is now replaced by the "slope of the line tangent to the curve at a particular time." (b) Velocity is the slope of the position-time graph. (c) Acceleration is the slope of the velocity-time graph.

in going from average velocity to instantaneous velocity and from average acceleration to instantaneous acceleration.

In Figure 2.27(a), a straight line has been drawn tangent to the position-time curve at $t = 3.0$ s. For this straight line, the position changes by $\Delta s = 27$ m $- 3$ m $= 24$ m during a time interval of $\Delta t = 6.2$ s $- 2.4$ s $= 3.8$ s. This is shown by the little triangle near the curve at $t = 3.0$ s. These data mean a slope (change in vertical over change in horizontal, or rise over run) of

$$\text{Slope} = \frac{\Delta s}{\Delta t} = \frac{24 \text{ m}}{3.8 \text{ s}} = 6.3 \text{ m/s}$$

Velocity is change in position divided by change in time. That is, *velocity is the slope of the position-time graph.* The velocity is 6.3 m/s at $t = 3.0$ s. In a similar manner, at $t = 10.0$ s we find the velocity to be 20 m/s. The line tangent to the curve and measurements of Δs and Δt for its slope are shown in the figure.

These two data points—$v = 6.3$ m/s at $t = 3.0$ s and $v = 20$ m/s at $t = 10.0$ s—are marked on the velocity-time graph in Figure 2.27(b). For this particular set of data, if we were to find the velocity at several different times, we would find that all velocities lie on this straight line. The velocity is increasing linearly with time for this case. If the velocity changes with time, then it is interesting to ask about the slope of this velocity-time graph. Since the curve is a straight line, there is only one slope. As shown in the figure, between $t = 2.0$ s and $t = 8.0$ s, with $\Delta t = 6.0$ s, the velocity changes from 4.0 m/s to 16.0 m/s, and so $\Delta v = 12.0$ m/s. The slope is always the rise, Δv in this case, over the run, Δt here:

$$\text{Slope} = \frac{\Delta v}{\Delta t} = \frac{12.0 \text{ m/s}}{6.0 \text{ s}} = 2.0 \text{ m/s}^2$$

And this slope is the acceleration. The acceleration is constant at 2.0 m/s². This constant acceleration is shown in Figure 2.27(c) as a horizontal line at 2.0 m/s² on an acceleration-time graph.

So you have seen that **velocity is the value of the slope of the curve on a position-time graph and acceleration is the value of the slope of the curve on a velocity-time graph.** If the graph is not a straight line, then the "slope of the curve" is the "slope of the straight line *tangent to* the curve" at the point of interest. We can put this information to good use in a still more interesting case, shown in Figure 2.28. We can first use the position-time graph to determine the velocity-time graph and from that derive the acceleration-time graph. Or we can go the other way.

For the first 2.0 s—up to "a" in the figure—the acceleration is constant at 1.0 m/s². That means the velocity increases linearly during that time. If the body starts at rest, after 2.0 s it is moving at 2.0 m/s. The position will change as shown in Figure 2.28(c).

Between "a" and "b" the acceleration is zero. That means the velocity is constant; at time "a" ($t = 2.0$ s) the velocity was 2.0 m/s, so it remains that while the acceleration is zero. Constant velocity means the position-time graph is a straight line as shown, linearly increasing with a slope of 2.0 m/s.

From "b" to "c" the acceleration is -1.0 m/s². Negative acceleration means the positive velocity is decreasing. Each second it decreases by 1.0 m/s.

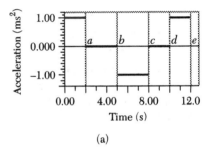

(a)

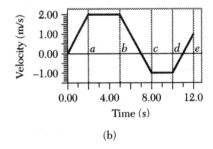

(b)

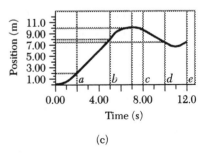

(c)

Figure 2.28
Graphs of position, velocity, and acceleration as functions of time are closely related to each other.

In this case it starts at 2.0 m/s and gets smaller and smaller until it first vanishes (at $t = 7.0$ s) and then becomes negative. The position-time graph is a curve as shown; when the velocity is zero, the slope of the position-time graph is zero, too. As the velocity becomes negative, the position-time graph curves down with a negative slope as shown.

Between "c" and "d" the acceleration is again zero, which means the velocity remains constant, which in turn means the position continues to change at a constant rate. At "c" ($t = 8.0$ s) the velocity is -1.0 m/s, and so the position-time graph continues downward with that as its slope.

Between "d" and "e" the acceleration is again 1.0 m/s². At "d" the velocity is -1.0 m/s, and so it must become first less negative, then zero, and then positive. The position-time graph changes accordingly; remember that the slope of the position-time graph is always equal to the velocity.

All of this is an illustration that velocity is the value of the slope of a position-time graph and acceleration is the value of the slope of a velocity-time graph.

SUMMARY

Kinematics is the *description* of motion. For motion in one dimension, a single number, the *displacement* s, completely locates an object. As an object moves, its velocity is a description of how rapidly it moves:

$$v = \frac{\Delta s}{\Delta t} \tag{2.1}$$

Acceleration describes how rapidly the velocity changes and is defined by

$$a = \frac{\Delta v}{\Delta t} \tag{2.3}$$

When determining instantaneous velocity and instantaneous acceleration, the time interval Δt must approach zero if the quantities that change do not change uniformly.

For the case of constant acceleration, three very useful relationships were developed:

$$v = v_i + at \tag{2.5}$$

$$s = s_i + v_i t + \frac{at^2}{2} \tag{2.6}$$

$$v^2 = v_i^2 + 2a(s - s_i) \tag{2.7}$$

In the absence of air resistance, all objects falling freely near Earth's surface have the same acceleration, 9.80 m/s² downward.

Velocity is the slope of the line tangent to the curve at every point on a position-time graph. Acceleration is the slope of the line tangent to the curve at every point on a velocity-time graph.

QUESTIONS

1. Can the distance an object travels ever be less than the object's displacement? Can it ever be greater?
2. Can the average speed of a car over some time interval ever be greater than any value of the instantaneous speed during that time?
3. If the average velocity over some time interval is zero, does that mean the instantaneous velocity was always zero during that time?
4. If the average speed over some time interval is zero, does that mean the instantaneous speed was always zero during that time?

5. Does a car's speedometer measure speed, velocity, or acceleration?

6. Can two objects have different velocities but the same speed?

7. Can an object's velocity vary if its acceleration is constant?

8. A physics student standing atop the physics building throws a stone straight up and watches as it goes up, stops, falls back down, passes beside her, and hits the sidewalk below. What is the stone's acceleration (including sign) (a) immediately after leaving her hand, (b) at the top of its path, and (c) just before it strikes the sidewalk?

9. Describe a situation in which a car has (a) positive velocity and negative acceleration,
(b) positive velocity and positive acceleration,
(c) negative velocity and positive acceleration,
(d) negative velocity and negative acceleration.

10. Explain the following statement: Acceleration is to velocity what velocity is to position.

11. Is motion at constant velocity a special case of motion at constant acceleration?

12. When an object is thrown upward, its speed first decreases, then goes to zero, then increases. How can the acceleration be constant throughout the motion?

13. When an object is thrown upward, its velocity is initially upward and decreasing in magnitude, goes to zero, and then is downward and increasing in magnitude. What happens to the acceleration throughout this motion?

14. How does air resistance affect the acceleration of an object dropped from a tower?

15. An object is thrown straight up. What are its velocity and acceleration at the top of its path?

16. Sketch a position-time graph for an object moving with constant velocity.

17. Sketch a velocity-time graph for an object moving with constant acceleration.

18. Sketch a position-time graph for an object moving with constant acceleration.

PROBLEMS

2.1 Kinematics: Describing Motion

2.2 Distance and Displacement

2.1   A truck, a house, and a car are located along a straight road as sketched in Figure 2.29. (a) What is the distance between truck and car? What are the following displacements: (b) truck from house, (c) house from truck, (d) car from truck, (e) truck from car (f) tree from house, (g) car from tree, (h) truck from tree?

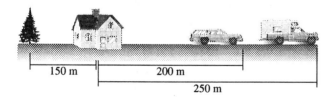

Figure 2.29

2.2   A lion paces back and forth along the front of its cage, which is 12 m wide. As a visitor watches, the lion starts at the left side of the cage, paces to the right, then back to the left, and back to the right again. (a) What total distance has the lion paced? (b) What is the lion's displacement?

2.3   During an automobile rally, a driving team travels 5 mi north, turns around and travels 3 mi south, and then travels 7 mi north again. (a) What distance has the team traveled? (b) What is its displacement?

2.3 Speed and Velocity

2.4   A long-distance runner completes a 10-km race in 48 min. What is her average speed?

2.5   You drive for 40 min at 60 km/h. Then traffic conditions force you to reduce your speed to 40 km/h for 20 min. What is your average speed?

2.6   You drive for 45 min at 80 km/h. Then heavy traffic forces you to reduce your speed to 50 km/h for 30 min. What is your average speed?

2.7   Out for a Sunday drive, you drive at 60 km/h for 1 h and 15 min to a picnic site and there stop for 2 h. Then you continue on at 80 km/h for 45 min. What is your average speed?

2.8   A car travels 10 km at 80 km/h, 20 km at 40 km/h, stops for 30 min, and continues on for 10 km at 50 km/h. What is its average speed?

2.9   A certain bacterium swims with a speed of 3.5 μm/s. How long will it take this bacterium to swim across a petri dish whose diameter is 8.4 cm? (1 μm = 10^{-6} m)

2.10   A bicyclist travels with an average velocity of 15 km/h north for 20 min. What is his displacement?

2.11   A jogger runs with an average speed of 12 km/h for 15 min. What distance does she cover?

O2.12   The speed of sound in air is about 330 m/s. You observe a lightning bolt strike a tree 1.5 km away. How much time elapses between your seeing the lightning bolt and your hearing the thunder that accompanies it? Why do you not need to take into account the speed of light?

O2.13   The speed of sound in air is about 330 m/s. You observe a lightning bolt and then hear thunder 4.5 s later. How far away was the lightning bolt? Why do you not need to take into account the speed of light?

O2.14   An arrow is fired with a horizontal velocity of 58.7 m/s and hits a target on an archery range 80 m away. How long did it take the arrow to travel this distance, assuming constant horizontal velocity?

2.4 Acceleration

2.15   A car accelerates from rest to 100 km/h in 9.3 s. What is its average acceleration in meters per second squared?

2.16   A car accelerates from rest to 90 km/h in 8.8 s. What is its average acceleration in meters per second squared?

2.17   A bicyclist starting from rest has a velocity of 16 m/s 8 s later. What is her average acceleration?

2.18   A jet aircraft launched from an aircraft carrier reaches a velocity of 230 km/h in 1.25 s. What is the aircraft's average acceleration in meters per second squared?

2.19   A jet aircraft landing on an aircraft carrier is brought to a complete stop from a velocity of 215 km/h in 2.7 s. What is its average acceleration in meters per second squared?

2.20   A car brakes to a stop from 90 km/h in 5.8 s. What is its average acceleration in meters per second squared?

2.21   A certain sports car has an acceleration of 3.2 m/s^2, assumed constant. Starting from rest, how long does the car require to reach a velocity of 30 m/s? How far does it travel while attaining that velocity?

2.22   A particular car has an acceleration of 3.8 m/s^2, assumed constant. Starting from rest, how long does the car require to reach a velocity of 100 km/h? How far does it travel while reaching that velocity?

2.23   A certain car has an acceleration of 2.4 m/s^2, assumed constant. Starting from rest, how long does the car require to reach a velocity of 90 km/h? How far does it travel while reaching that velocity?

2.24   A freight train has an acceleration of 0.05 m/s^2, assumed constant. How much time is required to accelerate from $v = 0$ to $v = 25$ m/s?

2.25   At normal speeds a certain car is capable of accelerating at 2.1 m/s^2. How much time does this car require to accelerate from 10 m/s to 30 m/s? How far does the car travel during this time?

2.26   A motorcyclist initially moving with a velocity of 25 km/h accelerates for 4.8 s. If the motorcycle accelerates at 1.25 m/s^2, how far does it travel during this time?

2.27   A rocket-driven sled is used to investigate the effects of extremely large accelerations. Starting from rest, such a sled is capable of attaining a speed of 1000 km/h in 18 s. What is its acceleration? How far does it travel during that time?

2.28   High-speed rocket sleds like the one described in the previous problem are often slowed very rapidly by the insertion of vanes into water channels running parallel to the sled's track. Using such a water brake, a sled is capable of coming to a stop from 500 km/h in 8.5 s. What is the acceleration? How far does the sled move in coming to rest?

O2.29 A truck initially traveling at 100 km/h slows down at a uniform acceleration of -2.5 m/s^2. How far has it traveled when its speed is half its initial value?

O2.30 A speedboat increases its speed at the rate of 1.8 m/s^2. How much time is required for it to increase its speed from 9 m/s to 18 m/s? How far does it travel during that time?

O2.31 A football is given a forward velocity of 15 m/s by a passer whose arm moves through a horizontal distance of 1.2 m. Compute the average forward acceleration of the football.

O2.32 A driver in a car going 30 m/s applies the brakes and causes an acceleration of -6.2 m/s^2. How far does the car travel before coming to a stop?

O2.33 A car accelerates from rest and travels 0.50 km in 9.0 s. What is the car's speed at that time? Assume constant acceleration.

O2.34 An airplane needs to reach a takeoff speed of 120 km/h before it reaches the end of a 2.5-km runway. What acceleration (assumed constant) must it have to do that?

O2.35 A jet aircraft reaches a takeoff speed of 54 m/s by the time it reaches the end of an 80-m catapult on an aircraft carrier. What is its acceleration? How much time does the launch require?

O2.36 A tennis ball with a speed of 10.0 m/s is thrown perpendicular to a wall. After striking the wall, the ball rebounds in the opposite direction with a speed of 8.0 m/s. If the ball is in contact with the wall for 12.0 ms, what is its average acceleration while in contact with the wall?

O2.37 To land on shorter runways, some airplanes use a "drag chute" in addition to normal brakes and flaps. From an initial velocity of 200 km/h, such a parachute-equipped airplane is able to come to rest in 0.25 km. What is its acceleration? How long does this landing require?

O2.38 In passing a slow truck, you must accelerate from 50 km/h to 90 km/h. What must your acceleration be so that you can complete your passing maneuver in 1.0 km?

O2.39 In passing a car, you must accelerate from 40 km/h to 100 km/h. What must your acceleration be so that you can pass in 0.50 km?

O2.40 An arrow being shot from a bow is accelerated to a velocity of 70 m/s over a distance of 0.75 m. Find its acceleration.

O2.41 In a television picture tube, an electron is accelerated to a velocity of 30 000 km/s over a distance of 12 cm. What average constant acceleration does the electron experience?

O2.42 A freight train 400 m long is moving on a straight track with a speed of 84.2 km/h. The engineer applies the brakes as the engine passes a crossing. Later, as the caboose passes the crossing, its speed has slowed to 16.4 km/h. Assuming constant acceleration, how long did the train block the crossing?

O2.43 Skid marks 32 m in length are found at the scene of an accident. Estimate the car's initial speed. Assume a uniform braking acceleration of -7.2 m/s^2.

O2.44 A freight train moving at a constant 80 km/h passes a passenger train stopped on a parallel track. As the freight train passes, the passenger train starts to move with a constant acceleration of 0.5 m/s^2. Where and when does the passenger train catch up with the freight train?

O2.45 Just as the light changes, a truck moving at a constant 50 km/h passes a car stopped at an intersection. At that moment, the car starts to move with a constant acceleration of 3.5 m/s^2. Where and when will the car overtake the truck?

O2.46 A car speeding at 130 km/h passes a police car at rest. Just as the speeding motorist passes the police car, the police car begins pursuit. If the police car maintains a constant acceleration of 5.8 m/s^2, when and where will the speeding motorist be overtaken?

●2.47 A particular car has a maximum positive acceleration of $+2$ m/s^2 and a maximum negative acceleration of -4 m/s^2. The speed limit is 16 m/s (what is that in kilometers per hour and miles per hour?). What is the shortest legal time in which the driver can go from one stop sign to another, a distance of 320 m?

●2.48 A late commuter, sprinting at 8 m/s, is 30 m away from the rear door of a commuter train when it starts to pull out of the station with an acceleration of 1 m/s^2. Can the commuter catch the train? (Assume the platform is more than long enough.)

●2.49   A car speeding at 150 km/h passes a police car at rest. Just as the speeding motorist passes the police car, the police car begins pursuit. If the police car maintains a constant acceleration and overtakes the speeding motorist in a distance of 1.25 km, what is the acceleration of the police car?

●2.50   Just as a traffic light changes, a truck moving at a constant 60 km/h passes a car stopped at the light. At that moment, the car starts to move with a constant acceleration. What is the car's acceleration if the car overtakes the truck in 0.75 km?

● 2.51   While standing still, a police car is passed by a speeding car traveling at a constant 120 km/h. The police officer waits 2 s before deciding to pursue. What must the acceleration of the police car be in order to catch the speeding car within 3 km?

●2.52   A driver is traveling at 80 km/h when he sees an obstacle in the road in front of him. His reaction time is 0.75 s. That is, 0.75 s passes before he applies the brakes. Thereafter, the car comes to a stop with an acceleration of -6.5 m/s^2. How far does the car travel after the driver first sees the obstacle?

●2.53   Generalize the previous problem. Given an initial velocity v_i, a reaction time t_r, and a braking acceleration a, show that the total stopping distance is $s_{stop} = v_i t_r - \frac{1}{2}v_i^2/a$. Remember that a will be negative.

●2.54   The yellow caution light on a traffic signal should stay on long enough to allow a driver to either pass through the intersection or safely stop before reaching the intersection. That is, if a car is more than the stopping distance of the previous problem away from the intersection, it can stop. If the car is less than this stopping distance from the intersection, the yellow light should stay on long enough to allow the car to pass entirely through the intersection. Show that the yellow light should stay on for a time $t_{light} = t_r - (v_i/2a) + (s_i/v_i)$, where t_r is a driver's reaction time, v_i is the velocity of any cars approaching the intersection (in other words, the speed limit), a is the braking accelera-

tion, and s_i is the length of the intersection.

●2.55   As city traffic planner, you expect cars to approach an intersection 16 m wide with a velocity of 60 km/h. Be cautious and assume a reaction time of 1.1 s to allow for a driver's indecision. Find the length of time the yellow light should remain on. Use a braking acceleration of -2 m/s^2.

●2.56   What of slower moving traffic? Consider a car approaching the intersection of the previous problem at 40 km/h. For such a car, how long should the yellow light remain on? As city traffic planner, which time would you choose?

2.5 Free Fall

(Ignore air resistance in all these problems.)

2.57   A pebble dropped into a well requires 1.3 s to hit the surface of the water. How far down is the water level?

2.58   A coin dropped down a long ventilation shaft requires 1.8 s to hit bottom. How long is the shaft?

2.59   A camera accidently dropped from the top of an observation tower takes 3.7 s to hit the ground. How high is the tower?

2.60   An ice cream cone dropped from the top of a stopped Ferris wheel takes 2.1 s to hit the ground. How high is the Ferris wheel?

2.61   A sky diver falls for 12 s before opening her parachute. How far has she fallen and how fast is she going when her parachute opens?

○2.62   How fast must a stone be thrown straight up in order to just reach the top of a building that is 10 m high?

○2.63   A shingle is dropped from the roof of a tall building. (a) What is its velocity at the end of 2 s? How far does it fall (b) in 2 s? (c) in 3 s? (d) What is its average velocity between $t = 2$ s and $t = 3$ s? (e) What is its velocity at the end of 3 s?

○2.64   How long does it take a ball to hit the ground when dropped from a height of 10 m? How fast is it going when it hits the ground?

○2.65   How long does it take a rivet to hit the ground when dropped from a height of

100 m? How fast is it going when it hits the ground?

O2.66 Cliff divers in Acapulco dive from rocks about 35 m above the water. What is their velocity as they hit the water?

O2.67 How fast must a stone be thrown straight up in order to just reach the top of a building that is 10 m high?

O2.68 A baseball is thrown vertically upward with a velocity of 20 m/s. How high does it go? How long is it in the air? How fast is it going when it strikes the ground? Assume it leaves the thrower's hand 2.0 m above the ground.

O2.69 Romeo stands directly below Juliet's second-story window. With what velocity must a stone leave his hand to just reach the window, 4 m above his hand?

O2.70 With what initial velocity does a kangaroo leave the ground if it jumps 2.3 m high?

O2.71 A nurse squeezes a syringe, and the liquid squirts 5.3 cm high. With what speed did the liquid emerge from the syringe?

O2.72 Killer whales at Sea World routinely jump 7.5 m above the water. What is their minimum speed as they leave the water?

(Copyright © 1993 Sea World of California, Inc. All rights reserved. Reproduced by permission.)

O2.73 A ball is thrown upward with a velocity of 12.8 m/s. What is its velocity (including sign) (a) after 1 s and (b) after 2 s?

O2.74 A construction worker drops a wrench from the fifth story of a building, 18 m above the sidewalk. How fast is the wrench going when it hits the sidewalk?

O2.75 A falling bolt hits a construction worker's hard hat with a speed of 21 m/s. From what height did the bolt fall?

O2.76 A baseball is thrown straight down with a speed of 4.9 m/s from the top of a stadium, 15 m above the parking lot. (a) How fast is the ball moving when it hits the ground of the parking lot? (b) How long does it take the ball to hit the ground?

O2.77 The acceleration due to gravity is 1.67 m/s^2 on the surface of the Moon. If a rock is thrown up from the Moon's surface with an initial velocity of 15 m/s, determine how high it will go and how much time elapses before it strikes the Moon's surface.

O2.78 A lunar astronaut slips from his ladder, 12 m above the Moon's surface. When he hits the surface, how fast will he be going? The acceleration due to gravity at the Moon's surface is 1.67 m/s^2.

O2.79 On the surface of Mars, the acceleration due to gravity is 3.92 m/s^2. With what initial velocity must a Martian throw a stone straight up so that it reaches a maximum height of 15 m?

O2.80 A hot-air balloon is ascending at a uniform speed of 4 m/s. At a height of 200 m, a sandbag is dropped. How much time is required for the sandbag to hit the ground? How fast is the sandbag going when it hits the ground?

O2.81 A ball is thrown vertically upward at 12.5 m/s. It leaves the thrower's hand 2.0 m above the ground. (a) How high does it rise? (b) How long does it take to reach its maximum height? (c) What is its acceleration at the top of its path? (d) What is its speed when it strikes the ground?

O2.82 A stone is dropped from a bridge, and 1 s later another stone is dropped. How far apart are the two stones when the first one has reached a speed of 19.6 m/s?

O2.83 Reaction time can be estimated by the "dollar bill catch." A dollar bill is held by con artist A with its long dimension vertical. Observer B holds her fingers at the bottom edge of the bill, 15.0 cm below the top. Then A releases the bill without warning, and B can keep the bill if she catches it. What must

B's reaction time be to catch the bill? It is very, very rare that A ever has to pay. (This is an interesting experiment to try yourself.)

Problem 2.83.

○2.84   Using a jet pack, a lunar astronaut is moving upward at a uniform velocity of 3 m/s. At a height of 15 m, she drops a wrench. How much time is required for the wrench to hit the Moon's surface? How fast is it going when it hits? How high is the astronaut then? The acceleration due to gravity near the Moon's surface is 1.67 m/s².

○2.85   In 1865 Jules Verne proposed sending people to the Moon by firing a space capsule from a cannon 220 m long with a final velocity of

Problem 2.85. A French postage stamp commemorates Jules Verne and the 1902 science fiction movie about his voyage to the Moon, produced by George Melies. (*The Granger Collection, New York*)

10.97 km/s. What would be the (unrealistically large) acceleration experienced by space travelers during launch?

○2.86   In the mid-1960s, McGill University launched high-altitude weather sensors by firing them from a cannon made from two World War II Navy cannons bolted together for a total length of 18 m. It was proposed that this arrangement be used to launch a satellite. The orbital speed of a satellite is about 29 000 km/h. What would the average acceleration throughout the 18 m length of the cannon have to be in order to have a muzzle velocity of 29 000 km/h for the satellite?

○2.87   A coin is dropped into a well 30 m deep. How long after it is released will one hear the sound of the splash? Sound travels about 330 m/s.

○2.88   A tennis ball is thrown upward from the top of an 80-m building. The ball goes up, comes back down beside the thrower, and continues down to the sidewalk below. The total time of the trip—up, back, and to the sidewalk—is 5.5 s. What was the ball's initial velocity?

○2.89   During the last second of its fall from rest, a certain body travels one half of its total fall distance. From what height did it fall?

○2.90   Two stones are released from the same height, 1.0 s apart. How long after the first one is released will they be 15 m apart?

○2.91   Two students are on a balcony 19.6 m above the street. One throws a ball vertically downward at 14.7 m/s. At the same moment the other throws a ball vertically upward at 14.7 m/s. The second ball just misses the balcony on its way down. (*a*) How long after the first ball strikes the street does the other ball strike the street? (*b*) With what velocity does each ball strike the street? (*c*) How far apart are the balls 1 s after being thrown?

●2.92   A falling flower pot takes 0.25 s to pass by a window 1.5 m high. How far above the top of the window was the balcony from which the flower pot fell?

●2.93   How fast must a ball be thrown upward to have a velocity of 5 m/s as it passes a rooftop observer 12 m above the thrower?

●2.94 A ball is dropped from a bridge 50 m above the water. One second later another ball is thrown downward. Both strike the water at the same time. What was the initial velocity of the second ball?

●2.95 Using a rocket pack with full throttle, a lunar astronaut accelerates upward from the Moon's surface with an acceleration of 2.0 m/s^2. At a height of 5 m, a bolt comes loose. (The acceleration due to gravity on the Moon's surface is about 1.67 m/s^2.) (a) How fast is the astronaut moving at that time? (b) When will the bolt hit the Moon's surface? (c) How fast will it be moving then? (d) How high will the astronaut be when the bolt hits? (e) How fast will the astronaut be traveling then?

●2.96 Poor Wyl E Coyote is at it again. Attempting to catch the Road Runner, he falls from the top of a 400-m cliff. He is wearing an Acme Fireworks rocket but requires 6 s to light the fuse and ignite the rocket. The burning rocket gives him an upward acceleration so that his velocity reaches zero just as he reaches the bottom. (a) How far has he fallen before the rocket starts? (b) What is his upward acceleration in order to have this gentle landing?

But, alas, as Edsel Murphy would have it, the rocket does not shut off. Nor is Wyl E able to release himself from the rocket. Therefore, the rocket now carries him upward with this same acceleration for an additional 5 s before it depletes its fuel. (c) How high is he as the fuel runs out? (d) How fast is he moving then? (e) How high above the canyon floor does he eventually go? (f) How fast is he going as he hits the canyon floor this time?

2.6 Graphical Interpretation

2.97 Figure 2.30 is a position-time graph of a bicycle moving along a straight sidewalk outside the physics lab. Determine the time at which the bicycle (a) has its greatest veloc-

ity, (b) has its greatest acceleration, (c) has constant velocity, (d) is stopped, (e) is moving backwards.

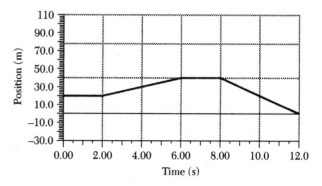

Figure 2.30
Problem 2.97.

2.98 Figure 2.31 is a position-time graph for several trains traveling along a straight track. Identify the graphs that correspond to (a) forward movement only, (b) backward movement only, (c) constant velocity, (d) greatest constant velocity, (e) no movement at all.

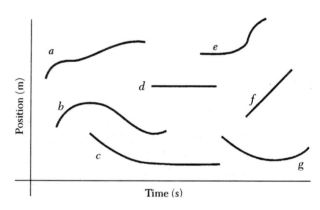

Figure 2.31
Problem 2.98.

2.99 Figure 2.32 is a position-time graph for a model train traveling along a straight track in a physics lab. Determine the time(s) at which the train (a) has its greatest displacement, (b) has its greatest velocity, (c) is moving backwards.

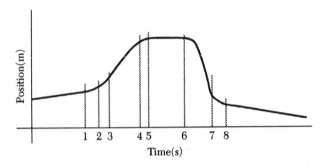

Figure 2.32
Problem 2.99.

○2.100 Figure 2.33 is a position-time graph. From it, construct the corresponding velocity-time and acceleration-time graphs.

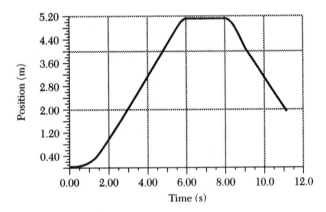

Figure 2.33
Problem 2.100.

○2.101 Figure 2.34 is a position-time graph for a cart traveling along a straight track in a physics lab. Determine the time(s) at which the cart (a) has its greatest velocity, (b) has constant velocity, (c) is moving backwards.

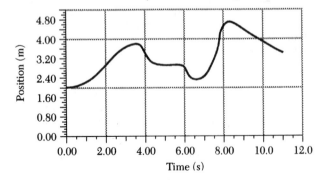

Figure 2.34
Problem 2.101.

○2.102 Figure 2.35 is a position-time graph. From it, construct the corresponding velocity-time and acceleration-time graphs.

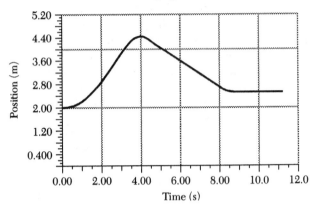

Figure 2.35
Problem 2.102.

○2.103 Figure 2.36 is a velocity-time graph of a bicycle moving along a straight sidewalk outside the physics lab. Determine the time(s) at which the bicycle (a) has its greatest velocity, (b) has its greatest acceleration, (c) has constant velocity, (d) is stopped, (e) is moving backwards.

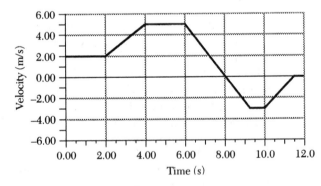

Figure 2.36
Problem 2.103.

○2.104 Figure 2.37 is a velocity-time graph. Use it to construct the corresponding position-time and acceleration-time graphs. Let the car start at $x = 0$ for $t = 0$.

○2.105 Figure 2.38 is a velocity-time graph. Use it to construct the corresponding position-time and acceleration-time graphs. Let the car start at $x = 0$ for $t = 0$.

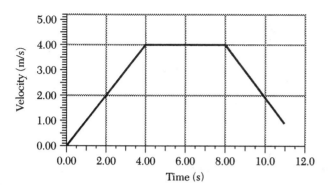

Figure 2.37
Problem 2.104.

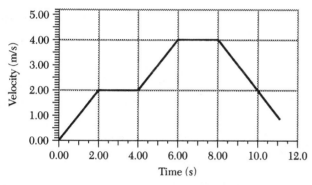

Figure 2.38
Problem 2.105.

○2.106 Figure 2.39 is an acceleration-time graph. Use it to construct the corresponding position-time and velocity-time graphs. Start the car from rest, $v_i = 0$, at $x = 0$ for $t = 0$.

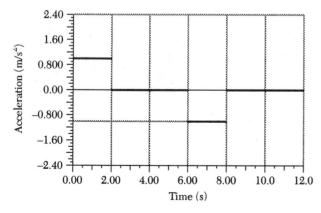

Figure 2.39
Problem 2.106.

○2.107 Figure 2.40 is an acceleration-time graph. Use it to construct the corresponding velocity-time and position-time graphs. Start the car from rest at the origin at time $t = 0$.

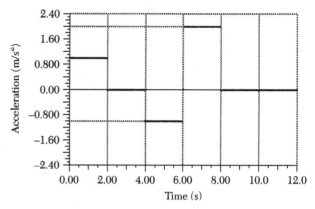

Figure 2.40
Problem 2.107.

General Problems

●2.108 When he was with the California Angels, Nolan Ryan crossed home plate with a speed of 45.06 m/s. Home plate is 18.5 from the pitcher's mound. How long did it take the baseball to travel this distance, assuming constant speed?

Nolan Ryan pitching. *(Phil Huber/Black Star) (Problem 2.108)*

PHYSICS AT WORK

At a busy airport such as LaGuardia International in New York City, air traffic controllers direct pilots to the correct runways with the language of vectors. They must take into account the paths of other airplanes and the speed and direction of the wind, which are also vector quantities. Imagine yourself in the control tower, giving directions to the pilot of the jet near the bottom of the picture. If that jet is going from New York to Miami, what vector should you tell the pilot to follow? What if the flight is bound for Chicago? What laws of vector math apply if the wind is blowing from west to east? What if it is blowing from north to south?

(Photo © Russ Kinne/Comstock)

3

Vectors and Motion in Two Dimensions

Key Issues

- Vectors provide a useful method of describing the motion of an object free to move in two dimensions. The displacement of an object from an origin or reference point is a good example of a vector. Velocity and acceleration are also vectors. A vector is represented by an arrow whose length is proportional to the magnitude of the vector and whose direction is the same as the vector.

- Vectors may be added graphically by making a careful drawing with each vector, beginning where the previous one has ended. The vector sum—called the resultant—of adding these vectors is a vector that begins where the first vector began and ends where the final vector ended.

- Vectors may be resolved into components along perpendicular axes. These components may then be added as scalars to find the components of the resultant. The resultant may then be found or constructed from these components.

- The motion of an object once it is free to move with only the force of gravity acting upon it is called *projectile motion*. Projectile motion is a combination of horizontal motion with constant velocity and vertical motion with constant acceleration.

By decade's end many automobiles will come equipped with on-board computers that can display on an electronic map the driver's current position as well as direction of motion. The driver will view changes in the car's position as the car moves because the on-board computer will receive constant input from global positioning satellites.

In the previous chapter we examined straight-line motion. That was an important first step, but most of the motion in our world is not along a single straight line. How do we locate an object that is not on a convenient reference line? How do we describe motion that does not follow a straight-line path? A yacht in the America's Cup race may be located by giving its longitude and latitude. Locating something that lies somewhere off a straight reference line requires vectors. What of the motion of a baseball going into the stands? Describing such motion requires the ideas of projectile motion. Vectors and projectile motion are ideas we seek to understand in this chapter.

3.1 Scalars and Vectors

CONCEPT OVERVIEW

- Scalars are fully specified by a number and a unit.

- Vectors require a direction (in addition to a number and a unit) to be fully specified.

An important distinction for physical quantities is whether they are scalars or vectors. As we learned in Chapter 2, scalars are fully specified by a number (how many?) and a unit (of what?). Some examples are temperature ($-5°C$), mass (65 kg), and time (24 h). Vectors, however, require a direction in addition to a number and unit to be fully specified. Some examples are displacement (150 km, due north) and velocity (80 km/h, 30° east of north). Without the direction, a vector quantity is not fully described; an important piece of information is missing.

Diagrams are always useful in understanding physics. Where vectors are concerned, however, diagrams are essential! We can represent a vector in a diagram by drawing an arrow. The **magnitude** of the vector, which is the number and unit (the scalar) part of the vector, is represented by the length of the arrow. The arrow is drawn pointing in the same *direction* as the vector it represents. In this text, to indicate that a quantity is a vector, we shall write it in boldface type, **B**. In handwritten notes, a vector is indicated either by an arrow over it, $\vec{B}$, or by a squiggly line underneath it, B̰. The magnitude alone is a scalar, and we shall write it simply as an italic *B* without an arrow, squiggle, or boldface type. Please note that **B** and *B* are *not* the same thing at all. If we want to emphasize that we are talking about just the magnitude—the scalar part—of a vector, we can indicate that with absolute value bars:

$$B = |\mathbf{B}| \tag{3.1}$$

Figure 3.1 shows several vector quantities represented by arrows.

As we learned in Chapter 2, distance is a scalar quantity. Suppose we say, for example, that the distance between Denver and Cheyenne is 150 km. This information simply tells us that Cheyenne is somewhere on the circle drawn in Figure 3.2. Any point on that circle is 150 km from Denver. The vector diagram in Figure 3.3 gives more information: Cheyenne is 150 km miles *north* of Denver. We can say that the displacement of Cheyenne from Denver is 150 km, due north. Displacement is a vector quantity. Distance is the magnitude, the scalar information, of displacement.

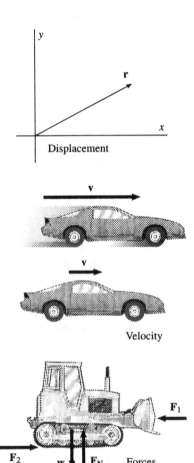

Figure 3.1
Common vector quantities. Direction is an important characteristic required to specify these quantities.

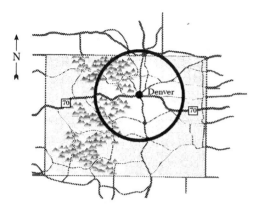

Figure 3.2
Stating that Cheyenne is 150 km from Denver means that Cheyenne lies somewhere on this circle.

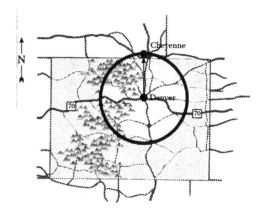

Figure 3.3
Specifying that Cheyenne is 150 km *due north* of Denver gives enough information to locate the city. This extra information, the direction, is needed because displacement is a vector.

A vector can be multiplied by a scalar, as shown in Figure 3.4. If **A** is some vector, we can multiply it by a scalar *s* (just a number, if you like) to get a new vector **C**. This operation is written

$$\mathbf{C} = s\mathbf{A} \tag{3.2}$$

The product **C** is a vector that has the same direction as the initial vector **A** (if *s* is positive), and its magnitude is just *s* times the magnitude of **A**:

$$C = sA \tag{3.3}$$

Note that Equations 3.2 and 3.3 are *not* the same thing; the distinction is very important.

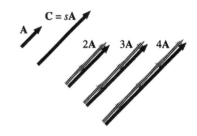

Figure 3.4
A vector **A** may be multiplied by a scalar *s* to create a new vector **C** = *s***A**. The new vector has the same direction as **A**, and its magnitude is *s* times greater than the magnitude of **A**.

3.2 Graphical Addition of Vectors

CONCEPT OVERVIEW

■ Vectors can be added graphically from a carefully drawn diagram.

The instructions on a treasure map—ten paces north, five paces southeast—provide a clear example of displacement vectors. We can write these two instructions as vectors,

$$\mathbf{A} = 10 \text{ paces, north}$$
$$\mathbf{B} = 5 \text{ paces, southeast}$$

and draw them as in Figure 3.5. A scale is chosen; perhaps one pace will be represented by 0.50 cm. The length of each vector is drawn accordingly. Here the length of vector **A** is 5.0 cm; that of **B**, 2.5 cm. The arrows are carefully drawn in the same directions as the vectors they represent, **A** pointing north and **B** pointing southeast.

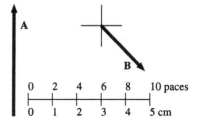

Figure 3.5
Vectors **A** and **B** represent displacements that are directions on a treasure map.

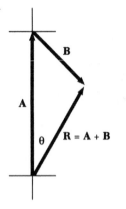

Figure 3.6
Vectors can be added graphically. The vector sum of **A** and **B** is called the resultant: **R = A + B**.

Walking ten paces north and then five paces southeast is just adding these two vectors together. What is the result of following these instructions—of adding these two vectors? We can write the answer symbolically as

$$R = A + B \qquad (3.4)$$

We call vector **R** the **resultant** of adding vector **A** and vector **B**. Figure 3.6 shows exactly what happens as we follow the instructions. From our origin, we take ten paces north. That is represented by vector **A**. The tip of **A** is our present location. Then from the tip of **A** we take five paces southeast. From the tip of **A** the second vector, **B**, is drawn. The tip of this second vector gives our final position.

How could we have gotten to the same place in a single displacement? The resultant vector **R** answers that question. Measuring the length of **R** shows it to be 3.7 cm long. The length of this vector is the magnitude of our final displacement:

$$R = 3.7 \text{ cm}\left(\frac{1 \text{ pace}}{0.50 \text{ cm}}\right) = 7.4 \text{ paces}$$

With a protractor, carefully measure angle θ in the diagram to determine the direction of **R**. This angle is a little less than 30°. Graphically, we have added vectors **A** and **B** and found their resultant to be

$$R = 7.4 \text{ paces, } 29° \text{ east of north}$$

Suppose the instructions on our treasure map had been given in the opposite order—go five paces southeast and then ten paces north. The path we would travel for those instructions is indicated in Figure 3.7. From the origin we would walk along vector **B** until we reached its tip. There we would turn and walk along vector **A**. Where would we be? At exactly the same place as before! We can write that location as

$$R = B + A \qquad (3.5)$$

The length and direction for the resultant vector **R** we find in Figure 3.7 are exactly the same as what we found in Figure 3.6. And that will always be the case. It makes no difference in what order vectors are added:

$$A + B = B + A \qquad (3.6)$$

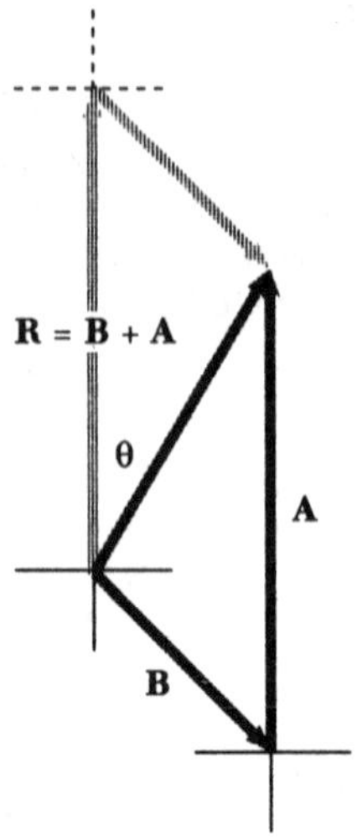

We describe this property by saying that vector addition is *commutative*. We have looked at the addition of two displacements, but what we have learned is true in general. *All vectors* add in this same way. Any vector can be represented graphically by drawing an arrow in the *direction* of the vector with its *length* proportional to the magnitude of the vector. Vectors are then added graphically by drawing one arrow with its tail at the origin and drawing the next arrow with its tail starting at the tip of the first. The resultant of adding vectors is a vector whose tail starts at the origin along with the first vector and whose tip ends at the tip of the last vector. This is illustrated in Figure 3.8 where two other vectors **A** and **B** are shown in Figure 3.8(a). Figure 3.8(b) shows the resultant **R = A + B**. Figure 3.8(c) shows **R = B + A**. Either of those is sometimes referred to as addition by completing a vector triangle. Figure 3.8(d) is just a

Figure 3.7
Vector addition is commutative: **A + B = B + A**.

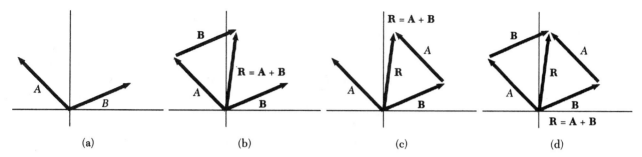

Figure 3.8
Another example of vector addition by the graphical method.

superposition of those two. Notice that the two vectors **A** and **B** form two sides of a parallelogram with **R** being the diagonal between them. Using a diagram like this is called the parallelogram method of vector addition.

Just as we can add more than two numbers, we can add more than two vectors, as illustrated in Figure 3.9. One vector is drawn to scale with its tail at the origin. The next vector is drawn to scale with its tail at the tip of the previous vector. This process is repeated until all the vectors to be added are used. The resultant—the sum of all the vectors—is then drawn with its tail at the origin and its tip at the tip of the final vector added. The magnitude and direction of the resultant can then be measured from the diagram. Since vector addition is commutative, the order in which these vectors are drawn does not matter. Obviously, vector-addition diagrams must be careful diagrams, not rough sketches.

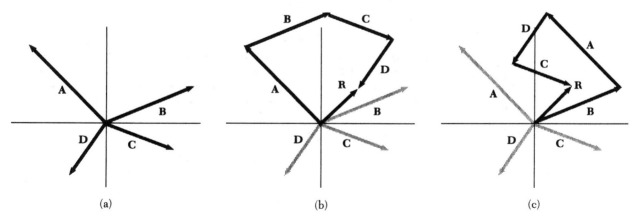

Figure 3.9
Vector addition need not be limited to only two vectors, and the multiple vectors, shown in (a), can be added in any order. In (b) the order of addition is **A** + **B** + **C** + **D**, while in (c) it is **B** + **A** + **D** + **C**. Regardless of the order, the resultant is **R**.

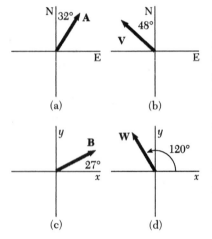

Figure 3.10
Diagrams are of great help in determining the direction of a vector.

When stating directions in terms of north, south, east, and west, it is common practice to measure the angle from north: vector **A** points 32° east of north or vector **V** points 48° west of north. That is shown in Figure 3.10(a) and (b), where x and y have been replaced by E and N as the coordinate labels. However, it is also quite common to measure angles from the x axis when directions are stated in terms of the x and y axes. Figure 3.10(c) shows a vector **B** that is directed at an angle of 27° counterclockwise from the x axis, and Figure 3.10(d) shows a vector **W** directed at an angle of 120° counterclockwise from the x axis.

There is no single "right" way to measure angles. You must read every physics problem carefully and always be sure that it is clear in your answer from which direction your angles are measured. Diagrams are extremely useful in making these directions clear.

Problem-Solving Tip

— Diagrams are essential for vector addition.

Example 3.1

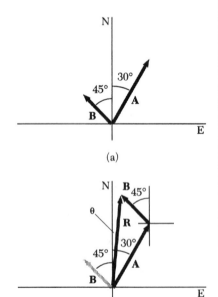

Figure 3.11
These displacements are **A** = 20 m, 30° east of north, **B** = 10 m, 45° west of north; **R** = **A** + **B** = 25 m, 6.8° east of north.

Add the following displacements:

$$A = 20 \text{ m}, 30° \text{ east of north}$$
$$B = 10 \text{ m}, 45° \text{ west of north}$$

Reasoning

A careful drawing—with measured lengths and measured angles—is essential. We shall add these vectors by first drawing vector **A** with its tail at the origin and then drawing vector **B** with its tail at the tip of **A**. The resultant **R** is the vector that begins at the tail of **A** and ends at the tip of **B**.

Solution

Begin by drawing a set of coordinate axes and the vectors to be added, as shown in Figure 3.11(a). Then redraw vector **B** at the tip of vector **A**. It usually helps to draw a new set of coordinate axes for this second step. The resultant vector **R** begins at the origin, at the tail of **A**, and ends at the tip of **B**. This is shown in Figure 3.11(b). Careful measurement shows that the length R is 25 m and the angle θ is 6.8°. We then write this result as

$$R = 25 \text{ m}, 6.8° \text{ east of north}$$

Practice Exercise Add the following displacements:

$$A = 10 \text{ m}, 45° \text{ east of north}$$
$$B = 15 \text{ m}, 60° \text{ west of north}$$
$$R = A + B = ?$$

Answer $R = 15.7 \text{ m}, 22°$ west of north

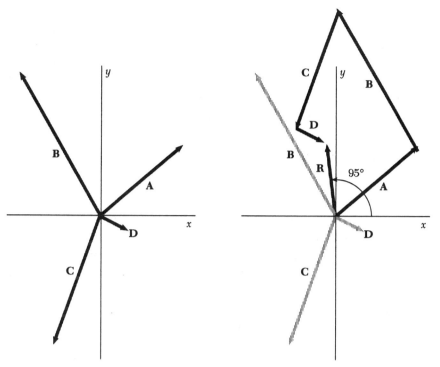

Figure 3.12
The four displacements are **A** = 12 m, 40°, **B** = 18 m, 120°, **C** = 15 m, 250°, **D** = 5 m, 330°; **R** = **A** + **B** + **C** + **D** = 7 m, 95° counterclockwise from the *x* axis.

Example 3.2

Add the following displacements (all angles are measured counterclockwise from the *x* axis): **A** = 12 m, 40°; **B** = 18 m, 120°; **C** = 15 m, 250°; **D** = 5 m, 330°.

Reasoning

As always, begin by drawing the vectors. Then make a new drawing with one vector originating at the origin. Choose a scale—such as 1 cm on your drawing represents 1 m for the displacements. One by one, draw the remaining vectors, each originating at the tip of the previous one. This is shown in Figure 3.12. The resultant vector **R** is the vector drawn from the origin to the tip of the last vector you drew. (Remember that the order in which you add the vectors does not matter.)

Solution

If the individual vectors have been carefully drawn to scale and in the correct direction, then we measure the length *R* to be 7 cm (meaning 7 m for the displacement) and the angle between **R** and the *x* axis to be 95°. We can write this resultant as

$$\mathbf{R} = 7 \text{ m, } 95° \text{ counterclockwise from the } x \text{ axis}$$

Practice Exercise Add the following displacements (all angles are measured counterclockwise from the x axis):

$$\mathbf{A} = 10 \text{ m}, 30°$$
$$\mathbf{B} = 12 \text{ m}, 130°$$
$$\mathbf{C} = 15 \text{ m}, 240°$$
$$\mathbf{D} = 10 \text{ m}, 300°$$
$$\mathbf{R} = \mathbf{A} + \mathbf{B} + \mathbf{C} + \mathbf{D} = ?$$

Answer $\mathbf{R} = 7.62 \text{ m}, 258.3°$

3.3 Components of Vectors and Vector Addition

CONCEPT OVERVIEW

- Any vector can be resolved into components along perpendicular axes.
- Components along either axis can be added as scalars to find the components of the resultant vector.
- The resultant components can then be used to reconstruct the resultant vector—to find both its magnitude and its direction.

Figure 3.13(a) shows a vector $\mathbf{V}$. You may think of this vector as the resultant of two vectors added together, such as the $\mathbf{V} = \mathbf{V}_1 + \mathbf{V}_2$ illustrated in Figure 3.13(b). Figure 3.13(c) shows five sets of two vectors each that could be added to give $\mathbf{V}$ as their resultant. In fact, there are an infinite number of sets of two

Figure 3.13
Any vector $\mathbf{V}$ can be described as the resultant of two other vectors. In particular, we can write $\mathbf{V} = \mathbf{V}_x + \mathbf{V}_y$, where $\mathbf{V}_x$ points in the direction of the x axis and $\mathbf{V}_y$ points in the direction of the y axis. The vectors $\mathbf{V}_x$ and $\mathbf{V}_y$ are called the x and y components of $\mathbf{V}$.

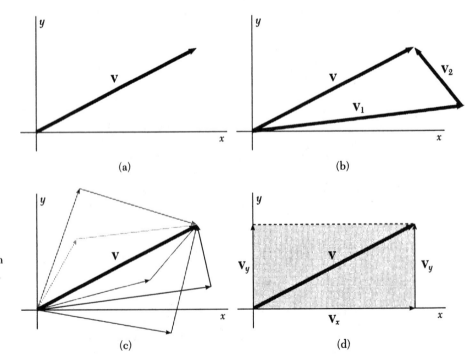

vectors that have **V** as their resultant. We are interested here in one particular set because it will make our work with and understanding of vectors easier. Figure 3.13(d) shows two vectors $\mathbf{V}_x$ and $\mathbf{V}_y$ that give **V** as their resultant when they are added. The vector $\mathbf{V}_x$ is parallel to the x axis, and $\mathbf{V}_y$ is parallel to the y axis. These two vectors are called the **vector components** of **V**. Any vector can be resolved into components that are parallel to the x and y axes.

Figure 3.14 shows two vectors and their vector components,

$$\mathbf{A} = \mathbf{A}_x + \mathbf{A}_y$$
$$\mathbf{B} = \mathbf{B}_x + \mathbf{B}_y$$

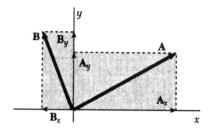

To add these two vectors, we can write their resultant **R** as

$$\mathbf{R} = \mathbf{A} + \mathbf{B} = (\mathbf{A}_x + \mathbf{A}_y) + (\mathbf{B}_x + \mathbf{B}_y)$$

Remember, from Equation 3.6, that the order of adding vectors is not important; vector addition is *commutative*. So let us add first $\mathbf{A}_x$ and $\mathbf{B}_x$ and then $\mathbf{A}_y$ and $\mathbf{B}_y$:

$$\mathbf{R} = \mathbf{A} + \mathbf{B} = (\mathbf{A}_x + \mathbf{B}_x) + (\mathbf{A}_y + \mathbf{B}_y)$$

The sum $\mathbf{A}_x + \mathbf{B}_x$ is a vector along the x axis; it is the vector component of the resultant **R** along the x axis:

$$\mathbf{R}_x = \mathbf{A}_x + \mathbf{B}_x$$

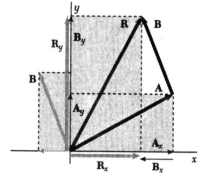

The sum $\mathbf{A}_y + \mathbf{B}_y$ is a vector along the y axis; it is the vector component of **R** along the y axis:

$$\mathbf{R}_y = \mathbf{A}_y + \mathbf{B}_y$$

All this is illustrated in Figure 3.14.

Since $\mathbf{A}_x$ and $\mathbf{B}_x$ both lie along the same direction, adding them is simple. Their resultant $\mathbf{R}_x$ is another vector in the same direction with its length equal to the sum of their lengths:

$$R_x = A_x + B_x \tag{3.7}$$

A_x or B_x will be positive if it points to the right and negative if it points to the left. $\mathbf{R}_x$ points to the right if R_x is positive; to the left if R_y is negative.

Likewise, the scalar part of $\mathbf{R}_y$, R_y, is equal to the sum of the lengths of the components in the y direction:

$$R_y = A_y + B_y \tag{3.8}$$

A_y or B_y will be positive if it points up and negative if it points down. $\mathbf{R}_y$ points up if R_y is positive; down if R_y is negative. With these components, we can now reconstruct the resultant vector:

$$\mathbf{R} = \mathbf{R}_x + \mathbf{R}_y \tag{3.9}$$

This procedure allows us to use scalar addition. The idea is as follows: Resolve each vector into its x and y components. All the x components add as ordinary numbers—positive if pointing right and negative if pointing left. All the y components add as ordinary numbers—positive if pointing up and negative if pointing down. These sums then form the x and y components of the resultant.

So far, we have always given a vector in terms of its length (magnitude) and

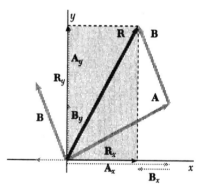

Figure 3.14
Vectors can be added by resolving them into components along the x and y axes, adding the components, and reconstructing the resultant.

direction. How do we go from there to finding its components? And how do we
calculate the resultant using the ideas summarized in Equations 3.7, 3.8, and
3.9? These questions are best answered by examples. In these examples we
shall use the trigonometry functions sine, cosine, and tangent, which are re-
viewed in Appendix.

Problem-Solving Tip

■ Diagrams are essential for vector addition even if the addition is done by the
  component method.

Example 3.3

Use the component method to add the following vectors:

$$\mathbf{A} = 30 \text{ units, } 30° \text{ counterclockwise from the } x \text{ axis}$$
$$\mathbf{B} = 20 \text{ units, } 140° \text{ counterclockwise from the } x \text{ axis}$$

Reasoning

Even though we are going to solve this problem with components, a clear
diagram is still essential. Each vector will be resolved into its x and y
components. This is shown in Figure 3.15(a). The components will then
be added and used to construct the resultant $\mathbf{R}$.

   We shall compare this numerically obtained result with our diagram.
It is all too easy to ask a calculator for a cosine instead of a sine or add
where subtraction is called for. Comparing the graphical result with the
numerical result is quick and easy and very useful for catching errors.

Solution

Finding the components of $\mathbf{A}$ is illustrated in Figure 3.15(b). The x component is the side of the right triangle adjacent to the 30° angle, and so its
value is

$$A_x = A \cos 30° = (30 \text{ units})(0.866) = 26 \text{ units}$$

The y component is the side opposite the 30° angle, and so its value is

$$A_y = A \sin 30° = (30 \text{ units})(0.500) = 15 \text{ units}$$

   Figure 3.15(c) illustrates the same process for $\mathbf{B}$. We use $\theta = 140°$
and proceed exactly as we did above with vector $\mathbf{A}$:

$$B_x = B \cos 140° = (20 \text{ units})(-0.766) = -15 \text{ units}$$

$$B_y = B \sin 140° = (20 \text{ units})(0.643) = 13 \text{ units}$$

   For angles greater than 90°, there is an alternate method you may
want to use. Vector $\mathbf{B}$ is the hypotenuse of a right triangle with a *reference angle* of $\phi = 40°$. The y component of $\mathbf{B}$ is the side of the right
triangle opposite this reference angle:

$$B_y = B \sin 40° = (20 \text{ units})(0.643) = 13 \text{ units}$$

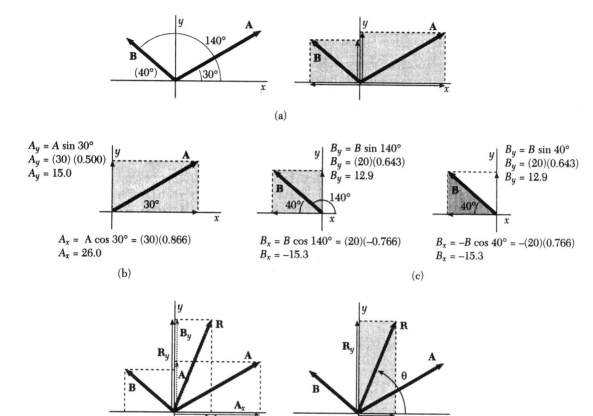

$A_y = A \sin 30°$
$A_y = (30)\,(0.500)$
$A_y = 15.0$

$A_x = A \cos 30° = (30)(0.866)$
$A_x = 26.0$

(b)

$B_y = B \sin 140°$
$B_y = (20)(0.643)$
$B_y = 12.9$

$B_x = B \cos 140° = (20)(-0.766)$
$B_x = -15.3$

$B_y = B \sin 40°$
$B_y = (20)(0.643)$
$B_y = 12.9$

$B_x = -B \cos 40° = -(20)(0.766)$
$B_x = -15.3$

(c)

(d)

Figure 3.15
Vector addition by the component method.

For the x component, we must be careful. Vector $\mathbf{B}_x$ points to the left. Therefore the component B_x is negative. The side of the right triangle adjacent to the reference angle is the hypotenuse multiplied by the cosine, but we must include a negative sign here to indicate that $\mathbf{B}_x$ points to the left along the negative x axis:

$$B_x = -B \cos 40° = -(20 \text{ units})(0.766) = -15 \text{ units}$$

We now add the x components together to obtain the x component of the resultant:

$$R_x = A_x + B_x = 26 \text{ units} + (-15 \text{ units}) = 11 \text{ units}$$

Likewise, the y components are added to give the y component of the resultant:

$$R_y = A_y + B_y = 15 \text{ units} + 13 \text{ units} = 28 \text{ units}$$

We could stop right here since these two components completely define the resultant that is the vector sum of our original two vectors. However, we may choose to determine the resultant in terms of its magnitude R and its direction θ. This is illustrated in Figure 3.15(d). The magnitude R

is the length of the hypotenuse of the right triangle that is formed when **R** is resolved into its x and y components. This length is given by the Pythagorean theorem:

$$R = \sqrt{R_x^2 + R_y^2} = \sqrt{11^2 + 28^2} = \sqrt{905} = 30 \text{ units}$$

The angle θ can be found from calculating the tangent:

$$\tan \theta = \frac{\text{opposite}}{\text{adjacent}} = \frac{R_y}{R_x} = \frac{28}{11} = 2.5$$

From a calculator, we find that $\theta = 68°$. Now we can write the resultant vector as

$$\mathbf{R} = 30 \text{ units, } 68° \text{ counterclockwise from the } x \text{ axis}$$

And our diagram does, indeed, look reasonably similar. Numbers are more accurate than sketches, but sketches show what is going on and either give you confidence in the numbers or else show you the numbers are incorrect.

Example 3.4

An airplane travels, in turn, along three horizontal displacements: $\mathbf{D}_1 = 100$ km, 30.0° east of north; $\mathbf{D}_2 = 150$ km, 60.0° west of north; $\mathbf{D}_3 = 120$ km, 35.0° west of south. What is its final (resultant) displacement? That is, what is the resultant when these three vectors are added?

Reasoning

As always, begin with a diagram, such as the one in Figure 3.16. Notice that the directions are stated in terms of angles measured from the y axis. This choice for measuring the angles makes sense because the y axis is the north-south direction on most maps and our angles in this example are given in terms of how much they deviate from the north-south axis.

Solution

Figure 3.16 contains the information needed to find the components of $\mathbf{D}_1$. Remember that the directions have been stated in terms of angles from north or south (from the y axis, not from the x axis). The component D_{1x} is the side opposite the 30.0° angle, so

$$D_{1x} = D_1 \sin 30.0° = 50.0 \text{ km}$$

The component D_{1y} is the side adjacent to the 30.0° angle, so

$$D_{1y} = D_1 \cos 30.0° = 86.6 \text{ km}$$

The figure provides the same information for vector $\mathbf{D}_2$. Again, be careful of the details:

$$D_{2x} = -D_2 \sin 60.0° = -130 \text{ km}$$

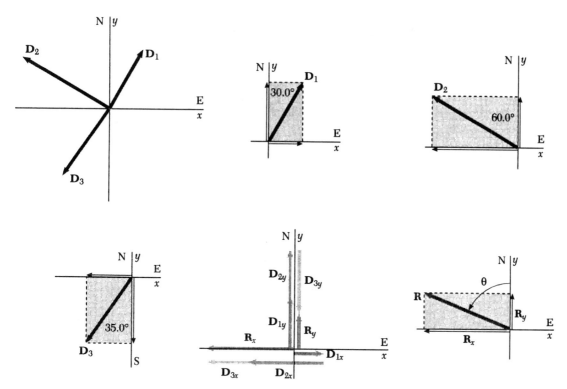

Figure 3.16
Three displacements being added by the component method.

Notice the negative sign. This is a displacement to the west (or to the left).

$$D_{2y} = D_2 \cos 60.0° = 75.0 \text{ km}$$

The components of $\mathbf{D}_3$ are illustrated in the figure also. South is just the negative y axis.

$$D_{3x} = -D_3 \sin 35.0° = -68.8 \text{ km}$$

$$D_{3y} = -D_3 \cos 35.0° = -98.3 \text{ km}$$

Notice that both components of $\mathbf{D}_3$ carry a negative sign. The D_{3x} is negative because it points to the west, and the D_{3y} is negative because it points south. *Be careful with the signs and the sines.*

Now we have all the information necessary to find the components of the resultant:

$$
\begin{aligned}
R_x &= D_{1x} + D_{2x} + D_{3x} \\
&= 50.0 \text{ km} + (-130 \text{ km}) + (-68.8 \text{ km}) = -149 \text{ km}
\end{aligned}
$$

$$
\begin{aligned}
R_y &= D_{1y} + D_{2y} + D_{3y} \\
&= 86.6 \text{ km} + 75.0 \text{ km} + (-98.3 \text{ km}) = 63.3 \text{ km}
\end{aligned}
$$

At this point **R** is fully specified because its components are known. We could therefore write **R** as

$$\mathbf{R} = (-149 \text{ km}, 63.3 \text{ km})$$

Instead, we shall find the magnitude and direction of the resultant. The magnitude comes from the Pythagorean theorem:

$$R = \sqrt{R_x^2 + R_y^2} = 162 \text{ km}$$

Since all directions were initially given as deviations from north or south, we shall continue this format. The tangent of angle θ in Figure 3.16 is

$$\tan \theta = \frac{\text{opposite}}{\text{adjacent}} = \frac{|R_x|}{R_y} = \frac{149 \text{ km}}{63.3 \text{ km}} = 2.35$$

where we use only the absolute value of R_x, ignoring the negative sign. We can do this because θ is essentially a reference angle. We must supply the rest of the information by specifying if θ is measured east or west from north or south. From a calculator, we find that a tangent of 2.35 corresponds to an angle of 66.9°. Therefore

$$\mathbf{R} = 162 \text{ km}, 66.9° \text{ west of north}$$

Example 3.5

Two men pull on the ends of a rope wrapped around a tree, as in Figure 3.17. Each exerts a force of 120 lb in the directions shown; that is,

$$\mathbf{F}_1 = 120 \text{ lb}, 30.0° \text{ counterclockwise from the } x \text{ axis}$$
$$\mathbf{F}_2 = 120 \text{ lb}, 307° \text{ counterclockwise from the } x \text{ axis}$$

Figure 3.17
Because forces are vectors, they follow the rules of vector addition.

(Notice that the direction of $\mathbf{F}_2$ might also be stated as 53.0° clockwise from the x axis.) Find the resultant force.

Reasoning

Forces are vectors, so this is another vector addition problem. We need the resultant $\mathbf{R} = \mathbf{F}_1 + \mathbf{F}_2$, which is best obtained by resolution into components followed by recombination of the components.

Solution

$$F_{1x} = F_1 \cos 30.0° = 104 \text{ lb}$$

$$F_{1y} = F_1 \sin 30.0° = 60.0 \text{ lb}$$

$$F_{2x} = F_2 \cos 53.0° = 72.2 \text{ lb}$$

$$F_{2y} = -F_2 \sin 53.0° = -95.8 \text{ lb}$$

$$R_x = F_{1x} + F_{2x} = 104 \text{ lb} + 72.2 \text{ lb} = 176 \text{ lb}$$

$$R_y = F_{1y} + F_{2y} = 60.0 \text{ lb} + (-95.8 \text{ lb}) = -35.8 \text{ lb}$$

Be sure to watch the sign carefully on F_{2y}.

To state $\mathbf{R}$ in terms of magnitude and direction, we can continue with

$$R = \sqrt{R_x{}^2 + R_y{}^2} = 180 \text{ lb}$$

$$\tan \theta = \frac{-35.8 \text{ lb}}{176 \text{ lb}} = -0.203$$

$$\theta = -11.5°$$

Notice that $\tan \theta = -0.203$ could mean $\theta = -11.5°$ or $\theta = 168.2°$. From Figure 3.17—since R_x is positive and R_y is negative—we determine that it is $-11.5°$ or $11.5°$ clockwise from the x axis. Therefore, the resultant is

$$\mathbf{R} = 180 \text{ lb, } 11.5° \text{ clockwise from the } x \text{ axis}$$

or

$$\mathbf{R} = 180 \text{ lb, } 349° \text{ counterclockwise from the } x \text{ axis}$$

3.4 Vector Subtraction

CONCEPT OVERVIEW

■ Vector subtraction is essentially vector addition with the direction of the vector to be subtracted reversed.

Just as you can subtract ordinary numbers, you can also subtract vectors. You may think of ordinary subtraction as addition of a negative number—

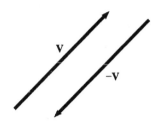

Figure 3.18
The vector $-\mathbf{V}$ is equivalent to $(-1)\mathbf{V}$.

subtracting $100 from your checking account is the same as adding $100 to your debit. We shall use this same idea when dealing with vectors.

As you know from Section 3.1, a vector can be multiplied by a scalar. The product is another vector pointing in the same direction with its magnitude multiplied by the scalar. Figure 3.18 shows a vector $\mathbf{V}$ and the same vector multiplied by -1. We shall simply call this new vector $-\mathbf{V}$:

$$-\mathbf{V} = (-1)\mathbf{V} \tag{3.10}$$

The vector $-\mathbf{V}$, by definition, has the same magnitude as vector $\mathbf{V}$ and points in the opposite direction. You can describe these two vectors as being *anti*parallel and having the same length.

Subtracting one vector from another is done by *adding* the negative of the vector that is being subtracted. Consider two vectors $\mathbf{A}$ and $\mathbf{B}$. We can define vector $\mathbf{C}$ as their difference:

$$\mathbf{C} = \mathbf{A} - \mathbf{B} = \mathbf{A} + (-\mathbf{B}) \tag{3.11}$$

The vector $-\mathbf{B}$ is just an ordinary vector, and so this new vector addition problem proceeds exactly as before, as shown in Figure 3.19. Each component of $-\mathbf{B}$ is just the negative of the corresponding component of $\mathbf{B}$, and so the components of $\mathbf{C}$ are

$$C_x = A_x - B_x$$
$$C_y = A_y - B_y$$

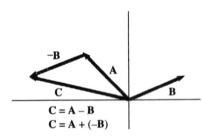

Figure 3.19
Vector subtraction is the same process as vector addition.

Example 3.6

Subtract the two vectors in Figure 3.20 to find $\mathbf{C} = \mathbf{A} - \mathbf{B}$, where

$\mathbf{A} = 12.0$ units, $30.0°$ counterclockwise from the x axis
$\mathbf{B} = 10.0$ units, $65.0°$ counterclockwise from the x axis

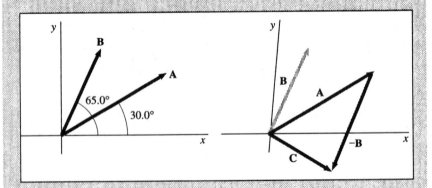

Figure 3.20
The vector subtraction $\mathbf{C} = \mathbf{A} - \mathbf{B}$ is the same as the vector addition $\mathbf{C} = \mathbf{A} + (-\mathbf{B})$.

Reasoning
Vector subtraction is essentially the same operation as vector addition.

Solution

Resolve **A** and **B** into components:

$$A_x = A \cos 30.0° = 10.4 \text{ units}$$
$$A_y = A \sin 30.0° = 6.00 \text{ units}$$

$$B_x = B \cos 65.0° = 4.23 \text{ units}$$
$$B_y = B \sin 65.0° = 9.06 \text{ units}$$

$$C_x = A_x - B_x = 10.4 \text{ units} - 4.23 \text{ units} = 6.2 \text{ units}$$
$$C_y = A_y - B_y = 6.00 \text{ units} - 9.06 \text{ units} = -3.06 \text{ units}$$

We can stop here and fully specify **C** in terms of its components:

$$\mathbf{C} = (6.2, -3.06)$$

Or we can find the magnitude and direction of **C**:

$$C = \sqrt{C_x{}^2 + C_y{}^2} = 6.9 \text{ units}$$

$$\tan \theta = \frac{-3.06}{6.2} = -0.49$$

$$\theta = -26°$$

$$\mathbf{C} = 6.9 \text{ units, } 26° \text{ clockwise from the } x \text{ axis}$$

or

$$\mathbf{C} = 6.9 \text{ units, } 334° \text{ counterclockwise from the } x \text{ axis}$$

3.5 Projectile Motion

CONCEPT OVERVIEW

■ Projectile motion is a combination of horizontal motion with constant velocity and vertical motion with constant downward acceleration, $-g$.

How can we describe the motion of a basketball after it leaves a player's hands? What path does a baseball follow after it is struck by a bat? What about the path of a football being kicked toward the goalposts? All of these are examples of **projectile motion,** the motion of an object that is thrown, or "projected," with only its initial velocity and gravity determining its further movement. We shall neglect air resistance and look only at motion that stays close to Earth's surface.

Think of what will happen if you gently release a ball and at exactly the same instant throw an identical ball horizontally. Which will hit the floor first? Figure 3.21(a), a stroboscopic photograph of two balls treated in exactly this way, shows that their vertical motions are the same. Figure 3.21(b) is a sketch of these same motions, showing the velocity vectors for the two balls. Both balls

An understanding of projectile motion was applied initially to the construction of more effective cannons. Even modern artillery, designed to be accurate at very long range, must still fire projectiles according to the laws of physics. Efforts are under way to adapt this weapons technology to design and build "super guns" that can hurl cargo or scientific instruments into low-earth orbit more cheaply than can rockets.

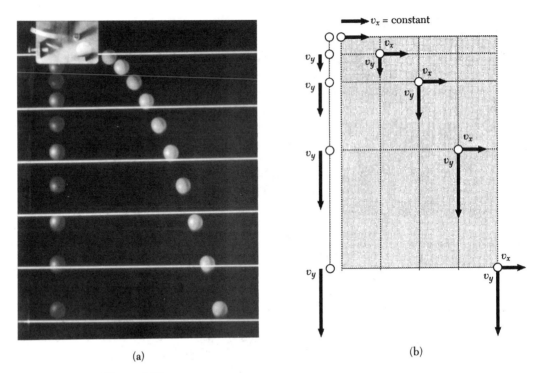

(a)   (b)

Figure 3.21
The vertical motion of a projected object is independent of its horizontal motion.
(Photograph © Richard Megna, 1990, Fundamental Photographs, NY)

experience the acceleration due to gravity in the vertical direction, and therefore both hit the floor at the same time.

Figure 3.22 shows a ball thrown from the top of a tall building. What path does it take? We can answer this question by using two observers, one looking at the horizontal motion without regard to the vertical and the other looking at the vertical motion without regard to the horizontal. What does the observer hovering overhead in the helicopter see? Whether the ball is high or low and whether it is rising or falling will not be clear to this observer, but its *horizontal* motion is quite clear: the ball moves horizontally with constant velocity. If we call the horizontal direction the x direction, then we can immediately write

$$a_x = 0 \tag{3.12}$$

$$v_x = v_{xi} = \text{constant} \tag{3.13}$$

$$x = x_i + v_{xi} \, t \tag{3.14}$$

These are just Equations 2.5 and 2.6 with $a = 0$. The helicopter observer sees the ball move *horizontally with a constant velocity* v_{xi}; this is the horizontal component of the initial velocity $\mathbf{v}_i$.

What does the observer far at the right see through the binoculars? This observer sees the ball move *vertically with constant acceleration*, 9.8 m/s² downward. This vertical motion is nothing more than the free fall discussed in

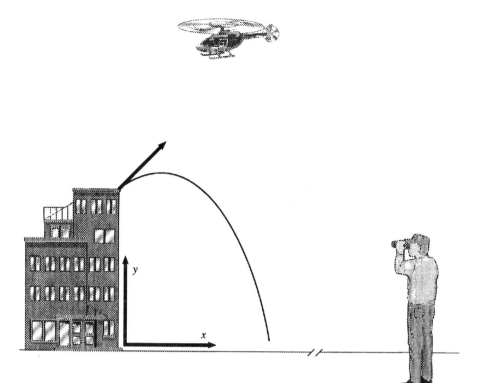

Figure 3.22
When observing projectile motion, a helicopter-based observer sees horizontal motion
with constant velocity while a distant observer sees vertical motion with constant downward acceleration.

Section 2.5. We shall call the vertical component of the initial velocity v_{yi}. The
vertical, or y-direction, motion seen by the far observer can be described by

$$a_y = -g = -9.8 \text{ m/s}^2 \qquad\qquad\qquad \textbf{(3.15)}$$

$$v_y = v_{yi} + a_y t = v_{yi} + (-g)t \qquad\qquad \textbf{(3.16)}$$

$$y = y_i + v_{yi}t + \frac{a_y t^2}{2} = y_i + v_{yi}t + \frac{(-g)t^2}{2} \qquad \textbf{(3.17)}$$

In writing $a_y = -9.8 \text{ m/s}^2$, we have defined up as the positive y direction.
These are just Equations 2.5 and 2.6 with $a = -g$. That is, the far observer sees
the ball move *vertically with a constant acceleration* $a_y = -g$.

Projectile motion is the combination of horizontal motion with constant
velocity and vertical motion with constant acceleration; the horizontal and vertical motions are independent of each other. We have already discussed both
types of motion separately. Now we must put them together. The variable that
connects the two sets of equations is the time t.

A numerical example may help make this clearer. From the edge of a

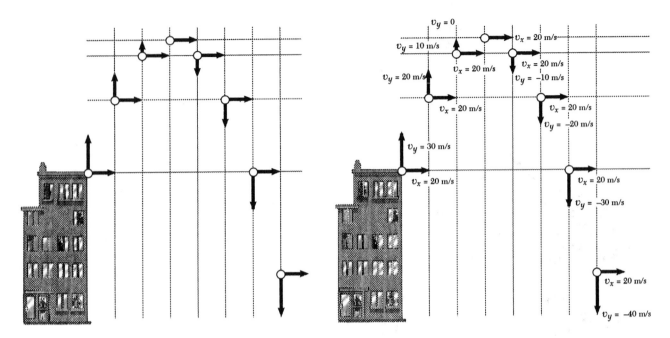

Figure 3.23
Projectile motion is horizontal motion with constant velocity ($v_x = 20$ m/s in this case) and vertical motion with constant downward acceleration. (These calculations use the approximation $a_y \approx -10$ m/s².)

building, throw a ball with a velocity of 36 m/s at 56.3° above the horizontal, as in Figure 3.23. The components of this initial velocity are

$$v_{xi} = v_i \cos 56.3° = (36 \text{ m/s})(0.555) = 20 \text{ m/s}$$
$$v_{yi} = v_i \sin 56.3° = (36 \text{ m/s})(0.832) = 30 \text{ m/s}$$

Figure 3.23 sketches the velocity components at 1-s intervals, using the approximation $g \approx 10$ m/s² so that we can do the arithmetic in our heads and concentrate on the motion. One second after being released, the ball still has a horizontal velocity of $v_x = v_{xi} = 20$ m/s but the vertical velocity has been reduced to $v_y = v_{yi} + (-g)t = 20$ m/s. At $t = 2$ s, the horizontal velocity remains $v_x = 20$ m/s while the vertical velocity is $v_y = 10$ m/s. At $t = 3$ s, the vertical velocity has vanished: $v_y = 0$. At this instant, the ball is at the top of its trajectory and falls downward after that. At $t = 4$ s, it is coming down with a velocity $v_y = -10$ m/s. It continues to accelerate downward with v_y increasing in the downward direction. Nothing happens to the horizontal velocity; v_x remains constant at 20 m/s.

The motion seen by the far observer of Figure 3.22 is shown at the right of Figure 3.24. This is exactly what would be seen if the ball had been thrown straight up; it is straight-line motion with constant acceleration. The motion seen by the helicopter observer of Figure 3.22 is shown at the top of Figure 3.24; it is straight-line motion with constant velocity.

The path, or **trajectory,** taken by such a projectile in free fall is a *parabola*. We can see this explicitly by eliminating the parameter t and writing an

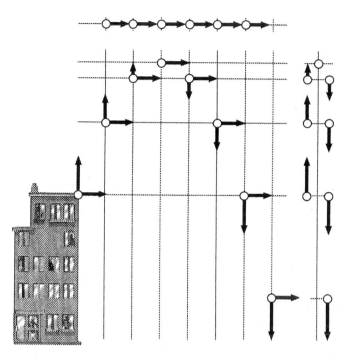

Figure 3.24
The vertical and horizontal motions of Figure 3.23 shown separately.

Water coming from the lawn sprinkler's jets follows a parabolic path. (© *Colin Molyneux/The Image Bank*)

equation for the path in terms of y as a function of x. For convenience, we shall establish a reference frame in which $x_i = y_i = 0$. From Equation 3.14, we have

$$x = x_i + v_{xi}t = v_{xi}t$$

which we can solve for the time:

$$t = \frac{x}{v_{xi}}$$

This value for t can then be substituted into Equation 3.17:

$$y = v_{yi}t + \frac{(-g)t^2}{2}$$

$$= v_{yi}\left(\frac{x}{v_{xi}}\right) + \frac{(-g)\left(\dfrac{x}{v_{xi}}\right)^2}{2}$$

$$= \left(\frac{v_{yi}}{v_{xi}}\right)x + \left(\frac{-g}{2v_{xi}^{2}}\right)x^2$$

$$= Ax + Bx^2$$

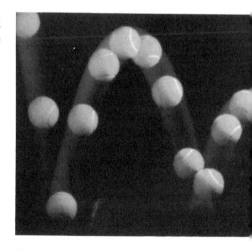

This stroboscopic image of a bouncing tennis ball clearly shows the parabolic nature of its trajectory. (© *Richard Megna, 1992, Fundamental Photographs, NY*)

This result, $y = Ax + Bx^2$, is the equation for a parabola, where in our particular case

$$A = \frac{v_{yi}}{v_{xi}}$$

$$B = \frac{-g}{2v_{xi}^2}$$

Problem-Solving Tip

- Projectile motion can be understood by examining the horizontal and vertical motions separately. Each is straight-line motion with constant acceleration, a situation you have seen before.

Example 3.7

A baseball is thrown from the top of a 50.0-m building with an initial horizontal velocity $v_{xi} = 9.00$ m/s and an initial vertical velocity $v_{yi} = 12.0$ m/s. (a) How high does the ball go? (b) How far from the side of the building does it strike the ground? (c) How fast is it moving just before it hits the ground?

(a) Reasoning

Diagrams are essential. Figure 3.25 will help as we answer these questions. How high the ball rises does not depend upon its horizontal motion. This is the type of question we handled in the previous chapter. At

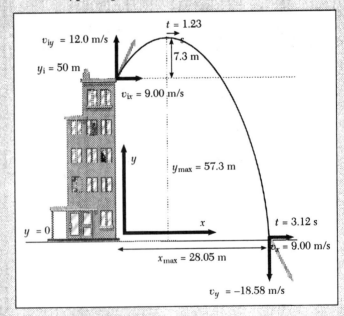

Figure 3.25
A baseball thrown off a building is an example of projectile motion.

the top of its flight, the ball's vertical velocity is zero. For what time do we find $v_y = 0$?

Solution

Use Equation 2.5 with $a = -g$:

$$v_y = v_{yi} + (-g)t$$

$$0 = 12.0 \text{ m/s} + (-9.80 \text{ m/s}^2)t$$

$$t = 1.22 \text{ s}$$

Now we know when the maximum height is reached, at $t = 1.22$ s, but *where* does this occur? To answer, we can use the position-time equation (2.6):

$$y = y_i + v_{yi}t + \frac{(-g)t^2}{2}$$

$$= 50.0 \text{ m} + (12.0 \text{ m/s})(1.22 \text{ s}) + 0.5(-9.80 \text{ m/s}^2)(1.22 \text{ s})^2$$

$$= 57.4 \text{ m}$$

Notice that the vertical distance has been measured from the ground. The ball starts from the top of the building with an initial position of $y_i = 50.0$ m.

The maximum height is 57.4 m above the ground (or 7.4 m above the top of the building). Remember that the horizontal velocity at this point is *still* $v_x = 9.00$ m/s.

(b) Reasoning

Where does the ball strike the ground? Part of that is easy to answer—at $y = 0$! It is the x coordinate we are really concerned with, but we can use the y coordinate to find the time and then the time to find the x coordinate.

Solution

Again, we use the position-time equation (2.6):

$$y = y_i + v_{yi}t + \frac{(-g)t^2}{2}$$

$$0 = 50.0 \text{ m} + (12.0 \text{ m/s})t + 0.5(-9.80 \text{ m/s}^2)t^2$$

If we remember that t is measured in seconds, we can drop the units and write

$$0 = 50.0 + 12.0\,t - 4.90\,t^2$$

This is a quadratic equation, and we can solve it using the usual quadratic formula:

$$t = \frac{-b \pm \sqrt{b^2 - 4ac}}{2a}$$

For our equation, $a = 4.90$, $b = -12.0$, $c = -50.0$. Therefore

$$t = \frac{-(-12.0) \pm \sqrt{(-12.0)^2 - 4(4.90)(-50.0)}}{2(4.90)} = \frac{12.0 \pm 33.5}{9.80}$$

As usual with quadratic equations, there are two possible mathematical answers:

$$t = 4.64 \text{ s} \quad \text{and} \quad t = -2.19 \text{ s}$$

The *physical* answer must be the positive time, $t = 4.64$ s.

What is the x coordinate at this time?

$$x = v_{xi}t = (9.00 \text{ m/s})(4.64 \text{ s}) = 41.8 \text{ m}$$

The ball hits the ground 41.8 m from the building.

(c) Reasoning

How fast is the ball traveling then? Its x velocity has remained constant at $v_x = 9.00$ m/s, but its y velocity has changed; we can calculate the velocity in the y direction from Equation 2.5.

Solution

$$v_y = v_{yi} + (-g)t$$
$$= 12.0 \text{ m/s} + (-9.80 \text{ m/s}^2)(4.64 \text{ s}) = -33.5 \text{ m/s}$$

That means its speed v at impact is

$$v = \sqrt{v_x^2 + v_y^2}$$
$$= \sqrt{(9.00 \text{ m/s})^2 + (-33.5 \text{ m/s})^2} = 34.7 \text{ m/s}$$

The ball hits the ground traveling with a speed of 34.7 m/s. Of course, this calculation neglects air resistance.

Example 3.8

A stone is kicked so that it leaves the ground with an initial velocity of 10.0 m/s at an angle of 37.0°. The stone is kicked toward a loading dock that is 3.00 m away and 1.00 m high. (a) What are the components of the stone's initial velocity? (b) How high does it go? (c) How far from the edge does it strike the dock? (d) How fast is it moving just before it hits the dock?

(a) Reasoning

Figure 3.26 will help as we answer these questions.

Solution

The components of the initial velocity are

$$v_{xi} = v_i \cos 37.0° = (10.0 \text{ m/s})(0.799) = 7.99 \text{ m/s}$$
$$v_{yi} = v_i \sin 37.0° = (10.0 \text{ m/s})(0.602) = 6.02 \text{ m/s}$$

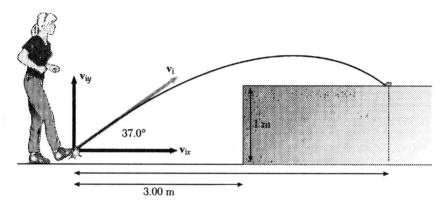

Figure 3.26
A kicked stone is another example of projectile motion.

(b) Reasoning

How high the stone rises does not depend upon its horizontal motion at all. "How high?" is exactly the type of question we handled in the previous chapter. At the top of its flight, the stone's vertical velocity is zero. Equation 2.5 will give us the time at which $v_y = 0$.

Solution

$$v_y = v_{yi} + (-g)t$$

$$0 = 6.02 \text{ m/s} + (-9.80 \text{ m/s}^2)t$$

$$t = 0.614 \text{ s}$$

Now we know when the maximum height is reached, at $t = 0.614$ s, but *where* does it occur? To answer this question, we can use the position-time equation (2.6):

$$y = y_i + v_{yi}t + \frac{(-g)t^2}{2}$$

$$y = (6.02 \text{ m/s})(0.614 \text{ s}) + 0.5(-9.80 \text{ m/s}^2)(0.614 \text{ s})^2$$

$$= (3.70 - 1.85)\text{m} = 1.85 \text{ m}$$

The maximum height is 1.85 m above the ground. Notice that the vertical distance has been measured from the ground; the stone starts from the ground at $y_i = 0$. Remember that the horizontal velocity when the stone reaches maximum height is still $v_x = 7.99$ m/s.

(c) Reasoning

Where does it land on the dock? The vertical part of that is easy to answer—at $y = 1.00$ m. It is the x coordinate that we must calculate. We can use the y coordinate to find the time and the time to find the x coordinate.

Solution

$$y = y_i + v_{yi}t + \frac{(-g)t^2}{2}$$

$$1.00 \text{ m} = 0 + (6.02 \text{ m/s})t + 0.5(-9.80 \text{ m/s}^2)t^2$$

If we keep in mind that t is measured in seconds, we can drop the units and write

$$4.90\,t^2 - 6.02\,t + 1.00 = 0$$

This is a quadratic equation and we can solve it using the usual quadratic formula with $a = 4.90$, $b = -6.02$, and $c = 1.00$:

$$t = \frac{-(-6.02) \pm \sqrt{(-6.02)^2 - 4(4.90)(1.00)}}{2(4.90)} = \frac{6.02 \pm 4.08}{9.80}$$

$$t = 1.03 \text{ s} \quad \text{and} \quad t = 0.198 \text{ s}$$

Both answers correspond to real physical values. The stone passes the point $y = 1.00$ m on the way up at $t = 0.198$ s. It continues up to $y_{max} = 1.83$ m and then comes back down to $y = 1.00$ m at $t = 1.03$ s. This second time, $t = 1.03$ s, is the one of interest to us now. What is the x coordinate at this time? The answer comes from Equation 2.6 with $x_i = 0$ and $a = 0$:

$$x = v_{xi}t = (7.99 \text{ m/s})(1.03 \text{ s}) = 8.23 \text{ m}$$

The stone hits the dock 8.23 m from the launch point; that is 5.23 m from the edge of the dock.

(d) Reasoning

How fast is the stone traveling then? Its x velocity has remained constant at $v_x = 7.99$ m/s, but its y velocity has changed. We can calculate the new y velocity using Equation 2.5 with $t = 1.03$ s.

Solution

$$v_y = v_{yi} + (-g)t$$

$$= 6.02 \text{ m/s} + (-9.80 \text{ m/s}^2)(1.03 \text{ s}) = -4.07 \text{ m/s}$$

This value for v_y means that the stone's speed v at impact is

$$v = \sqrt{v_x^2 + v_y^2}$$

$$= \sqrt{(7.99 \text{ m/s})^2 + (-4.07 \text{ m/s})^2} = 8.97 \text{ m/s}$$

The stone hits the dock moving at a speed of 8.97 m/s.

These examples show us that all projectile problems are really very similar. Projectile motion is a combination of horizontal motion with constant velocity and vertical motion with constant acceleration (free fall).

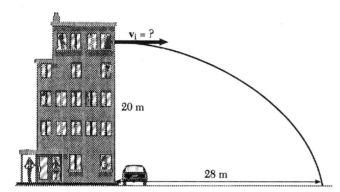

Figure 3.27
What is the initial velocity?

Example 3.9

A baseball is thrown horizontally out a window 20 m above a parking lot. The ball hits the parking lot 28 m from the base of the building, as shown in Figure 3.27. Find the velocity with which the ball was thrown.

Reasoning

If the ball is thrown horizontally, we know its initial velocity in the y direction is zero: $v_{yi} = 0$. Its vertical motion, described by Equations 2.5 and 2.6, which is motion with constant acceleration with $a = -g$, is the same as that of an object dropped from rest from a height of 20 m. If we call the ball's initial y position zero, $y_i = 0$, then it hits the ground at $y = -20$ m. We can find the time it takes for the ball to hit the ground from Equation 2.6 and then use the time to find the initial horizontal velocity.

Solution

$$y = y_i + v_{yi}t + \frac{at^2}{2}$$

$$-20 \text{ m} = 0.5(-9.8 \text{ m/s}^2)t^2$$

$$t^2 = 4.1 \text{ s}^2$$

$$t = 2.0 \text{ s}$$

In this 2.0-s interval, the ball moves 28 m horizontally at constant horizontal velocity. We can use Equation 2.6 with $a_x = 0$ to find the horizontal velocity $v_x = v_{xi}$:

$$x = x_i + v_{xi}t$$

$$28 \text{ m} = v_{xi}(2.0 \text{ s})$$

$$v_{xi} = 14 \text{ m/s}$$

Practice Exercise   If the baseball hits the parking lot 38 m from the base of the building, what is its initial velocity?

Answer   19 m/s

It seems easy enough to think of a baseball or a stone falling as it moves through the air, but what of something much faster, like a rifle bullet? Now the motion is much faster, and the distance the bullet falls is far shorter than the distance it travels horizontally. Despite these differences, the bullet's behavior is the same as that of the slower moving objects we have considered so far.

Example 3.10

A .22-caliber bullet leaves a rifle with a muzzle velocity of 350 m/s. If the target is 100 m away, how far does the bullet fall (Fig. 3.28)?

Reasoning

To travel the horizontal distance from $x_i = 0$ to $x = 100$ m requires a time t given by Equation 2.6 with $a_x = 0$. During this time of flight, the bullet will drop, and we can calculate the distance it drops from Equation 2.6 with $a = -g$.

Solution

$$x = x_i + v_{xi}t$$

$$100 \text{ m} = (350 \text{ m/s})t$$

$$t = 0.286 \text{ s}$$

During this time interval, the bullet falls a distance y given by Equation 2.6 with $a = -g$:

$$y = y_i + v_{yi}t + \frac{(-g)t^2}{2}$$

$$= 0.5 \, (-9.80 \text{ m/s}^2)(0.286 \text{ s})^2 = -0.401 \text{ m}$$

If the target were only 50 m away, the bullet would fall only one fourth this 0.401-m distance. That is why the sights on a gun must be adjusted for distance.

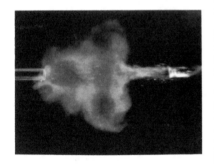

(a)

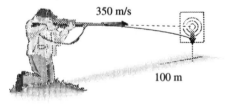

350 m/s

100 m

(b)

Figure 3.28
(a) Even a high-speed projectile such as this shotgun shell will fall as it moves forward. What do you think the motion of the pellets contained in the shell will be once they emerge from the casing? We see the shell 4.7×10^{-3}s after the shell was detonated. *(Stephen Dalton/Photo Researchers, Inc.)* (b) Even a high-speed bullet falls vertically as it moves horizontally.

As you solve projectile-motion problems, you are free to choose any point you like to be the origin of your coordinate system. That is, you can choose x_i and y_i to be whatever you like or whatever is convenient for the problem you are solving. The physics and the mathematics of the situation are not changed.

A common lecture demonstration involves the fable illustrated in Figure 3.29(a). A naturalist with a tranquilizer dart has taken careful aim at a monkey in a tree some distance away. The dart gun points directly at the monkey. The

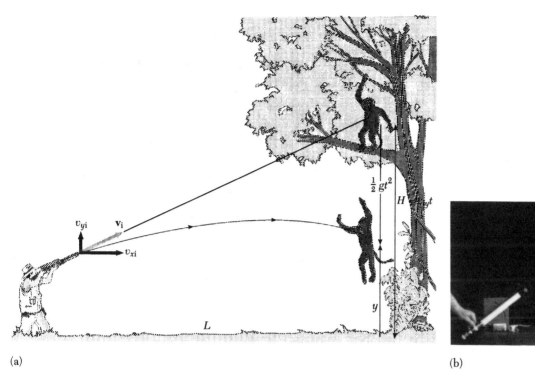

(a)

(b)

Figure 3.29

The naturalist will always hit the monkey. *(Courtesy of Central Scientific Company)*

monkey sees this and thinks of what action to take. He knows a little about projectile motion and so decides to wait until he sees the dart leave the gun and at that instant drop from the tree. He expects the dart to pass over his head as he safely falls to the ground. What will happen?

The monkey does not know quite enough physics and so will be hit by the dart. And this will happen regardless of the initial speed of the dart. As a lecture demonstration, an electromagnet usually holds a stuffed monkey or a yellow ball as in Figure 3.29b and the projectile breaks an electric circuit as it leaves the gun. This demonstration is often run two or three times with different initial velocities. The monkey is always hit. How can this be?

Figure 3.29(a) shows the path the dart would take if we could turn gravity off. The monkey is at height H, and the dart is aimed directly at him. In the absence of gravity, the dart would move horizontally with constant speed $v_x = v_{xi}$ while it moved vertically with constant speed $v_y = v_{yi}$. The vertical velocity v_{yi} (which would remain constant if gravity were turned off), the height H, and the time t are related by $H = v_{yi}t$. With no gravity, the dart moves along the straight line shown in the figure; the monkey remains in the tree and is hit. With gravity turned back on, the monkey falls a distance $gt^2/2$ from the initial height H. At time t he is located at height $y_{monkey} = H + (-gt^2/2)$. With gravity turned back on, the dart continues to fall away from the straight line. Its y position is $y_{dart} = v_{yi}t + (-gt^2/2)$. After a time t, the dart has deviated from its straight-line (no-gravity) path by a distance $gt^2/2$, which is exactly the distance the monkey has fallen from his initial position. The dart will always hit the monkey. The only exception is if the monkey is so far away that it and the dart both hit the ground before they have a chance to collide.

We can see more explicitly that the monkey is always hit by writing the equations for the motion:

$$x_{dart} = v_{xi}t \qquad\qquad x_{monkey} = L = \text{constant}$$

$$y_{dart} = v_{yi}t + \frac{(-g)t^2}{2} \qquad y_{monkey} = H + \frac{(-g)t^2}{2}$$

Since the monkey falls straight down with $x_{monkey} = L = \text{constant}$, he can be hit only when $x_{dart} = x_{monkey} = L$. That will be at time

$$t = \frac{x_{dart}}{v_{xi}} = \frac{L}{v_{xi}}$$

From the diagram, we can see that $v_{yi}/v_{xi} = H/L$. Now use the two y equations:

$$y_{monkey} = H - \frac{gt^2}{2}$$

$$y_{dart} = v_{yi}\left(\frac{L}{v_{xi}}\right) + \frac{(-g)t^2}{2}$$

$$= L\left(\frac{v_{yi}}{v_{xi}}\right) + \frac{(-g)t^2}{2}$$

$$= L\left(\frac{H}{L}\right) + \frac{(-g)t^2}{2}$$

$$= H + \frac{(-g)t^2}{2}$$

$$y_{dart} = y_{monkey}$$

That is, by the time $x_{dart} = x_{monkey}$, it will always also be true that $y_{dart} = y_{monkey}$. The monkey will always be hit!

SUMMARY

Scalars are fully specified by a number and a unit. *Vectors* require a *direction* in addition to a number and unit to be fully specified. A vector is represented in a diagram by an arrow. The length of the arrow is proportional to the *magnitude* of the vector. The arrow is drawn pointing in the same *direction* as the vector it represents.

Vectors can be added graphically. One vector is drawn with its tail at the origin. The next vector is drawn with its tail at the tip of the previous one. This is repeated until all the vectors to be added are used.

The *resultant*—the sum of all the vectors thus added—is then drawn with its tail at the origin and its tip at the tip of the final vector added. The magnitude and direction of the resultant can then be measured from the diagram.

Any vector can be resolved into *components* that are parallel to the x and y axes. These components are extremely useful because their directions are known. When adding vectors by the component method, the x components are added together as ordinary numbers and the y components are added together as ordinary numbers. That is, for

$$R = A + B + C + \cdots$$

the components of **R** are

$$R_x = A_x + B_x + C_x + \cdots$$
$$R_y = A_y + B_y + C_y + \cdots$$

Trigonometric functions are used to calculate the components and to determine the magnitude and direction of the resultant **R**.

Projectile motion is a combination of horizontal motion with constant velocity and vertical motion with constant acceleration. This constant acceleration is the downward acceleration due to gravity.

QUESTIONS

1. If vectors **A** and **B** are added to give resultant **R**, what must be true if $R = A + B$? What must be true if $R = 0$? (Remember, the scalar R is the magnitude of the vector **R**.)
2. Two displacement vectors of magnitudes 100 m and 40 m are added. What is the minimum possible value of the magnitude of the resultant? The maximum possible value?
3. Can a vector with nonzero magnitude have a component that is zero?
4. Can a vector **A** ever have an x or y component larger than A?
5. Can the magnitude of a vector be negative?
6. A scalar can multiply a vector. Can a scalar be added to a vector?
7. Can two vectors with different magnitudes be directed in such a way that their resultant is zero?
8. Can three vectors with different magnitudes be directed in such a way that their resultant is zero?
9. If **C** = **A** + **B**, under what conditions can $C < A$ and $C < B$ hold true?
10. If **C** = **A** + **B**, under what conditions can $C > A$ and $C > B$ hold true?
11. What factors are important to a broad-jumper? To a high-jumper?
12. Which will require longer to reach the ground—a ball dropped from the window of a moving car or a ball dropped from the window of a car at rest?
13. A wrench is dropped from the top of a 10-m mast on a sailing ship while the ship is traveling at 20 knots. Where will the wrench hit the deck?
14. Usually the trajectory of a projectile is a parabola. Is it possible to throw an object in such a manner that its trajectory will be an arc of a circle instead?
15. At what point or points in its trajectory does an object have its minimum speed? Its maximum speed?
16. Compare the speeds that an object has at the following three points: as it is thrown from ground level, at the top of the trajectory, and as it hits the ground.
17. Can you throw an object at such an angle as to achieve both maximum horizontal distance and maximum height for a particular speed v_i?
18. In projectile motion, maximum horizontal distance is achieved when the launch angle is 45°. What angle, measured from horizontal, gives the minimum horizontal distance for a given initial speed v_i?

PROBLEMS

3.1 Scalars and Vectors

3.1 Construct a vector diagram showing a displacement vector **D** = 250 km, 30° east of north. Use a scale of 50 km = 1 cm.

3.2 Construct a vector diagram showing a displacement vector **D** = 150 km, 53° west of north. Use a scale of 50 km = 1 cm.

3.3 Construct a vector diagram showing a displacement vector **D** = 75 paces, 37° east of south. Use a scale of 25 paces = 1 cm.

3.4 Construct a vector diagram showing a displacement vector $\mathbf{D} = 125$ m, $60°$ counterclockwise from the x axis. Use a scale of 25 m = 1 cm.

3.5 Construct a vector diagram showing a displacement vector $\mathbf{D} = 80$ m, $250°$ counterclockwise from the x axis. Use a scale of 10 m = 1 cm.

3.2 Graphical Addition of Vectors

3.6 Construct a vector diagram to add the displacements $\mathbf{A} = 100$ km, due north and $\mathbf{B} = 150$ km, due south.

3.7 Construct a vector diagram to add the displacements $\mathbf{A} = 100$ paces, due north and $\mathbf{B} = 150$ paces, due east.

3.8 Three displacements are $\mathbf{A} = 100$ m, due north; $\mathbf{B} = 150$ m, due east; $\mathbf{C} = 100$ m, $30°$ west of north. Construct a separate diagram for each of the following possible ways of adding these vectors: $\mathbf{R}_1 = \mathbf{A} + \mathbf{B} + \mathbf{C}$; $\mathbf{R}_2 = \mathbf{A} + \mathbf{C} + \mathbf{B}$; $\mathbf{R}_3 = \mathbf{B} + \mathbf{A} + \mathbf{C}$; $\mathbf{R}_4 = \mathbf{C} + \mathbf{A} + \mathbf{B}$.

3.9 Construct a vector diagram to add the vectors $\mathbf{V}_1 = 10$ units, $37°$ counterclockwise from the x axis and $\mathbf{V}_2 = 15$ units, $53°$ counterclockwise from the x axis.

3.10 Construct a vector diagram to add the vectors $\mathbf{V}_1 = 20$ units, $37°$ counterclockwise from the x axis and $\mathbf{V}_2 = 10$ units, $127°$ counterclockwise from the x axis.

3.3 Components of Vectors and Vector Addition

3.11 Construct a vector diagram showing a displacement vector $\mathbf{D} = 75$ paces, $37°$ east of south. Use a scale of 25 paces = 1 cm. Show the components of $\mathbf{D}$ and carefully measure to find D_x (or D_{east}) and D_y (or D_{south}).

3.12 Construct a vector diagram showing a displacement vector $\mathbf{D} = 125$ m, $60°$ counterclockwise from the x axis. Use a scale of 25 m = 1 cm. Show the components of $\mathbf{D}$ and carefully measure to find D_x and D_y.

3.13 Construct a vector diagram showing a displacement vector $\mathbf{D} = 175$ m, $250°$ counterclockwise from the x axis. Use a scale of 25 m = 1 cm. Show the components of $\mathbf{D}$ and carefully measure to find D_x and D_y.

Determine the magnitude of the x and y components of the displacements in Problems 3.14 through 3.19.

3.14 $\mathbf{A} = 100$ km, due north; $\mathbf{B} = 150$ km, due east; $\mathbf{C} = 100$ km, $30°$ west of north.

3.15 $\mathbf{A} = 10$ km, $60°$ west of north; $\mathbf{B} = 25$ km, $37°$ east of south; $\mathbf{C} = 10$ km, $53°$ west of south.

3.16 $\mathbf{A} = 20$ m, $53°$ counterclockwise from the x axis; $\mathbf{B} = 15$ m, $30°$ counterclockwise from the x axis; $\mathbf{C} = 10$ m, $45°$ counterclockwise from the x axis.

3.17 $\mathbf{A} = 12$ paces, $210°$ counterclockwise from the x axis; $\mathbf{B} = 15$ paces, $330°$ counterclockwise from the x axis; $\mathbf{C} = 25$ paces, $135°$ counterclockwise from the x axis.

3.18 $\mathbf{A} = 12$ units, $120°$ counterclockwise from the x axis; $\mathbf{B} = 15$ units, $60°$ counterclockwise from the x axis; $\mathbf{C} = 25$ units, $250°$ counterclockwise from the x axis.

3.19 $\mathbf{A} = 32$ units, $225°$ counterclockwise from the x axis; $\mathbf{B} = 45$ units, $135°$ counterclockwise from the x axis; $\mathbf{C} = 20$ units, $270°$ counterclockwise from the x axis.

3.20 A girl delivering newspapers covers her route by traveling three blocks west, four blocks north, and then six blocks east. What is her final displacement?

3.21 Figure 3.30 illustrates the difference in average proportions of the male and female anatomy. The displacements $\mathbf{d}_{1m}$ and $\mathbf{d}_{1f}$ from the bottom of the feet to the navel have magnitudes of 104 cm and 84 cm, respectively. The displacements $\mathbf{d}_{2m}$ and $\mathbf{d}_{2f}$ have magnitudes of 50 cm and 43 cm, respectively. (*a*) Find the vector sum of $\mathbf{d}_1$ and $\mathbf{d}_2$ for each body. (*b*) The male figure is 180 cm tall; the female, 168 cm. Rescale the displacements of each figure to a common height of 200 cm and recalculate the resultant vectors as in (*a*). (*c*) Find the vector difference between the two values in (*b*).

Use the component method to calculate the resultant $\mathbf{R}$ of adding the three vectors given in Problems 3.22 through 3.27. In each case give $\mathbf{R}$ in terms of its components and in terms of its magnitude and its direction.

3.22 The vectors of Problem 3.14.

3.23 The vectors of Problem 3.15.

3.24 The vectors of Problem 3.16.

3.25 The vectors of Problem 3.17.

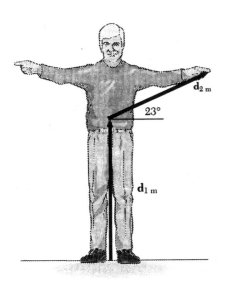 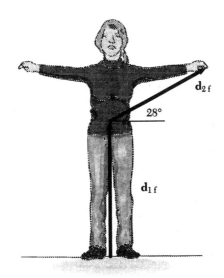

Figure 3.30
Problem 3.21.

3.4 Vector Subtraction

3.26   Both graphically and by the component
method, calculate vector **C**, where **C** =
A − **B**, **A** = 25 units, 35° counterclockwise
from the x axis, and **B** = 30 units, 75° counterclockwise from the x axis.

3.27   For the vectors in Figure 3.31, find
(a) **A** + **B**;   (b) **A** − **B**;   (c) **B** − **A**;
(d) 2**B** − **A**;   (e) 2**A** − 3**B**.

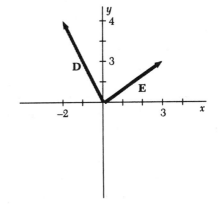

Figure 3.32
Problem 3.28.

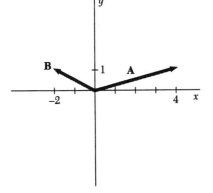

Figure 3.31
Problem 3.27.

3.28   For the vectors in Figure 3.32,
(a) **D** + **E**;   (b) **D** − **E**;   (c) **E** − **D**;
(d) 1.5**D** − 2.5**E**;   (e) 2.25**E** − 3.5**D**.

3.29   For the vectors in Figure 3.33, find
(a) **V** + **W**;   (b) **V** − **W**;   (c) **W** − **V**;
(d) 2.5**V** − 3.5**W**;   (e) 5**W** − 7**V**.

Figure 3.33
Problem 3.29.

3.5 Projectile Motion

3.30   A stone is thrown horizontally from a cliff
30 m high. The initial velocity is 6.0 m/s.
How far from the base of the cliff does the
stone strike the ground? What is the velocity
of the stone just before it strikes the ground?

3.31   A baseball is thrown horizontally from the
top of a building 10 m high. If the ball has
an initial velocity of $v_i = 9.00$ m/s,   (a) what
are its initial velocity components?   (b) How
long does it take to strike the ground?
(c) How far from the side of the building
does it strike the ground?   (d) What is its
velocity just before it hits the ground?   (e)
What is its speed just before it hits the
ground?

3.32   A package of medical supplies is released
from a small plane flying a relief mission
over an isolated jungle settlement. The plane
flies horizontally with a speed of 20 m/s at an
altitude of 20 m. Where will the package
strike the ground? That is, how far horizontally from where it was released, will the
package strike the ground?

3.33   A Ping-Pong ball rolls with a speed of
0.60 m/s off the edge of a table 0.75 m above
the floor. Neglect air resistance.   (a) How
long is the ball in the air?   (b) How far out
from the edge of the table does it hit the
floor?

3.34   (a) With what vertical component of velocity
must a jumping frog take off if it is to be in
the air for 0.60 s?   (b) How high does the
frog jump?   (c) How far ahead of its starting
point does the frog land if its takeoff is 35°
above the horizontal?

3.35   A stone is thrown at an angle of 30° above
the horizontal from the top of a cliff 15 m
above a wide river. If the initial velocity is
$v_i = 5.00$ m/s,   (a) what are the horizontal
and vertical components of the initial
velocity?   (b) How long does the stone take
to hit the water?   (c) How far from the side
of the cliff does it hit the water?   (d) What
is its velocity just before it hits the water?
(e) What is its speed just before it hits the
water?

○3.36   A rock is kicked so that it leaves the ground
with an initial velocity of 5.00 m/s at an angle

of 20.0° above the horizontal. The rock is
kicked toward a low platform that is 0.10 m
high and is 1.00 m high.   (a) What are the
components of the rock's initial velocity?
(b) How high does it go?   (c) How far from
the edge does it strike the platform?   (d)
How fast is it moving just before it hits the
platform?

○3.37   Water leaves a particular water hose and nozzle with a speed of 8.0 m/s. The nozzle is
held at ground level, and the stream of water
is directed at 60° above the horizontal.   (a)
What are the components of the water's initial velocity?   (b) How high does the water
go?   (c) Where does it strike the ground?
(d) How fast is it moving just before it hits
the ground?

○3.38   A supply airplane flies in level flight at a
speed of 300 km/h at an altitude of 100 m.
How far (horizontally) before the plane is
directly over a clearing should packages be
released if they are to land in the clearing?

○3.39   An airplane, traveling in level flight at a
speed of 200 km/h and an altitude of 200 m,
carries bales of hay to a herd of snow-bound
cattle. How far (horizontally) before the airplane is directly over the cattle should the
bales of hay be thrown out if they are to
land very near the cattle?

○3.40   A football is thrown with a velocity of 24 m/s
at 30° above the horizontal. The receiver
catches the ball at the same height as the
quarterback released it. Calculate   (a) the
maximum height of the ball;   (b) the time in
the air;   (c) the horizontal distance to the
point where the ball is caught.

○3.41   A book is thrown horizontally out a dorm
window 7.00 m above the ground. The book
lands on the grass 8.00 m from the base of
the dorm. Find the velocity with which the
book was thrown.

○3.42   A stunt driver plans to make her car go over
eight cars (each 2.5 m wide) parked side by
side below a platform as sketched in Figure
3.34. The height of the parked cars is 1.5 m.
The vertical height of the platform is 3.2 m,
and the stunt car leaves the platform horizontally. What minimum speed must the car
have as it leaves the platform if it is to clear
the last car?

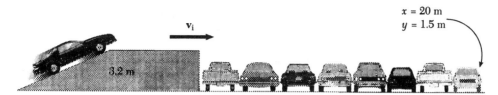

$x = 20$ m
$y = 1.5$ m

3.2 m

v_i

Figure 3.34
Problem 3.42.

3.43 Cliff divers at Acapulco jump into the sea from a cliff 36 m high (Fig. 3.35). At the level of the sea, there is a rock that sticks out a horizontal distance of 6.0 m. With what minimum horizontal velocity must the cliff divers leave the top of the cliff if they are to miss this rock?

Figure 3.35
Problem 3.43. Acapulco cliff divers.

3.44 A bullet leaves a rifle with a muzzle velocity of 400 m/s. If the target is 40.0 m away, how far does the bullet fall?

3.45 A person throws a ball horizontally at 8 m/s from a window. The ball is caught by a friend on the street, 10 m from the base of the building. How high is the window?

3.46 A cat jumps off a piano that is 1.2 m high with an initial velocity of 2.5 m at 37° above the horizontal. How far out from the piano does the cat land?

3.47 On level ground a ball is thrown forward and upward. The ball is in the air 2.0 s and is caught 35 m from the thrower. With what velocity (speed and direction) is the ball thrown?

3.48 A rock is thrown horizontally from the top of a building 60.0 m high. When the rock strikes the ground, it is moving at an angle of 48.0° from the horizontal. (a) What was its initial velocity? (b) What is its speed just before it hits the ground? (c) How far is it from the building when it hits the ground?

3.49 A pitcher throws a baseball, releasing it horizontally, with a speed of 35.0 m/s from a height of 1.84 m above the ground. The pitcher is 18.5 m from home plate. What is the height of the ball when it crosses home plate?

3.50 A long-jumper takes off at an angle of 20° with the horizontal and reaches a maximum height of 0.55 m at mid-flight. (a) What is the forward component of her velocity? (b) How far does she jump in the forward direction?

3.51 A fire department needs to get water to a fire that is burning on top of a building, 15 m above the street. The center of the fire is 10 m in from the edge of the building, and the firefighters are 12 m away from the building. The fire hose is aimed at 50° above the horizontal. What must be the initial speed of the water as it leaves the hose if it is to hit the fire?

3.52 A diver takes off with a speed of 8.00 m/s from a diving board 3.00 m high, at an angle of 35.0° above the horizontal. (a) How much later does she strike the water? (b) Where does she strike the water? (c) What is her velocity just before she strikes the water?

3.53 Attempting to gain altitude, a pilot releases his external fuel tanks. At the time of release, he is 120 m above the ground and traveling upward at an angle of 15° above the horizontal with a speed of 98 m/s. How long are the tanks in the air and where do they strike the ground?

○3.54 A mountain climber is stranded on a ledge 300 m above the ground (Fig. 3.36). Rescuers on the ground want to shoot a projectile to him with a rope attached to it. If the projectile is directed upward at an initial angle of 35° and fired at the ledge from a horizontal distance of 500 m, determine the initial speed the projectile must have in order to land on the ledge.

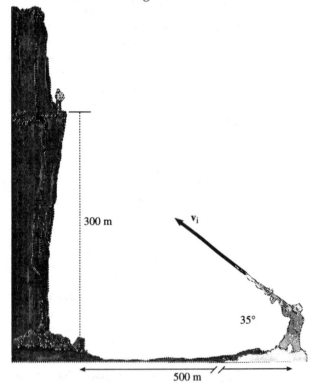

Figure 3.36   Problem 3.54.

●3.55 Show that the maximum height reached by a projectile with initial speed v_i and initial angle θ, measured relative to the horizontal, is

$$y_{max} = \frac{v_i^2 \sin^2 \theta}{2g}$$

●3.56 Show that the maximum horizontal distance reached by a projectile with initial speed v_i and initial angle θ, measured relative to the horizontal, is

$$x_{max} = \frac{v_i^2 \sin \theta \cos \theta}{2g}$$

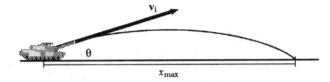

Figure 3.37   Problem 3.56.

This maximum horizontal distance reached by an object projected from a horizontal surface as shown in Figure 3.37 is called the **range** of the projectile.

●3.57 If a quarterback can throw a football with a maximum speed of 24 m/s, what is the greatest distance he can throw it?

●3.58 If a batter can hit a baseball so that it leaves the bat with a maximum speed of 32 m/s, what is the greatest distance he can hit it?

●3.59 A hay baler throws each finished bale up into the air so that it can land 2.5 m high on a trailer 5.0 m behind the baler. What must be the velocity (speed and direction) with which the bales are launched?

●3.60 A large gun on a battleship has a range (see Problem 3.56) of 60 km. What is its muzzle velocity? Neglect air resistance.

●3.61 According to ancient Greek records, a certain catapult was capable of hurling a huge stone 730 m. What must have been the minimum initial speed of the stone as it left the catapult?

●3.62 A gardener is just barely able to send water 12 m to the far end of a garden. With what speed does the water leave the hose?

●3.63 The green on a certain hole of a golf course is 240 m from the tee. A golfer hits the ball so that it leaves the tee at 15.0° above the horizontal. What must be the initial speed of the ball if it is to reach the green?

●3.64 A man can throw a ball a maximum distance R. If he can throw vertically upward with the same speed, what is the maximum height he can throw the ball?

●3.65 If a pitcher whose hand is 2.0 m above the ground throws a baseball with a velocity of 25 m/s at an angle of 35° above the horizontal, where does it strike the ground (Fig. 3.38)?

●3.66 Show that the **range**—the maximum horizontal distance—of a projectile fired from and onto a horizontal surface is the same for an angle θ and for $90° - \theta$. That is, show

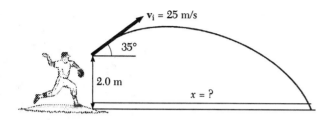

Figure 3.38
Problem 3.65.

that baseballs thrown at 30° and 60° with the same initial speed strike the ground at the same place. Or that water from a garden hose held at ground level directed at 20° and 70° will water the same patch of grass. Neglect air resistance.

●3.67 A projectile is fired with an initial velocity of 15 m/s at 53° above the horizontal from the foot of a ramp inclined 20° above the horizontal, as in Figure 3.39. How far up the ramp does the projectile strike the ramp?

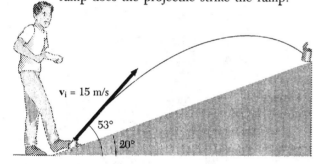

Figure 3.39
Problem 3.67.

●3.68 A projectile is fired with initial velocity v_i at angle ϕ above the horizontal at the foot of a ramp inclined θ above the horizontal, as in Figure 3.40. How far up the ramp does the projectile strike the ramp?

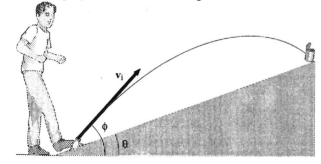

Figure 3.40
Problem 3.68.

●3.69 A basketball player shoots toward a basket that is 6.00 m horizontally from him and 3.048 m above the floor. The ball leaves his hand 2.400 m above the floor at an angle of 30° above the horizontal. What speed should the player give the ball?

General Problems

3.70 Use the component method to calculate the resultant **R** of adding the following vectors: **A** = 12 units, at 120° counterclockwise from the x axis, **B** = 15 units, at 60° counterclockwise from the x axis, and **C** = 25 units, at 250° counterclockwise from the x axis. State the resultant **R** in terms of its components and in terms of its magnitude and its direction.

3.71 Use the component method to calculate the resultant **R** of adding the following vectors: **A** = 32 units, 225° counterclockwise from the x axis, **B** = 45 units, 135° counterclockwise from the x axis, and **C** = 20 units, 270° counterclockwise from the x axis. State the resultant **R** in terms of its components and in terms of its magnitude and its direction.

3.72 Both graphically and by the component method, calculate vector **C**, where **C** = **A** − **B**, **A** = 125 units, 135° counterclockwise from the x axis, and **B** = 75 units, 60° counterclockwise from the x axis.

○3.73 A stunt driver leaves a horizontal platform that is 2 m high with speed v_i and lands on the ground 25 m away. What must v_i be?

○3.74 An archer shoots an arrow with a horizontal velocity component of 40.0 m/s and a vertical velocity component of 19.6 m/s. It strikes a target at the same height as the bow. Calculate (a) the initial speed of the arrow, (b) its time in the air, (c) the horizontal distance it travels, (d) its maximum height, (e) its final speed.

○3.75 A tennis ball is hit toward the net with an initial velocity of 24.4 m/s at an angle of 3.00° downward with respect to the horizontal. The net is 12.2 m away and 0.915 m high. The ball just clears the net. (a) How high was the ball when struck by the racket? (b) How far does the ball go beyond the net before striking the ground?

PHYSICS AT WORK

According to Newton's Second Law of Motion, a net force on an object causes the object to accelerate. In order to produce even a small acceleration of something as massive as this Boeing 737 airliner, the force applied by Malaysian strongman R. Letchemanah must be extremely large. Even more astonishing, the harness is attached to his hair!

Newton's Third Law says that whenever one object exerts a force on another object, the second object exerts on the first a force of equal magnitude but in the opposite direction. What does this imply about the force exerted on the strongman's hair? List all the forces that combine to produce the net force on the airplane and the net force on the strongman, and use them to draw a free-body diagram of the situation.

(Photo © Reuters/Bettmann)

4

Dynamics: Newton's Laws of Motion

Key Issues

- While kinematics describes motion, dynamics explains the cause of motion. The ideas of dynamics were first understood by Sir Isaac Newton—also in the 17th century, although a little later than Galileo.

- Dynamics can be explained by Newton's three laws of motion:

- Newton's First Law of Motion, also known as the Law of Inertia, states that, in the absence of an unbalanced force, an object will continue doing what it is doing. An object at rest will remain at rest, and an object moving with some velocity will continue to move with that velocity (in the same direction).

- Newton's Second Law explains that an unbalanced force is the cause of a change in motion. If an unbalanced force acts on an object, that object will accelerate in the direction of the unbalanced force, and the magnitude of the acceleration will be proportional to the unbalanced force and inversely proportional to the mass of the object. This law can be summarized in the equation $\mathbf{F} = m\mathbf{a}$.

- Newton's Third Law, sometimes called the Law of Action and Reaction, says that when object #1 exerts a force on object #2, object #2 exerts an equal force back on object #1 but in the opposite direction. This can be written as the following equation: $\mathbf{F}_{12} = -\mathbf{F}_{21}$.

So far we have *described* motion; this is called **kinematics.** Now we move on to **dynamics,** which is the explanation of the *causes* of motion or, more precisely, the causes of acceleration. Forces will be of prime importance in our study of dynamics. Dynamics relies heavily on Newton's laws of motion.

A **law** of physics is a basic statement that is true under a wide range of conditions. It is a generalization based upon a great number of observations. We expect it to be true for any and every experiment. Throughout this course, we shall spend most of our time and effort looking at many applications of a few laws of physics. In this chapter we introduce Newton's three laws of motion. These laws—these general statements about force and motion—will help us understand what causes objects to move as they do. Newton's laws of motion form the foundation of all of mechanics.

In addition to using the term *law* for general statements that are applicable to all situations, physicists refer to other, very specific observations as laws also. Hooke's law describes the behavior of springs. Ohm's law describes the behavior of metallic conductors at constant temperature. Language—even in physics—is often imprecise. Newton's laws of motion are fundamental to mechanics. They are far more important than some of the other statements commonly referred to as laws.

The importance of Sir Isaac Newton to physics—indeed, to science—is difficult to overstate. His contributions are not limited to the laws of motion we shall study in this chapter. He also invented differential and integral calculus. He contributed greatly to the field of optics, inventing the reflecting telescope that still forms the basis of modern astronomical telescopes. Newton's Law of Universal Gravitation accurately describes the motion of such diverse bodies as the satellites that bring us weather photos, comets, and the planets in our solar system. Indeed, it explains motion throughout the Universe.[1] An amazing indication of Newton's brilliance is that he developed most of these ideas when he returned home, at age 22, from Cambridge University when the university was closed because of the bubonic plague that was devastating England.

Though Newton's description of gravity has now been incorporated into Einstein's more encompassing General Relativity theory, it was an understanding of Newton's laws of motion that permitted humanity's spacecraft first to orbit the Earth and then to journey to the Moon and planets in our solar system.

4.1 Inertia: Newton's First Law of Motion

CONCEPT OVERVIEW

- If the vector sum of forces acting is zero, an object at rest will remain at rest and an object in motion will remain in motion with the same velocity—along the same straight line with the same speed.

If there are no outside forces acting on an object at rest or if the net force on the object at rest is zero, we expect it to remain at rest. For instance, if a glass is sitting on a table with nothing but the table touching it or acting upon it, we expect the glass to remain there. We have all observed this; if we look closely under controlled conditions, it remains true.

---

[1] Newton's Law of Universal Gravitation explains most astronomical motion remarkably well. However, it cannot describe precisely very large velocities (near the speed of light) or the motion of small bodies near such massive bodies as stars and black holes. Not until the early 20th century did Albert Einstein and his special and general theories of relativity provide the necessary framework for us to understand more precisely motion under such extreme conditions.

What if something is already moving but there are no outside forces acting upon it? Now we must be careful. Think of a crate sliding across a concrete floor, a book sliding across a table, and a hockey puck sliding across a rink, all illustrated in Figure 4.1. Large forces of friction stop the crate almost immediately, friction is certainly acting on the book, and a little friction is even acting on the hockey puck. We have to work hard—or, at times, just imagine—to produce a demonstration that has no friction force at all. Think again of the crate, book, and puck. As the friction force is reduced, each object has a greater tendency to continue to move without change in its motion. If we can reduce to zero the friction force on a moving object, we expect the object to continue to move with constant velocity—in the same direction and with constant speed.

It was Galileo who first realized all these details about motion. However, we usually refer to a formal statement of these ideas as Newton's first law of motion, for Newton restated Galileo's observations and expanded upon them. This **first law of motion** may be stated as

- *An object at rest will remain at rest and an object in motion will remain in motion with the same velocity—along the same straight line with the same speed—unless an unbalanced force acts upon it.*

The net force acting on something is the vector sum of all the forces acting on it. The key idea in Newton's first law of motion is that an object continues to move with constant velocity unless some net force from outside the object causes it to change its velocity. Think of the hockey puck sliding across the ice. Once it is moving, no additional agent is needed to keep it moving. It continues to move because it is already moving! This property is called **inertia,** and the first law of motion is also known as the law of inertia.

Mass is a description of inertia. As illustrated in Figure 4.2, it is easier to kick a football than a lead shot (from a shot-put event). The shot has more mass—more inertia.

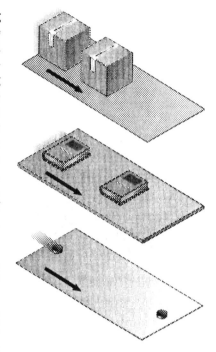

Figure 4.1

Common friction forces cause most objects in motion to come to rest.

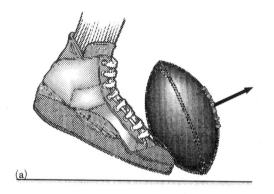

(a)

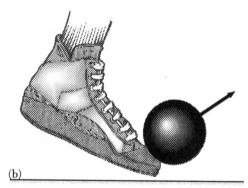

(b)

(c)

Figure 4.2

Mass is a measure of inertia. (a) A football is easier to move than (b) a lead shot because the football has less mass and therefore less inertia.

(c)   (© *Joe Strunk and Loren Winters*)

4.2 Force, Mass, and Acceleration: Newton's Second Law of Motion

CONCEPT OVERVIEW

━ If there is a net force on an object, that object will accelerate. The acceleration will be in the direction of the net force and will be directly proportional to that force and inversely proportional to the mass of the object.

Though both are "boats," this heavy cruiser and canoe offer a striking contrast in size and, more importantly, mass. The more massive cruiser has a much greater inertia and is harder to accelerate than the lightweight canoe. *(Courtesy of Homer Dodge, First President of AAPT)*

What if the net force on something is *not* zero? If you push on a car initially at rest, it moves. If you pull on a wagon initially at rest, it moves. If the car or wagon is already moving, then your push or pull will change its velocity—either speed, direction, or both. A bigger push or pull causes a bigger change. And you know it is much easier to push a sports car than a semitrailer.

These ideas are all contained in Newton's **second law of motion:**

If there is a net force on an object, that object will accelerate. The acceleration will be in the direction of the net force and will be directly proportional to that force and inversely proportional to the mass of the object.

Remember, it is the *net* force we must use here, the vector sum of all the forces acting on the object. That the acceleration is *directly* proportional to the net force means that a larger force causes a larger acceleration. Think of pushing a car to accelerate it from rest; it is easier to do this with more people (exerting more force). That the acceleration is *inversely* proportional to the mass means that, for a given force, the acceleration decreases as the mass increases. As a truck is loaded with more and more cargo, it becomes more and more difficult to accelerate. This relationship we call Newton's second law of motion can be written as

$$a \propto \frac{1}{m} F$$

Equations are more useful than mere proportionalities. With a proper choice of units, Newton's second law can be written

$$a = \frac{1}{m} F \tag{4.1}$$

where a is the acceleration resulting from a net force F acting on an object of mass m. If all the motion takes place along a straight line, Equation 4.1 is all we need. Remember, though, that both acceleration and force are vectors—their directions are important. For the general case, then, we must write

$$\mathbf{a} = \frac{1}{m} \mathbf{F} \tag{4.1'}$$

where we have explicitly indicated that **a**, the acceleration, and **F**, the net force, are vectors. Either of these equations is, effectively, a definition of **mass.** Often they are written in the form

$$F = ma \tag{4.2}$$

or

$$\mathbf{F} = m\mathbf{a} \tag{4.2'}$$

Equation 4.2′ is the most important equation in all of mechanics.

It is common and easy to write $F = ma$, but it is very important to remember that the F in that equation is the *net* force, the sum of all the forces acting on mass m. To emphasize this, we shall often write Newton's second law as

$$\Sigma \mathbf{F} = m\mathbf{a} \qquad (4.3)$$

or

$$F_{\text{net}} = ma \qquad (4.3')$$

where the Greek letter Σ (sigma) means the *sum* of all the forces.

As we learned in earlier chapters, the SI unit of mass is the kilogram and the units of acceleration are meters per second per second. That means the units of force must be

$$[\text{Force}] = [\text{kg}]\left[\frac{m}{s^2}\right] = \text{kg·m/s}^2$$

This unit of force is called the newton (N):

$$1 \text{ N} = 1 \text{ newton} = 1 \text{ kg·m/s}^2$$

A force of 1 N will cause a 1-kg mass to accelerate at 1 m/s^2.

Problem-Solving Tip

■ If it is not zero, the net force—the vector sum of all the forces acting— causes an object to accelerate. Newton's second law is $\Sigma F = ma$. Be sure all of the forces acting on mass m have been included when you write this sum. Be sure only the forces acting on mass m have been included.

Example 4.1

A skateboard in a physics experiment has a mass of 5.0 kg. What net force must be applied to it in order to observe an acceleration of 1.5 m/s^2?

Reasoning

Newton's second law of motion connects mass, acceleration, and force.

Solution

We shall use Equation 4.2, Newton's second law.

$$F = ma = (5.0 \text{ kg})(1.5 \text{ m/s}^2) = 7.5 \text{ kg·m/s}^2$$

$$= 7.5\frac{\text{kg·m}}{s^2}\left(\frac{1 \text{ N}}{1 \text{ kg·m/s}^2}\right) = 7.5 \text{ N}$$

Practice Exercise What net force must the skateboard experience to have an acceleration of 2.5 m/s^2?

Answer 12.5 N

Example 4.2

The engine and drive train of a 1-metric-ton ($m = 1000$ kg) car provide an average net force of 850 N. What will be the acceleration of the car?

Reasoning

Again, we start with Newton's second law, $F = ma$. This is a very powerful summary of the cause of motion and of the relationship among force, mass, and acceleration.

Solution

Solving Equation 4.2 for the acceleration, we get

$$a = \frac{F}{m} = \frac{850 \text{ N}}{1000 \text{ kg}}$$

$$= 0.850 \, \frac{\text{N}}{\text{kg}} \left(\frac{1 \text{ kg·m/s}^2}{1 \text{ N}} \right) = 0.850 \text{ m/s}^2$$

Practice Exercise   If the net force is increased to 950 N, what will be the acceleration?

Answer   0.950 m/s^2

Practice Exercise   What is the net force if the acceleration is 1.2 m/s^2?

Answer   1200 N

Example 4.3

The brakes and tires of a van provide a maximum braking force of 11 200 N (Fig. 4.3). The van comes to a stop with an acceleration of -5.6 m/s^2. What is the mass of the van?

Reasoning

The negative acceleration means the van is slowing down, coming to a stop—in other words, the acceleration is in the direction opposite the van's velocity. If the direction of motion is considered positive, then the braking force is also negative, $F = -11\,200$ N. We start with Newton's second law, $F = ma$, Equation 4.2, and solve for m, the mass of the van.

Solution

$$F = ma$$

$$m = \frac{F}{a} = \frac{-11\,200 \text{ N}}{-5.6 \text{ m/s}^2} = 2000 \, \frac{\text{N}}{\text{m/s}^2}$$

$$= 2000 \left(\frac{\text{N·s}^2}{\text{m}} \right) \left(\frac{1 \text{ kg·m/s}^2}{1 \text{ N}} \right) = 2000 \text{ kg}$$

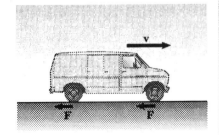

Figure 4.3
Braking forces, in the direction opposite the velocity direction, bring the van to rest.

Notice that the units always "take care of themselves." This little bit of extra effort with the units helps us form a habit that will prove quite useful in more complex problems.

Problem-Solving Tip

- In using Newton's second law, $\Sigma F = ma$, you may first have to find the acceleration as in a kinematics problem.

Example 4.4 _____

A performer in a circus is fired from a cannon as a "human cannonball" and leaves the cannon with a speed of 18 m/s (Fig. 4.4). The performer's mass is 80 kg. The cannon barrel is 9.2 m long. Find the average net force exerted on the performer while he is being accelerated inside the cannon.

Reasoning

This example is more involved; we cannot immediately plug numbers into a single equation. Newton's second law, $F = ma$, provides the relationship among the net force we want, the mass we know, and the acceleration, but how can we find the acceleration? Nothing is said about time, but we do know the distance and speeds involved. Equation 2.7 relates distance, speed, and acceleration, so we can use that expression to find the acceleration. Once we know the acceleration, Newton's second law will give us the force.

Solution

Begin with Equation 2.7:

$$v^2 = v_i{}^2 + 2a(s - s_i)$$

$$(18 \text{ m/s})^2 = 0^2 + 2a(9.2 \text{ m})$$

$$324 \text{ m}^2/\text{s}^2 = (18.4 \text{ m})a$$

$$a = \frac{324 \text{ m}^2/\text{s}^2}{18.4 \text{ m}} = 17.6 \text{ m/s}^2 \approx 18 \text{ m/s}^2$$

This is the average acceleration of the performer. Therefore, the average net force acting on him must be given by Newton's second law, Equation 4.2:

$$F = ma$$

$$F = (80 \text{ kg})(18 \text{ m/s}^2) = 1400 \text{ kg·m/s}^2\left(\frac{1 \text{ N}}{1 \text{ kg m/s}^2}\right) = 1400 \text{ N}$$

Figure 4.4
How does this human cannonball avoid serious injury from the explosive charge propelling him from the cannon? (© *J. Yulsman/The Image Bank*)

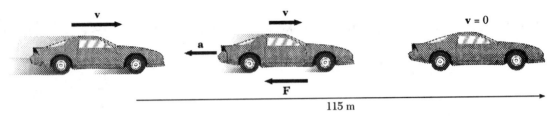

115 m

Figure 4.5
Knowing distance and velocity, we can find acceleration. Then we can apply $F = ma$ to find the net force.

Example 4.5

A 940-kg car comes to a stop from 80.0 km/h in a distance of 115 m (Fig. 4.5). What is the average net force supplied by the braking system?

Reasoning

As in the previous example, before we can use Newton's second law, we must find the acceleration. Again, Equation 2.7 relates distance, velocity, and acceleration, so we can use that. Everything will be easier if we first convert the initial velocity to meters per second.

Solution

$$v_i = 80.0 \frac{\text{km}}{\text{h}} \left(\frac{1000 \text{ m}}{1 \text{ km}} \right) \left(\frac{1 \text{ h}}{3600 \text{ s}} \right) = 22.2 \text{ m/s}$$

Now use first Equation 2.7 and then Equation 4.2:

$$v^2 = v_i^2 + 2a(s - s_i)$$

$$0^2 = (22.2 \text{ m/s})^2 + 2a(115 \text{ m})$$

$$a = \frac{-(22.2 \text{ m/s})^2}{230 \text{ m}} = -2.14 \text{ m/s}^2$$

$$F = ma$$

$$= (940 \text{ kg})(-2.14 \text{ m/s}^2) = -2010 \text{ kg·m/s}^2 = -2010 \text{ N} = -2.01 \text{ kN}$$

Because we have defined the direction of the car's velocity as being the positive direction, the negative sign means the stopping force is directed opposite the velocity.

4.3 Mass and Weight

CONCEPT OVERVIEW

■ Mass is related to the inertia of an object—a measure of how difficult it is to accelerate that object.

■ Weight is the force that gravity exerts on an object.

If you drop your physics book or any other object near the surface of Earth, the object accelerates downward. That means there is a net force acting on it. We call this force exerted by gravity the **weight** of the object.

Mass is a measure of the inertia an object has—how difficult it is to move the object or to stop it from moving. Mass and weight are related, but they are not the same thing. Near Earth's surface, we already know that all objects fall with the same acceleration, $g = 9.80$ m/s^2. For an object with mass m, its weight w (the force of gravity with which Earth pulls on it) must be given by Newton's second law:

$$F = ma$$

$$w = mg \qquad \text{(4.4)}$$

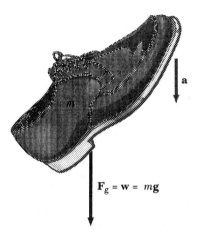

Figure 4.6
The force exerted by gravity on an object near Earth's surface equals the object's mass m times the acceleration due to gravity g. We call this force the weight of the object: $w = mg$.

Figure 4.6 shows this relationship in the form of a force diagram for a freely falling object.

An object has the same mass wherever it is—in Death Valley, atop Mount Fuji, in an orbiting space shuttle, or even on the Moon. No matter where we put the object, it will be just as difficult to move or to stop from moving. Its weight, however, varies with the local value of the acceleration due to gravity. An object weighs slightly more in Death Valley than atop Mount Fuji. In an orbiting space shuttle, it seems weightless. And on the Moon it has a weight one sixth its weight on Earth.

Example 4.6

What is the weight of a 1.0-kg mass?

Reasoning

To solve this problem, we need only our definition of weight, Equation 4.4.

Solution

$$w = mg = (1.0 \text{ kg})(9.8 \text{ m/s}^2) = 9.8 \text{ kg·m/s}^2 \left(\frac{1 \text{ N}}{1 \text{ kg·m/s}^2} \right) = 9.8 \text{ N}$$

Any object with a mass of 1.0 kg has a weight of 9.8 N.

4.4 Free-Body Diagrams

CONCEPT OVERVIEW

■ Free-body diagrams provide an effective means of ensuring that all forces acting on a body are accounted for.

We shall often write Newton's second law of motion as

$$\Sigma \mathbf{F} = m\mathbf{a}$$

with the Σ there to remind us that what we need is the vector sum of *all* external forces acting on mass m. The best way to ensure that we have correctly

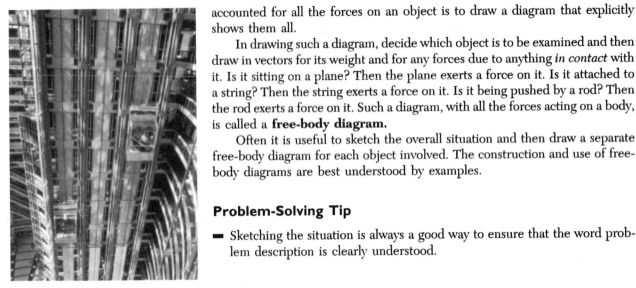

(a)

accounted for all the forces on an object is to draw a diagram that explicitly shows them all.

In drawing such a diagram, decide which object is to be examined and then draw in vectors for its weight and for any forces due to anything *in contact* with it. Is it sitting on a plane? Then the plane exerts a force on it. Is it attached to a string? Then the string exerts a force on it. Is it being pushed by a rod? Then the rod exerts a force on it. Such a diagram, with all the forces acting on a body, is called a **free-body diagram.**

Often it is useful to sketch the overall situation and then draw a separate free-body diagram for each object involved. The construction and use of free-body diagrams are best understood by examples.

Problem-Solving Tip

- Sketching the situation is always a good way to ensure that the word problem description is clearly understood.

Example 4.7

An elevator has a mass of 1500 kg. What is the tension in its supporting cable when it is moving up at a constant 2.5 m/s?

Reasoning

The sketch to the left in Figure 4.7 illustrates the elevator and cable. To the right is a free-body diagram showing just the two forces acting on the elevator: its weight w acting downward and the force of the cable (equal to the tension T) acting upward. The tension is the magnitude of the force exerted by the cable that supports the elevator. We already know how to calculate the weight, $w = mg$. The vector sum of these forces is the net force and, according to Newton's second law, must equal the product of mass times acceleration. To move at a constant velocity, the elevator has *zero acceleration*, which means the net force is zero. In other words, the upward force (the tension T) must equal the downward force (the weight mg).

Solution

$$w = mg = (1500 \text{ kg})(9.80 \text{ m/s}^2) = 14\,700 \text{ kg·m/s}^2 \left(\frac{1 \text{ N}}{1 \text{ kg·m/s}^2} \right)$$

$$= 14\,700 \text{ N} = 14.7 \text{ kN}$$

Since the elevator has zero acceleration, the net force on it must also be zero:

$$\Sigma F = ma = T - w = 0$$

$$T = w = 14\,700 \text{ N}$$

Figure 4.7
(a) A modern hotel elevator with its shaft and cable open to view. (© *Brent Jores/Stock, Boston*) (b) A sketch (left) helps us to understand what is going on. A free-body diagram (right) shows all the forces acting on an object.

The tension, which pulls upward, just equals the weight, which pulls downward. Notice that the speed of the elevator is irrelevant. It is the acceleration that determines the forces involved.

Example 4.8

An elevator has a mass of 1500 kg. What is the tension in its supporting cable when it is moving upward with an upward acceleration of 1.50 m/s²?

Reasoning

We begin with a free-body diagram like the one on the right in Figure 4.8. There are two forces: the upward force exerted by the cable (and equal to the tension) and the weight of the elevator, exerted downward. Since forces are vectors, we must be careful with signs—up will be positive and down negative. We set the net force equal to the product of mass and acceleration.

Solution

$$\Sigma F = T - w = T - (1500 \text{ kg})(9.80 \text{ m/s}^2) = T - 14\,700 \text{ N}$$

This net force causes the 1500-kg elevator to accelerate upward at 1.50 m/s². We can use Newton's second law to get a value for ΣF:

$$\Sigma F = ma = (1500 \text{ kg})(1.50 \text{ m/s}^2) = 2250 \text{ kg·m/s}^2\left(\frac{1 \text{ N}}{1 \text{ kg·m/s}^2}\right) = 2250 \text{ N}$$

These two expressions for ΣF must equal each other:

$$\Sigma F = T - 14\,700 \text{ N} = 2250 \text{ N}$$

$$T = 16\,950 \text{ N} \approx 17\,000 \text{ N} = 17.0 \text{ kN}$$

The tension in the cable, $T = 17\,000$ N, is greater than it would be if the elevator were moving with zero acceleration. The additional tension provides a net upward force that causes the elevator to accelerate upward.

Since velocity and acceleration are both positive, the elevator is speeding up, as it does when the doors close and it leaves for a higher floor. If velocity and acceleration were in opposite directions, the elevator would be slowing down, as it does when it comes to a stop before the doors open.

Practice Exercise What is the tension if the elevator has an upward acceleration of 1.20 m/s²?

Answer 16 500 N

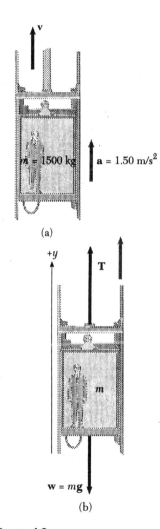

(a)

(b)

Figure 4.8
A sketch (a) and free-body diagram (b) for Example 4.8. Free-body diagrams show all the forces acting on a body. These forces can then be used in Newton's second law, $\Sigma F = m\mathbf{a}$.

Example 4.9

An elevator has a mass of 1500 kg. What is the tension in its supporting cable when it is moving upward but accelerating downward with an acceleration of 1.10 m/s²?

Reasoning

We can use the weight calculated in Example 4.7 and the same free-body diagram (Fig. 4.9). As before, we shall use the forces in the free-body diagram to calculate the net force ΣF and then set that equal to the mass times the acceleration. Force and acceleration are vectors, and that means that signs (directions) are important. We shall again take up as the positive y direction, so that a downward acceleration of 1.10 m/s² means $a_y = -1.10$ m/s².

Solution

$$\Sigma F = T - w = T - (1500 \text{ kg})(9.80 \text{ m/s}^2) = T - 14\,700 \text{ N}$$

We can now use Newton's second law:

$$\Sigma F = ma = (1500 \text{ kg})(-1.10 \text{ m/s}^2) = -1650 \text{ kg·m/s}^2\left(\frac{1 \text{ N}}{1 \text{ kg·m/s}^2}\right)$$

$$= -1650 \text{ N}$$

The negative sign means that the net force is downward. Of course, it must be downward because the acceleration is downward.

These two expressions for ΣF must equal each other:

$$\Sigma F = T - 14\,700 \text{ N} = -1650 \text{ N}$$

$$T = 13\,050 \text{ N} \approx 13\,000 \text{ N} = 13.0 \text{ kN}$$

Again notice that the speed of the elevator is not needed, nor does it matter whether the elevator is moving up (with $v > 0$) or down (with $v < 0$). It is the acceleration that determines the forces involved.

___

Practice Exercise What is the tension if the elevator has a downward acceleration of 1.20 m/s²?

Answer 12 900 N

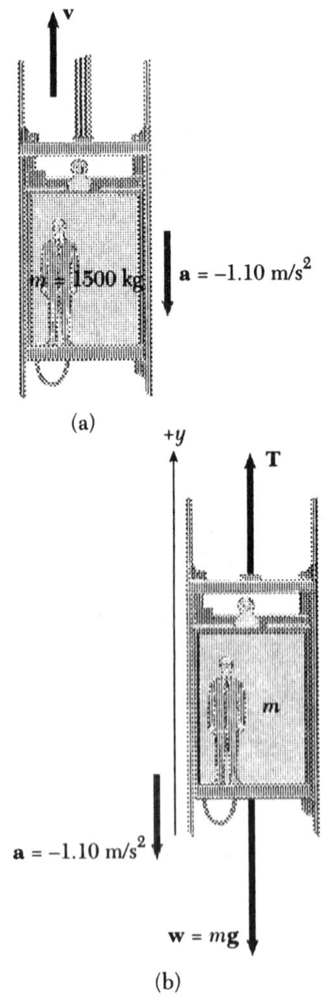

Figure 4.9
A sketch (a) and free-body diagram (b) for Example 4.9. Note that here the acceleration is downward and so carries a negative sign.

Example 4.10

An *Atwood's machine* consists of two masses attached by a string that runs over a pulley, as shown in Figure 4.10. For an Atwood's machine with $m_1 = 0.250$ kg and $m_2 = 0.300$ kg, find the acceleration of the masses. For a frictionless, lightweight pulley (which we assume here), the tension in the string on either side of the pulley is the same.

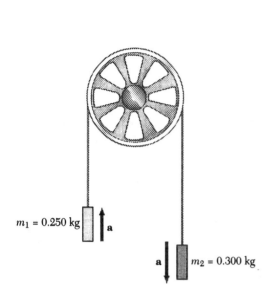

Figure 4.10
An Atwood's machine consists of two masses
attached by a string that runs over a pulley.

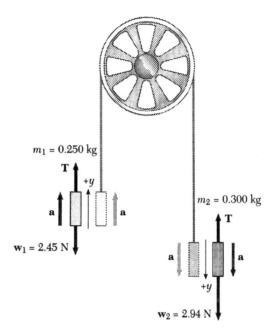

Figure 4.11
A separate free-body diagram is required for each
body in the system.

Reasoning

Before writing Newton's second law, we draw a separate free-body diagram for each mass, showing all the forces acting. This is done in Figure 4.11.

Then we apply Newton's second law to each mass separately.

Solution

For mass m_1, the forces are the tension T and the weight w_1

$$w_1 = m_1g = (0.250 \text{ kg})(9.80 \text{ m/s}^2) = 2.45 \text{ N}$$

Thus, taking the upward direction as positive, we have

$$\Sigma F = T - w_1 = m_1a$$

$$T - 2.45 \text{ N} = (0.250 \text{ kg})a$$

This equation has two unknowns (T and a), so we cannot immediately solve for either of them. We must find more information—that is, another equation.

We can find more information by looking at mass m_2. The forces that act on it are the tension T and its weight w_2

$$w_2 = m_2g = (0.300 \text{ kg})(9.80 \text{ m/s}^2) = 2.94 \text{ N}$$

For this mass, we expect the acceleration to be downward and have drawn it that way in Figure 4.11. We are free to choose the coordinate system however we want for this mass. If it is going to accelerate downward, then choosing down as positive will be convenient; then the weight is positive and the tension is negative. Now when we apply Newton's second law we have

$$\Sigma F = w_2 - T = m_2 a$$

$$2.94 \text{ N} - T = (0.300 \text{ kg})a$$

This is a second equation with our two unknowns (T and a). We now have enough information to solve for both. We can solve the first equation for the tension:

$$T = 2.45 \text{ N} + (0.250 \text{ kg})a$$

and substitute this value for T into the second equation:

$$2.94 \text{ N} - [2.45 \text{ N} + (0.250 \text{ kg})a] = (0.300 \text{ kg})a$$

$$0.49 \text{ N} = (0.550 \text{ kg})a$$

$$a = 0.89 \text{ m/s}^2$$

If we also want the tension in the string, we can find that by substituting this value of the acceleration into the equation for the tension:

$$T = 2.45 \text{ N} + (0.250 \text{ kg})(0.89 \text{ m/s}^2) = 2.67 \text{ N}$$

Notice that the tension is greater than the weight of m_1, so that mass accelerates upward. The tension is less than the weight of m_2, so that mass accelerates downward.

Practice Exercise   For an Atwood's machine with $m_1 = 0.350$ kg and $m_2 = 0.400$ kg, find the acceleration of the masses.
Answer   0.653 m/s^2

Example 4.11

A 0.250-kg glider sits on a frictionless air track and is attached to a string that runs over a pulley and is then attached to a 0.0500-kg hanging mass, as in Figure 4.12. Find the acceleration of the glider and of the mass (also often stated as "find the acceleration of the system"). As with Example 4.10, consider the pulley to be massless and frictionless; then the tension in the string is the same on either side of the pulley.

Reasoning

Before we apply Newton's second law, we construct a separate free-body diagram for each body, as in Figure 4.13. Call the glider m_1; three forces act on it. The tension supplied by the string pulls the glider to the right,

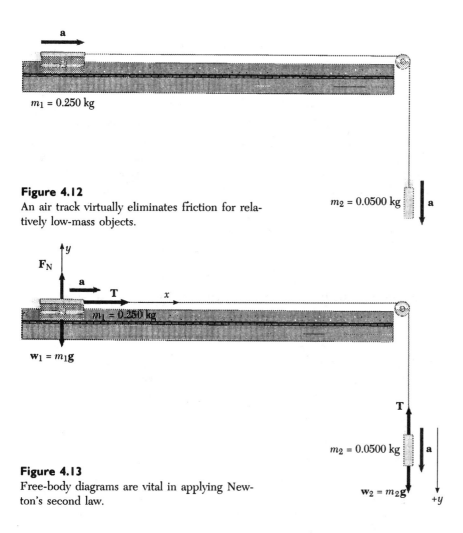

Figure 4.12
An air track virtually eliminates friction for relatively low-mass objects.

Figure 4.13
Free-body diagrams are vital in applying Newton's second law.

gravity pulls down on it with a force equal to its weight $w_1 = m_1g$, and the air track pushes up with what is called a "normal force" F_N.[2] Likewise, for hanging mass m_2, we show the tension T pulling up and its weight $w_2 = m_2g$ pulling down.

Solution

Using the forces in the free-body diagrams, we can now apply Newton's second law. Notice that the forces on the glider do not all lie along one line; therefore we must use the vector form of Newton's second law: $\Sigma \mathbf{F} = m\mathbf{a}$. As with all vector equations, this is really shorthand notation for *two* equations, $\Sigma F_x = ma_x$ and $\Sigma F_y = ma_y$.

[2] The word *normal* means "perpendicular"; the normal force is perpendicular to the surface exerting it. Even if a surface has no friction, it always provides a normal force. *Normal* does not always mean vertical, as we shall see later when we encounter inclined planes.

Take the horizontal direction for the x axis and the vertical direction for the y axis. Then only the tension acts in the x direction on the glider, and the acceleration in that direction is just a:

$$\Sigma F_x = T = ma_x = (0.250 \text{ kg})a$$

From the free-body diagram for m_2, the hanging mass, apply Newton's second law. There we have two forces: the tension T upward and the weight w_2 downward. As usual we are free to choose the coordinate system however we want. Since we know the hanging mass is going to move down and the acceleration is downward, it is convenient to choose down as the positive direction. Then

$$\Sigma F = ma = w_2 - T = m_2 a$$
$$0.490 \text{ N} - T = (0.0500 \text{ kg})a$$
$$T = 0.490 \text{ N} - (0.0500 \text{ kg})a$$

This value of the tension may be substituted into the earlier equation involving tension and acceleration:

$$T = (0.250 \text{ kg})a$$
$$0.490 \text{ N} - (0.0500 \text{ kg})a = (0.250 \text{ kg})a$$
$$0.490 \text{ N} = (0.300 \text{ kg})a$$
$$a = 1.63 \text{ m/s}^2$$

Problem-Solving Tip

■ In solving $\Sigma \mathbf{F} = m\mathbf{a}$, it is often necessary to resolve the forces into their x and y components and solve the two corresponding scalar equations, $\Sigma F_x = ma_x$ and $\Sigma F_y = ma_y$. Choose one axis along the direction of motion and the other perpendicular to the motion.

Example 4.12 _____

Find the acceleration of a glider sitting on an air track that is inclined $5.00°$ from the horizontal.

Reasoning

It may seem disconcerting that the mass of the glider is not given. Somehow, finding the acceleration of a 0.300-kg glider often seems easier than finding the acceleration of merely "a glider," with mass unknown. We simply label this unknown mass m and expect that it will drop out of our final answer.

Figure 4.14 is a sketch of the situation and a free-body diagram showing the forces acting on the glider. Two forces act: gravity pulls

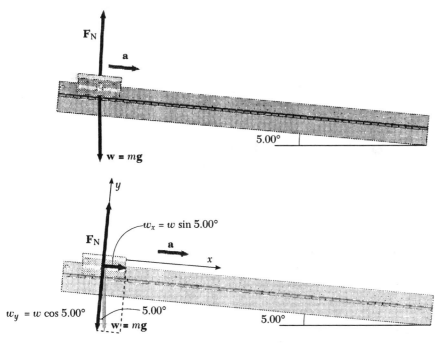

Figure 4.14
Sketch and free-body diagram for a glider on an inclined air track.

straight down with a force equal to the weight $w = mg$, and the air track pushes *perpendicular* to its surface with the normal force F_N. Notice that the normal force is *not* straight up; normal forces are always perpendicular to the surface exerting them.

Knowing these forces, we then apply Newton's second law in component form.

Solution

Take the direction along the air track for the x axis and the direction perpendicular to that for the y axis. The normal force acts only in the y direction, but the weight does not lie along either axis; we must therefore resolve the weight into components:

$$w_x = w \sin 5.00°$$
$$w_y = w \cos 5.00°$$

Now we can use these values in determining ΣF_x and ΣF_y. Because the glider has no motion in the y direction, it must be true that $\Sigma F_y = 0$. This means that the normal force $\mathbf{F_N}$ along the positive y axis exactly balances the force along the negative y axis, which is w_y. Thus we need only calculate ΣF_x in order to know F_{net}. Once we know $F_{net} = ma$, we can find a.

$$\Sigma F_x = ma_x = w_x = ma$$
$$w_x = w \sin 5.00° = 0.0872\, w = ma$$

Since $w = mg$ we can say

$$0.0872\ mg = ma$$

$$0.0872\ (9.80\ \text{m/s}^2) = a$$

$$a = 0.854\ \text{m/s}^2$$

Practice Exercise What is the acceleration of a glider sitting on an air track that is inclined 15° from the horizontal?

Answer 2.54 m/s^2

Note that the mass m drops out of the equations in Example 4.12. The fact that mass does not enter into these calculations means that different gliders with different masses will still have the same acceleration. How can that be? We have already seen the reason in free fall. To the extent that we can neglect friction due to air, all objects fall with the same acceleration regardless of any differences in mass. The weight of an object—whether in free fall or on an inclined air track—is proportional to the mass by the relationship $w = mg$. This proportionality carries right on through with Newton's second law in other contexts. If the net force is proportional to the mass, then the acceleration is independent of the mass. Thus in Example 4.12, the fact that $mg \sin \theta = \Sigma F = ma$ tells us that $a = g \sin \theta$ for any and all values of m.

4.5 Action and Reaction: Newton's Third Law of Motion

CONCEPT OVERVIEW

■ Whenever one object exerts a force on a second object, the second object exerts on the first a force of equal magnitude but in the opposite direction.

A force is a push or a pull, but it is always a push or pull on one object by another—a push on object 1 by object 2, a pull on body A by body B. A force is an interaction between two things; there are always *pairs* of forces involved. For example, a hammer hits a nail and drives it into the wall; the nail brings the hammer to rest. A bat hits a baseball and changes the ball's speed and direction; the baseball changes the motion of the bat. A receiver catches a pass, and the football comes to a stop from high speed; the football exerts a force on the receiver as he catches it.

No single force ever exists by itself. This is the basis of Newton's third law of motion:

Whenever one object exerts a force on a second object, the second object exerts on the first a force of equal magnitude but in the opposite direction.

It is useful to express this law as an equation. Look at the forces that objects A and B exert on each other in Figure 4.15. The force exerted *on* object A *by* object B is $\mathbf{F}_{AB}$, and the force exerted *on* object B *by* object A is $\mathbf{F}_{BA}$. In general, the double subscripts mean the force *on* the first object *by* the second.

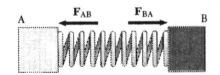

Figure 4.15
The equality $\mathbf{F}_{AB} = -\mathbf{F}_{BA}$ is a statement of Newton's third law. An example could be two blocks exerting forces on each other by means of a compressed spring.

Newton's third law of motion, then, can be written

$$F_{AB} = -F_{BA} \tag{4.5}$$

This law is sometimes stated, "For every action, there is an equal and opposite reaction," and the two forces are called an **action-reaction pair.**

It is important to remember that Equation 4.5 refers to forces that act on *different* objects at the same time. Every force has an agent and an object. The agent produces the force; the force acts upon the object. The forces in Equation 4.5 have different agents and act on different objects. Figure 4.16 illustrates a few examples of this. As you walk along the floor, you push back on the floor and the floor pushes forward on you. The tires of a car push back on the road as the road pushes forward on the tires. The propeller of an airplane pushes back on the air as the air pushes forward on the airplane. The spinning rotor of a helicopter is shaped so that it pushes down on the air, and the air in turn pushes up on the rotor. This upward push is called *lift*. When the (upward) lift equals the (downward) weight of the helicopter, the helicopter is able to hover in midair. When the lift is greater than the weight, the helicopter accelerates upward.

(a)

(b)

(c)

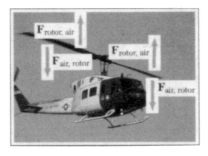

(d)

Figure 4.16
Forces always come in pairs: (a) a person walking, *(© Comstock, Inc.)* (b) a car on the road, *(© Comstock, Inc.)* (c) an airplane in flight, *(© Philip Wallick/FPG International Corporation)* (d) a helicopter in flight. *(© Comstock, Inc.)*

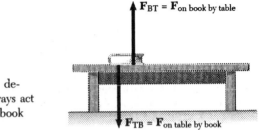

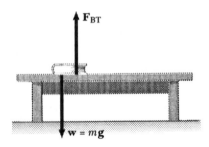

Figure 4.17
The action-reaction pair of forces described in Newton's third law always act on different objects, such as this book and table.

Figure 4.18
The forces acting on the book are the force exerted by gravity, which is the book's weight **w**, and the force exerted on the book by the table $\mathbf{F}_{BT}$, which is often called the normal force and labeled $\mathbf{F}_N$.

In Figure 4.17, which shows a book resting on a table, $\mathbf{F}_{TB}$ is the downward force on the table by the book and $\mathbf{F}_{BT}$ is the upward force on the book by the table. Since the book has an upward force acting on it, why does it remain resting on the table? Why doesn't it accelerate upward? To answer these questions, we must realize that we have not considered all the forces acting on the book, which are shown in Figure 4.18. In addition to the force on the book by the table $\mathbf{F}_{BT}$ from Figure 4.17, the book also experiences a gravitational force, its weight **w**. The book remains at rest because the net force on it, the vector sum $\Sigma \mathbf{F} = \mathbf{F}_{BT} + \mathbf{w}$, is zero.

In considering forces, it is important to note which object each force acts on. Do not overlook a force exerted by an inanimate object. If you push on a wall, as in Figure 4.19, it is easy to understand the force exerted on the wall by you $\mathbf{F}_{WY}$. There is also a force $\mathbf{F}_{YW}$, however, exerted on you by the wall. That force may be more difficult to understand. Think of placing a piece of paper between your hand and the wall, as in Figure 4.20. You push on the paper with force $\mathbf{F}_{PY}$. If that is the only force on the paper, it should accelerate. The paper

Figure 4.19
When you push on a wall, the wall pushes back on you.

Figure 4.20
These forces act on a piece of paper held between your hand and a wall. (Gravity and a friction force also act, but they have not been drawn in this diagram.)

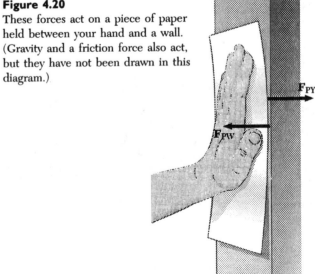

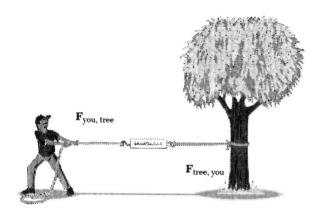

Figure 4.21
Inanimate objects can exert reaction forces.

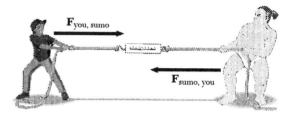

Figure 4.22
Reaction forces may be easier to visualize when you think of another person exerting them rather than an inanimate object.

does not accelerate, however, because the wall also exerts a force $\mathbf{F}_{PW}$ on it. This force $\mathbf{F}_{PW}$ has the same magnitude as $\mathbf{F}_{PY}$ but is directed in the opposite direction, as shown in Figure 4.20. Remove the paper and the force you exert acts on the wall instead of on the paper ($\mathbf{F}_{PY} = \mathbf{F}_{WY}$) and the force the wall exerts acts on you instead of on the paper ($\mathbf{F}_{PW} = \mathbf{F}_{YW}$).

Consider pulling on a rope tied to a tree, as in Figure 4.21. As you pull on the tree, the tree pulls back on you. This pair of forces can be measured by cutting the rope and splicing in a spring scale. If you exert a force of 500 N on your end of the rope, the rope does not accelerate. That means the tree must also be exerting a force of 500 N at the other end. The spring scale will read 500 N, the tension in the rope. At one end the rope exerts a force of 500 N to the right on you, while at the other end it exerts a force of 500 N to the left on the tree. Tension is the magnitude of these two forces acting along the rope.

How can an inanimate object *do* something like exert a force? To answer this question, it may help to look at Figure 4.22, where the tree has been replaced by an animate object—a Sumo wrestler! If you pull with a force of 500 N on your end of the rope and the wrestler pulls with 500 N of force on his end, the net force on the rope is zero and it remains at rest. The tension read by a scale spliced into the rope is still 500 N, exactly the same situation as when the rope is tied to the tree.

Newton's third law assures us that we cannot touch without being touched, whether the situation involves ropes and trees as in Figure 4.21 or the simpler situation of Figure 4.23.

4.6 Apparent Weight

CONCEPT OVERVIEW

■ The apparent weight of a body is equal to the upward force of support on the body. For accelerating bodies, this apparent weight is usually different from the true weight.

■ A body in free fall appears to be weightless.

Figure 4.23
We cannot touch without being touched. *(© Comstock, Inc.)*

Figure 4.24
Your apparent weight is the force you feel pushing or pulling up on you.

$\mathbf{F}_{you, trapeze}$

$\mathbf{F}_{you, chair}$

$\mathbf{F}_{you, scale}$

$\mathbf{w}$

$\mathbf{w}$

$\mathbf{w}$

An astronaut orbiting the Earth experiences "weightlessness" because she and her spacecraft are in free fall around the Earth. Lacking an apparent weight, astronauts on long space flights suffer from bone loss, muscle atrophy, and occasionally nausea. Normally living in an environment where we sense our weight, humans are not very comfortable being "weightless."

$a = g$　　　$w = mg$

Figure 4.25
A skydiver in free fall feels weightless even though it is his weight that is causing him to fall. We call this condition apparent weightlessness.
(F. Rickard-Artdia/Agence Vandystadt/Photo Researchers, Inc.)

Weight is the force gravity exerts on a body, but what is it that you *feel* as the force of your own weight? If you sit on a chair, you feel the chair pushing up on you. If you stand on a bathroom scale, you feel the scale pushing up on you. If you hang from a trapeze, you feel the trapeze pulling up on you. These situations are sketched in Figure 4.24.

The skydiver in Figure 4.25 is in free fall for a brief moment before air resistance becomes significant and before his parachute opens. During this interval of free fall, he still has weight—the force of gravity pulling him down—but he *feels* weightless because there is no opposing force pushing up on him. In other words, what you feel as your body weight is not the downward pull of gravity but rather the force of the floor, chair, whatever, pushing up on you. Your apparent weight is what a set of bathroom scales would read under your feet at any given moment. If the skydiver carried a set of bathroom scales, they would be falling along with him and would read zero; his apparent weight would be zero.

Example 4.13

A 70-kg man is in an elevator that is accelerating upward at 1.2 m/s². What are his true weight and his apparent weight on the elevator?

Reasoning

We can use Equation 4.4, $w = mg$, to find his true weight since we know his mass.

Figure 4.26 shows a sketch of the man in the elevator and a free-body diagram of the forces on him. The net force is the vector sum of his true weight w and the normal force of the elevator floor F_N. This net force causes his mass to have an upward acceleration of 1.2 m/s². The normal force F_N is his apparent weight. His true weight is given by Equation 4.4.

Solution

$$w = mg = (70 \text{ kg})(9.8 \text{ m/s}^2) = 686 \text{ kg}\cdot\text{m/s}^2 = 686 \text{ N} \approx 690 \text{ N}$$

When we use the free-body diagram to apply Newton's second law, Equation 4.3 gives[3]

$$\Sigma F = ma$$
$$\Sigma F = F_N - 686 \text{ N} = (70 \text{ kg})(1.2 \text{ m/s}^2)$$
$$F_N - 686 \text{ N} = 84 \text{ kg}\cdot\text{m/s}^2 = 84 \text{ N}$$
$$F_N = 770 \text{ N}$$

The man's apparent weight is 770 N. He feels pushed down into the floor as if his weight were 770 N, and this is just what a set of bathroom scales would read if he stood on them as he accelerated upward in the elevator.

___

Practice Exercise What is the man's apparent weight if the elevator accelerates downward at 0.8 m/s²?

Answer 630 N

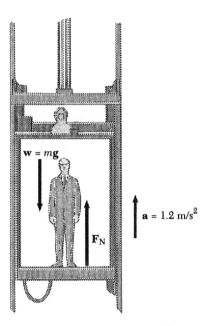

Figure 4.26
The two forces on a person in an elevator are the person's weight, pulling down, and the normal force from the floor of the elevator, pushing up.

Example 4.14

At an amusement park, the parachute tower drops guests with an initial downward acceleration of 7.2 m/s² (Fig. 4.27). What is the apparent weight of a 50-kg woman as she experiences this acceleration?

Reasoning

Figure 4.28 shows the forces acting on the woman. Her weight w acts downward, and the normal force F_N of the floor of the cage attached to the parachute acts upward. From the free-body diagram, we can apply Newton's second law and solve for the normal force, which is her apparent weight.

Solution

The forces on the woman are

$$\Sigma F = F_N - w = F_N - (50 \text{ kg})(9.8 \text{ m/s}^2) = F_N - 490 \text{ N}$$

Figure 4.27
The parachute drop amusement park ride at Coney Island, New York, in the 1940's. *(The Bettmann Archive)*

___

[3] Our given data, 70 kg and 1.2 m/s², limit our answer here to two significant figures. In working physics problems, however, it is a good idea to carry one uncertain figure through the calculations and then at the end round the answer to the correct number of significant figures. Thus in this example the answer to the true-weight question is "690 N," but we use the value 686 N, with the "6" uncertain, in subsequent calculations. The apparent weight is reported to only the two significant figures allowed by the input data. You should get in the habit of treating significant figures in this way.

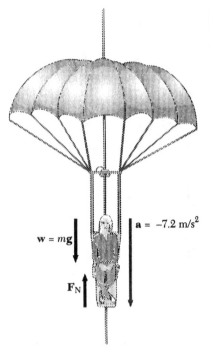

Since her acceleration is downward, we must use $a = -7.2$ m/s^2 in Equation 4.3:

$$\Sigma F = ma = F_N - 490 \text{ N}$$

$$F_N - 490 \text{ N} = (50 \text{ kg})(-7.2 \text{ m/s}^2)$$

$$F_N = 490 \text{ N} - 360 \text{ N} = 130 \text{ N}$$

Her apparent weight is only 130 N (remember, her true weight is 490 N). She feels much lighter. If her downward acceleration were a full 9.8 m/s^2, she would feel weightless.

4.7 Friction

CONCEPT OVERVIEW

- To a good approximation, the force of friction between two surfaces is proportional to the normal force pushing the surfaces together.

- The force of friction opposes the relative motion of the two surfaces (or the relative motion they would have if friction were not present).

Figure 4.28
The forces on an amusement park guest as she experiences "controlled free fall" are her weight w and the normal force F_N exerted by the floor of the cage attached to the parachute.

What happens when you try to slide a crate across the floor? There is resistance to your force. The heavier the crate, the greater the resistance; the rougher the floor or the bottom of the crate, the greater the resistance. This resistance is known as **friction.**

It is easy to see why there is resistance—friction—if a heavy, rough crate slides across a rough floor. The two surfaces are not smooth; rather they consist of small hills and valleys. The hills and valleys of the crate bottom are pushed down into the hills and valleys of the floor. Rough spots—hills—are squashed and mashed and deformed as the crate moves across the floor. There may even be momentary "welding" of high spots on the two surfaces. The same sort of thing occurs even if the crate and floor appear smooth. A microscopic view would show that the surfaces are still rough even if they appear smooth to the eye or touch, as illustrated in Figure 4.29.

There is much we can measure and understand about friction even though all of the details are not fully understood. Figure 4.30 shows a crate sitting on a floor. What is the force of friction between floor and crate? This straightforward question cannot be answered without additional information. The friction force may even be zero. If the man exerts no external force on the crate, then the friction force is zero. If he pushes to the right with a force of 1 N, the friction force $\mathbf{F}_f$ is 1 N to the left. If he pulls on the crate with a force of 5 N to the left, the friction force is 5 N to the right. Until some maximum value $F_{f,max}$ is reached, the friction force is whatever force is necessary to prevent movement. Friction forces are directed such that they always *oppose the motion* that would result if they were not present.

Figure 4.30 also has a free-body diagram of the forces on the crate. While it is at rest, it is in equilibrium. The sum of the forces in the y direction must vanish (because there is no motion in this direction), so the normal force F_N is equal to the weight w (notice that these are now *magnitudes*). The sum of the

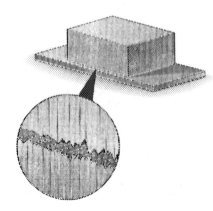

Figure 4.29
What appear to be smooth surfaces are not at all smooth when viewed on a microscopic scale.

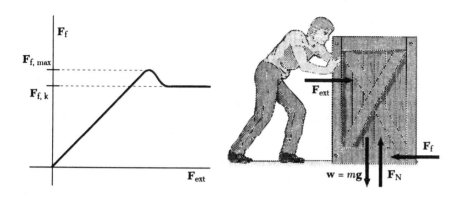

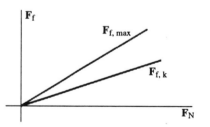

Figure 4.30
The friction force depends upon other forces acting on an object.

forces in the x direction must vanish, and so the friction force F_f is equal to the horizontal external force F_{ext}.[4]

As the horizontal force F_{ext} increases, so does the friction force F_f, but there is a maximum value $F_{f,max}$. While the crate is at rest, we refer to the friction force as *static friction*. The static friction force may be any value from zero to some maximum value which depends upon the other forces involved. If the horizontal external force becomes great enough, the crate will begin to move. Once it starts to move, the force of friction becomes a little less than what it was. The force is now referred to as *kinetic friction*. This decrease in the friction force can be seen on the graph in Figure 4.30.

What determines the value of the kinetic friction force $F_{f,k}$ and the maximum value of the static friction force $F_{f,max}$? If rocks are added to the crate, increasing its weight and the normal force, these friction forces will increase. If we do such an experiment and plot the data, we find a graph like that in Figure 4.31. For a moving object, we find that the force of kinetic friction $F_{f,k}$ is *directly proportional* to the normal force. That means, for instance, that doubling the normal force will double the friction force. Decreasing the normal force by 20 percent will decrease the friction force by 20 percent. This direct proportionality can be written

$$F_{f,k} = \mu_k F_N \tag{4.6}$$

where $F_{f,k}$ is the kinetic friction force, μ_k (the Greek letter mu) is a quantity known as the coefficient of kinetic friction, and F_N is the normal force between the two surfaces. Likewise, if we push on the crate until it just begins to move and find the maximum static friction force $F_{f,max}$, we find that this force, too, is *directly proportional* to the normal force, and we can write

$$F_{f,max} = \mu_s F_N \tag{4.7}$$

where μ_s is the coefficient of static friction.[5]

Figure 4.31
The force of kinetic friction $F_{f,k}$ is proportional to the normal force F_N between the two surfaces. The maximum value of the force of static friction $F_{f,max}$ is also proportional to the normal force between the two surfaces.

---

[4] Note that these are again magnitudes; while it is true that $F_f = F_{ext}$, it is also true that $\mathbf{F}_f = -\mathbf{F}_{ext}$. Always be careful to distinguish scalars from vectors.

[5] When it is clear from the context which kind of friction we are describing, it is common to drop the modifiers *static* and *kinetic*, talk of "the friction force" or "the coefficient of friction," and write F_f or $F_f = \mu F_N$.

TABLE 4.1
Selected List of Coefficients of Friction

	μ_s	μ_k
Skis on wet snow	0.15	0.12
Oiled metal on metal	0.15	0.12
Wood on wood	0.4	0.3
Metal on metal	0.6	0.5
Rubber on concrete	2	1.7

Many skiers wax their skis to reduce frictional resistance between the ski bottom and the surface of the snow. Why do skiers prefer slopes covered with "powder" rather than ice, which offers less resistance? (© Grafton M. Smith/The Image Bank)

The coefficient of friction depends upon the two surfaces in contact. A few approximate values are given in Table 4.1. It must be emphasized that any such list contains only approximate values. The coefficient of friction for pine sliding on pine, say, is different from that of maple sliding on oak. Also, the details of surface preparation are very important. Have the wood surfaces been sanded or waxed? How wet is the snow? How cold? What kind of a metal surface? What kind of oil? Is the concrete smooth or rough? What about the tire tread?

What about surface area? Will the friction force for long, thin, cross-country skis be different from the friction force for short, wide, jumping skis? Will some change with surface area, but that change is very small for hard surfaces. If we return to the crate in Figure 4.30 and turn it on its side—giving a different surface area but the same type of surface—and fill it with rocks again as we measure its weight, the normal force, and maximum friction force, we would obtain a graph nearly identical to the one in Figure 4.31. As long as the surfaces remain of the same type, the amount of surface area in contact has very little effect. The effect is so small that we shall ignore it and always use Equation 4.6 for kinetic situations and Equation 4.7 for static situations. This is a very useful approximation, valid under a wide range of situations. If you design skis for Yamaha or tires for Michelin, you will need to look at friction more carefully, but for most other purposes, this approximation will do.

Problem-Solving Tip

- In solving $\Sigma F = ma$, it is often necessary to resolve the forces into components and solve the equations $\Sigma F_x = ma_x$ and $\Sigma F_y = ma_y$. Choose one axis along the direction of motion and the other axis perpendicular to that.

- Friction always opposes the relative motion that would occur without friction. Assume a direction for the motion and draw the friction force in the opposite direction. If the results are inconsistent or unreasonable, then your assumed direction for motion was incorrect and you must start again.

Example 4.15

A wooden block sits on a wooden plane inclined 30° from the horizontal. Find the acceleration of the block down the plane. The coefficient of kinetic friction between block and plane is $\mu_k = 0.30$.

Reasoning

We can use Figure 4.32 as our free-body diagram, find all the forces, and then solve Newton's second law for the acceleration. Since the mass of the wooden block is not given, we shall simply represent it by M and expect it to drop out of our final answer.

The weight w of the block is

$$w = Mg = M \ (9.8 \ \text{m/s}^2)$$

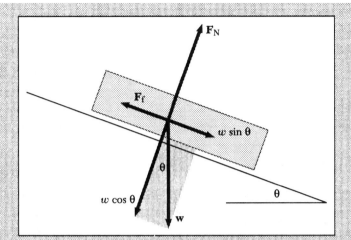

Figure 4.32
Forces on a block sitting on an inclined plane.

We shall take the x axis parallel to the plane; the block will move along this axis, and the y axis will be perpendicular to it. In terms of these directions, the components of the weight are

$$w_x = w \sin 30° = M \ (9.8 \text{ m/s}^2)(0.500) = M \ (4.90 \text{ m/s}^2)$$

$$w_y = -w \cos 30° = -M \ (9.8 \text{ m/s}^2)(0.866) = -M(8.49 \text{ m/s}^2)$$

The negative sign means this component lies in the direction of the negative y axis. In addition to those components, we have the normal force F_N in the positive y direction and the friction force F_f in the negative x direction.

Solution

Now we can apply Newton's second law in component form:

$$\Sigma F_x = w \sin 30° - F_f = Ma_x$$

$$M \ (4.90 \text{ m/s}^2) - F_f = Ma_x$$

We know the friction force in terms of the normal force, $F_f = \mu F_N$, but we do not yet know F_N. We can find that from the other component of $\Sigma F = Ma$:

$$\Sigma F_y = Ma_y$$

$$= F_N - w \cos 30° = Ma_y$$

There is no motion in the y direction (the box stays on the plane); therefore $a_y = 0$ and

$$F_N = w \cos 30° = M(8.49 \text{ m/s}^2)$$

Now we can solve for the friction force:

$$F_f = \mu F_N = (0.30)M(8.49 \text{ m/s}^2)$$

and the x-component equation is now complete:

$$\Sigma F_x = M \ (4.90 \text{ m/s}^2) - (0.30)M(8.49 \text{ m/s}^2) = Ma_x$$

$$4.90 \text{ m/s}^2 - (0.30) \ (8.49 \text{ m/s}^2) = a_x$$

$$a_x = 2.0 \text{ m/s}^2$$

Practice Exercise   Find the acceleration of the block down the plane if the incline is increased to 35°.

Answer 3.2 m/s^2

Example 4.16

A wooden block of mass $m_1 = 0.250$ kg sits on a wooden horizontal plane. The coefficient of friction between block and plane is $\mu_k = 0.35$. A string attached to the block runs over a pulley and is attached to a hanging block of mass $m_2 = 0.150$ kg. Find the acceleration of the blocks.

Reasoning

First, look back at Example 4.11 and notice that this arrangement is very similar to that one except now friction is present. Figure 4.33 shows a sketch of the equipment and free-body diagrams for the two blocks. As always, such free-body diagrams are essential!

With the free-body diagrams completed, we are ready to apply Newton's second law to each block. We shall take the x axis along the direction of motion of m_1, parallel to the plane, and the y axis perpendicular to that. For each block, we shall apply Newton's second law to determine the acceleration.

Solution

For m_1 we have

$$\Sigma F_x = T - F_f = m_1 a_x$$

$$T - F_f = (0.250 \text{ kg})a$$

We know that $a_y = 0$ since the block does not jump up off the plane or burrow down into it and we have set $a_x = a$. We do not yet know the

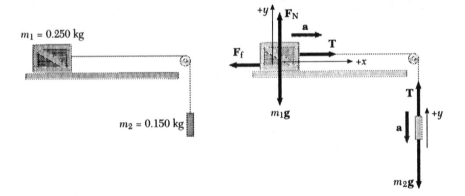

Figure 4.33
Friction is an important force to consider in drawing free-body diagrams and in applying Newton's second law.

friction force F_f, but we shall find that value after we write the y-component part of Newton's second law for m_1:

$$\Sigma F_y = F_N - m_1 g = m_1 a_y = 0$$

$$F_N - (0.250 \text{ kg})(9.8 \text{ m/s}^2) = 0$$

$$F_N = 2.4 \text{ N}$$

With the normal force known, we also have the friction force:

$$F_f = \mu F_N = (0.35)(2.4 \text{ N}) = 0.86 \text{ N}$$

Therefore,

$$T - 0.86 \text{ N} = (0.250 \text{ kg})a$$

and we still have two unknowns, T and a, meaning, as it always does, that more information is needed. We can find that additional information by applying $F = ma$ to m_2. If we take up as positive, the acceleration, which is downward, is negative: $a_y = -a$. Therefore

$$\Sigma F = T - m_2 g = m_2(-a) = m_2 a_y$$

$$T - (0.150 \text{ kg})(9.8 \text{ m/s}^2) = (0.150 \text{ kg})(-a)$$

$$T = 1.5 \text{ N} + (0.150 \text{ kg})(-a)$$

Now we can return to the x-component equation for m_1:

$$T - F_f = (0.250 \text{ kg})a$$

$$[1.5 \text{ N} + (0.150 \text{ kg})(-a)] - 0.86 \text{ N} = (0.250 \text{ kg})a$$

$$(0.400 \text{ kg})a = 0.64 \text{ N}$$

$$a = \frac{0.64 \text{ N}}{0.400 \text{ kg}} = 1.6 \text{ m/s}^2$$

Example 4.17

A wooden box of mass $m_1 = 0.500$ kg sits on a steel plane inclined 20° from the horizontal. The coefficient of friction between block and plane is $\mu_k = 0.10$. A string attached to this block runs over a pulley and is attached to a hanging block of mass $m_2 = 0.250$ kg, arranged to pull the wooden block up the plane. Find the acceleration of the blocks.

Reasoning

This resembles the previous example except that the plane is inclined this time, and that is an important difference. Do not start by writing equations; rather, start by drawing diagrams. Figure 4.34 shows a sketch of the equipment and free-body diagrams for the two masses.

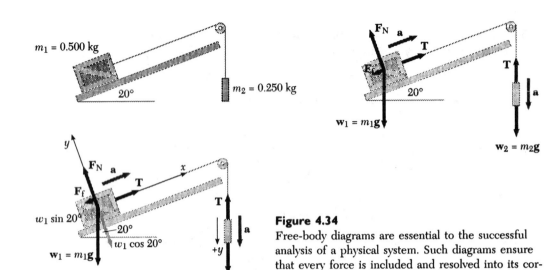

Figure 4.34
Free-body diagrams are essential to the successful analysis of a physical system. Such diagrams ensure that every force is included and resolved into its correct components.

In this example, forces act in several directions; so we must choose a set of coordinate axes for each block and resolve the forces into components along those axes. The hanging block is simple; the forces of tension T and weight w_2 lie only along the vertical. If we take down as positive, the acceleration is a. For the block on the plane, we choose the x axis along the plane, in the direction of the acceleration, and the y axis perpendicular to that.

Solution

With down as the positive direction, Newton's second law applied to the hanging block gives

$$\Sigma F_2 = -T + w_2 = m_2 a$$

$$(0.250 \text{ kg})(9.80 \text{ m/s}^2) - T = (0.250 \text{ kg})a$$

$$2.45 \text{ N} - T = (0.250 \text{ kg})a$$

As we have come to expect, this equation has two unknowns so additional information is needed.

Newton's second law for m_1 in vector form is

$$\Sigma \mathbf{F}_1 = \mathbf{T} + \mathbf{F}_N + \mathbf{F}_f + \mathbf{w}_1 = m\mathbf{a}$$

but we need it in component form to be able to solve it. With our choice of axes described above, the tension points in the positive x direction, the normal force in the positive y direction, and the friction force in the negative x direction. The weight does not lie along either axis, so we must resolve it into components: $w_1 \sin 20°$ in the negative x direction and $w_1 \cos 20°$ in the negative y direction. The acceleration along the x axis is just a, the acceleration of the block, and the acceleration along the y axis is zero.

Now we can proceed:

$$\Sigma F_{1x} = T - F_f - w_1 \sin 20° = m_1 a_x$$

$$T - \mu F_N - (0.500 \text{ kg})(9.80 \text{ m/s}^2)(0.342) = (0.500 \text{ kg})a$$

$$\Sigma F_y = F_N - w_1 \cos 20° = m_1 a_y = 0$$

$$F_N = w_1 \cos 20° = (0.500 \text{ kg})(9.80 \text{ m/s}^2)(0.940) = 4.60 \text{ N}$$

$$F_f = \mu F_N = (0.10)(4.60 \text{ N}) = 0.46 \text{ N}$$

$$T - 0.46 \text{ N} - 1.68 \text{ N} = (0.500 \text{ kg})a$$

$$T - 2.14 \text{ N} = (0.500 \text{ kg})a$$

$$T = 2.14 \text{ N} + (0.500 \text{ kg})a$$

Now that we have T in terms of a, we can substitute for T in the equation for the hanging block we obtained earlier:

$$2.45 \text{ N} - T = (0.250 \text{ kg})a$$

$$2.45 \text{ N} - [2.14 \text{ N} + (0.500 \text{ kg})a] = (0.250 \text{ kg})a$$

$$0.31 \text{ N} = (0.750 \text{ kg})a$$

$$a = 0.413 \text{ m/s}^2 \approx 0.41 \text{ m/s}^2$$

Notice that the normal force $F_N = 4.60$ N on the wooden block in Example 4.17 is *not* equal to the weight $w_1 = 4.90$ N. There are just enough special cases where the normal force happens to equal the weight to make students sometimes (and wrongly) assume that the normal force and weight are always the same. Be careful, for that is simply not true!

Example 4.18

In Figure 4.35 a heavy, rough crate sits on an inclined plane. The weight of the crate is 60 N. The plane is inclined 37° from the horizontal, and the coefficient of friction between plane and crate is $\mu_k = 0.8$. Find the force needed to just keep the crate moving up the plane. Then find the force needed to just keep the crate moving down the plane. What happens to the friction force in the two cases?

Reasoning

Figure 4.35 is a free-body diagram for the crate moving up the plane. The force of friction is down the plane since friction always opposes the motion. The external force $\mathbf{F}_{ext}$ is directed up the plane, and the weight $\mathbf{w}$ acts vertically downward. The plane exerts two forces: the friction force $\mathbf{F}_f$ parallel to its surface and directed down the plane, opposite the crate's motion, and the normal force $\mathbf{F}_N$ perpendicular to the plane surface. Of course, $F_f = \mu F_N$. We shall resolve the weight into its components paral-

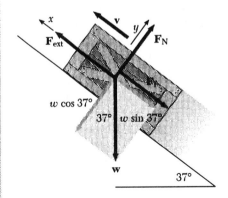

Figure 4.35
When the block moves up the plane, the friction force is directed down the plane.

lel and perpendicular to the plane (along the x and y axes in the diagram) and apply Newton's second law. We choose up the plane as the positive x direction and above the plane as the positive y direction.

Solution

From the diagram we find all the forces to use in Newton's second law. Since we need only "just keep the crate moving," we know that $a = 0$.

$$\Sigma F_x = ma = F_{ext} - F_f - w\sin 37° = 0$$

$$F_{ext} - \mu F_N - (60\text{ N})(0.60) = 0$$

$$F_{ext} = 0.8F_N + 36\text{ N}$$

$$\Sigma F_y = ma = F_N - w\cos 37° = 0$$

$$F_N = (60\text{ N})(0.80) = 48\text{ N}$$

$$F_{ext} = (0.8)(48\text{ N}) + 36\text{ N} = 74\text{ N} \approx 70\text{ N}$$

Reasoning

Figure 4.36 is a free-body diagram for the crate moving down the plane. The external force is now directed down the plane, and the force of friction is directed up the plane. The normal force remains perpendicular to the plane and the weight remains vertically downward. Again, we use Newton's second law. Because we have already chosen up the plane as our positive x direction, we must remember to pay special attention to the sign we assign to each force.

Solution

$$\Sigma F_x = ma_x = F_f - F_{ext} - w\sin 37° = 0$$

$$\mu F_N - F_{ext} - (60\text{ N})(0.60) = 0$$

$$0.8\,F_N - F_{ext} - 36\text{ N} = 0$$

$$\Sigma F_y = ma_y = F_N - w\cos 37° = 0$$

$$F_N = (60\text{ N})(0.80) = 48\text{ N}$$

$$(0.8)(48\text{ N}) - F_{ext} - 36\text{ N} = 0$$

$$F_{ext} = 2.4\text{ N} \approx 2\text{ N}$$

It is interesting to examine this example with a different coefficient of friction. This time, sand the bottom of the crate so that it is smoother and $\mu = 0.4$. The problem seems the same. The free-body diagrams are the same. Proceed just as before.

For moving the crate up the plane, we have

$$\Sigma F_x = F_{ext} - F_f - w\sin 37° = 0$$

$$F_{ext} - \mu F_N - (60\text{ N})(0.60) = 0$$

$$F_{ext} = 0.4\,F_N + 36\text{ N}$$

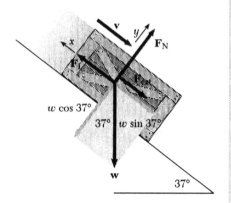

Figure 4.36
The direction of the friction force is opposite the direction of the velocity as the crate slides along the plane. When the crate moves down the plane, the force of friction is directed up the plane.

$$\Sigma F_y = F_N - w \cos 37° = 0$$

$$F_N = (60 \text{ N})(0.80) = 48 \text{ N}$$

$$F_{ext} = (0.4)(48 \text{ N}) + 36 \text{ N} = 55 \text{ N}$$

So far, so good; this is exactly what we did with the rougher crate. However, look at the case of sliding down the plane. The free-body diagram looks the same as for the $\mu = 0.8$ crate.

$$\Sigma F_x = F_f - F_{ext} - w \sin 37° = 0$$

$$\mu F_N - F_{ext} - (60 \text{ N})(0.60) = 0$$

$$0.4 \, F_N - F_{ext} - 36 \text{ N} = 0$$

$$\Sigma F_y = F_N - w \cos 37° = 0$$

$$F_N = (60 \text{ N})(0.80) = 48 \text{ N}$$

$$(0.4)(48 \text{ N}) - F_{ext} - 36 \text{ N} = 0$$

$$F_{ext} = -17 \text{ N}$$

How do we interpret a negative value for the external force? When we drew Figure 4.36 for the crate moving down the plane, we *assumed* F_{ext} would be directed down the plane. A negative value means our assumption was incorrect! With its smoother surfaces, the crate will now accelerate down the plane if left alone. Therefore, to keep it "just moving down" with constant velocity, we must exert a force up the plane.

Getting a negative answer when you don't expect to usually means that something—usually a direction—assumed in drawing the diagrams or setting up the equations is the opposite of what you assumed. One point you should remember is that, if the direction of the velocity is different from your initial assumption, you must also change the direction of the friction force.

Equations are useful and therefore important, but never simply plop numbers into an equation, solve for an answer, and leave without asking what that answer means. Equations are a *summary* of ideas and understanding. They must never become a *substitute* for ideas and understanding.

Example 4.19

A car initially traveling at 30 m/s comes to an emergency stop and leaves skid marks for 82 m (Fig. 4.37). What is the coefficient of kinetic friction between tires and road?

Reasoning

We know the initial velocity and the distance traveled, and from that information we hope to determine the coefficient of friction. Forces

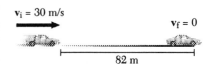

Figure 4.37
The braking car leaves skid marks for 82 m. What is the coefficient of friction between tires and road?

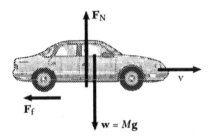

Figure 4.38
Free-body diagram showing the friction force that supplies the braking force required to bring the car to a stop.

cause accelerations, so looking for the force of friction means finding the acceleration of the car.

Solution

Distance traveled, velocity, and acceleration are related by Equation 2.7:

$$v^2 = v_i^2 + 2a(s - s_i)$$

$$0^2 = (30 \text{ m/s})^2 + 2a(82 \text{ m})$$

$$2a(82 \text{ m}) = -900 \text{ m}^2/\text{s}^2$$

$$a = \frac{-900 \text{ m}^2/\text{s}^2}{164 \text{ m}} = -5.5 \text{ m/s}^2$$

Notice that the acceleration is negative. A braking acceleration is usually negative.

Knowing the acceleration, we can now use Newton's second law to find the force. The mass of the car is not given; such omissions may seem unnerving sometimes. If you think more information is needed, use a symbol for the information you think you need and see if the symbol somehow drops out of the picture. We shall call the mass of the car M.

Figure 4.38 is a free-body diagram of the forces on the car while it brakes to a stop. The coefficient of friction involves the normal force, and the normal force involves the weight of the car. We write $w = Mg$ and continue with the problem, expecting that the M's will later disappear.

From the free-body diagram, you can see that the upward normal force exerted by the road on the car is just equal to the downward weight. Perhaps you can see this detail by inspection, but just to be sure let's determine the y component of Newton's second law:

$$\Sigma F_y = F_N - w = Ma_y = 0$$

From inspection or from solving this equation, we have

$$F_N = w = Mg$$

The net force on the car is just the force of friction; since this force points to the left, we attach a negative sign to it:

$$\Sigma F_x = -F_f = Ma$$

$$F_f = \mu F_N = \mu Mg$$

$$-\mu Mg = Ma$$

$$-\mu(9.8 \text{ m/s}^2) = (-5.5 \text{ m/s}^2)$$

$$\mu = \frac{5.5 \text{ m/s}^2}{9.8 \text{ m/s}^2} = 0.56$$

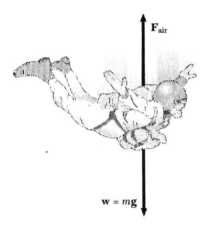

Figure 4.39
A skydiver in free fall experiences two opposing forces: the resistance of the air and the pull of gravity. The upward force of air resistance increases with the skydiver's downward speed.

The force of static friction is always greater than the force of kinetic friction. It takes greater force to start something moving than to keep it moving. While automobile tires roll along the road, the part of the tire touching the pavement at any instant is at rest with respect to the pavement. The friction between road and tire is static friction. This is true whether the car is accelerating, coasting, or braking—as long as the wheels are rolling normally. However, if the brakes are applied strongly enough to lock the wheels so that the tires skid along the pavement, then only the force of kinetic (sliding) friction is present. A car skidding with locked brakes requires a greater distance to come to a stop. Therefore, more and more cars now come equipped with antilock brakes.

As a driver begins to apply the brakes in a car equipped with an antilock braking system on wet or icy pavement, sensors report the amount of wheel lock to a computer that signals the brakes to reduce the pressure or pulse the pressure, thus preventing loss of control.

*4.8 Terminal Velocity

CONCEPT OVERVIEW

- As an object falls through the air, the upward force exerted on it by the air increases with the object's speed until the force finally equals the downward force exerted by gravity. With the upward and downward forces on the object balanced, it will no longer accelerate; it has reached its terminal velocity.

What happens to the forces on a skydiver after she leaves the airplane? No doubt you have held your hand out the window of a moving car. When you were stopped, the air exerted no force on your hand. As the car—and your hand—gained speed, the air colliding with your hand exerted a force on it. As the speed increased, so did this force. That is what happens to the skydiver. Figure 4.39 is a sketch of the forces acting on her as she falls. Her weight remains the same; the downward force exerted by gravity does not change. When the downward velocity is small, the force due to air resistance F_{air} is also small. There is a net force downward of $\Sigma F_y = w - F_{air}$, and the skydiver continues to accelerate. As the velocity increases, so does F_{air}. The net force—and hence the acceleration—decrease. The velocity will increase until F_{air} just equals w. When this happens, the net force will be zero. A net force of zero means zero acceleration and constant velocity. The speed at which this happens is known as **terminal velocity** or **terminal speed.**

The force of air resistance F_{air} depends upon the size and shape of the object in addition to its speed. Terminal velocity, then, depends upon object size and shape. A skydiver pulled up into a ball offers very little air resistance. Her terminal velocity therefore must be large to provide an air-resistance force large enough to balance her weight. Spreading her arms and legs increases her surface area and provides much greater air resistance. A smaller terminal velocity will now give an air-resistance force that balances her weight. Thus by changing size and shape, one skydiver can fall more slowly than another. This possibility allows skydivers to maneuver or catch up with each other and even form and fall in intricate formations, such as in Figure 4.40(a).

Of course, the parachute in Figure 4.40(b) offers much greater surface area and therefore much greater air resistance. Therefore the terminal velocity of a falling parachutist is much less than it would be without the parachute. After all, that is the whole idea behind using a parachute!

(a)

(b)

Figure 4.40
(a) Skydivers can slow down, speed up, and maneuver by changing their size and shape. Such skills allow experienced skydivers to control their fall and form intricate formations. (© *Guy Savage/Photo Researchers*) (b) The last thing skydivers do in making a safe descent is, of course, open their parachutes. Why do you think parachute shapes have changed over the past 50 years from the circular ones seen so often in old war movies to the rectangular modern ones shown here? (© *Peter M. Miller/The Image Bank*)

SUMMARY

Newton's three laws of motion are general statements about motion and its causes.

Newton's first law may be stated as follows:

An object at rest will remain at rest and an object in motion will remain in motion with the same velocity—along the same straight line with the same speed—unless an unbalanced force acts upon it.

A force is basically a push or a pull. A force is anything that causes a change in the motion of a body, that is, anything that causes an acceleration. This definition is described by Newton's second law:

If there is a net force on an object, that object will accelerate. The acceleration will be in the direction of the net force and will be directly proportional to that force and inversely proportional to the mass of the object.

In equation form Newton's second law can be written

$$\Sigma F = ma \tag{4.3}$$

or

$$F_{net} = ma \tag{4.3'}$$

No single force ever exists by itself. A force on one object by a second object brings into existence a force back on the first object. This is Newton's third law:

Whenever one object exerts a force on a second object, the second object exerts on the first a force of equal magnitude but in the opposite direction.

If we call the force exerted on body A by body B $\mathbf{F_{AB}}$ and the force exerted on body B by body A $\mathbf{F_{BA}}$, Newton's third law can be written

$$\mathbf{F_{AB}} = -\mathbf{F_{BA}} \tag{4.5}$$

Mass is a measure of the inertia of an object. *Weight* is the force that gravity exerts on an object. For any object, the relationship between these two quantities is

$$\mathbf{w} = m\mathbf{g} \tag{4.4}$$

An accelerating body appears to have a weight—an *apparent weight*—different from its true weight. A body in free fall has an apparent weight of zero.

Friction is a force parallel to a surface and opposes the relative motion of one surface past another. The friction force F_f is proportional to the normal force F_N pressing the two surfaces together. The maximum force of static friction is greater than the force of kinetic friction:

$$F_{f,k} = \mu_k F_N \tag{4.6}$$

$$F_{f,max} = \mu_s F_N \tag{4.7}$$

An object in free fall experiences an upward force exerted by the air. As the fall speed increases, this force increases until it just balances the object's weight. The object has then reached its *terminal velocity*.

QUESTIONS

1. If an object is at rest, can you conclude there are no forces acting on it?
2. If an object moves with constant velocity, can you conclude there are no forces acting on it?
3. If an object moves with constant acceleration, can you conclude there are no forces acting on it?
4. Why are cleats important on a baseball player's shoes?
5. Explain the trick shown in Figure 4.41, where a tablecloth is quickly pulled out from under some dishes without appreciably moving the dishes.
6. As an airplane accelerates down the runway from rest to takeoff speed, you feel yourself pushed back into your seat. Explain.
7. If you are in a car stopped in traffic and are hit from behind, you might suffer a "whiplash" injury because your head is thrown backward. Ex-

Figure 4.41
Question 5. How is the "magician" able to yank away the tablecloth without disturbing the dishes?

plain how a force to the car, directed forward, could cause your head to be thrown backward.

8. The iron ball in Figure 4.42 hangs by a thread; a similar thread is attached to the lower side of the ball and hangs loose. If the loose thread is given a slow and steady pull, the top thread will break. If the loose thread is given a sharp jerk, it will break. Explain.

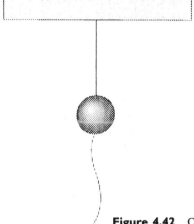

Figure 4.42   Question 8.

9. A box in a delivery van accelerates along with the van. What produces the forces on the box that cause its acceleration?
10. When the brakes are applied hard, a box in a delivery van slides toward the front. Explain.
11. If a pair of binoculars are dropped from the bridge of a cruise ship moving at constant velocity, where will they land on the deck below? If the same binoculars are dropped while the ship is accelerating, where will they land? If your two answers are different, explain.
12. If buying gold by weight, would you get a better

deal in Antarctica or Ecuador? At sea level in a coastal city or in the mountains of Nepal?
13. If buying gold by mass, would you get a better deal in Antarctica or Ecuador? At sea level in a coastal city or in the mountains of Nepal?
14. Two skaters—one massive and one lightweight— face each other and push themselves apart. Compare the forces on the two skaters. Compare their accelerations.
15. What are the action-reaction forces when a baseball is hit with a bat? What are the action-reaction forces while the baseball is in flight?
16. Identify all the action-reaction forces when you hold a book on your lap while sitting in a chair.
17. Identify all the action-reaction forces when you walk along pulling a child in a wagon.
18. A horse is pulling a heavy cart, and the cart and horse are both accelerating. Is the force exerted by the cart on the horse equal to and opposite that exerted by the horse on the cart? If your answer is "yes," explain how these equal and opposite forces give rise to the net force necessary to cause acceleration. If your answer is "no," explain whether or not Newton's third law is true while the cart is accelerating.
19. Why does your stomach sometimes feel funny as you ride on an express elevator in a tall building?
20. Consider giving a block an initial velocity up an inclined plane. Compare the accelerations as it slides up the plane and then as it slides back down.
21. You give a block an initial velocity up an inclined plane. It slows down and stops but does not slide back down. Why?
22. Why is it easier to pull a lawn mower than to push it?
23. Consider pushing a sled with a force **F** inclined at angle θ, as shown on the left in Figure 4.43, or pulling it with a force of the same magnitude

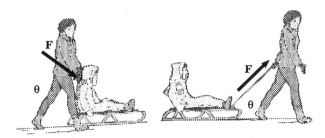

Figure 4.43   Question 23.

F inclined at the same angle θ, as shown on the right in the figure. Both forces have the same horizontal component. What difference do you expect in these two cases?

24. Consider pushing a mop with a force **F** directed along its handle, as shown in Figure 4.44. What happens as the angle θ changes? In particular, is it possible to choose an angle θ such that you

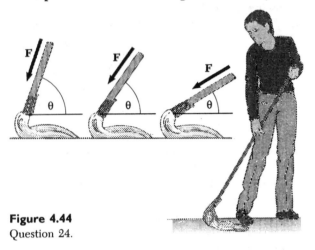

Figure 4.44
Question 24.

cannot move the mop, no matter how large the force is?

25. Is it possible to have a coefficient of friction greater than 1.0?

26. Why will pumping your brakes stop your car in a shorter distance than slamming them on with all your strength?

27. Why will applying your brakes gently stop your car in a shorter distance than slamming them on with all your strength?

28. What things determine the force of friction?

29. What things determine the coefficient of friction?

30. How can an adjustable inclined plane be used to measure the coefficient of friction?

31. Why would a very heavy skydiver want a larger parachute than a very lightweight skydiver?

32. If a stone and a cork fall with the same acceleration, how can skydivers control their jumps so that a group can fall together in a pattern even though the members jumped from the plane at different times?

PROBLEMS

4.1 Inertia: Newton's First Law of Motion

4.2 Force, Mass, and Acceleration: Newton's Second Law of Motion

4.1 A 1200-kg car moves to the right with a constant speed of 1.4 m/s. What is the net force on the car? A 1200-kg car moves to the right with a constant acceleration of 1.4 m/s². What is the net force on the car?

4.2 A model boat whose mass is 6.5 kg floats on a pond. If the boat experiences a net horizontal force of 2.5 N, what is its acceleration?

4.3 What net force will provide a 4.5-kg mass with an acceleration of 2.3 m/s²?

4.4 A net force of 13.2 N acts on a skateboard accelerating at 1.80 m/s². What is the mass of the skateboard?

O4.5 Advertisements for a 1280-kg automobile state that it can accelerate from zero to 90 km/h in 8.3 s. What is the average net force acting on the car?

O4.6 A 10-g rifle bullet has a muzzle velocity of 880 m/s. The length of the rifle barrel is 80 cm. What is the net force accelerating the bullet? Assume the force is constant.

O4.7 A billiard ball and a bowling ball are at rest on a frictionless table (Fig. 4.45). When the same net force is applied to both, the billiard ball accelerates with 32 times the acceleration of the bowling ball. The bowling ball has a mass of 7.3 kg. What is the mass of the billiard ball?

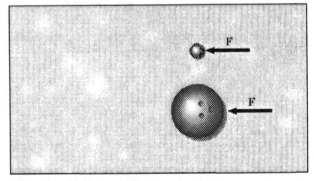

Figure 4.45 Problem 4.7.

○4.8   A 160-g baseball moving horizontally at 30 m/s toward the batter is struck by the bat and begins to move in the opposite direction at 45 m/s. The contact between ball and bat lasts for 1.0 ms. Find the average force exerted by the bat on the ball.

○4.9   An arrow of mass 0.15 kg is shot from a bow with an average force of 180 N applied through a distance of 0.6 m. (*a*) With what speed does the arrow leave the bow? (*b*) If it is fired straight up, how high will it go? (*c*) What maximum horizontal distance can it be shot? Assume it is shot from ground level and returns to the same level.

○4.10   During a test of seat belts, a car that has a mass of 1.4 metric tons (1400 kg) and is traveling at 12 m/s collides with a heavy truck. The car moves forward 3.0 m while coming to rest. (*a*) What average force is exerted on the car? (*b*) What average force is exerted by the seat belt on an 80-kg driver?

○❖4.11   Shown in Figure 4.46 are six rocks that have all been thrown straight up from the same initial height. The rocks all have the same size and shape but different masses, and they are all thrown with different speeds. Masses and speeds when released are given in the figure. Neglecting air resistance, rank these rocks according to maximum height attained.

Figure 4.46
Problem 4.11.

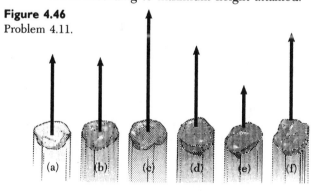

Mass:	1.0 kg	1.1 kg	1.3 kg	2.5 kg	2.8 kg	4.1 kg
v_i:	2.5 m/s	2.3 m/s	3.8 m/s	2.5 m/s	1.3 m/s	2.7 m/s

○4.12   A person is riding in a car that is struck from behind, causing a sudden and brief forward acceleration of 1215 m/s². In order to follow along with the person's body, the person's head, of mass 3.40 kg, must be acted upon by a large forward force exerted by the neck muscles and upper spine. What is the value of the force involved in this whiplash injury?

●❖4.13   Shown in Figure 4.47 are six laboratory carts of various masses. Each has the initial velocity and net force shown. Rank these carts according to velocity after 10 s.

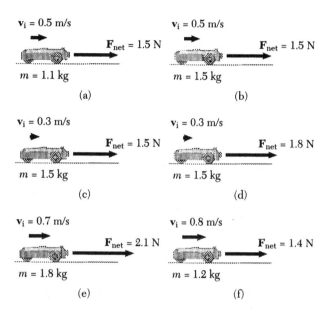

Figure 4.47   Problem 4.13.

●4.14   A basketball player wishes to pass the 0.62-kg ball to a teammate who is 5.0 m away, measured horizontally. The thrower draws his arm back to a position 5.8 m from the teammate and releases the ball after pushing it forward 0.8 m. The ball leaves the thrower's hand moving horizontally. (*a*) What force must he apply to the ball if it is to reach the teammate 0.25 s after release? (*b*) How far does the ball fall during its trajectory?

4.3 Mass and Weight

4.15   What is the weight of a 100-kg mass?

4.16   What is the mass of a 100-N weight?

4.17   A man has a mass of 80 kg; what is his weight?

4.18   A woman has a mass of 60 kg; what is her weight?

4.19   The acceleration of gravity on the Moon's surface is about one sixth what it is on Earth's surface. What is the weight on the Moon of an astronaut whose mass is 85 kg on Earth? What is the astronaut's mass on the Moon?

4.20 A typical human liver has a mass of 0.8 kg. What is its weight in newtons?

4.21 A typical human head has a weight of 35 N. What is its mass in kilograms?

4.4 Free-Body Diagrams

4.22 Two people pull as hard as they can on ropes attached to a boat that has a mass of 200 kg. If they pull in the same direction, the boat accelerates at 1.5 m/s^2 to the right. If they pull in opposite directions, the boat accelerates at 0.52 m/s^2 to the left. What is the force exerted by each person on the boat? (Disregard any other forces on the boat.)

4.23 An elevator weighs 39.2 kN, and the tension in its supporting cable is 35.2 kN. (*a*) What are the magnitude and direction of its acceleration? (*b*) If it starts from rest, how far will the elevator move in 2.5 s? In which direction will it move?

4.24 A 2000-kg truck is moving at 20 m/s. It is acted upon by two horizontal forces (assumed constant): a forward force of 800 N exerted by the engine (and the road) and a 300-N retarding force due to air resistance. What is the acceleration of the truck?

4.25 A block starts from rest and slides down a frictionless inclined plane that is inclined 20° from the horizontal and is 2.5 m long. How fast is the block moving just as it reaches the bottom?

4.26 During a rescue at sea following a boating accident, a 70-kg man dangles on the end of a rope attached to a Coast Guard helicopter. If the helicopter accelerates upward at 2.3 m/s^2, what is the tension in the rope?

4.27 A supertanker of mass 250 000 metric tons (1 metric ton = 1000 kg) coasts to a stop from an initial speed of 30 km/h in 25 min. Assume its acceleration is constant. (*a*) How far does it go in that time? (*b*) What resistance force is supplied by the water?

4.28 A tractor of mass 1000 kg is attached by means of a horizontal, massless cable to a log whose mass is 400 kg. The tension in the cable is 2000 N, and the force of friction exerted by the ground on the log is 800 N. How far will the log move in 2.00 s, starting from rest?

4.29 A 60-g golf ball is struck by a golf club and given a speed of 70 m/s during the impact, which lasts for 2.0×10^{-4} s. What force (assumed constant) is exerted on the ball?

4.30 After falling from a height of 30 m, a 0.5-kg ball rebounded upward, rising to 80 percent of its initial height. If the contact between ball and ground lasted for 2.0 ms, what average force was exerted on the ball while it was in contact with the ground?

4.31 Four airline baggage carts are coupled as shown in Figure 4.48 to make a train; each cart has a mass of 250 kg. Find the force **F** that must be applied to accelerate this train at 0.8 m/s^2. Find the tension in each coupling between carts.

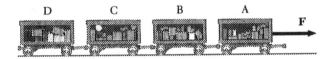

Figure 4.48 Problem 4.31.

4.32 A 75-kg skydiver acquires a velocity of 50 m/s and then opens her parachute. After she has fallen an additional 30 m, her velocity is reduced to 2.5 m/s. (*a*) What is her acceleration during that 30 m? (*b*) What average force does the parachute exert during that time interval?

4.33 A 5.0-kg iron mass is jerked upward by a rope; it rises 15 cm in 0.10 s. Assume uniform acceleration. What is the tension in the rope during this time interval?

4.34 A 50-ton (50 000-kg) rocket is acted on by an upward force of 600 kN. If the rocket is 25 m tall, how long does it require to rise off the launching pad a distance equal to its height?

4.5 Action and Reaction: Newton's Third Law of Motion

4.35 A tractor of mass 800 kg is attached by a horizontal chain (whose mass is negligible compared with the other masses involved) to a log whose mass is 400 kg. The force of friction exerted by the ground on the log is 500 N. The system (tractor, chain, and log) has a forward acceleration of 2.00 m/s^2.

(*a*) What is the tension in the chain?

(*b*) What is the forward force of the ground on the tractor?

O4.36 A 30-kg block is at rest on a horizontal table. The coefficient of static friction between block and table is 0.20; the coefficient of kinetic friction, 0.18. The block is attached to a string that runs over a lightweight, frictionless pulley and is then attached to a 10-kg hanging mass. (*a*) What is the tension in the string while the 30-kg block is at rest? (*b*) What is the tension when the block is released? (*c*) What is the acceleration of the block when it is released? (*d*) How long does it take the hanging mass to descend 0.60 m?

Static friction must be less than or equal to

$F_p = \mu_s F_N$

a)

O4.37 Objects with masses of 3 kg and 5 kg are connected by a light cord that passes over a lightweight frictionless pulley, forming an Atwood's machine. (*a*) How long is required for the 5-kg object to fall 1 m, starting from rest? (*b*) What is the tension in the cord?

●4.38 Two football linemen are locked in combat. Player A, on the right, has a mass of 102 kg; player B, on the left, 94 kg. The players accelerate to the right with a constant acceleration of 0.54 m/s². If A exerts a force of 320 N to the left on B, determine the force of the ground on each player.

●4.39 Two small balls of equal mass are connected by a string 1 m long and are laid out on a smooth (in other words, frictionless) table with ball A just at the edge of the table and ball B 1 m in from the edge. The table is 0.5 m high. Ball A is nudged gently over the edge of the table, and things begin to happen. (*a*) How much time is required for A to strike the floor? (*b*) How much time is required for B to strike the floor? (*c*) How far from the base of the table does B strike the floor?

●4.40 Three cars of a model railroad train, each of mass 0.8 kg, are connected together and pulled by an engine of mass 1.6 kg. The track supplies a forward force of 12 N on the engine. (*a*) What force does the engine exert on the track? (*b*) What is the acceleration of the train? (*c*) What is the tension in the coupling between the front and middle cars?

4.6 Apparent Weight

4.41 What is the apparent weight of a 60-kg woman standing in an elevator that is ascending at 2.4 m/s and has an upward acceleration of 1.2 m/s²? What is her true weight?

4.42 What is the apparent weight of an 80-kg man riding in an elevator that is ascending at 3.2 m/s and has a downward acceleration of 1.8 m/s²? What is his real weight?

4.7 Friction

4.43 A 25-kg box sits on a rough surface. The coefficient of sliding friction between box and floor is 0.30. A rope is attached to the box and makes an angle of 30° above the horizontal. What must the tension in the rope be if the box is to move along the floor at constant velocity?

4.44 A 60-kg crate sits on a rough surface. The coefficient of sliding friction between crate and floor is 0.35. A rope is attached to the crate and makes an angle of 37° above the horizontal. What must the tension in the rope be if the crate is to move along the floor at constant velocity?

4.45 A 7.0-kg package sits on a rough but level floor. The coefficient of sliding friction between package and floor is 0.27. A cord is attached to the package and makes an angle of 30° above the horizontal. What must the tension in the cord be if the package is to move along the surface with an acceleration of 1.0 m/s²?

4.46 A man pushes a 20-kg lawn mower at constant speed with a force of 80 N directed along the handle, which makes an angle of 37° with the horizontal (Fig. 4.49). Calculate (*a*) the horizontal and vertical components of the force exerted by the man, (*b*) the horizontal retarding force on the mower, (*c*) the normal force between mower and lawn (be sure to include the mower's weight), (*d*) the coefficient of friction between mower and lawn.

4.47 Consider the mower of the previous problem. What force must be exerted along the handle to give the mower an acceleration of 0.5 m/s²?

4.48 A block starts from rest and slides down a plane that is inclined 30° from the horizontal

Figure 4.49
Problem 4.46.

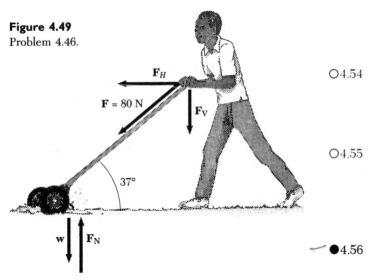

F = 80 N

F_H

F_V

37°

w F_N

and is 1.5 m long. The coefficient of friction between block and plane is 0.20. How fast is the block moving as it reaches the bottom?

O4.49   A 1.2-kg block sits on a level surface and is attached to a cord that runs over a pulley and then supports a 0.5-kg hanging mass. Find the acceleration of the system   (a) if the block and surface are frictionless,   (b) if the coefficient of friction between block and surface is 0.25.

O4.50   A 2.4-kg block sits on a level surface and is attached to a cord that runs over a pulley and then supports a 1.2-kg hanging mass. Find the acceleration of the system   (a) if the block and surface are frictionless,   (b) if the coefficient of friction between block and surface is 0.18.

O4.51   A 2.5-kg block sits on a plane that is inclined 37° from the horizontal. The coefficient of friction between plane and block is 0.22. Find the acceleration of the block down the plane.

O4.52   A 3.2-kg block sits on a plane inclined 30° from the horizontal. The block is attached by a cord to a 2.1-kg mass that hangs from a pulley at the upper end of the plane. Ignore friction.   (a) When released, will the 3.2-kg block slide down or up the plane?   (b) What acceleration will it have?

O4.53   A 2.7-kg block sits on a plane inclined 20° from the horizontal. The block is attached by a cord to a 2.4-kg mass that hangs from a pulley at the upper end of the plane. The coefficient of friction between block and

plane is 0.33.   (a) When released, will the 2.7-kg block slide down or up the plane?   (b) What acceleration will it have?

O4.54   A car can decelerate at −4.8 m/s² without skidding when coming to a stop on a level road. What will its deceleration be when the road inclines 15° uphill? Assume the same coefficient of friction in both cases.

O4.55   A 30-kg sled coasts with a constant velocity of 4 m/s over perfectly smooth, level ice. It then enters a rough stretch of ice 20 m long, where the coefficient of friction is 0.3. With what speed does the sled emerge from this rough stretch?

●4.56   A 3.0-kg block is given an initial speed of 1.8 m/s up a plane inclined 20° from the horizontal. The coefficient between block and plane is 0.30.   (a) How much time is required for the block to slide up the plane and stop?   (b) How much time is required for the block to slide back down the plane to its initial position?   (c) What speed does it have just as it returns to its initial position?

●4.57   A 2.0-kg block is given an initial speed of 2.5 m/s up a plane inclined 15° from the horizontal. The coefficient between block and plane is 0.25.   (a) How far up the plane does the block go before it stops?   (b) What speed does it have just as it returns to its initial position?

*4.8 Terminal Velocity

4.58   A parachutist who weighs 900 N is partially supported by a 700-N upward force of air resistance. What is his acceleration?

4.59   A skydiver who weighs 800 N falls at a constant speed of 4.5 m/s. What force is exerted by the parachute?

General Problems

●4.60   A 1000-kg, 3.5 m long, car traveling at 10 m/s is 15 m from the beginning of an intersection when the traffic light changes from green to yellow; then 6 s later the light changes to red. The intersection is 55 m wide. Can the driver avoid being in the intersection when the light changes to red? Available forces are 3000 N for braking to a quick stop and 1000 N for forward acceleration.

●4.61   The driver of a 600-kg sports car, heading directly for a railroad crossing 250 m away,

Figure 4.50   Problem 4.62.

applies the brakes in a panic stop. The car is moving (illegally) at 40 m/s, and the brakes can supply a force of 1200 N. (a) How fast is the car moving when it reaches the crossing? (b) Will the driver escape collision with a freight train that is 80 m away from the intersection and traveling at 23 m/s?

●4.62   Bob and Joe, two construction workers on the roof of a building, are about to raise a keg of nails from the ground by means of a rope passing over a pulley 10 m above the ground (Fig. 4.50). Bob weighs 900 N; Joe, 600 N; the keg, 300 N; the nails, 600 N. Both men slip off the roof, and the following unfortunate sequence of events takes place. The two men, hanging onto the rope together, strike the ground just as the keg hits the pulley. Unnerved by his fall, Bob lets go of the rope. The keg pulls Joe up to the roof, where he cracks his head against the pulley but gamely hangs on. However, the nails spill out of the keg when it strikes the ground, and the empty keg rises as Joe returns to the ground. Finally, Joe has had enough and lets

go of the rope and remains on the ground, only to be hit by the empty keg. Ignoring the possible (or probable) midair collisions that merely add insult to injury, how long did this industrial accident take to run its course?

●4.63   Objects of mass 2 kg and 3 kg are connected by a light cord that passes over a frictionless pulley 5 m above the floor. Initially the masses are at rest 1 m above the floor, with the cord taut. After the system is released, what maximum height above the floor does the 2-kg object reach? (*Hint:* The answer is *not* 2 m.)

●4.64   A motorcycle rider coasts along without engine power at a steady speed of 13 m/s and then enters a stretch of rough pavement where the coefficient of friction is 0.35 (Fig. 4.51). The strip of rough pavement is 15 m long. Will the rider be able to get through this section without having to start her engine? If so, what will her speed be as she returns to the smooth pavement on the other side of the rough patch?

Figure 4.51   Problem 4.64.

v = 13 m/s

15 m

The *Apollo 11* spacecraft made the first manned lunar landing on July 20, 1969. Astronauts Neil Armstrong and Edwin E. Aldrin, Jr., touched down in the lunar module *Eagle* while Michael Collins orbited the Moon in the command service module *Columbia*. Armstrong was the first to step on the surface; he and Aldrin spent a day collecting lunar rock samples and taking pictures. The two astronauts then launched the lunar module from the surface, and it is seen here returning to the orbiting *Columbia*.

What principles of circular motion and gravity were involved in orbiting the Moon, landing on the surface, and returning to rendezvous with the *Columbia*? Compare the mass of the Moon, 7.35×10^{22} kg, to the mass of the Earth listed in Table 5.1. What fraction of your Earth weight would you have on the Moon? What consequences do you think this fact had on the astronauts and on the amount of fuel their craft needed to lift off?

(NASA)

5

Circular Motion and Gravity

- Motion in a circle at constant *speed*—uniform circular motion—is motion with a changing *velocity* for velocity involves direction. Changing velocity means there is an acceleration. The acceleration of uniform circular motion is called "centripetal acceleration" and is directed toward the center of the circle. Its magnitude is given by $a_c = \dfrac{v^2}{r}$.

- From Newton's Second Law, we know that there must be an unbalanced force to cause the acceleration of an object. The net force on an object which causes it to go in a circle is known as the "centripetal force." Its magnitude is given by $F_c = \dfrac{mv^2}{r}$.

- There is a force of gravity between any two objects simply because they have mass. This force of gravity was first understood and explained by Sir Isaac Newton in his Law of Universal Gravitation and may be written in an equation as $F_{\text{gravity}} = G\dfrac{Mm}{r^2}$. This force of gravity supplies the centripetal force necessary to keep a satellite in its orbit.

- Johannes Kepler described the movements of the planets in terms of three general laws. Newton's Law of Universal Gravitation gained wide acceptance because it provided the explanation or cause behind Kepler's laws.

You probably first experienced circular motion on a merry-go-round.
(Courtesy of the author)

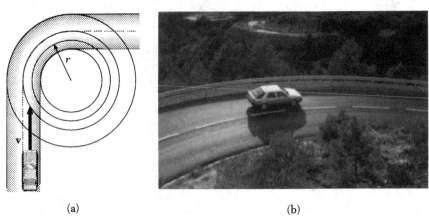

(a)   (b)

Figure 5.1
Highway curves are arcs, and so a car rounding a curve is undergoing circular motion.

Circular motion is interesting to observe, investigate, and explain. It is interesting because it is all around us. Amusement parks—where we experience physics demonstrations on a grand scale—use circular motion in many rides. You probably had your first experience with circular motion on a playground merry-go-round. Cars rounding a curve undergo circular motion (Fig. 5.1). The motion of Earth and the other planets around the Sun is nearly circular. (Though the Earth's orbit is an ellipse, the eccentricity is only 0.017—a very good approximation to a circle.) The communications satellites that bring news from around the world into our living rooms circle Earth in orbits that have a period of 24 h. (**The period** is the time needed to complete one orbit.) Gravity, of course, holds these satellites in orbit—the same gravity that pulls an apple toward the ground and holds galaxies together. We investigate all these ideas in this chapter.

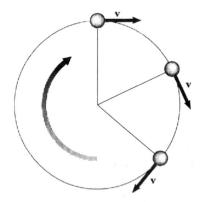

Figure 5.2
Uniform circular motion is circular motion with constant *speed* but not constant velocity. The velocity changes because the direction changes.

Any body moving along a curved path experiences acceleration because the direction of the velocity vector is always changing.

5.1 Circular Motion

CONCEPT OVERVIEW

- An object moving in uniform circular motion experiences a centripetal acceleration having a magnitude $a_c = v^2/r$ and directed toward the center of the circle.

- The net force acting on an object in uniform circular motion, often called the centripetal force $F_c = mv^2/r$, causes the centripetal acceleration.

When an object moves in a circle at constant speed, as in Figure 5.2, we describe it as undergoing **uniform circular motion.** Its speed is constant, but its velocity is not because velocity includes direction and the object's direction clearly changes.

A changing velocity means an acceleration, but what is the cause of this acceleration? As you know from Newton's second law, $\mathbf{F} = m\mathbf{a}$, force and acceleration are closely related. Figure 5.3 shows a person swinging a ball on the end of a string above his head in a horizontal circle. The ball is traveling in

Figure 5.3
What happens to the ball if the string breaks?

uniform circular motion and the string exerts a force on it. If the string breaks, the ball flies off in a straight line. It is the force of the string that causes the acceleration in this example of uniform circular motion.

Figure 5.4 is a more general sketch of an object in uniform circular motion. Initially the object is at position A with velocity $\mathbf{v}_0$. At a time Δt later, it is at position B with velocity $\mathbf{v}_1$. Remember that the speed is constant, which means that

$$v_0 = v_1 = v$$

The triangle made by the two velocities and their difference $\Delta \mathbf{v}$ is geometrically similar to triangle AOB. That is, the angle $\Delta \theta$ between the velocities is the same as the angle $\Delta \theta$ between the radii OA and OB. This is true because velocity $\mathbf{v}_0$

As the toy airplanes "fly" around the earth in a circular path, they constantly change direction. (© *L. W. Black*)

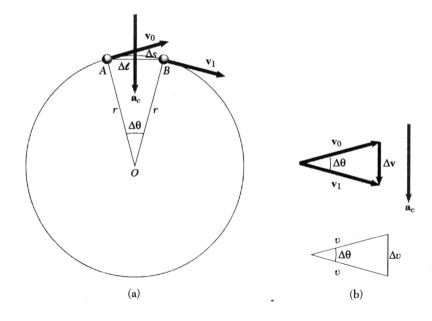

Figure 5.4
(a) As the object moves from A to B, the direction of its velocity vector changes from $\mathbf{v}_0$ to $\mathbf{v}_1$. (b) The construction for determining the direction of the change in velocity $\Delta \mathbf{v}$, which is directed toward the center of the circle.

is perpendicular to radius OA and velocity $\mathbf{v}_1$ is perpendicular to radius OB. The ratios of corresponding sides of similar triangles are equal, and so

$$\frac{\Delta v}{v} = \frac{\Delta l}{r}$$

where Δl is the chord length from A to B and r is the radius of the circle of motion. For small intervals of time Δt and consequently small angles $\Delta \theta$, Δl is very nearly equal to the arc length Δs along which the object actually travels. As Δt approaches zero, Δl and Δs become equal, and so we can write

$$\frac{\Delta v}{v} = \frac{\Delta s}{r}$$

$$\Delta v = \frac{\Delta s}{r} v$$

where Δs is the distance the object travels in time Δt at constant speed v. Therefore, from Equation 2.1 we have

$$\Delta s = v\,\Delta t$$

$$\Delta v = \frac{v\,\Delta t}{r} v = \frac{v^2\,\Delta t}{r}$$

Notice from Figure 5.4 that $\Delta \mathbf{v}$ is directed toward the center of the circle; this will always be the case.

Since acceleration is $\Delta v / \Delta t$ (Equation 2.3), we can say

$$a = \frac{\Delta v}{\Delta t} = \frac{v^2\,\Delta t}{r}\frac{1}{\Delta t} = \frac{v^2}{r}$$

This acceleration is called the **centripetal acceleration,** and we shall write it a_c:

$$a_\mathrm{c} = \frac{v^2}{r} \tag{5.1}$$

The centripetal acceleration is always directed toward the center of the circle of motion (*centripetal* means "center-seeking"). Note that the centripetal acceleration is perpendicular to the tangential velocity. It is this centripetal acceleration that is responsible for the change in the *direction* of the velocity; the magnitude of the velocity remains constant.

We know from Newton's second law, $\mathbf{F}_\mathrm{net} = m\mathbf{a}$, that an object undergoing an acceleration is experiencing a net force. The net force on an object undergoing uniform circular motion is called the **centripetal force $\mathbf{F}_\mathrm{c}$.** Combining Newton's second law and Equation 5.1, we can say

$$F_\mathrm{c} = \frac{mv^2}{r} \tag{5.2}$$

This centripetal force is necessary for an object of mass m to travel with constant speed v in a circle of radius r. The centripetal force is the net force; it is not some additional force. The centripetal force always points toward the center of the circle about which the object moves with uniform speed.

Example 5.1

Calculate the centripetal acceleration of our Moon. It is in a nearly circular orbit of radius $r = 3.84 \times 10^8$ m and completes one orbit every 27.32 days.

Reasoning

Our Moon moves in a circle, so its direction changes continuously. Thus, it has an acceleration because its velocity changes. Knowing the radius of the orbit and the period, we can find the speed. Then, to calculate the centripetal acceleration, we can use Equation 5.1.

Solution

First we must find the Moon's speed from the total distance traveled d divided by the time required t:

$$v = \frac{d}{t} = \frac{2\pi r}{T} \quad \text{— circumference} \quad \Rightarrow \text{period}$$

where T is the symbol we shall use for the period. For this case, we have

$$v = \frac{2\pi(3.84 \times 10^8 \text{ m})}{27.32 \text{ days}}\left(\frac{1 \text{ day}}{24 \text{ h}}\right)\left(\frac{1 \text{ h}}{3600 \text{ s}}\right) = 1022 \text{ m/s}$$

$$a_c = \frac{(1022 \text{ m/s})^2}{3.84 \times 10^8 \text{ m}}$$

$$= 2.72 \times 10^{-3} \text{ m/s}^2 = 2.72 \text{ mm/s}^2$$

The Earth-Moon system as imaged by the Galileo spacecraft in 1990. The two images were each taken separately and overlapped to show relative sizes. *(NASA)*

2360448

Example 5.2

A rubber ball with a mass of 50 g is attached to a string 1.0 m in length. What is the tension in the string if the ball moves in a circle on a horizontal, frictionless table with a constant speed of 4.2 m/s?

Reasoning

The vertical forces—the downward force of gravity (weight) and the upward normal force of the table—cancel each other. The tension in the string provides the net force on the ball. That net force causes the ball to move in a circle; it is the centripetal force $F_c = mv^2/r$.

Solution

This is a direct application of Equation 5.2 since the tension in the string is the centripetal force:

$$F_c = \frac{mv^2}{r} = \frac{(0.050 \text{ kg})(4.2 \text{ m/s})^2}{1.0 \text{ m}} = 0.88 \text{ N}$$

Practice Exercise   What is the tension in the string if the constant speed is 3.1 m/s?

Answer   0.48 N

5.2 Motion on a Curve

CONCEPT OVERVIEW

- The net force on a car traveling around a curve is the centripetal force.
- For a level (unbanked) curve, the centripetal force is supplied by the friction force between tires and roadway; a banked curve can supply the centripetal force without relying on friction.

As a car travels around a curve, the net force on the car must be the centripetal force, directed toward the center of the circle the curve is part of (Fig. 5.5). On a flat, level curve, this centripetal force is supplied by the friction between tires and roadway. If the tires are worn smooth or the roadway is icy or oily, this friction force will not be available. The car will not be able to move in a circle; it will keep going in a straight line and therefore go off the roadway.

Example 5.3

What must the coefficient of friction between tires and level roadway be to allow a car of mass M to make a curve of radius $r = 350$ m at a speed of 80.0 km/h?

Reasoning

For a level curve, the force of friction is the only horizontal force on the car and provides the centripetal force. This can be seen from the free-body diagram of Figure 5.5(b) or 5.6.

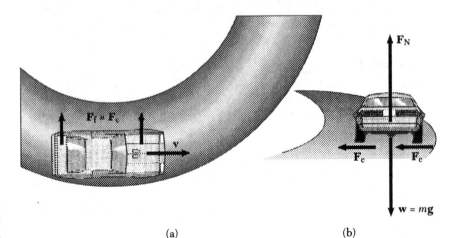

Figure 5.5
A net horizontal force is necessary in order for the car to travel in a circle; this net force is the centripetal force.

(a)   (b)

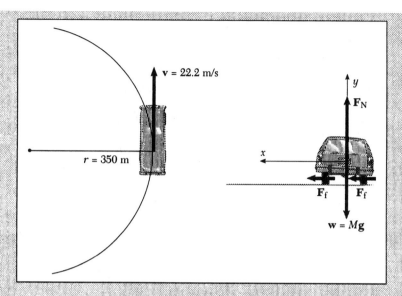

Figure 5.6
The net force must be horizontal—pointing toward the center of the circle—and only the friction force is available to provide it.

Solution

Calculations are easier if the speed is expressed in meters per second, and so we shall convert that first:

$$v = 80.0 \, \frac{\text{km}}{\text{h}} \left(\frac{1000 \text{ m}}{1 \text{ km}} \right) \left(\frac{1 \text{ h}}{3600 \text{ s}} \right) = 22.2 \text{ m/s}$$

From Equation 5.2, we can calculate the necessary centripetal force:

$$F_c = \frac{Mv^2}{r} = M \frac{(22.2 \text{ m/s})^2}{350 \text{ m}} = M(1.41 \text{ m/s}^2)$$

Do not be disturbed that the mass of the car M appears in this equation and a value has not been given for it. Perhaps we do not need it; we must just wait and see.

Figure 5.6 is a sketch and free-body diagram showing the forces on the car. Nothing exciting happens in the vertical direction since the upward and downward vertical forces cancel each other. For this flat (unbanked) curve, the centripetal force is supplied by the friction force $F_f = \mu F_N$. From the free-body diagram we see that $F_N = Mg = M(9.80 \text{ m/s}^2)$, and so we have

$$F_f = \mu M(9.80 \text{ m/s}^2)$$

Because $F_f = F_{\text{net}} = F_c$, we can write

$$\mu M(9.80 \text{ m/s}^2) = M(1.41 \text{ m/s}^2)$$
$$\mu = 0.144$$

This is the minimum value for μ. Any coefficient of friction greater than 0.144 will keep the car from slipping.

Note that the mass of the car does cancel. The centripetal force F_c is directly proportional to M, as are the weight, the normal force, and the friction force. Since all forces are proportional to M, the actual value of M does not matter.

Notice also that the force of friction is perpendicular to the velocity but in the direction necessary to oppose the motion that would occur without friction. In Figure 5.6, the car will slide to the right if it encounters a patch of low-friction surface (like ice or oil); therefore the friction force is directed toward the left.

Practice Exercise What must the coefficient of friction be if the radius of the curve is increased to 400 m?

Answer 0.126

Banked curves in a velodrome allow the riders of fast racing bicycles to maintain high speeds on the turns without sliding. (© Co Rentmeester/The Image Bank)

How can a car ever travel at all on a slippery highway? Some curves are banked to compensate for slippery conditions—ice on a highway, say, or oil on a racetrack. Figure 5.7 shows a car making a banked turn. In addition to any friction force that may or may not be present, the road still exerts a normal force perpendicular to its surface. Also, the downward force of the car's weight is present. Those two forces add as vectors to provide a net force $\mathbf{F}_{net}$ that points toward the center of the circle; this is the centripetal force. Note that it points toward the center of the circle; it is not parallel to the banked roadway. Figure 5.8 shows the forces on the car resolved into components. Since we are interested in the force that points toward the center of the circle, we choose a coordinate axis that lies along that direction. There is no acceleration in the y direction, so the sum of the forces in the y direction must be zero.

$$F_{net,y} = \Sigma F_y = F_N \cos \theta - w = 0$$

$$F_N \cos \theta = w$$

$$F_N = \frac{w}{\cos \theta}$$

$$F_{net,x} = \Sigma F_x = F_N \sin \theta$$

Because the only force in the x direction is the centripetal force, we can say

$$F_c = \frac{mv^2}{r} = F_{net,x} = F_N \sin \theta = \left(\frac{w}{\cos \theta}\right) \sin \theta = w \tan \theta = mg \tan \theta$$

$$\frac{v^2}{r} = g \tan \theta$$

$$\tan \theta = \frac{v^2}{gr} \tag{5.3}$$

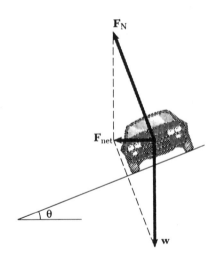

Figure 5.7
On a slippery banked curve, the only forces acting on a car are the weight of the car and the normal force exerted by the road.

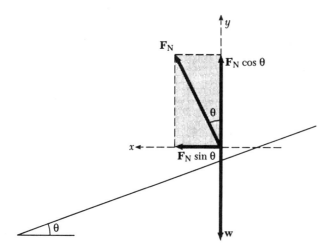

Figure 5.8
The net force acting on a car traveling in a circle must point toward the center of the circle.

This equation gives the banking angle that allows a car to travel in a curve of radius r with constant speed v and require no friction force.

A banked curve is designed for one specific speed. If the banked curve is icy so that there is no friction force at all, then traveling at a speed higher than design speed means the car will slide out, up, and over the edge. Traveling at a speed lower than design speed means the car will slide in, down, and off the bank.

Problem-Solving Tip

- The vector sum of all the forces acting on a body in uniform circular motion provides the centripetal force in $F_{net} = ma = F_c = mv^2/r$; the centripetal force is *not* an extra force.

Example 5.4

At what angle should a curve of radius 200 m be banked so that no friction is required when a car travels at 60.0 km/h around the curve?

Reasoning

For a banked curve with no friction, the only forces acting on the car are the normal force and the weight. We have just looked at this situation in Figures 5.7 and 5.8. The centripetal force must be supplied by the horizontal component of the normal force. We have just analyzed this situation in arriving at Equation 5.3.

Solution

While it is comfortable to think of a car traveling at 60 km/h, it will be easier in our calculations to use velocity in meters per second, so we begin by changing units:

$$v = 60.0 \frac{\text{km}}{\text{h}} \left(\frac{1000 \text{ m}}{1 \text{ km}} \right) \left(\frac{1 \text{ h}}{3600 \text{ s}} \right) = 16.7 \text{ m/s}$$

Now we can apply Equation 5.3 directly:

$$\tan \theta = \frac{v^2}{gr} = \frac{(16.7 \text{ m/s})^2}{(9.80 \text{ m/s}^2)(200 \text{ m})} = 0.142$$

$$\theta = 8.10°$$

Practice Exercise   If the radius of the curve is increased to 300 m, what angle is required?

Answer 5.42°

5.3 Vertical Circles

CONCEPT OVERVIEW

▬ At the top and bottom of a vertical circular path, the centripetal force is supplied by the resultant of the weight and a supporting force (often the normal force).

What are the forces on a roller coaster victim—I mean, passenger—at the very top of the loop-the-loop shown in Figure 5.9? As the free-body diagram of Figure 5.10 shows, the passenger's weight $\mathbf{w} = m\mathbf{g}$ is present and so is the

Figure 5.9
The motion of a loop-the-loop on a roller coaster is circular. *(Courtesy of Busch Gardens, Tampa, Florida)*

normal force $\mathbf{F}_N$ that the seat exerts on him. The net force is just the sum of these two forces:

$$\mathbf{F}_{net} = \mathbf{F}_N + m\mathbf{g}$$

Since the motion is circular, this net force must also be the centripetal force $F_c = mv^2/r$. Therefore, looking at the magnitudes of the forces, we have at the top of the loop

$$F_{net} = F_N + mg = \frac{mv^2}{r} = F_c$$

$$F_N = \frac{mv^2}{r} - mg \qquad (5.4)$$

Notice that this expression means that the magnitude of the normal force—the force you feel on the seat of your pants—can be positive, zero, or negative. A negative value for F_N means the passenger has to be strapped in, with the straps exerting a force *upward*. Such a situation would be very dangerous, and roller coaster designers avoid that. A normal force of zero means the passenger seems to be weightless as well as upside-down.

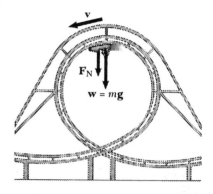

Figure 5.10
The force of gravity (the passenger's weight) and the normal force of the seat provide the centripetal force.

Problem-Solving Tip

- At the top or bottom of a vertical circular path, the weight and the normal force (or an equivalent supporting force) are the only forces acting on an object; the vector sum of these two forces provides the centripetal force. Use a free-body diagram to determine the directions of these forces.

Example 5.5

A roller coaster similar to the one in Figure 5.9 has a speed of 11.5 m/s at the top of a loop that has a radius of 8.1 m. What force does the seat exert on an 82-kg passenger?

Reasoning

At the top and bottom of a vertical circle, the centripetal force is supplied by only the normal force and the weight. At the top, both these forces lie in the same direction—downward. So the net force is the normal force plus the weight and this net force is the centripetal force.

Solution

We can use Equation 5.4 directly

$$F_N = \frac{mv^2}{r} - mg$$

$$= \frac{(82 \text{ kg})(11.5 \text{ m/s})^2}{8.1 \text{ m}} - (82 \text{ kg})(9.8 \text{ m/s}^2) = 535 \text{ kg·m/s}^2$$

$$= 535 \text{ N} \approx 540 \text{ N}$$

The seat exerts a downward force of 540 N. The passenger's usual weight is 804 N.

Practice Exercise If the speed at the top of the loop is 13.0 m/s (about 15% faster), what force is exerted on the passenger?

Answer 910 N

Practice Exercise A car goes over the crest of the hill with a speed of 9.4 m/s. The top of the hill has a radius of 22.3 m. What upward force does the seat exert on a 72-kg driver?

Answer 420 N

Practice Exercise A car goes over the crest of the hill with a speed of 12.5 m/s. The top of the hill has a radius of 17.5 m. What upward force does the seat exert on a 72-kg driver?

Answer 63 N

(a)

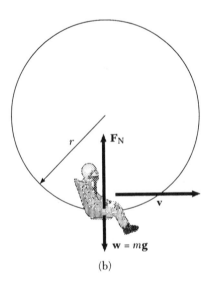

(b)

Figure 5.11
As an airplane comes out of a dive, it may be analyzed in terms of circular motion. *(Photo Crown copyright 1991. Reproduced by permission of the Controller of Her Majesty's Stationery Office.)*

The forces on an airplane pilot at the bottom of a dive can be quite large. Figure 5.11 shows airplanes coming out of a dive and the forces involved. Gravity exerts a weight force **w** downward and the seat exerts its usual normal force $\mathbf{F}_N$, this time upward. The net force is the vector sum of these:

$$\mathbf{F}_{net} = \mathbf{F}_N + \mathbf{w} = \mathbf{F}_N + m\mathbf{g}$$

or, looking at only the magnitudes,

$$F_{net} = F_N - mg$$

where the minus sign on the weight term indicates that we have chosen up as the positive direction. (In writing this equation, we have omitted the thrust of the engine and the drag of air resistance; we have assumed that those two forces just balance each other.) Since the airplane is traveling along the arc of a circle, the net force is also the centripetal force. Therefore,

$$F_{net} = F_N - mg = \frac{mv^2}{r} = F_c$$

$$F_N = \frac{mv^2}{r} + mg \qquad (5.5)$$

For this case—at the bottom of a dive—the normal force can only be positive, must be greater than the weight mg, and can become quite large. A roller coaster at the bottom of a hill provides the same situation. The mass appears in both terms on the right of Equation 5.5, so the normal force is always proportional to the mass of the pilot. A pilot with a mass of 120 kg will experience twice as much force as a co-pilot with a mass of 60 kg. Because of this difference, these situations are often expressed in terms of the normal force divided by the mass, which is an acceleration:

$$a = \frac{F_N}{m}$$

When we substitute F_N from Equation 5.5, this expression for acceleration gives

$$a = \frac{v^2}{r} + g \qquad (5.5')$$

If the normal force were twice the weight, the acceleration would be $2g$, twice normal gravity. This acceleration is expressed in terms of g's (pronounced "gees") and you can immediately see that the pilot would feel a force of two times or four times or 7.5 times her or his own weight if the acceleration were $2\,g$'s or $4\,g$'s or $7.5\,g$'s.

Example 5.6

An acrobatic pilot comes out of a dive with her airplane traveling at 180 km/h. At the bottom of the dive, the airplane is traveling along the arc of a 150-m circle in a vertical plane. How many g's does the pilot feel?

Reasoning

This situation is just like the one we have discussed, and the forces involved are illustrated in Figure 5.12. At the top and bottom of a vertical circle, the centripetal force is supplied by only the normal force and the weight. At the bottom, these forces lie in opposite directions. So the net force is the normal force (up) minus the weight (down), and this net force is the centripetal force.

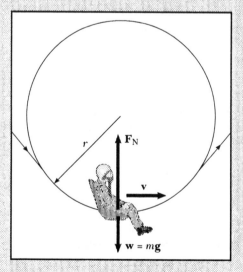

Figure 5.12
The net force on a pilot coming out of a dive must provide a centripetal force.

Solution

We can use Equation 5.5', which we have developed to handle this situation. Since we know $g = 9.80$ m/s^2, we shall first change 180 km/h to meters per second:

$$v = 180\frac{\text{km}}{\text{h}}\left(\frac{1000 \text{ m}}{1 \text{ km}}\right)\left(\frac{1 \text{ h}}{3600 \text{ s}}\right) = 50 \text{ m/s}$$

$$a = \frac{v^2}{r} + g$$

$$= \frac{(50 \text{ m/s})^2}{150 \text{ m}} + 9.80 \text{ m/s}^2 = 26.5 \text{ m/s}^2$$

$$= 26.5 \text{ m/s}^2\left(\frac{1 \text{ g}}{9.80 \text{ m/s}^2}\right) = 2.70 \text{ g's}$$

The pilot is pushed into the seat with 2.7 times his or her weight. Everything feels 2.7 times as heavy. The pilot's hands feel this heavy, making control of the airplane a little more difficult. The pilot's blood feels this heavy, making circulation more difficult on the heart. Accelerations of eight or nine or ten g's make it difficult for the circulatory system to get enough blood to the brain and a pilot may black out in such circumstances. Pressure suits that squeeze on the legs push blood back into the rest of the body, including the brain, and help prevent blackouts.

Please do not memorize the equations we have developed in this section. They are good only in very limited situations. Rather, remember what we have done. For each situation we have made a free-body diagram of all the forces involved, found the net force, set that force equal to mv^2/r, which is the centripetal force, and applied Newton's second law.

5.4 Inertial Reference Frames

CONCEPT OVERVIEW

- An inertial reference frame is one in which the law of inertia—Newton's first law of motion—is valid; a noninertial reference frame is one in which the law of inertia is not valid.

- Accelerating reference frames are noninertial; this includes rotating reference frames.

- In an inertial reference frame, Newton's second law is also valid; in a noninertial reference frame, additional forces have to be assumed to explain Newton's second law.

You already know that all measurements—of position, say, or velocity or acceleration—are made with respect to some reference frame. A reference frame in which Newton's first law, the law of inertia, is valid is known as an **inertial reference frame**. If this law is valid, Newton's other two laws of motion will also be valid, and so we can apply $\mathbf{F} = m\mathbf{a}$ and understand the motion as usual.

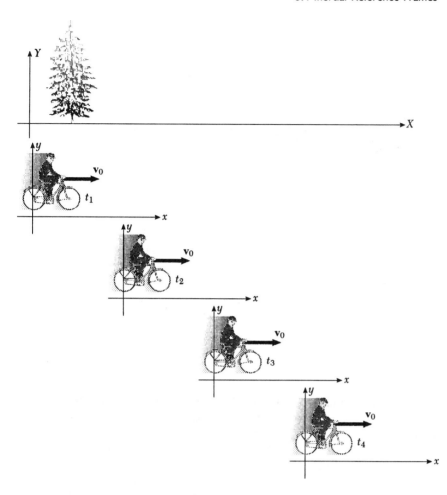

Figure 5.13
The tree is at rest in the XY reference frame. With zero net force acting on it, the tree remains at rest and the XY reference frame is an inertial reference frame.

Figure 5.13 shows a tree sitting still in an XY reference frame. The net force on the tree is zero, and its acceleration is zero—it remains at rest, the law of inertia holds, and therefore the XY frame is an inertial frame. A bicycler comes riding by along a straight line with constant velocity $\mathbf{v}_0$ to the right. The bicycler constitutes another reference frame; we shall call it the xy frame. The bicycler in his xy frame sees the tree move with a constant velocity $-\mathbf{v}_0$ to the left in the xy frame; this is sketched in Figure 5.14. The net force on the tree is zero and its acceleration is zero—it is moving with constant velocity—and therefore the xy frame is also an inertial frame. In general, any reference frame moving at constant velocity with respect to an inertial frame is also an inertial frame. The same forces are present and the same laws of motion are valid in all inertial frames.

The best possible inertial frame might be one at rest with respect to the average position of the "fixed" stars. Earth moves relative to such a frame—it rotates about its axis and revolves in its orbit about the Sun. We can, however, use Earth as an inertial reference frame for most situations. What, then, is a noninertial reference frame? If the bicycle of Figure 5.13 and its xy frame had been accelerating, then observations of the tree from that frame would have shown the tree to be accelerating; in other words, a tree experiencing zero net force would be observed to accelerate! That violates the law of inertia, so the accelerating xy frame would be a noninertial frame. In general, any frame that accelerates relative to an inertial frame is noninertial.

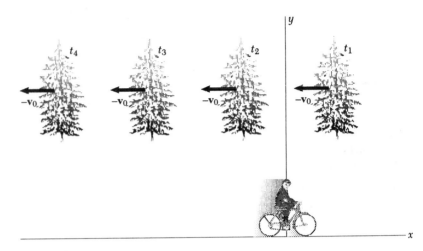

Figure 5.14
The bicycler carries the xy reference frame. Because the tree moves with constant velocity relative to it, the xy reference frame is an inertial frame.

It is easy to think of an acceleration along a straight line, such as an accelerating bicycle. Rotation also involves acceleration, however, so a rotating reference frame is noninertial. Observations taken from a rotating reference frame violate $\mathbf{F}_{net} = m\mathbf{a}$, and we must introduce additional forces to make Newton's second law appear valid.

Picture yourself as a passenger in a car making a left turn, your sunglasses on the dashboard. What happens? You feel pushed to the outside of the curve—to the right—and your sunglasses slide across the dashboard to the right. Are not both these motions caused by a force directed away from the center of the circle? No. You and your sunglasses are doing what is described by Newton's first law as you and the sunglasses try to continue in a straight line, as shown in Figure 5.15. The car is turning to the left beneath you and the sunglasses. The car is a noninertial reference frame because it is accelerating—not speeding up but accelerating because it is turning. The seat or the car door exerts a force on you toward the center of the turning circle, to the left; that is the centripetal force that causes *you* and your sunglasses to move in a circle. Likewise, in order for the car to travel in a circle, it must experience a centripetal force also. These are real forces that are observable by an outside observer.

An outside observer watching all this from the inertial reference frame of Earth sees you and the sunglasses both trying to continue in a straight line as the car turns, but remember—inside the car you see the sunglasses accelerate to the right. Acceleration means force. Either $\mathbf{F}_{net} = m\mathbf{a}$ is not valid in the car as it turns, or else there is some unexpected force acting toward the right. Such a rotating reference frame is a noninertial frame, as discussed above, but we want to keep on believing $\mathbf{F}_{net} = m\mathbf{a}$, and so we devise the fiction that there is a force directed toward the outside of the curve; this is called the *centrifugal force*. It is a fictitious force (sometimes called an inertial force) necessary to explain motion observed in a noninertial reference frame.

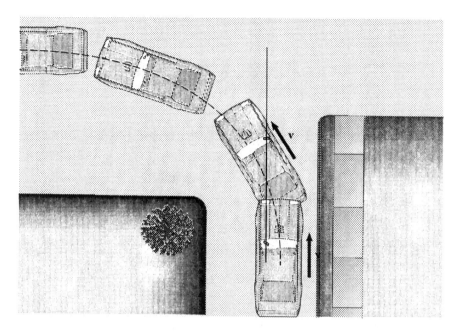

Figure 5.15
To an outside observer, the sunglasses on the dashboard attempt to continue in a straight line as the car curves to the left. From your point of view inside the turning car, the sunglasses move off to the right as if there were a force on them, directed to the outside of the curve (to the right).

In this book we shall limit our observations and calculations to real forces in inertial reference frames. Centrifugal forces, because they are not real, will not be referred to again.

5.5 Gravity

CONCEPT OVERVIEW

■ The gravitational force between any two objects is proportional to the product of their masses and inversely proportional to the square of the distance between them. The force is attractive and acts along the line joining the two objects.

Sir Isaac Newton not only developed the three laws of motion we studied in the previous chapter but also developed a law of universal gravitation telling us that the force causing an apple to drop from a tree and the force holding the Moon in orbit are one and the same. Newton's **Law of Universal Gravitation** can be stated as

The gravitational force between two objects is proportional to the product of their masses and inversely proportional to the square of the distance between them. The force is attractive and acts along the line joining the two objects.

For any two objects with masses M and m a distance r apart, we can write

$$F \propto \frac{Mm}{r^2}$$

Any proportionality can be written as an equation by including a constant of proportionality and equations are always more useful. Therefore, we can write

Newton's Law of Universal Gravitation as

$$F = G\frac{Mm}{r^2}$$

(5.6)

where G is called the **universal gravitational constant** and has a measured value of

$$G = 6.67 \times 10^{-11} \text{ N·m}^2/\text{kg}^2$$

The importance of this law is that gravitation is truly universal. A book falls to the floor if dropped. The Moon is held in a nearly circular orbit around Earth. The planets are held in orbits around the Sun. Binary stars rotate around each other. Galaxies are attracted to each other. Gravity is the force in each case.

For two tiny objects or for objects separated by astronomical distances, there is no question as to what value to use for r. For spherical objects—our Earth and Moon, for example—r is the distance from one center to the other. Let us apply the law of gravitation to an apple located near the surface of Earth. We shall call the mass of the apple m and the mass of Earth $M_E = 5.98 \times 10^{24}$ kg. The distance between the center of Earth and the apple is just the radius of Earth $R_E = 6.38 \times 10^6$ m. Putting these values into Equation 5.6, we have the force of gravity on an object of mass m near the surface of Earth:

$$F = \frac{(6.67 \times 10^{-11} \text{ N·m}^2/\text{kg}^2)(5.98 \times 10^{24} \text{ kg})m}{(6.38 \times 10^6 \text{ m})^2}$$

$$F = (9.80 \text{ m/s}^2)m$$

Of course, this is exactly what we expect since the force of gravity is just the weight of an object $w = mg$. We have just used the law of gravitation to calculate the acceleration of free fall at Earth's surface, 9.80 m/s^2.

TABLE 5.1   Solar System Data

Planet	Radius (km)	Mass ($\times 10^{24}$ kg)	Orbital Radius (AU)[a]	Period (days)[b]
Mercury	2439	0.33	0.39	87.96
Venus	6052	4.87	0.72	224.68
Earth	6378	5.98	1.00	365.25
Mars	3397	0.64	1.52	686.95
Jupiter	71 398	1901	5.20	4337
Saturn	60 000	561	9.54	10 760
Uranus	26 071	87.1	19.2	30 700
Neptune	24 764	103	30.1	60 200
Pluto	1150	0.01	39.5	90 780
Moon	1740	0.0736	3.84×10^8 m	27.32
Sun	6.96×10^5	1.99×10^{30} kg	—	—

[a] AU = astronomical unit = 1.5×10^8 km
[b] The **period** is the time necessary to complete one orbit around the Sun.

Table 5.1 gives data on the planets in our solar system. From those data we can calculate the force of gravity at the surface of any planet and thus the acceleration of free fall there. The next example illustrates this.

Example 5.7

Take our proverbial apple of mass m to the surface of Mars and calculate the force of gravity that Mars exerts on the apple. What will its acceleration be if we drop the apple near the surface of Mars?

Reasoning

Newton's law of universal gravitation describes the force between any two objects—and that includes the force between Mars and an apple. We need the mass and radius of Mars from Table 5.1 to find the force on the apple. Newton's second law will then provide the apple's acceleration when dropped.

Solution

We use Equation 5.6 with m standing for the (unknown) mass of the apple:

$$F = G\frac{Mm}{r^2}$$

$$= \frac{(6.67 \times 10^{-11}\ \text{N·m}^2/\text{kg}^2)(0.64 \times 10^{24}\ \text{kg})m}{(3.397 \times 10^6\ \text{m})^2} = (3.7\ \text{m/s}^2)m$$

The Martian force of gravity on a 1-kg apple will be 3.7 N. If the apple is dropped, this is the only—and therefore net—force acting on it, so that

$$F = ma = m(3.7\ \text{m/s}^2)$$
$$a = 3.7\ \text{m/s}^2$$

Any object dropped near the surface of Mars will accelerate downward at $3.7\ \text{m/s}^2$.

It is useful to express Equation 5.6 in terms that apply to Earth only. The force of gravity at Earth's surface on an object of mass m is

$$F = G\frac{M_E m}{R_E^2}$$

where M_E is the mass of Earth and R_E the radius. We have already expressed this force of gravity as the weight of an object,

$$F = w = mg$$

Setting these two expressions for the same force equal to each other provides

$$mg = G\frac{M_E m}{R_E^2}$$

or

$$g = G\frac{M_{\text{E}}}{R_{\text{E}}^2}$$

Note again how very small G is. The gravitational force between most ordinary objects is exceedingly small. On an astronomical scale, however, where the masses involved are enormous, gravity is the dominant force.

Example 5.8

What is the gravitational force of attraction between two airplanes whose masses are both 1000 metric tons ($M = m = 1.0 \times 10^6$ kg) when they are 100 m away from each other?

Reasoning

Newton's law of universal gravitation applies to all objects—including airplanes.

Solution

The gravitational force between the two airplanes is given by Equation 5.6:

$$F = \frac{(6.67 \times 10^{-11}\ \text{N·m}^2/\text{kg}^2)(1.0 \times 10^6\ \text{kg})^2}{(100\ \text{m})^2} = 6.7 \times 10^{-3}\ \text{N}$$

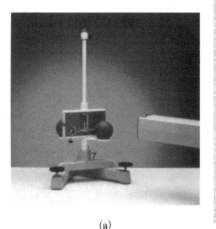

(a)

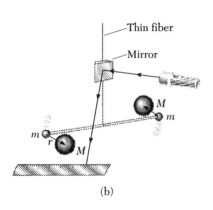

(b)

Figure 5.16
(a) A Cavendish balance for measuring the small gravitational force between ordinary masses. *(Courtesy of PASCO Scientific)* (b) An optical lever is used to measure the very small rotation caused by the gravitational attraction between the masses.

Newton's law of gravitation provides one equation with several unknowns. The universal gravitational constant G must be experimentally determined. That involves measuring the two masses M and m, the distance r, and the force F exerted by the two masses. Values for the mass of Earth, the Sun, and other planets were calculated only after G was experimentally determined. This was first done in 1798 by Henry Cavendish using an apparatus like that illustrated in Figure 5.16, which is now known as a *Cavendish balance*. A thin rod with two known small masses m, one at each end, is placed near two known large masses M positioned a carefully measured distance r away. The rod is supported by a thin fiber that twists slightly as a result of the very small gravitational force between the small masses and the large ones. That small rotation is measured by shining a light onto a mirror mounted on the fiber and measuring the deflection of the light. This arrangement is known as an *optical lever* and is useful for measuring very small displacements or rotations.

A scientific report titled "Measuring the Universal Gravitational Constant" would have been of interest only to Cavendish and a few colleagues. Knowing, however, that once the value of G was established, the mass of Earth could be calculated immediately, Cavendish caused great interest by titling his work "Weighing the Earth"!

In Example 5.1 we calculated the Moon's centripetal acceleration due to its circular motion around Earth to be 2.72×10^{-3} m/s^2. The net force on the

Moon must therefore be

$$F_{\text{net}} = m_M a_M = m_M(2.72 \times 10^{-3} \text{ m/s}^2)$$

where m_M is the mass of the Moon. From the law of gravitation, we can calculate the force of gravity exerted on the Moon by Earth:

$$F = G\frac{M_E m_M}{r^2}$$

$$= \frac{(6.67 \times 10^{-11} \text{ N·m}^2/\text{kg}^2)(5.98 \times 10^{24} \text{ kg})m_M}{(3.84 \times 10^8 \text{ m})^2}$$

$$= (2.70 \times 10^{-3} \text{ m/s}^2)m_M$$

where we have used the radius of the Moon's orbit, given in Example 5.1, as our value for the Earth-Moon distance. The force of gravity that Earth exerts on the Moon provides the centripetal force necessary to keep the Moon moving in uniform circular motion.

5.6 Kepler's Laws of Planetary Motion

CONCEPT OVERVIEW

- Kepler described planetary motion in three statements based upon observational data.

Newton presented his law of universal gravitation in 1687, and its greatest test or verification was its explanation of the motion of the planets in our solar system. Earlier in the same century, the German astronomer Johannes Kepler had determined that the motion of the planets could be described in three statements now known as Kepler's laws of planetary motion:

1. *Each planet travels around the Sun in an ellipse with the Sun at one focus of the ellipse.*
2. *A line drawn from the Sun to a planet sweeps out equal areas in equal time intervals.*
3. *The periods and average distances of the planets from the Sun are such that the ratio of the square of the period to the cube of the average distance is the same for every planet.*

Kepler's first law requires that the force between the Sun and a planet be an inverse-square force just as Newton's law of gravitation predicts. Proof of this is beyond the scope of this text. Kepler's second law (equal areas in equal times), illustrated in Figure 5.17, is true for any force acting along the line between two bodies. Kepler's third law can be written

$$\frac{T_1^2}{r_1^3} = \frac{T_2^2}{r_2^3}$$

where T is the period of any planet, r its average distance from the Sun, and the subscripts refer to any two planets; this ratio is the same constant for all the planets orbiting the Sun.

It is both easy and interesting to demonstrate Kepler's third law if we restrict our attention to circular orbits. For circular orbits, there is only a single

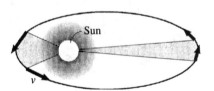

Figure 5.17
In orbit around the Sun, a planet moves more slowly when it is farther away and faster when it is nearer. The radius of the orbit sweeps out equal areas in equal times, as shown by the shaded areas. (Note that real planetary orbits are much less elliptical than what is shown here.)

radius. We shall use Newton's Law of Universal Gravitation to calculate the gravitational force exerted by the Sun, of mass M_S, on a planet of mass m, in a circular orbit of radius r and then set that force equal to the planet's mass multiplied by the centripetal acceleration $a_c = v^2/r$:

$$F = G\frac{M_S m}{r^2} = ma_c = m\frac{v^2}{r}$$

Since $v = $ distance/time $= $ circumference/period $= 2\pi r/T$, we have

$$\frac{GM_S m}{r^2} = \frac{4\pi^2 rm}{T^2}$$

$$\frac{T^2}{r^3} = \frac{4\pi^2}{GM_S} = \text{constant}$$

The ratio of the square of the period T to the cube of the average radius r is a constant for every planet orbiting the Sun.

Kepler discovered his laws of planetary motion after 16 years of looking for some pattern in the motion of the planets. His work was based upon meticulous astronomical data taken by his mentor, Tycho Brahe, throughout the latter's lifetime. Brahe carefully sighted stars and planets using a sextant. He was the last of the great naked-eye astronomers, doing all his work before the invention of the telescope. Galileo, whom we have met before, was the first astronomer to use a telescope to look at the stars and planets.

*5.7 Satellite Orbits

CONCEPT OVERVIEW

■ A satellite is held in a circular orbit because the force of gravity supplies the necessary centripetal force.

Earth satellites are now commonplace and satellite receiving antennae like the one in Figure 5.18 are seen everywhere. Weather photographs from satellites show us how a cloud cover has traveled throughout the day. Satellites beam television signals into millions of homes. Satellites provide photographs like the ones in Figure 5.19 to help identify pollution or other environmental problems.

What keeps a satellite in orbit? Why does it not fall from the sky or move away from Earth along some straight-line path? We shall answer these questions for satellites in orbit around Earth, but both the questions and their answers apply just as well to the other planets orbiting our Sun. Planets are satellites of the Sun.

In order for any object to move in a circle, there must be a net force on the object directed toward the center of the circle; this is the centripetal force we have already studied. For a satellite in orbit around Earth, this centripetal force is simply the gravitational force exerted by Earth. A satellite does not fall because it is moving; it has been given a tangential velocity by the rocket that launched it. It does not travel off in a straight line because Earth's gravity pulls

Figure 5.18
A large satellite receiver-transmitter dish in Cornwall, England. Smaller satellite dishes have become common sights on building roofs as more and more individuals and businesses use satellites to send messages, entertainment, and data to distant places.
(Martin Bond/Science Photo Library/Photo Researchers)

(a)   (b)

Figure 5.19
Observations of the Earth from orbiting satellites and manned spacecraft help us predict the weather, detect pollution, and determine more easily the extent of changes to the ecosphere. (a) Hurricane Elens, 1985 *(© Telegraph Colour Library/FPG International Corporation)* (b) Sediment from the mouth of the Amazon River flowing into the Atlantic Ocean. *(© Telegraph Colour Library/FPG International Corporation)*

it toward Earth. Earth's radius is about 6400 km. A satellite in a low orbit may have an altitude of 150 km. That means it is 6550 km from the center of Earth. The force of gravity at that altitude will not be very different from the force of gravity at Earth's surface. We may investigate such a low-orbit satellite by using $w = mg$ as a reasonable estimate of the gravitational force on the satellite and setting this equal to the centripetal force:

$$mg = \frac{mv^2}{r}$$

We can solve for the orbital speed v:

$$v^2 = gr = (9.80 \text{ m/s}^2)(6.55 \times 10^6 \text{ m})$$

$$v = 8010 \frac{\text{m}}{\text{s}} \left(\frac{3600 \text{ s}}{1 \text{ h}} \right) \left(\frac{1 \text{ km}}{1000 \text{ m}} \right) = 28\ 800 \text{ km/h}$$

To place a satellite into a low orbit around Earth, a rocket must give the satellite a tangential speed of 28 800 km/h (about 18 000 mi/h).

The period T may be of more interest than the orbital velocity; we can easily determine the period from the relationship

$$v = \frac{2\pi r}{T} = \frac{2\pi R_{\text{E}}}{T}$$

$$T = \frac{2\pi R_{\text{E}}}{v} = \frac{2\pi (6.55 \times 10^3 \text{ km})}{28\ 800 \text{ km/h}} = 1.4 \text{ h} = 86 \text{ min}$$

Figure 5.20
If a satellite in a low orbit around Earth is used for communications, it must be tracked with movable antennae.

The satellite "dish" receiving antennas seen in so many yards get a free ride from our rotating earth.

In the early days of satellite communications (the early 1960s), only low-orbit satellites existed to send radio and television signals. You can see from Figure 5.20 the difficulty in using satellites with a period such as the 86 min we just calculated. Both the antenna that broadcasted to the satellite and the antenna that received the signal from the satellite had to be steerable, and for only a few minutes during each orbit was the satellite "above the horizon" for both sender and receiver. Figure 5.21 illustrates a satellite in a *geosynchronous* orbit. Such a satellite is placed in an orbit above the equator with a period of 24 h. Thus, it revolves once around its orbit in exactly the time it takes Earth to rotate once about its axis. The result is that the satellite appears to remain stationary above Earth. We can find the radius necessary for this orbit since we know the period. This will clearly not be a low orbit, so we must use the correct value for the gravitational force rather than the $w = mg$ estimate we used earlier. Therefore we begin with Newton's second law:

$$F = ma$$

$$G\frac{M_E m}{r^2} = m\frac{v^2}{r}$$

$$r = \frac{GM_E}{v^2}$$

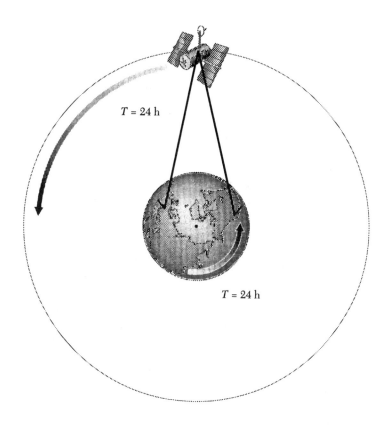

$T = 24$ h

$T = 24$ h

Figure 5.21
A satellite in a geosynchronous orbit has a period of 24 h and so appears to remain stationary to an Earth-based observer. Antennae tracking such a satellite can be stationary on the ground because Earth is doing the moving.

We do not know the orbital velocity yet; we know only the period $T = 24$ h:

$$v = \frac{2\pi r}{T}$$

$$r = \frac{GM_E}{(2\pi r/T)^2} = \frac{GM_E T^2}{4\pi^2 r^2}$$

$$r^3 = \frac{GM_E T^2}{4\pi^2}$$

Pause for a moment and notice that this is Kepler's third law for Earth satellites; the ratio r^3/T^2 is a constant, $GM_E/4\pi^2$. Now we can continue with numerical values to find that

$$r^3 = \frac{(6.67 \times 10^{-11}\ \text{N·m}^2/\text{kg}^2)(5.98 \times 10^{24}\ \text{kg})(24\ \text{h})^2}{4\pi^2}$$

$$= 5.82 \times 10^{15} \frac{\text{N·m}^2}{\text{kg}} \text{h}^2 \left(\frac{1\ \text{kg·m/s}^2}{1\ \text{N}}\right)\left(\frac{3600\ \text{s}}{1\ \text{h}}\right)^2$$

$$= 7.54 \times 10^{22}\ \text{m}^3 = 75.4 \times 10^{21}\ \text{m}^3$$

$$r = 4.22 \times 10^7\ \text{m} = 42\ 200\ \text{km}$$

*5.8 Types of Forces

CONCEPT OVERVIEW

- There are four known fundamental forces in nature: gravity, electromagnetism, the weak nuclear force, and the strong nuclear force.

- The everyday forces that we often refer to as "contact forces" are the result of electromagnetic forces between atoms of one object interacting with atoms of another object.

When the ideas of force, mass, acceleration, and Newton's second law were introduced in this book, a force was described as a push or a pull. And it is. A force is anything that causes an acceleration, a change in motion; but what different kinds of forces are there?

All the forces that are encountered in all of nature can be described in terms of four fundamental forces (Table 5.2). We have just studied one of them in detail, the **gravitational force.** Inside the atomic nucleus, there are two kinds of forces: the **strong nuclear force** and the **weak nuclear force.** The fourth fundamental force is the **electromagnetic force.** Most of our common, ordinary forces—our usual pushes and pulls—are electromagnetic forces. When you hold a pencil, the electrons in the atoms at the surface of your hand interact with the electrons in the atoms at the surface of the pencil. We sometimes describe a force that acts in this way as a *contact force.* As you walk across the room, you are held up by the normal force, or contact force, that the floor exerts on your feet. Ultimately this force is a manifestation of the electromagnetic forces between the electrons in your feet and those in the floor. For most purposes, it is easier and clearer to describe such a force as a contact force, and we shall do so throughout this book.

TABLE 5.2 Summary of Fundamental Forces

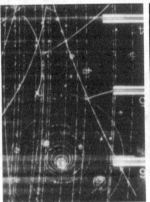

(Courtesy of Central Scientific Company) *(University of California/ Lawrence Berkeley Laboratory)* *(NASA)*

Force	Strong	Electromagnetic	Weak	Gravitational
Relative Strength	1	10^{-2}	10^{-6}	10^{-38}
Range	Short (1×10^{-15} m)	Long ($\propto 1/r^2$)	Short (1×10^{-15} m)	Long ($\propto 1/r^2$)

The electromagnetic force and the weak nuclear force can be described as two aspects of a single, more fundamental force called the **electroweak force.** *Grand Unified Theories* (GUTs) seek to combine the electroweak force with the strong and gravitational forces, hoping that all forces can finally be understood in terms of a single "ultrafundamental" force.

A fifth and sixth fundamental force, if they exist (and that is probably the biggest IF in physics), both have to do with gravity, and proof of their existence will force us to rethink Newton's gravitation law.

SUMMARY

An object traveling in a circle of radius r with uniform speed v accelerates because its direction changes. This acceleration, called the centripetal acceleration a_c, is equal to

$$a_c = \frac{v^2}{r} \tag{5.1}$$

and points toward the center of the circle. The net force on such an object is the force necessary to produce this acceleration. Called the centripetal force F_c, it is equal to

$$F_c = \frac{mv^2}{r} \tag{5.2}$$

and also points toward the center of the circle.

An inertial reference frame is one in which Newton's first law is valid. A reference frame moving at a constant velocity with respect to another inertial frame is also an inertial frame. An accelerating frame is never an inertial frame.

Newton's law of universal gravitation can be written as

$$F = G\frac{Mm}{r^2} \tag{5.6}$$

where G is the universal gravitational constant and has a value

$$G = 6.67 \times 10^{-11} \ \mathrm{N \cdot m^2/kg^2}$$

Kepler's three laws describe planetary motion in our solar system:

1. *Each planet travels around the Sun in an ellipse with the Sun at one focus.*
2. *A line drawn from the Sun to any planet sweeps out equal areas in equal time intervals.*
3. *For any two planets, $T_1^2/r_1^3 = T_2^2/r_2^3$.*

There are four fundamental forces in nature: strong nuclear force, weak nuclear force, electromagnetic force, and gravitational force. The second and third of these are two aspects of what is called the electroweak force.

QUESTIONS

1. Is it possible for an object to be accelerated when its speed is constant?
2. Suppose you swing a yo-yo in a horizontal circle above your head and the string breaks. Describe the motion of the yo-yo after the break.
3. A marble rolls around the edge of a circular pan that has a hole cut out of its side as shown in Figure 5.22. Which of the four paths shown does the marble follow when it reaches the hole?

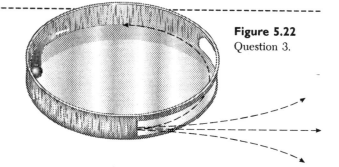

Figure 5.22
Question 3.

4. On a high-speed roller coaster, are you more likely to be thrown out of your seat at the top of a hill or at the bottom?

5. Why are curves banked on a racecar speedway?

6. Explain how the spin-dry cycle on an automatic washer removes water from clothes (a) from the point of view of someone standing in the laundry room watching the clothes go around and (b) from the point of view of someone inside the washer with the clothes.

7. Explain how you can swing a bucket of water in a vertical circle and keep the water in the bucket even at the top of the path, when the bucket is upside-down.

8. Because of the inverse-square nature of the law of gravitation, the force of gravity decreases but is still present even hundreds of thousands of kilometers from Earth. This being so, how can astronauts feel weightless in an orbiting space shuttle?

9. Television coverage of space shuttle astronauts often shows pens and notebooks floating around in the cabin. We describe this by saying the astronauts are "weightless" in the orbiting craft. How were astronauts able to walk on the Moon? Did they require heavy boots to stay on the Moon's surface?

10. Why might a geological survey satellite be put into an orbit that passes over Earth's north and south geographic poles?

11. Can a communications satellite be placed in a "stationary" position in a polar orbit?

12. Earth moves faster around the Sun during the northern hemisphere's winter than during the summer. Based upon this information, is Earth closer to the Sun during winter or during summer?

PROBLEMS

5.1 Circular Motion

5.1 A horse moves with a tangential speed of 1.9 m/s when it is 8.5 m from the center of a carousel. Calculate its centripetal acceleration. If a 70-kg person sits on the horse, what is the net force on the person?

5.2 A 0.050-kg ball on the end of a string is revolving at a uniform rate in a vertical circle of radius 1.2 m. If the speed is a constant 2.8 m/s, find the tension in the string when the ball is at the top and bottom of the circle.

5.3 Tarzan plans to cross a deep ravine by swinging in an arc on a vine. His hands and arms can exert a force of 1500 N. His mass is 80 kg, and the vine is 5.0 m long. What is the maximum speed with which he can go through the lowest point of his swing?

5.2 Motion on a Curve

5.4 What is the centripetal force on a car of mass 800 kg traveling at 24 m/s around a curve of radius 60 m?

○5.5 A centripetal force of 12 kN acts on a truck of weight 15 kN as the truck rounds a level, unbanked curve of radius 60 m. What is the truck's speed?

5.6 A coin sits 15.0 cm from the center of a variable-speed turntable. The coin remains in place as the speed of the turntable increases until the speed reaches a rate of 60 rev/s and then starts to slide. What is the coefficient of friction between coin and turntable?

5.7 A standard turn for level flight with a propeller-driven airplane is one in which the plane makes a complete circular turn in 2 min. If the speed of the plane is 120 m/s, find the radius of the circle and the centripetal acceleration of the plane during such a turn.

5.8 A car goes around a curve at 12 m/s; the radius of the curve is 50 m. Calculate the centripetal acceleration of the car.

5.9 A fly of mass 0.3 g is sunning itself on a phonograph turntable at a location 8 cm from the axis. The turntable rotates at 33 rev/min. What is the centripetal acceleration of the fly? What force is necessary to keep the fly from slipping?

5.10   A 1200-kg car takes an unbanked curve of radius 60 m at a speed of 15 m/s. Will the car make the turn if (a) the pavement is dry and has a coefficient of friction of 0.65? (b) there is oil on the pavement and the coefficient of friction is only 0.20?

5.11   A car of mass *m* goes around an unbanked curve at 20 m/s. If the radius of the curve is 60 m, what is the smallest coefficient of friction that will allow the car to negotiate the curve without skidding?

5.12   What is the maximum speed with which a 1200-kg car can take a curve of radius 85 m on a flat road if the coefficient of friction between tires and road is 0.60?

5.13   What is the minimum coefficient of friction necessary to allow a car to take a curve of radius 105 m at a speed of 80 km/h on a flat road?

5.14   What is the minimum coefficient of friction necessary to allow a car to take a curve of radius 85 m at a speed of 70 km/h on a flat curve?

5.15   What should be the radius of a flat curve that is to carry traffic at 70 km/h if the coefficient of friction between road and tires is 0.45?

5.16   At what angle should a curve of radius 200 m be banked for cars and trucks traveling at 24.6 m/s so that there need be no reliance on friction?

5.17   What is the proper speed for a car to go around a slippery curve of radius 50 m if the road is banked at an angle of 15°?

5.18   At what angle should a curve of radius 130 m be banked to allow cars to travel at 70 km/h on the curve even if it is icy or slick?

5.19   What should be the bank angle for a curve with a radius of 200 m to allow cars to safely travel at 80 km/h on the curve without need of frictional forces?

○❖ 5.20   The table below lists data for six level highway curves. Find the maximum speed at which these curves may be traveled and rank them from slowest to fastest.

Curve	A	B	C	D	E	F
Radius (m)	330	350	280	420	180	250
Friction coefficient	1.2	1.8	1.6	1.8	2.1	2.2

○❖ 5.21   The table below lists data for six banked highway curves. Find the maximum speed at which these curves may be traveled without depending upon friction and rank them from slowest to fastest.

Curve	A	B	C	D	E	F
Radius (m)	330	350	280	420	180	250
Bank angle (°)	5	3	11	5	8	10

5.3 Vertical Circles

5.22   In terms of the radius of the circle *r*, what is the minimum speed a bucket must have if it is to swing in a vertical circle with no water falling out?

○5.23   A pilot comes out of a dive by flying along an arc of radius 600 m. What is his speed if he experiences 6.3 *g*'s at the bottom of the dive?

5.24   A 60-kg pilot comes out of a dive by flying along an arc of radius 800 m with a speed at the bottom of 120 km/h. What is her apparent weight?

5.25   A roller coaster car travels at 102 km/h at the bottom of the first hill. There the track forms part of a circle that has a radius of 10.6 m. With what force is a 58.0-kg passenger pushed down into the seat? How many *g*'s acceleration is this?

5.26   A roller coaster car travels at 50 km/h as it goes along the top of a hill that forms part of a circle. What must the radius of that circle be if the passengers are to feel weightless at the top of the hill? This feeling means the normal force between car and passengers is zero.

5.27   At the top of a loop-the-loop of radius 9.3 m, what is the minimum speed a roller coaster car must have so that passengers do not fall out?

5.28   What must be the minimum speed at the top of a roller coaster loop of radius 9.0 m if the passengers are to be pushed into their seats with a force of one fourth their weight?

5.29   The local highway department is building a highway across hilly terrain. The highway is to carry automobile traffic at 105 km/h. What is the smallest radius of curvature the road can have on the crest of a hill so that cars do not lose contact with the road?

○5.30 A 1000-kg car goes over the crest of a hill whose radius of curvature is 40 m measured in a vertical plane. (a) What is the force of the car on the road surface if the car's speed is 15 m/s? (b) Calculate the magnitude and direction of the necessary force between car and road if the speed at the top of the crest is 20 m/s. Explain your answer.

●5.31 The roller coaster car in Figure 5.23 leaves the top of the 30-m hill with negligible speed. (a) How fast is it going when it reaches the center of the 5-m valley? At that valley, the track forms part of a circle of radius 8 m. (b) What force ("apparent weight") does a 65-kg passenger feel at the valley?

●5.32 The roller coaster car in Figure 5.23 leaves the top of the 30-m hill with negligible speed. (a) How fast is it going at the top of the 24-m hill? At that hill, the track forms part of a circle of radius 12 m. (b) What force ("apparent weight") does a 65-kg passenger feel at the top of the 20-m hill?

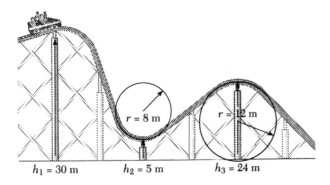

Figure 5.23
Problems 5.31 and 5.32.

●5.33 The **Highland Fling** amusement park ride is pictured in Figure 5.24. The cars are free to revolve around the circle, and at high speed they move as shown, with the passengers' feet toward the outside of the circle. The radius of the circle is 7 m at the location of the passengers' center of gravity. (a) What is the linear speed of the passengers if they feel weightless at the top? (b) What is the apparent weight of a 60-kg passenger at the bottom?

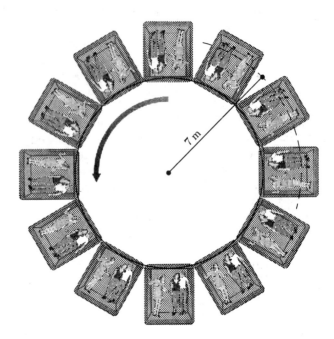

Figure 5.24
Problem 5.33.

5.4 Inertial Reference Frames

●5.34 In the **Rotor** amusement park ride, passengers are pressed against the inside vertical wall of a rotating drum 2.5 m in radius. This drum takes 2.0 s to make a complete rotation; that is, it rotates at 30 rev/min. (a) Viewed from an Earth-based frame of reference outside the drum, what are the magnitude and direction of the force exerted by the wall on a rider whose mass is 70 kg? Initially the rider's weight is supported by the floor. After the drum is at full speed, the floor is removed. (b) If the coefficient of static friction between passenger and wall is 0.6, will the passenger slip or be held pressed against the vertical wall? (c) Describe the situations—including the forces—from the reference frame of the passenger.

5.5 Gravity

5.35 Calculate the gravitational attraction between two physics books, each with mass 2.1 kg, when the center-to-center distance is 20 cm.

5.36 Calculate the gravitational attraction between a 60-kg person and an 80-kg person when their center-to-center distance is 30 cm.

5.37 Calculate the gravitational attraction between a 3-kg copper block and a 2-kg aluminum block when their center-to-center distance is 1.0 m.

5.38 How far apart are two objects, each of mass 1 metric ton (1000 kg) if the gravitational force of attraction between them is 1 nN?

5.39 Calculate the gravitational attraction between Earth and the Sun and between Earth and the Moon.

5.40 Calculate the gravitational force between a proton and an electron in a hydrogen atom. The mass of a proton is 1.67×10^{-27} kg; the mass of an electron is 9.11×10^{-31} kg, and the radius of the electron orbit is 5.29×10^{-11} m.

○5.41 Calculate the mass of Jupiter, given that its moon Callisto has a mean orbital radius of 1.88×10^6 km and an orbital period of 16 days, 16.54 h.

○5.42 Calculate the mass of the Sun, given that the length of the year is 365.25 days and Earth's mean orbital radius is 1.5×10^8 km.

○5.43 The mass of the Moon is 7.4×10^{22} kg, and its radius is 1740 km. Calculate the acceleration of free fall at the Moon's surface.

○5.44 Starting with the Moon's period of 27.3 days, calculate the radius of its orbit.

●5.45 A 1000-kg spacecraft is three fourths of the way from Earth to our Moon during a new Moon, when Earth, the Moon, and the Sun are aligned as shown in Figure 5.25. Find the net gravitational attraction on the space-craft.

●5.46 A 1000-kg spacecraft is three fourths of the way from Earth to our Moon during a half

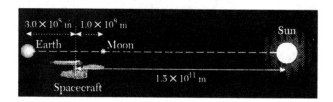

Figure 5.25
Problem 5.45. (Not to scale.)

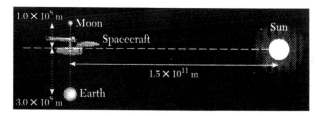

Figure 5.26
Problem 5.46. (Not to scale.)

Moon, when Earth, the Moon, and the Sun are aligned as shown in Figure 5.26. Find the net gravitational attraction on the space-craft.

●5.47 The acceleration of a falling body near Earth's surface, at a distance $R_E = 6.38 \times 10^3$ km from Earth's center, is 9.80 m/s². (a) Use a suitable proportion to calculate the acceleration toward Earth of a falling body that is $60R_E$ from Earth's center. (b) Our Moon is in an orbit of radius $60R_E$, with a period of revolution of 27.26 days. Show, as did Newton, that the centripetal acceleration of the moon toward Earth agrees with your answer from part (a).

5.6 Kepler's Laws of Planetary Motion

○5.48 Use Kepler's second law to show that $v_A/v_B = r_B/r_A$, where A and B are two positions in a planet's orbit as shown in Figure 5.27.

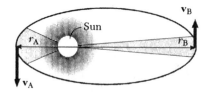

Figure 5.27
Problem 5.48.

5.7 Satellite Orbits

5.49 What orbital radius should a weather satellite have if it is to have a period of 6.0 h?

5.50 What would be the period of a satellite placed in a low orbit around the Moon?

Without an atmosphere to hinder its motion, such a satellite need have an altitude only high enough to clear the lunar mountains.

○5.51 A certain binary star system has two identical stars that are 6.5×10^5 km apart. They revolve about the midpoint of a line connecting their centers of mass with a period of 11.3 years. What is the mass of each star?

○5.52 A space probe is placed into a circular orbit about the Sun with an orbital radius of 149×10^6 km. Find its period.

○5.53 Consider placing a satellite in orbit by firing it horizontally from a high-flying airplane or an extremely high mountaintop. (a) What speed must the satellite have? (b) What will its period be?

○5.54 A particular asteroid revolves about the Sun with a period of 10 years. What is its average distance from the Sun?

General Problems

○5.55 At what angle should the well-known cyclist in Figure 5.28 lean inward if he is negotiating a curve of radius 16 m at a speed of 8 m/s?

●5.56 A car goes around an unbanked curve of radius R. The coefficient of static friction is μ_S. Derive a formula for the maximum speed v_{max} in terms of μ_S, g, and R. Does this speed depend upon the mass of the car?

●5.57 An airplane comes out of a dive, turning upward in a curve whose center of curvature is 1200 m above the plane. The plane's speed is 300 m/s. (a) Calculate the centripetal acceleration and compare it with g. (b) Calculate the upward force of the seat on the 80-kg pilot. (c) Calculate the upward force on a 1-g sample of blood in the pilot's brain. What happens if the heart muscles and circulatory system cannot supply this force?

●5.58 A proposed space station consists of a large, rotating circular tube (like a huge donut), as shown in Figure 5.29. The tube would have a radius of about 1.5 km. Once the tube was rotating, people could walk inside it, heads toward the center and feet radially outward, very much as if gravity were present. For a point on the edge of the space station, find the speed that will give the same effect as gravity at Earth's surface.

Figure 5.28
Problem 5.55.

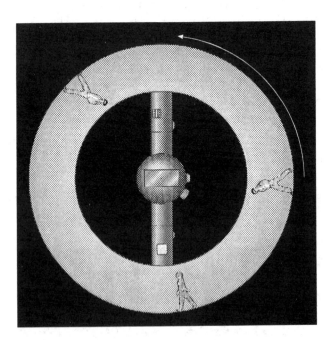

Figure 5.29
Problem 5.58. (Not to scale.)

●5.59 A prototype station similar to the one in the previous problem is built with a radius of 150 m. What tangential speed is necessary to give occupants the same effective gravity as at Earth's surface?

●5.60 From the data in Table 5.1, calculate the acceleration of free fall on the surface of (*a*) Jupiter, (*b*) Saturn, and (*c*) our Moon.

●5.61 At what distance from Earth on its way to the Moon would a spacecraft experience exactly equal gravitational force from both Earth and the Moon?

PHYSICS AT WORK

It takes work to make a sailboat go—not just in the everyday sense of human effort, but in the scientific sense of exerting a force to move an object through a distance. The wind does work on the sails; some of that work overcomes the friction between the boat and the water, and the rest is converted into the boat's kinetic energy. (Where does the wind's energy come from?)

Because the wind usually doesn't blow exactly in the direction the skipper wants to go, the sails can be turned at an angle to the boat's keel (a vertical board that runs along the boat's bottom and helps to keep the boat from drifting sideways). Then the force of the wind and the boat's displacement are separated by an angle θ, which appears in the equation for work. List some of the experiments you would do to find ways to improve a sailboat's performance (its speed, maneuverability, or ability to withstand storms).

(Photo © Ken Fraser/FPG, International Corporation)

6

Work, Power, and Energy

- In physics "work" has a very specific definition. Work is done to an object when it moves through a distance with a force acting on it—a force with a component in the direction of its displacement. In terms of an equation, this may be written as $W = F_{||}s$ or $W = Fs \cos \theta$.

- Work and energy are two forms of the same thing.

- Kinetic energy is the energy an object has because it is in motion; in terms of an equation the kinetic energy may be written as $KE = \frac{1}{2}mv^2$. The net work done on an object causes its kinetic energy to change by an amount equal to the net work done.

- Work may be done on a body moving it against the force of gravity or the force of a spring. This work changes the gravitational potential energy or the spring potential energy.

- In the absence of outside forces—in a closed system—the total amount of energy remains constant. Energy may change from one form to another—from kinetic energy to gravitational potential energy and back again to kinetic energy—but the sum of all the different forms of energy remains constant. This is known as conservation of energy.

We have been able to apply Newton's laws of motion, especially $\mathbf{F} = m\mathbf{a}$, and predict the motion of an object by knowing the forces acting on it. Sometimes, however, it is difficult to know the detailed nature of the forces on an object; this is especially true when objects collide. In this chapter we look at the idea of *energy*, and in the following chapter we look at *momentum*. These ideas provide another way of investigating motion.

Energy and momentum are important physical quantities because they are *conserved*. To be conserved means that the total energy and total momentum remain constant in a closed system. A closed system is one that has no outside forces acting upon it. That energy and momentum are conserved is an especially useful concept in understanding collisions.

In this chapter we study work, power, and energy, all useful concepts throughout physics as well as in other areas both scientific and nonscientific.

6.1 Work

CONCEPT OVERVIEW

■ Work depends upon force, displacement, and the angle between them and is defined by $W = Fs \cos \theta$.

■ For a variable force, work is the area under the curve on a force-distance graph.

In normal, everyday conversation, the word *work* often means something difficult or not enjoyable. We commonly say that we work hard at our jobs or that it required a lot of work to rake all the leaves or even that solving physics problems requires a lot of work. However, in physics—as well as in all other areas of science—we mean something very specific by *work*.

Look at Figure 6.1, where a man is pushing on a brick wall; nothing happens to the wall. He then pushes on a skateboard; as a force is exerted on the skateboard, it moves across the room. What is the difference between the two cases? The answer to this question is that no work is done on the wall but work is done on the skateboard. You may become very tired from pushing on a wall because biological work is being done in your body—blood is moving to supply oxygen and nutrients to your muscles, work is being done in your cells, and so on—but no *mechanical* work is done on the wall.

It is possible to work with all your might and still have your physics instructor say you did zero work.

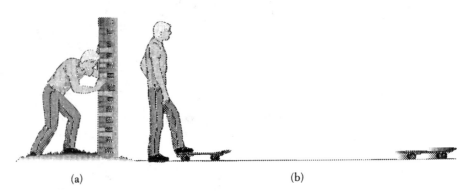

(a)   (b)

Figure 6.1
What is the difference between pushing on a brick wall and pushing on a skateboard?

Work is done on an object when a force **F** is applied to the object as it moves through a displacement **s**, as sketched in Figure 6.2. We can state this explicitly as follows:

*For a constant force **F** acting on an object as the object moves through a displacement **s**, the amount of **work** done is defined as*

$$W = Fs \cos \theta \qquad (6.1)$$

Figure 6.2
Work is done when a force **F** moves an object through displacement **s**.

where θ is the angle between the force vector and the displacement vector.

Both force and displacement are vector quantities, but only their magnitudes F and s appear in this definition of work. This means that *work is a scalar quantity*.

You can see from Equation 6.1 that the units of work are units of force multiplied by units of distance. In the SI, therefore, work is measured in newton-meters (N·m). This unit occurs so frequently that we give it a special name, **joule** (J):

$$1\,J = 1\,joule = 1\,N{\cdot}m$$

This unit of work is named after James Prescott Joule (1818–1889), who did a great deal of work establishing the connections among work, energy, and heat—including careful temperature measurements at the top and bottom of a waterfall he and his bride visited on their honeymoon!

Example 6.1

As sketched in Figure 6.3, a force of 40 N acts on a block at an angle of 30° to the horizontal. If the block moves through a horizontal displacement of 3.0 m, how much work is done on it?

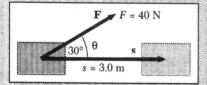

Figure 6.3
Work is done when a force acts through a displacement; the amount of work done is $W = Fs \cos \theta$.

Reasoning
We can use the definition of work, Equation 6.1, directly with $F = 40$ N, $s = 3.0$ m, and $\theta = 30°$.

Solution

$$
\begin{aligned}
W &= Fs \cos \theta \\
&= (40\,\text{N})(3.0\,\text{m})(0.87) \\
&= 104\,\text{N·m} \approx 100\,\text{N·m} = 100\,\text{J}
\end{aligned}
$$

Notice that, even when a force is exerted, the work can still be zero (if either $s = 0$ or $\theta = 90°$) and that, even when a body moves through a displace-

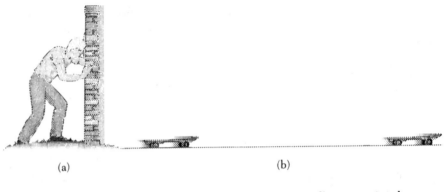

Figure 6.4
No work is done on the brick wall
because it does not move. No work
is done on the moving skateboard
because no net force is exerted on it.

(a)   (b)

ment, the work can still be zero (if either $F = 0$ or $\theta = 90°$). Figure 6.4 shows a
man exerting great force on a wall, but the wall's displacement is zero and so
the work done on it is zero. The figure also shows a skateboard moving along
the sidewalk. No net force is exerted on the board, so the work done on it is
zero. Even if you consider the downward force of gravity or the upward normal
force of the sidewalk, there is still no work done on the skateboard because
both those forces are perpendicular to the displacement so $\theta = 90°$.

It is useful to look at the work done by the components of the force, as
sketched in Figure 6.5. Look first at $F_{\parallel}$, the component parallel to the displacement. We can rearrange Equation 6.1 to obtain

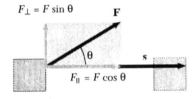

Figure 6.5
The force on an object can be resolved into components parallel and
perpendicular to the displacement.

$$W = (F \cos \theta)s$$
$$W = F_{\parallel}s \qquad\qquad (6.2)$$

This expression tells us that work is equal to the component of force *in the
direction of the motion* multiplied by the displacement. Now look at $F_{\perp}$, the
perpendicular component. The angle between $F_{\perp}$ and the displacement is 90°,
and $\cos 90° = 0$ so that *no work* is done by this component. In general, any
force perpendicular to the displacement does no work.

Example 6.2

Figure 6.6 shows a block of mass $m = 2.00$ kg being pulled along a horizontal table a distance $s = 4.00$ m by an external force $F_{ex} = 15.0$ N directed 37.0° above the horizontal. The coefficient of friction between
block and table is 0.350. Determine the magnitude and direction of all
the forces acting on the block, the work done by each, and the total work
done.

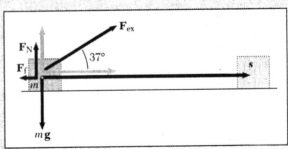

Figure 6.6
All forces, includ-
ing those of fric-
tion, are able to
do work.

Reasoning

We start with a free-body diagram showing *all* the forces on the block. Since the block moves horizontally, we take the x axis to be in this direction and the y axis perpendicular to it. After resolving the forces into components along these directions, we can use Newton's second law in the y direction to solve for the normal force. (Remember, the normal force need not be equal to the weight.) We can then use the normal force to find the friction force.

After using Equation 6.1 to calculate the work done by each force separately, we shall add all the work values algebraically to determine the total work done.

Solution

Since there is no motion in the y direction, we know the acceleration in the y direction must be zero:

$$\Sigma F_y = ma_y = 0$$
$$\Sigma F_y = F_N + F_{ex} \sin 37.0° - mg = 0$$
$$F_N + (15.0 \text{ N})(0.602) - (2.00 \text{ kg})(9.80 \text{ m/s}^2) = 0$$
$$F_N = 10.6 \text{ N}$$

$$F_f = \mu F_N = (0.350)(10.6 \text{ N}) = 3.71 \text{ N}$$

Now that all the forces are known, we can find the work done by each. *No work* is done by the forces in the y direction because they are perpendicular to the direction of motion:

$$F_N: \text{ no work done}$$
$$mg: \text{ no work done}$$

Therefore only the external force and the friction force do work:

$$F_{ex}: \quad W_{ex} = F_{ex}s \cos \theta_{ex}$$
$$= (15.0 \text{ N})(4.00 \text{ m})(0.799) = 47.9 \text{ N·m} = 47.9 \text{ J}$$

$$F_f: \quad W_f = F_f s \cos \theta = (3.71 \text{ N})(4.00 \text{ m})(\cos 180°)$$
$$= -14.8 \text{ N·m} = -14.8 \text{ J}$$

Notice that this last work calculated is *negative* since the direction of the friction force is opposite the direction of the displacement.

The total work done on the block is

$$W_{tot} = 47.9 \text{ J} - 14.8 \text{ J} = 33.1 \text{ J}$$

Work may be either positive or negative. Figure 6.7 shows a force and a displacement whose directions are different by more than 90°. Here the force component along the direction of the displacement is directed *opposite* the displacement. This configuration gives a *negative* value for work. Equivalently, Equation 6.1 applied to this configuration gives a negative value since θ is greater than 90° so $\cos \theta$ is negative.

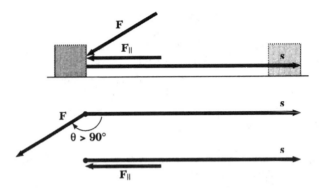

Figure 6.7
The work done is negative if θ is greater than 90°.

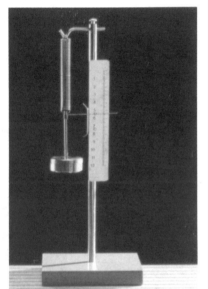

An apparatus for investigating Hooke's law. *(Courtesy of Central Scientific Company)*

So far we have considered only constant forces, but many forces are not constant. Figure 6.8 illustrates the force exerted by a spring as it is stretched. This force is found to be proportional to how far the spring has been stretched or compressed:

$$F_s = -kx \qquad\qquad (6.3)$$

where k is known as the **spring constant** and describes how stiff or how strong the spring is and x is the distance the spring has been moved from its equilibrium position. The negative sign indicates that the spring opposes its displacement from equilibrium. If x is positive, indicating, perhaps, a stretch to the right, then the spring force is negative, meaning it is directed to the left. If x is negative, indicating the spring has been compressed and moved to the left of

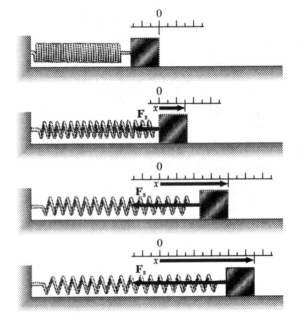

Figure 6.8
The force exerted by a spring is $F_s = -kx$, Hooke's law.

Hooke's Law

the equilibrium position, then the force is positive and directed to the right. Equation 6.3 is known as **Hooke's law,** although, as noted in the opening of Chapter 4, it is not a grand *law of physics* in the sense of Newton's laws of motion or his law of universal gravitation.

Figure 6.9 is a graph of the spring force as a function of the distance the spring is stretched. The graph is a straight line, and its slope is the spring constant of the spring. The force $F_{ex} = |F_s|$ is the external force applied to the spring; $F_{ex} = kx$. Even though the force varies smoothly and continuously, we can think of it as being constant over some small distance and then jumping to a new value and remaining constant over some small distance, and so on. This noncontinuous aspect is indicated by the rectangles in Figure 6.9, one of which is highlighted. Notice that the area of that rectangle is force times distance, which is just the work done while the spring moves that small distance.

The total work done in stretching or compressing a spring is the area under the entire curve of Figure 6.9. You could draw many little rectangles, find the area of each, and add all the areas together, as shown in Figure 6.10. For a spring, the area under the graph is the area of a triangle. When the spring has been stretched an amount x_{max}, the work that has been done on it is equal to the area under the graph, the area of a triangle:

$$W_s = \text{area} = \tfrac{1}{2}\,\text{base} \times \text{height}$$
$$= \tfrac{1}{2}\,(x_{max})(F_{max})$$
$$= \tfrac{1}{2}\,(x_{max})(kx_{max})$$
$$W_s = \tfrac{1}{2}\,kx_{max}^2 \tag{6.4}$$

In general, the work done by a variable force is the area under the curve on a force-distance graph. Another example is shown in Figure 6.11, which shows us what to do when the area is not something familiar like that of a triangle. We approximate the area by drawing many small rectangles, finding the area of each, and adding all the areas together.

6.2 Power

CONCEPT OVERVIEW

▬ Power is the rate at which work is done.

▬ Watts and horsepower are two units for measuring power.

In addition to how *much* work is done, we often want to know how *fast* it is done. Practically any car can climb Pikes Peak if given enough time, but we expect a Porsche to climb it more quickly than a Ford Escort. What is the difference?

Power is the *rate* at which work is done:

$$P = \frac{\Delta W}{\Delta t} \tag{6.5}$$

This ratio of amount of work done ΔW divided by time Δt taken to do that work is the *average* power. For the *instantaneous* power—the value of the power at a particular moment—we must take this ratio over a very small time interval Δt, that is, in the limit as Δt approaches zero.

In physics, as in the world at large, there are laws, and then there are *laws.*

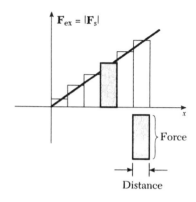

Figure 6.9
Work done is the area under the curve on a force-distance graph.

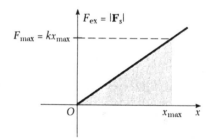

Figure 6.10
For a triangle, area = $\tfrac{1}{2}$ base × height.

Figure 6.11
Even when the curve is irregular, work done is still the area under the curve on a force-distance graph. The summed area of all the thin rectangles gives a good approximation of the total area under the curve.

In the SI, power is measured in joules per second, and this unit is given a special name, the **watt** (W), in honor of James Watt, inventor of the steam engine:

$$1 \text{ W} = 1 \text{ watt} = 1 \text{ J/s}$$

We can solve Equation 6.5 for the work done:

$$\Delta W = P \, \Delta t$$

This expression tells us that, if power is used at a constant rate P for a time Δt, an amount of work ΔW will be done. In this sense, we can describe a joule as the amount of work done if power is used at the rate of 1 W for a time interval of 1 s:

$$1 \text{ J} = 1 \text{ W·s}$$

Example 6.3

Your electric bill lists your monthly usage in terms of a unit of energy or work called a kilowatt-hour. How many joules of work is done by a kilowatt-hour of electricity?

Reasoning
This is another case of converting units. We start with 1.0 kW·h and multiply by fractions that are equal to unity.

Solution

$$1.0 \text{ kW·h}\left(\frac{10^3 \text{ W}}{1 \text{ kW}}\right)\left(\frac{1 \text{J/s}}{1 \text{ W}}\right)\left(\frac{3600 \text{ s}}{1 \text{ h}}\right) = 3.6 \times 10^6 \text{ J}$$

James Watt was somewhat optimistic in his measurements of British workhorses. These Clydesdales are among the very few breeds of horses that can actually put out one horsepower. (© *Robin Smith/FPG International Corporation*)

U.S. automotive engineers "think metric" and use watts in their calculations when designing car engines.

*Horsepower

In British engineering units (now used only in the United States), force is measured in pounds, distance in feet, and time in seconds. Thus, work is measured in foot-pounds and power in foot-pounds per second. No special name is given to this unit. However, we do often use the British unit **horsepower** (hp):

$$1 \text{ hp} = 1 \text{ horsepower} = 550 \text{ ft·lb/s}$$

Originally 1 hp was intended to be the power produced by a typical British workhorse. Watt introduced this idea in order to explain how powerful his steam engines were. However, only a very fine, strong workhorse can continuously produce a full 1 hp.

Horsepower and watts measure the same thing—the rate of doing work. It is only because of tradition that we talk about the power of an automobile in horsepower and the power of an electric motor in kilowatts (Fig. 6.12). Automobile power in Europe and Japan is usually given in kilowatts, and automotive engineers even in the United States do their internal measurements, calcula-

(a)

(b)

(c)

Figure 6.12
It is only because of tradition that we think of automobile power in horsepower and electrical power in kilowatts. (a) A modern, eight cylinder automobile engine capable of producing nearly 300 horsepower. *(Courtesy of Cadillac Motor Division, General Motors Corp.)* (b) An electric motor. *(Courtesy of General Motors Corp.)* (c) An old-fashioned carbon filament incandescent lamp. Such lamps are usually rated in watts. What happens to the energy (in the form of electricity) entering the lamp?
(Courtesy of Central Scientific Company)

tions, and designs with engine power measured in kilowatts. Values can easily be converted from one system of units to the other:

$$1 \, \text{hp} = 746 \, \text{W} = 0.746 \, \text{kW}$$
$$1 \, \text{kW} = 1.34 \, \text{hp}$$

Example 6.4

A 75.0-kg student runs up the stairs in the Science Hall in 23.0 s. If his final elevation is 21.0 m greater than his initial elevation, what average power does he produce?

Reasoning

The student does work against gravity as he lifts his weight through a vertical distance of 21.0 m. The amount of work done is his weight multiplied by the vertical distance. Dividing this amount of work by the time required will give the average power produced.

Solution

$$w = mg = (75.0 \, \text{kg})(9.80 \, \text{m/s}^2) = 735 \, \text{N}$$

$$\Delta W = Fs = (735 \, \text{N})(21.0 \, \text{m}) = 15 \, 400 \, \text{N·m} = 15.4 \, \text{kJ}$$

Power is the rate at which this work is done:

$$P = \frac{\Delta W}{\Delta t} = \frac{15 \, 400 \, \text{J}}{23.0 \, \text{s}} = 670 \, \text{W}$$

It is more fun to express this in horsepower:

$$P = 670 \, \text{W}\left(\frac{1 \, \text{hp}}{746 \, \text{W}}\right) = 0.898 \, \text{hp}$$

You may want to calculate your own power output by this method.

> **Practice Exercise**   If the student requires 29.0 s to run up the stairs, what is the average power?
>
> **Answer**   532 W

There is an interesting and useful expression relating force, power, and speed. Consider a constant force F exerted on an object moving at speed v. An example might be the thrust of airplane engines or the friction force between a road and the tires of an automobile. In time Δt, the object moves a distance $\Delta s = v\Delta t$, so an amount of work $\Delta W = F\Delta s = Fv\Delta t$ is done. The power supplied is therefore

$$P = \frac{\Delta W}{\Delta t} = \frac{F v \Delta t}{\Delta t}$$

$$P = Fv \qquad\qquad (6.6)$$

Example 6.5

> In a particular car, the net power delivered to the wheels is 28 hp when the car is traveling at a constant 90 km/h. What net force of air resistance and rolling friction must the engine overcome at that speed?
>
> **Reasoning**
> Force, power, and speed are related by Equation 6.6, which we may solve to find the force.
>
> **Solution**
> It will make our calculation easier if we first change the power to kilowatts:
>
> $$P = 28 \text{ hp}\left(\frac{1 \text{ kW}}{1.34 \text{ hp}}\right) = 21 \text{ kW}$$
>
> $$F = \frac{21 \text{ kW}}{90 \text{ km/h}}\left(\frac{1 \text{ km}}{1000 \text{ m}}\right)\left(\frac{1000 \text{ W}}{1 \text{ kW}}\right)\left(\frac{3600 \text{ s}}{1 \text{ h}}\right)\left(\frac{1 \text{ J/s}}{1 \text{ W}}\right)\left(\frac{1 \text{ N·m}}{1 \text{ J}}\right) = 840 \text{ N}$$

6.3 Kinetic Energy

CONCEPT OVERVIEW

- Kinetic energy, which is energy due to motion, is defined as $KE = \frac{1}{2}mv^2$.
- Work done on an object changes the kinetic energy of the object.

Kinetic energy is energy due to motion. Figure 6.13 shows a glider of mass m sitting on a frictionless air track. Initially, the glider is moving with velocity $\mathbf{v_i}$ when a net force $\mathbf{F}$ is applied to it as shown. The motion is along a straight line,

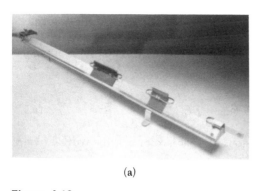

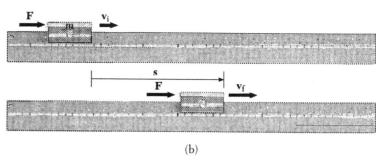

(a)   (b)

Figure 6.13

An air track glider goes faster as work is done on it. *(Photo courtesy of Central Scientific Company)*

and $\mathbf{F}$ is parallel to the velocity. While $\mathbf{F}$ is applied, the glider moves through a displacement $\mathbf{s}$, meaning work $W = Fs$ has been done. While $\mathbf{F}$ is applied, the glider accelerates, with $\mathbf{a} = \mathbf{F}/m$, to a final velocity $\mathbf{v}_f$. From Equation 2.7 (with $s_i = 0$), we can determine this final velocity,

$$v_f^2 = v_i^2 + 2as$$
$$= v_i^2 + 2\left(\frac{F}{m}\right)s = v_i^2 + \frac{2Fs}{m}$$

Since the Fs factor is, by definition, the work done on the glider, this expression can be solved for Fs ($=W$) to give

$$W = \tfrac{1}{2} mv_f^2 - \tfrac{1}{2} mv_i^2 \qquad \textbf{(6.7)}$$

The net work done is equal to the *change* in the quantity $\tfrac{1}{2} mv^2$. If an object of mass m is initially at rest (so that $v_i = 0$), then the work done to get it moving with speed v is equal to $\tfrac{1}{2} mv^2$. This means that the quantity $\tfrac{1}{2} mv^2$ is a measure of the work done to an object of mass m moving with speed v or a measure of the work that an object of mass m moving with speed v could do to something else as the object returned to rest. Therefore, we can define the **kinetic energy** *KE* by

$$KE = \tfrac{1}{2} mv^2 \qquad \textbf{(6.8)}$$

Note that kinetic energy depends upon mass and speed but not direction. Energy of any form, like work, is a scalar quantity and is measured in joules or equivalent units. Thus a 1.0-kg mass traveling at 1.0 m/s has 0.50 J of kinetic energy, according to the definition given by Equation 6.8.

Notice that kinetic energy depends upon the mass in a linear way but upon the square of the speed. Figure 6.14 shows two trucks traveling with the same speed v_i, but one truck is loaded so that it has twice the mass of the other. The more massive truck has twice as much kinetic energy. If the brakes are applied and the two trucks brought to a stop, twice as much work will be done in stopping the more massive truck. Figure 6.15 shows two identical trucks, one traveling at twice the speed of the other. The faster truck has *four* times the

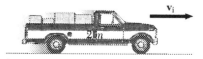

Figure 6.14

The more massive truck has the greater kinetic energy.

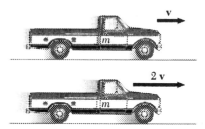

Figure 6.15

The faster truck has the greater kinetic energy.

kinetic energy of the slower one. If the brakes are applied and the two trucks brought to a stop, four times as much work will be done in stopping the faster truck. Figure 6.16 shows typical kinetic energies for a variety of physical events. We can rewrite Equation 6.7 as

$$W = KE_f - KE_i \qquad\qquad (6.7')$$

Equations 6.7 and 6.7′ are two versions of what is known as the **work-energy theorem:**

The net work done on an object is equal to the change in kinetic energy of the object.

The work-energy theorem is a powerful tool for solving motion problems too difficult to be handled with Newton's motion laws.

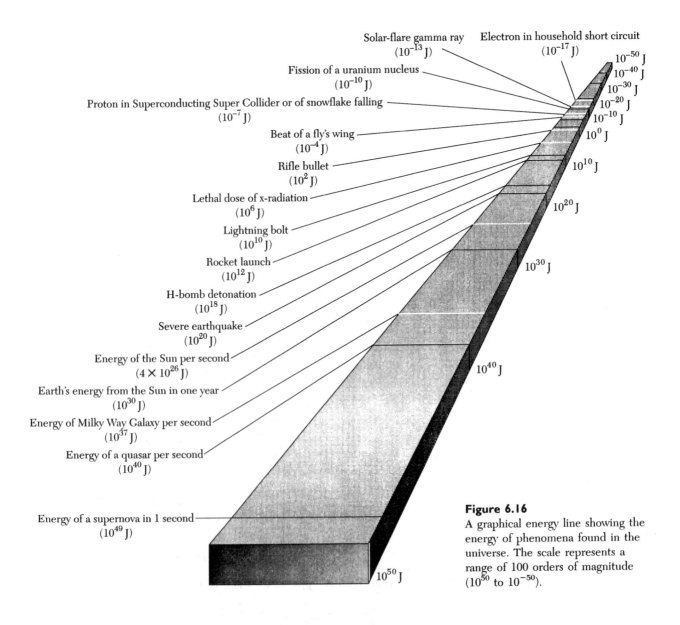

Figure 6.16
A graphical energy line showing the energy of phenomena found in the universe. The scale represents a range of 100 orders of magnitude (10^{50} to 10^{-50}).

Example 6.6

Calculate the kinetic energy of a 1200-kg car traveling at 10 km/h and at 100 km/h.

Reasoning

Since the mass and speed are known, both parts of this problem are a direct application of the definition of kinetic energy, Equation 6.8.

Solution

It will make our calculations easier if we convert the speed to meters per second:

$$v = 10\frac{km}{h}\left(\frac{1000\ m}{1\ km}\right)\left(\frac{1\ h}{3600\ s}\right) = 2.8\ m/s$$

$$KE = \tfrac{1}{2}mv^2 = \tfrac{1}{2}(1200\ kg)(2.8\ m/s)^2 = 4700\frac{kg{\cdot}m^2}{s^2}\left(\frac{1\ J}{1\ kg{\cdot}m^2/s^2}\right)$$

$$= 4700\ J = 4.7\ kJ$$

$$v = 100\frac{km}{h}\left(\frac{1000\ m}{1\ km}\right)\left(\frac{1\ h}{3600\ s}\right) = 28\ m/s$$

$$KE = \tfrac{1}{2}(1200\ kg)(28\ m/s)^2 = 470\ 000\ J = 470\ kJ$$

Increasing the speed of the car by a factor of ten increases its kinetic energy by a factor of 100. Kinetic energy depends upon the *square* of the speed.

Practice Exercise What is the car's kinetic energy when it travels 20 km/h (twice as fast)?

Answer 18 000 J = 18 kJ

Table 6.1 shows some interesting data on typical distances needed to stop a car at various speeds. Notice the first part of the distance, the *reaction distance*. This is how far the car travels from the time the driver sees a need to stop until the foot gets to the brake pedal—in other words, how far the car travels during the driver's reaction *time*. Reaction distance is *linear* with respect to the velocity, for it is just $s = vt$. If it requires 0.75 s to get the driver's foot on the brake, then the car travels 4.2 m at 20 km/h, 8.4 m at 40 km/h, 16.8 m at 80 km/h.

Now look at the second part, the *braking distance*. This is how far the car travels after the brakes are applied. Because the brakes must do work in order to stop the car, braking distance is a function of kinetic energy, so it is proportional to the *square* of the speed. When the brakes are applied, a force **F** is applied to the tires by the road, and this force is applied over a distance **s** in bringing the car to a stop. We may justifiably consider this braking force to be constant. This force and the car's displacement as it comes to a stop are in opposite directions, so the work done is negative (see Figure 6.7 if you need to review the concept of negative work). Also, the change in kinetic energy, from

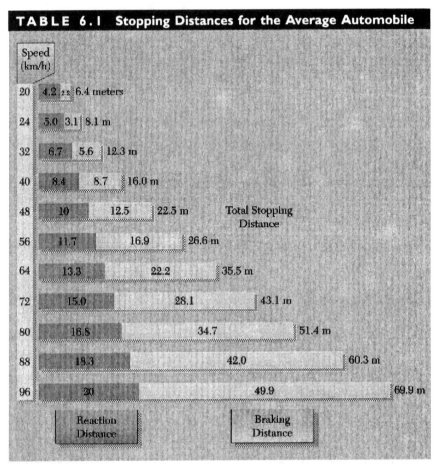

TABLE 6.1 Stopping Distances for the Average Automobile

Speed (km/h)	Reaction Distance	Braking Distance	Total Stopping Distance
20	4.2	2.2	6.4 meters
24	5.0	3.1	8.1 m
32	6.7	5.6	12.3 m
40	8.4	8.7	16.0 m
48	10	12.5	22.5 m
56	11.7	16.9	26.6 m
64	13.3	22.2	35.5 m
72	15.0	28.1	43.1 m
80	16.8	34.7	51.4 m
88	18.3	42.0	60.3 m
96	20	49.9	69.9 m

Adapted from Day, Maryott, and Stiska, *In the Driver's Seat*, Boston, Houghton Mifflin Co., 1978.

$\frac{1}{2} mv^2$ to zero, is negative. The work done is just equal to this change in kinetic energy. Look at the braking distance in Figure 6.16 for 20 km/h, 40 km/h, and 60 km/h. In doubling the speed from 20 km/h to 40 km/h, the kinetic energy is increased by a factor of four. This means the work that must be done to stop a car traveling at 40 km/h is four times as great as the work needed to stop a car traveling at 20 km/h. For a constant braking force **F**, the braking distance will also be four times as great. In tripling the speed, from 20 km/h to 60 km/h, the kinetic energy is increased by a factor of *nine*. This means the work needed to stop a 60-km/h car is nine times the work needed to stop a 20-km/h car. For a constant braking force **F**, the braking distance is also nine times as great!

Example 6.7

A 0.60-kg hammer head that is moving with a speed of 8.0 m/s hits a nail and drives it 1.0 cm into a wall. Find the kinetic energy of the hammer head and the average force exerted by the nail on the wall.

Reasoning

Calculating the kinetic energy is a direct application of Equation 6.8. All this kinetic energy is given up as work done by the nail exerting a force F through a distance s as the nail is driven into the wall. We shall set the kinetic energy equal to the work and solve for F.

Solution

$$KE = \tfrac{1}{2}mv^2 = \tfrac{1}{2}(0.60 \text{ kg})(8.0 \text{ m/s})^2 = 19 \text{ J}$$

This much kinetic energy is reduced to zero by the work done by the nail exerting an average force F through a distance of 1.0 cm = 0.010 m:

$$W = Fs = KE$$
$$Fs = F(0.010 \text{ m}) = 19 \text{ J}$$

$$F = \frac{19 \text{ J}}{0.010 \text{ m}}\left(\frac{1 \text{ N·m}}{1 \text{ J}}\right) = 1900 \text{ N}$$

This simple analysis ignores the force that the carpenter continues to exert as the nail is hit and the mass and energy of the carpenter's hand and the hammer handle, but it is a reasonable approximation.

6.4 Gravitational Potential Energy

CONCEPT OVERVIEW

- Near Earth's surface, the gravitational potential energy, which is energy due to position, is $PE_g = mgy$.

- When the only work done on an object is work by the gravitational force, the sum of the object's kinetic energy and gravitational potential energy remains constant.

- Only *changes* in potential energy have meaning in physics.

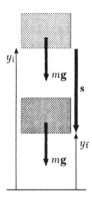

Figure 6.17
As the block moves from y_i to y_f, the force of gravity mg does work $W_g = mg(y_i - y_f)$ on it.

Figure 6.17 shows a block of mass m moving from an initial position y_i to a final position y_f under the influence of the force of gravity. This force, $F_g = mg$, acts on the block while it moves a distance $s = y_i - y_f$, and so the force does work W_g on the block:

$$W_g = F_g s$$
$$W_g = mg(y_i - y_f) \tag{6.9}$$

This is the net work because F_g is the only force acting on the block. According to the work-energy theorem, Equation 6.7 or 6.7′, this net work equals the change in kinetic energy:

$$mg(y_i - y_f) = KE_f - KE_i$$

Rearranging terms, we get

$$mgy_i + KE_i = mgy_f + KE_f \tag{6.10}$$

The sum, $mgy + KE$, or $mgy + \tfrac{1}{2}mv^2$, remains constant whenever the only work done on an object is work done by gravity.

Water flowing over the Shohola Dam in northeastern Pennsylvania has gravitational potential energy. (© L. W. Black)

You can pick any point as your origin in solving problems on gravitational potential energy.

We define the quantity mgy as the **gravitational potential energy:**

$$PE_g = mgy \qquad (6.11)$$

Gravitational potential energy is the potential energy of a system composed of the object with mass m and Earth; however, it is common to speak of the potential energy as belonging to the object only.

Equation 6.10 can now be written

$$PE_{gi} + KE_i = PE_{gf} + KE_f \qquad (6.12)$$

The sum of potential energy and kinetic energy is the total energy of the system,

$$E = PE_g + KE \qquad (6.13)$$

Equation 6.12 states that the initial energy equals the final energy. In other words, energy is conserved, as mentioned in the opening paragraph of this chapter. Energy may change from kinetic to potential or vice versa, but the total energy remains constant.

The potential energy defined by Equation 6.11 depends upon our choice for the origin used in measuring the height y. However, it is only the *difference* in height $y_f - y_i$ that enters into the form of the work-energy theorem we used above to obtain Equations 6.10 and 6.12. We are free to choose whatever reference position is convenient for calculating potential energy, for only the difference is ever used. An example will make this clear.

Problem-Solving Tip

- It is useful to define the potential energy to be zero for some position and then measure the potential energy at other positions relative to that zero position.

Example 6.8

Figure 6.18 shows a 1.2-kg book initially at rest on a table at position A. The book is lifted to a height of 0.80 m above the table at position B. From there it is dropped and lands on the floor, 1.0 m beneath the table. Discuss the changes in its gravitational potential energy as it moves from A to B to C.

Reasoning

To answer the question, what is the change in gravitational potential energy as the book moves from A to B?, we use Equation 6.11. To answer the question, what is the total change in gravitational potential energy as it moves from A to C?, we use Equation 6.11 again. At C, the book has *less* potential energy than it had initially; therefore the change in potential energy is negative. In the solution that follows, notice that we use values for $y_B - y_A$ and $y_C - y_A$ and have no need for the absolute values y_A, y_B, and y_C.

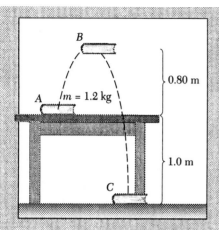

Figure 6.18
Work is done on a book in moving
it up and down.

Solution

First we go from A to B. The change in potential energy is that at B minus
that at A:

$$PE_{gB} = mgy_B$$

$$PE_{gA} = mgy_A$$

$$\Delta PE_g = PE_{gB} - PE_{gA} = mg(y_B - y_A)$$
$$= (1.2 \text{ kg})(9.8 \text{ m/s}^2)(0.80 \text{ m}) = 9.4 \text{ J}$$

We use Equation 6.11 again to find the potential energy at C:

$$PE_{gC} = mgy_C$$

$$\Delta PE_g = PE_{gC} - PE_{gA} = mg(y_C - y_A)$$
$$= (1.2 \text{ kg})(9.8 \text{ m/s}^2)(-1.0 \text{ m}) = -12 \text{ J}$$

The negative value means that the book has *less* gravitational potential
energy at C than at A.

Another way of looking at the situation described in Example 6.8 is to ask,
what *is* the potential energy at positions A, B, and C? To answer this, we still
need Equation 6.11, but now we must be careful to ask, what is the potential
energy at A, B, and C *relative to* some location? Let us begin by asking for the
potential energy relative to the table top (relative to y_A), which means we
measure heights from the table top. Therefore, $y_A = 0$, $y_B = +0.80$ m, and
$y_C = -1.0$ m. Then,

$$PE_{gA} = 0$$

$$PE_{gB} = (1.2 \text{ kg})(9.8 \text{ m/s}^2)(0.80 \text{ m}) = 9.4 \text{ J}$$

$$PE_{gC} = (1.2 \text{ kg})(9.8 \text{ m/s}^2)(-1.0 \text{ m}) = -12 \text{ J}$$

We could just as well ask for the potential energy relative to the floor, however,
which means we measure heights from the floor. Now, $y_A = 1.0$ m, y_B
$= 1.8$ m, and $y_C = 0$. Then,

$$PE_{gA} = (1.2 \text{ kg})(9.8 \text{ m/s}^2)(1.0 \text{ m}) = 12 \text{ J}$$

$$PE_{gB} = (1.2 \text{ kg})(9.8 \text{ m/s}^2)(1.8 \text{ m}) = 21 \text{ J}$$

$$PE_{gC} = 0$$

The numbers are quite different in the two cases, but notice that the *change* in potential energy is the same in both cases. In the first case, we have

$$PE_{gB} - PE_{gA} = 9.4 \text{ J} - 0 = 9.4 \text{ J}$$

$$PE_{gC} - PE_{gA} = -12 \text{ J} - 0 = -12 \text{ J}$$

In the second case we have

$$PE_{gB} - PE_{gA} = 21 \text{ J} - 12 \text{ J} = 9 \text{ J}$$

$$PE_{gC} - PE_{gA} = 0 - 12 \text{ J} = -12 \text{ J}$$

You may choose any place as an origin from which to measure the height y to calculate gravitational potential energy, but it is important that all heights are measured from the same origin.

Example 6.9

Figure 6.19 shows the same 1.2-kg book initially at position A on the table but moving upward with speed v_i. It rises to a height of 0.80 m above the table to position B. What is its initial speed v_i at A? The book then falls to the floor, 1.0 m below the table. What is its speed v_f just before striking the floor at C?

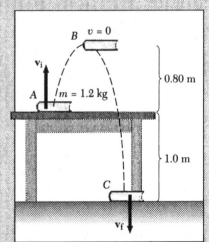

Figure 6.19
Kinetic and potential energy change as the book moves along its path, but the sum of kinetic and potential energy remains constant.

Reasoning
Kinetic and potential energy both change as the book moves along the path shown, but the sum of kinetic and potential energy does *not* change.

At A we know the position of the book so we can calculate the potential energy there. Because we do not know the speed, we write it symbolically as v_i in the kinetic energy equation.

At B we know the speed—it is zero at the top of the motion—so we can calculate the kinetic energy: $KE_B = 0$. We also know the height so can calculate the potential energy. Now we know the total energy at B. Since the total energy of the system does not change, we can set the total energy at B equal to the sum of kinetic energy and potential energy at A and solve for v_i, the speed at A.

At C we know the book's position so we can calculate the potential energy. Setting the total energy at C equal to the total energy at A or B will allow us to determine the kinetic energy at C and, from that, the speed v_f at C.

Solution

We start with position A, choosing it as our origin so $y_A = 0$. In making this choice, we have established the table top as the reference level and must ensure that y_B and y_C are also measured from the table top.

$$KE_i = KE_A = \tfrac{1}{2} mv^2 = (0.60 \text{ kg}) \, v_i^2$$

$$PE_{gA} = mgy_A = (1.2 \text{ kg})(9.8 \text{ m/s}^2)(0) = 0$$

The total energy at A is

$$E_A = KE_A + PE_{gA} = (0.60 \text{ kg}) \, v_i^2$$

Now we need the total energy at B:

$$KE_B = \tfrac{1}{2} mv_B^2 = \tfrac{1}{2}(1.2 \text{ kg})(0)^2 = 0$$

$$PE_{gB} = mgy_B = (1.2 \text{ kg})(9.8 \text{ m/s}^2)(0.80 \text{ m}) = 9.4 \text{ J}$$

$$E_B = 0 + 9.4 \text{ J} = 9.4 \text{ J}$$

Because energy is conserved,

$$E_A = E_B$$

$$(0.60 \text{ kg}) \, v_i^2 = 9.4 \text{ J}$$

$$v_i^2 = \frac{9.4 \text{ J}}{0.60 \text{ kg}} \left(\frac{1 \text{ N·m}}{1 \text{ J}} \right) \left(\frac{1 \text{ kg·m/s}^2}{1 \text{ N}} \right) = 16 \text{ m}^2/\text{s}^2$$

$$v_i = 4.0 \text{ m/s}$$

Now we do the same thing for position C:

$$KE_C = \tfrac{1}{2}(1.2 \text{ kg}) \, v_f^2 = (0.60 \text{ kg}) \, v_f^2$$

$$PE_{gC} = (1.2 \text{ kg})(9.8 \text{ m/s}^2)(-1.0 \text{ m}) = -12 \text{ J}$$

The negative sign here is important. On the floor the book has less potential energy than on the table top. Work must be done on the book to move it back to the position of zero potential energy.

The total energy at C is

$$E_C = (0.60 \text{ kg}) \, v_f^2 - 12 \text{ J}$$

Energy is conserved so

$$E_C = E_B$$

$$(0.60 \text{ kg}) \, v_f^2 - 12 \text{ J} = 9.4 \text{ J}$$

$$v_f^2 = \frac{21 \text{ J}}{0.60 \text{ kg}} = 35 \text{ m}^2/\text{s}^2$$

$$v_f = 5.9 \text{ m/s}$$

6.5 Elastic Potential Energy

CONCEPT OVERVIEW

- Stretching or compressing a spring requires the expenditure of energy, and this energy is stored by the spring and may be recovered; it is known as elastic potential energy.

- Stretching or compressing a spring an amount x stores elastic potential energy equal to $PE_{el} = \frac{1}{2}kx^2$.

- When a spring force is the only force acting on an object, the sum of the object's kinetic energy and elastic potential energy remains constant.

In Section 6.1 we looked at the force a spring exerts and the work it does. We determined in Equation 6.4 that the work done in stretching or compressing a spring an amount x is

$$W_s = \tfrac{1}{2}kx^2$$

where k is the spring constant for the spring. Figure 6.20 shows a spring with spring constant k attached to an object of mass m. Initially the spring is stretched a distance x_i from equilibrium, and finally it is stretched a distance x_f from equilibrium. The work done by the spring in moving from x_i to x_f is

$$W_s = \tfrac{1}{2}kx_i^2 - \tfrac{1}{2}kx_f^2$$

The force exerted by the spring is the net force acting on the object, so, according to the work-energy theorem, Equation 6.7 or 6.7′, the work done by the spring equals the change in the kinetic energy of the mass:

$$W_s = \tfrac{1}{2}kx_i^2 - \tfrac{1}{2}kx_f^2 = KE_f - KE_i$$

After rearranging terms, we get

$$\tfrac{1}{2}kx_i^2 + KE_i = \tfrac{1}{2}kx_f^2 + KE_f$$

We define the **elastic potential energy** of the spring as

$$PE_{el} = \tfrac{1}{2}kx^2 \qquad\qquad (6.14)$$

and then we have

$$PE_{el,\,i} + KE_i = PE_{el,\,f} + KE_f \qquad\qquad (6.15)$$

The elastic potential energy of the spring plus the kinetic energy of the mass is the total energy of the spring-mass system. Equation 6.15, which is an applica-

Both PE_g and PE_{el} are potential energy because they have to do with an object's position.

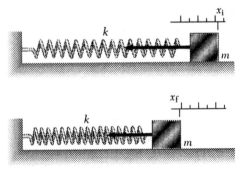

Figure 6.20
As a mass-spring system moves, energy changes from kinetic to elastic potential, but the sum of the two remains constant.

tion of the work-energy theorem, tells us that the total energy of the system remains constant; in other words, energy is conserved:

$$E_i = E_f = \text{constant}$$

As with gravitational potential energy, this conservation of energy is very useful and provides us with a new tool to use in understanding and predicting the motion of an object.

Example 6.10

A 0.500-kg mass is attached to a spring for which the spring constant is 7.2 N/m. The mass is pulled 12 cm from its equilibrium position and given an initial speed of 42 cm/s. What is its speed when it is 6.0 cm from equilibrium?

Reasoning

We shall calculate the initial energy—the sum of the elastic potential energy and the kinetic energy—and set that equal to the final energy. We can calculate the final elastic potential energy because we know the final stretch of the spring. Then we shall solve for the final kinetic energy of the mass and use that energy to determine the final speed.

Solution

The initial total energy is

$$E_i = PE_{el,\,i} + KE_i = \tfrac{1}{2}kx_i^2 + \tfrac{1}{2}mv_i^2$$
$$= \tfrac{1}{2}(7.2 \text{ N/m})(0.12 \text{ m})^2 + \tfrac{1}{2}(0.500 \text{ kg})(0.42 \text{ m/s})^2 = 0.096 \text{ J}$$

As the spring and mass continue their motion, the elastic potential energy and the kinetic energy change, but their sum always remains 0.096 J. For the final state, with $x_f = 6.0$ cm, we have

$$E_f = PE_{el,\,f} + KE_f = \tfrac{1}{2}Kx_f^2 + \tfrac{1}{2}mv_f^2$$
$$= \tfrac{1}{2}(7.2 \text{ N/m})(0.060 \text{ m})^2 + \tfrac{1}{2}(0.500 \text{ kg})(v_f)^2$$
$$E_f = 0.013 \text{ J} + (0.25 \text{ kg})v_f^2 = 0.096 \text{ J} = E_i$$
$$(0.25 \text{ kg})v_f^2 = 0.083 \text{ J}$$
$$v_f^2 = 0.33 \text{ m}^2/\text{s}^2$$
$$v_f = 0.58 \text{ m/s} = 58 \text{ cm/s}$$

We would not have been able to solve this problem using the kinematics equations we developed in Chapter 2 and Newton's second law. The force of the spring is not constant, so the acceleration of the mass is not constant. Therefore the tools and techniques we developed for uniform acceleration do not apply here.

Practice Exercise What is the speed of the mass as it passes through equilibrium?

Answer 62 cm/s

6.6 Energy Conservation

CONCEPT OVERVIEW
■ The total energy of a closed system remains constant.

As already mentioned several times in this chapter, energy is a **conserved** quantity. By *conserved* we mean that the total amount of energy in any closed system remains constant. (Again, a **closed system** is one that has no interaction with anything outside itself.) Let us now look at this concept of energy conservation in detail.

Consider a force acting on an object. The force is said to be a **conservative force** if all the work done against it is stored as energy and is available to do work later on. The gravitational force and the force exerted by a spring are two examples of conservative forces. The work done, for example, in lifting a box against the conservative force of gravity is stored as gravitational potential energy in the raised box and is available to do work. In contrast, friction is a **non-conservative force.** If you drag a wooden crate across a rough concrete floor, the work you do against friction is not available for recovery, as evidenced by the fact that the crate does not accelerate and return to its original position when released. Rather, the work you have done shows up as thermal energy; the crate and floor both get warmer.

For a conservative force F_c, we can define a change in potential energy ΔPE that is the negative of the work W_c done by the conservative force in moving from any initial position to a final position,

$$\Delta PE = PE_f - PE_i = -W_c \tag{6.16}$$

Notice that this is consistent with Equations 6.11 and 6.14, which define potential energy for the force of gravity and the elastic force of a spring.

If the net work done on an object is due to conservative forces, then Equation 6.16 and the work-energy theorem, Equation 6.7′, can be combined as follows:

$$W = KE_f - KE_i = W_c = -\Delta PE = PE_i - PE_f$$

$$PE_i - PE_f = KE_f - KE_i$$

$$PE_i + KE_i = PE_f + KE_f \tag{6.17}$$

The sum of potential energy plus kinetic energy is the **total mechanical energy** E of a system:

$$E = PE + KE \tag{6.18}$$

Then

$$E_i = E_f \tag{6.18'}$$

which means that, if only conservative forces are present, the total mechanical energy of a system remains constant; it is conserved. Energy may change from one form to another—from kinetic to potential or the other way around—but the total energy remains constant. That is what energy conservation means.

Figure 6.21 shows a child on a swing. As the child swings back and forth—once she is started and no one pushes from the outside—the forces acting on

At this water-powered ironworks, the potential energy of a mass of water at some height above the ironworks is changed to the kinetic energy of a mass of flowing water. As the mass of water flows downhill due to gravity, the potential energy of that water decreases as its kinetic energy increases. When the water hits the water wheel, the kinetic energy of the water is transferred to setting the wheel in motion. This motion of the wheel can then be transferred again to move other devices used in manufacturing iron products. If total mechanical energy is conserved in this system, why is a measurement of the *useful* energy available inside the ironworks always less than the energy of the water just as it hits the wheel? Where has the "missing" energy gone? (© Paul Bodine)

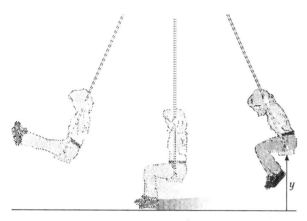

Stopped — no *KE*
High off the ground — high *PE*

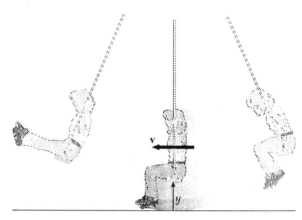

Moving fast — high *KE*
Small height — low *PE*

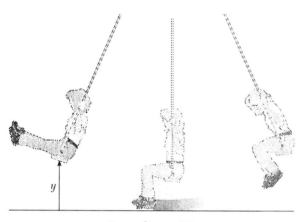

Stopped — no *KE*
Greater height — large *PE*

Figure 6.21
A playground swing illustrates conservation of energy.

her are the force of gravity and the force of the chains that hold the swing. Gravity is a conservative force. The force of the chains always acts perpendicular to her motion and so does no work. Therefore her total mechanical energy remains constant. At one extreme of the swing, she stops for just a moment. Being stopped, her kinetic energy is zero at this point. Here she is highest off the ground, so her potential energy is greatest. From that extreme she starts to move, gaining kinetic energy, and losing potential energy. At the bottom, she is moving fastest, so has maximum kinetic energy, and is at the lowest point, so has minimum potential energy. As she continues to move, she goes higher. Swinging higher increases her potential energy, meaning that her kinetic energy decreases and finally vanishes as she stops at the other extreme. Energy changes back and forth between potential and kinetic. This whole procedure repeats itself over and over and over again. Throughout, her total energy remains constant. Notice that this means she will swing as high on one side as on the other.

We have to plan well to conserve energy in its everyday sense in our everyday lives, but energy is conserved "automatically" in physics.

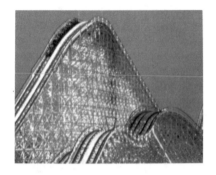

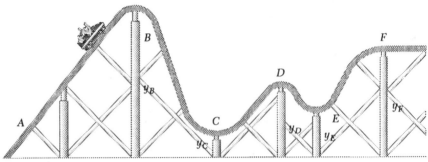

Figure 6.22
Energy conservation keeps a roller coaster going, such as Colossus at Magic Mountain in Valencia, California. (© *Comstock, Inc.*)

Roller coasters are delightful physics laboratories. Figure 6.22 illustrates the principle behind a roller coaster. As the cars come up the first hill from A to B, an external force is acting on them and work is therefore done to lift them to point B. From this point on, very little external work is done, and we can assume that the system is conservative. (Actually, of course, there is some friction, braking may occur at some places for safety, and, finally, work is done as brakes are applied to bring the cars to a stop.)

We start by writing the total energy E_B at position B:

$$E_B = KE_B + PE_B$$

At B the cars are barely moving, but they have great potential energy because they are at a great height y_B. We can write therefore

$$KE_B \approx 0$$
$$PE_B = mgy_B$$
$$E_B = mgy_B$$

Conservation of energy means that the energy remains constant throughout. Therefore, we can write

$$E = E_B = mgy_B$$

As the cars come down the hill from B to C, they lose potential energy (and height) while gaining kinetic energy (and speed). At C we can write

$$E_C = KE_C + PE_C$$
$$KE_C = \tfrac{1}{2} mv_C^2$$
$$PE_C = mgy_C$$
$$E_C = \tfrac{1}{2} mv_C^2 + mgy_C$$

Energy conservation lets us solve for, say, v_C:

$$E_C = E = mgy_B$$
$$\tfrac{1}{2} mv_C^2 + mgy_C = mgy_B$$
$$\tfrac{1}{2} mv_C^2 = mgy_B - mgy_C$$
$$\tfrac{1}{2} v_C^2 = gy_B - gy_C$$
$$v_C^2 = 2g(y_B - y_C)$$
$$v_C = \sqrt{2g(y_B - y_C)}$$

All the potential energy lost in descending from y_B to y_C goes into kinetic energy and an increased speed. We have found the velocity at C after the cars have descended from y_B to y_C, having started from rest at B.

As the cars go from C to D, kinetic energy is turned into potential energy. The cars go more slowly as they go higher. At D they have less potential energy than they did at B and more kinetic energy (they are going faster than at B). They speed up as they go on to E, turning potential energy into kinetic energy once more. As they go on to F, kinetic energy is once again turned into potential energy. At F the potential energy is greater than at D, so the kinetic energy is less than at D. The cars travel more slowly at F than they do at D.

Problem-Solving Tip

■ To use energy conservation, find the total energy—kinetic plus potential—for one position of the system and set that value equal to the total energy for some other position of the system.

Example 6.11

Return to the roller coaster in Figure 6.22. If point B is 25 m above the ground and point C is 3.0 m above the ground, find the speed of the roller coaster as it reaches C.

Reasoning

Notice that we cannot solve this problem using Newton's second law. We have solved similar problems with a car coming down a *straight* inclined plane because in that case we either knew or could find the constant acceleration. This roller coaster track is not straight, and we do not even know—or need to know—the horizontal distance between B and C.

To use energy conservation, we must find the total energy at some point. At B, as the ride begins, the cars move so slowly that we can consider them as starting from rest, $v_B = 0$. The mass of the cars is not given, so we write it symbolically as M and expect to get rid of it later. (You should recognize this as a standard technique by now.)

We shall determine the energy at two points in the roller coaster's run—B and C. Neglecting friction, the only forces exerted by the track are perpendicular to the cars' motion. Therefore no work is done on the cars by the track, and energy should be conserved.

Solution

At B,

$$KE_B = 0$$

$$PE_B = Mgy_B = M(9.8 \text{ m/s}^2)(25 \text{ m}) = M(245 \text{ m}^2/\text{s}^2)$$

$$E = E_B = M(245 \text{ m}^2/\text{s}^2)$$

At C,

$$KE_C = \tfrac{1}{2} M v_C{}^2$$
$$PE_C = Mgy_C = M(9.8 \text{ m/s}^2)(3.0 \text{ m}) = M(29 \text{ m}^2/\text{s}^2)$$
$$E = E_C = \tfrac{1}{2} M v_C{}^2 + M(29 \text{ m}^2/\text{s}^2)$$

Energy conservation means that the energy at B is the same as the energy at C. We therefore set these quantities equal to each other and solve for v_C:

$$E = E_C = E_B$$
$$E_C = \tfrac{1}{2} M v_C{}^2 + M(29 \text{ m}^2/\text{s}^2) = M(245 \text{ m}^2/\text{s}^2) = E_B$$
$$\tfrac{1}{2} M v_C{}^2 + M(29 \text{ m}^2/\text{s}^2) = M(245 \text{ m}^2/\text{s}^2)$$
$$\tfrac{1}{2} M v_C{}^2 = M(245 \text{ m}^2/\text{s}^2) - M(29 \text{ m}^2/\text{s}^2)$$
$$\tfrac{1}{2} v_C{}^2 = 216 \text{ m}^2/\text{s}^2$$
$$v_C = 21 \text{ m/s} = 47 \text{ mi/h}$$

Problem-Solving Tip

— Careful simplification of a complex system may be necessary to allow a solution.

Example 6.12

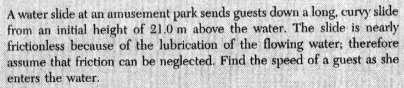

A water slide at an amusement park sends guests down a long, curvy slide from an initial height of 21.0 m above the water. The slide is nearly frictionless because of the lubrication of the flowing water; therefore assume that friction can be neglected. Find the speed of a guest as she enters the water.

Reasoning

At the top of the slide, the guest's kinetic energy is zero and her potential energy is $Mg(21.0 \text{ m})$. At the bottom, her kinetic energy is $\tfrac{1}{2} M v^2$ and her potential energy is zero. Since we expect her total energy to be conserved, we shall set the energy at the top equal to the energy at the bottom and solve for v. Notice that her mass has been included symbolically as M since we do not have a numerical value.

Solution

At the top,

$$KE_i = 0$$
$$PE_i = mgy = M(9.80 \text{ m/s}^2)(21.0 \text{ m}) = M(206 \text{ m}^2/\text{s}^2)$$
$$E_i = KE_i + PE_i = M(206 \text{ m}^2/\text{s}^2)$$

As the guest descends—even if there are hills or curves—her potential energy decreases while her kinetic energy increases and her total energy remains constant. At the bottom, as she goes into the water,

$$PE_f = Mgy = Mg(0) = 0$$
$$KE_f = \tfrac{1}{2} M v^2$$
$$E_f = KE_f + PE_f = \tfrac{1}{2} M v^2$$

A ride down a water slide is nearly frictionless. *(© Mike Valeri/FPG International Corporation)*

From energy conservation (since we have ignored friction), we know that

$$E_f = E_i$$

$$\tfrac{1}{2}Mv^2 = M(206 \text{ m}^2/\text{s}^2)$$

$$v^2 = 412 \text{ m}^2/\text{s}^2$$

$$v = 20.3 \text{ m/s} = 45.4 \text{ mi/h}$$

That is very fast!

Practice Exercise   If the height of the waterslide were 25.0 m, what would the speed of the guest be?

Answer   22.1 m/s

To understand complicated situations or complex systems, physicists often simplify them. For instance, it is reasonable to ignore friction in 6.12. The friction in a water slide is small, but it is not quite zero. (Adventurous sliders will note a difference in the thrill of the ride depending on whether they wear a cotton swim suit or a nylon swim suit, or longer swim trunks compared to briefer swim wear; these differences are clear evidence that friction is not totally absent.) Simplifications such as neglecting friction may be necessary for understanding and for a numerical solution. Remember, though, that the solution is only as accurate as the simplification. The speed 45.4 mi/h we calculated in the example is the speed only if there really is no friction. For a real person in the real world, the speed may be 40 mi/h or 43 or 38.

Problem-Solving Tip

- Once the total energy of a system is known for one time, that value can be used as the total energy of the system for all other times.

Example 6.13 _____

What is the minimum height h of the first hill for a roller coaster that does a loop-the-loop of radius r, as sketched in Figure 6.23, such that the car makes it through the loop and does not fall off?

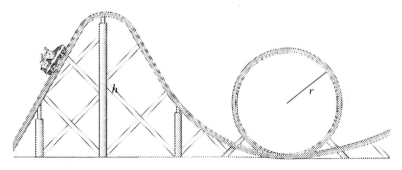

Figure 6.23
What minimum height h is required? The Scorpion loop-the-loop roller coaster.
(Courtesy of Busch Gardens, Tampa, Florida)

Reasoning

This example combines ideas we have discussed at different times. First we must find the minimum speed the car must have at the top of the loop to keep it from falling off the track; this calculation involves uniform circular motion. Then we can use energy conservation to find the height h of the hill that will provide that speed.

Let the speed at the top of the loop be v. Since the car is moving in a circle, we know that the net force on it must be the centripetal force mv^2/r (Equation 5.2). For the minimum speed v, the normal force of the track will just vanish. If the track exerts no normal force on the car, then the net force is only the gravitation force $w = mg$. Knowing this much, we can solve for the speed v.

Remember, v is the speed at the top of the loop, a distance $2r$ above the bottom. To determine what height h is necessary to provide that speed at that height, we use energy conservation.

Solution

At the top of the loop,

$$F_c = \frac{mv^2}{r} = mg = F_{net}$$
$$v^2 = gr$$
$$v = \sqrt{gr}$$

What initial height h will provide this speed when the car is at a height of $2r$? The potential and kinetic energies for the car at the top of the loop are

$$PE = mgy = mg(2r) = 2mgr$$
$$KE = \tfrac{1}{2} mv^2 = \tfrac{1}{2} m \left(\sqrt{gr}\right)^2 = \tfrac{1}{2} mgr$$
$$E = PE + KE = 2mgr + \tfrac{1}{2} mgr = (5/2)mgr$$

At the top of the hill, the car is hardly moving, so we consider its kinetic energy to be zero. The potential and kinetic energies for the car at the top of the hill are

$$PE = mgh$$
$$KE \approx 0$$
$$E = PE + KE = mgh$$

Now set the two expressions for total energy equal to each other since energy is conserved for a roller coaster:

$$mgh = (5/2)mgr$$
$$h = (5/2)r$$

This is the minimum height necessary. If the initial hill is lower than this value, the car's speed at the top of the loop will be less than necessary and the car will fall off the track (or need additional means of support). A higher hill will give the car a speed at the top of the loop faster than the minimum we calculated. Of course, a designer would want a speed somewhat faster than minimum in order to compensate for any possible energy losses due to friction.

Example 6.14

Figure 6.24 is a sketch of the "Convincer," a device used to demonstrate the importance of wearing seatbelts. A passenger slides down a track, reaching a speed of 25 km/h at the bottom and then stopping abruptly by running into a heavy frame. A seat/shoulder belt prevents injury. In order to reach a final speed of 25 km/h, what must the initial elevation of the passenger be?

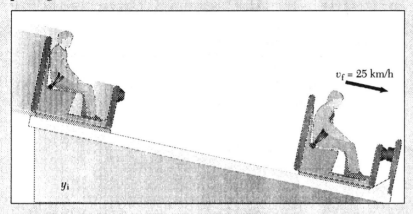

$v_f = 25$ km/h

y_i

Figure 6.24
The "Convincer," a device that shows the importance of wearing seatbelts.

Reasoning

Since the passenger starts from rest, his initial kinetic energy is zero. We evaluate the total energy at the beginning and at the end. Because energy is conserved, these two values are the same. Knowing the total energy, we can solve for the height y_i.

Solution

As in previous situations, we shall write the passenger's mass as M and expect that it will be eliminated from the equations.

$$KE_i = \tfrac{1}{2} M v_i^2 = 0$$
$$PE_i = M g y_i$$
$$E = E_i = M g y_i$$
$$KE_f = \tfrac{1}{2} M v_f^2$$
$$PE_f = M g y_f$$
$$E = E_f = \tfrac{1}{2} M v_f^2 + M g y_f$$

All we are interested in is the height of y_i above y_f, so we can take $y_f = 0$. We know the final velocity in kilometers per hour, and we know the acceleration due to gravity is 9.8 m/s². Therefore it will be helpful if we convert this velocity to meters per second:

$$v_f = 25 \, \frac{\text{km}}{\text{h}} \left(\frac{1 \, \text{h}}{3600 \, \text{s}} \right) \left(\frac{1000 \, \text{m}}{1 \, \text{km}} \right) = 6.9 \, \text{m/s}$$

Now, set the two expressions for the energy equal:

$$E = Mgy_i = \tfrac{1}{2} Mv_f^2 + Mgy_f$$

$$Mgy_i = \tfrac{1}{2} Mv_f^2 + Mgy_f = \tfrac{1}{2} Mv_f^2$$

$$gy_i = \tfrac{1}{2} v_f^2$$

$$y_i = \tfrac{1}{2} \frac{v_f^2}{g} = \frac{(6.9 \text{ m/s})^2}{2(9.8 \text{ m/s}^2)} = 2.4 \text{ m}$$

6.7 Nonconservative Forces

CONCEPT OVERVIEW

- In the presence of nonconservative forces, the sum of kinetic and potential energies is not constant. When thermal energy is included, the total energy remains constant even in these systems.

For a conservative force, like gravity, the amount of work required to move an object from some initial position to some final position is independent of the path taken; the work done depends only upon the initial and final positions. This is illustrated in Figure 6.25, where a block is shown being moved along three paths. Along each path the work done is mgh, where h is the vertical distance between the initial and final positions.

In contrast, Figure 6.26 is an overhead view of a wooden crate being pushed along a rough concrete floor. The work done in pushing the crate from some initial position to some final position depends drastically on the path taken. The work done is work against friction, and friction is a nonconservative force. The direction of the friction force is opposite the direction of the velocity. That means friction is a velocity-dependent force. All velocity-dependent forces are nonconservative. (Do not be misled by the seeming similarity of Figures 6.25 and 6.26. The former is a side view and depicts changes in height. The latter is an overhead view and depicts three pathways all at the same level, on the concrete floor.)

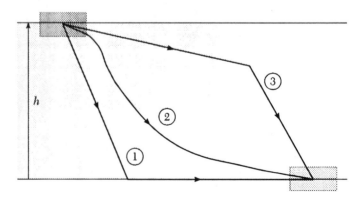

Figure 6.25
For a conservative force, the work done depends only on the initial and final positions.

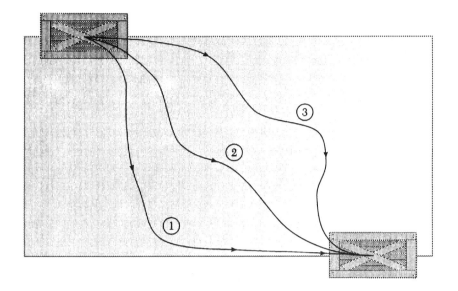

Figure 6.26
How much work is done in moving the crate depends on the path taken because the nonconservative friction force is involved.

We can still use the work-energy theorem with nonconservative forces. If both conservative and nonconservative forces are present, the total work W done on a system is the sum of the work due to conservative forces W_c and the work due to nonconservative forces W_{nc}:

$$W = KE_f - KE_i = W_c + W_{nc}$$

Since we know from Equation 6.16 that $W_c = -(PE_f - PE_i)$, we can substitute for W_c:

$$W = W_{nc} + PE_i - PE_f = KE_f - KE_i$$

$$W_{nc} = (PE_f + KE_f) - (PE_i + KE_i)$$

$$W_{nc} = E_f - E_i = \Delta E \qquad (6.19)$$

Work done by nonconservative forces changes the total mechanical energy of the system. This work usually appears in the form of heat, which is *thermal energy*. As the brakes of your car change the car's kinetic energy, the brake disks or drums become very hot. When nonconservative forces are involved, the total energy remains constant only if we look closely enough and count all forms of energy, including thermal.

Nonconservative forces can complicate conservation-of-energy calculations, but when we learn how to measure amounts of heat in Chapter 12, the complication will disappear.

*6.8 More on Energy

CONCEPT OVERVIEW
- Energy can take many forms; from $E = mc^2$ we find that mass and energy are equivalent.

Figure 6.27 illustrates examples of potential energy. The Sun exerts a force on Earth, so work is done as Earth moves away from the Sun. That this work is done means the potential energy of Earth with respect to the Sun increases as

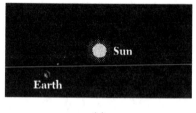

(a)

(b)

(c)

Figure 6.27

Examples of potential energy. (a) Because its orbit is an ellipse, Earth has the most potential energy when it is farthest from the Sun (a point known as aphelion). (b) Metal objects are attracted to permanent magnets; this attractive force provides the potential energy. *(PASCO Scientific)* (c) A pole vaulter has both kinetic energy and potential energy. *(© Harold E. Edgerton. Courtesy of Palm Press, Inc.)*

Earth moves away from the Sun. It requires work to pull a bar away from a magnet. Therefore there is potential energy as a result of the attractive force between bar and magnet. A pole vaulter stores elastic potential energy in his bent pole, energy that later helps him gain additional height and gravitational potential energy.

Energy is associated with matter through Einstein's mass-energy relationship,

$$E = mc^2 \qquad\qquad (6.20)$$

which states that the energy concentrated in an amount of matter having mass m is equal to that mass multiplied by the square of c, the speed of light. The speed of light is enormous:

$$c = 3.00 \times 10^8 \text{ m/s}$$

Example 6.15

The eraser on the end of a pencil has a mass of about 1 g. How much energy is concentrated in it?

Reasoning

This is an application of Equation 6.20,

Solution

$$E = mc^2 = (0.001 \text{ kg})(3 \times 10^8 \text{ m/s})^2$$

$$= 9 \times 10^{13} \frac{\text{kg} \cdot \text{m}^2}{\text{s}^2} = 9 \times 10^{13} \text{ J}$$

And that is an enormous amount of energy!

In nuclear reactions, some of the mass of the fuel is converted to energy, and the reactions that take place in nuclear power plants are examples of mass-energy equivalence. Chemical energy—released when, say, oxygen and hydrogen are combined to form water—also turns mass into energy, but the mass change for chemical reactions is so very small that it is usually ignored. All energy changes have corresponding mass changes, but with the conversion factor of c^2 connecting the two (in $E = mc^2$), the mass change is often difficult or impossible to measure because it is so small.

*6.9 Escape Speed

CONCEPT OVERVIEW

■ Escape speed is the speed an object must have in order to escape the effects of gravity.

We derived the equation $PE_g = mgy$ for the case where the force of gravity is $F_g = mg$. That is the case near Earth's surface, and we used this potential energy in understanding how a roller coaster works and how fast a book is going when it lands on the floor. But what is the gravitational potential energy of a communications satellite in Earth orbit or an interplanetary space probe as it approaches the orbit of Uranus? For such cases, the gravitational force clearly is *not* $F_g = mg$; rather, it is given by Newton's law of universal gravitation $F_g = GMm/r^2$ (Equation 5.6).

Gravitational potential energy is the amount of work required to move two objects from an initial separation r_i to a final separation r_f, as sketched in Figure 6.28. For a variable force, the work done is the area under the force-distance curve, as sketched in Figure 6.9. We could carefully construct such a graph every time we wanted to calculate a potential energy. Instead, we shall derive an equation that we can use in an approach that is far less tedious.

Work is force times distance, but what value of the force should we use as a mass m, initially a distance r_i away from a mass M, moves to a distance r_f away? Let us move from r_i to r_f in a series of very small increments, from r_i to r_1, from r_1 to r_2, and so on until r_f is finally reached; this is sketched in Figure

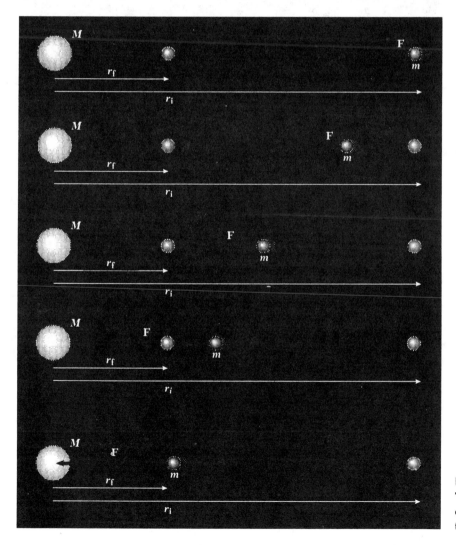

Figure 6.28
The change in gravitational potential energy is the work required to move the mass m from r_i to r_f.

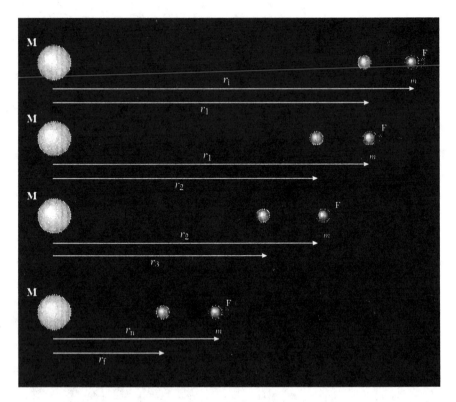

Figure 6.29
In calculating the work done in moving from r_i to r_f, we break the path up into several smaller paths—from r_i to r_1, from r_1 to r_2, and so on, finally from r_n to r_f.

6.29. During each small step, the distance does not change very much, so we can use a single value for r as an approximation. For the move from r_i to r_1, we use

$$\frac{1}{r^2} = \frac{1}{r_i r_1}$$

for the distance-squared term in Equation 5.6, with similar approximations for all the other steps. The work done in moving from r_i to r_1 is

$$\Delta W_{i1} = F_{i1}s_{i1} = G\frac{Mm}{r_i r_1}(r_1 - r_i) = GMm\left(\frac{1}{r_i} - \frac{1}{r_1}\right)$$

and the work done during each increment from each position to the next is given by a similar expression:

$$\Delta W_{12} = F_{12}s_{12} = G\frac{Mm}{r_1 r_2}(r_2 - r_1) = GMm\left(\frac{1}{r_1} - \frac{1}{r_2}\right)$$

$$\Delta W_{23} = F_{23}s_{23} = G\frac{Mm}{r_2 r_3}(r_3 - r_2) = GMm\left(\frac{1}{r_2} - \frac{1}{r_3}\right)$$

$$\cdots$$

$$\Delta W_{nf} = F_{nf}s_{nf} = G\frac{Mm}{r_n r_f}(r_f - r_n) = GMm\left(\frac{1}{r_n} - \frac{1}{r_f}\right)$$

The change in the potential energy is the total work required to change the separation of the masses from r_i to r_f, which is the sum of the work done in all the little increments:

$$\Delta PE_{if} = \Delta W_{if} = \Delta W_{i1} + \Delta W_{12} + \Delta W_{23} + \cdots + \Delta W_{nf}$$

Notice that part of each term cancels part of the one before; only a part of the first and last terms remains:

$$\Delta PE_{if} = GMm\left(\frac{1}{r_i} - \frac{1}{r_f}\right) \qquad \textbf{(6.21)}$$

This is comparable to Equation 6.9, which gives the work done by the force of gravity near Earth's surface as an object moves from height y_i to height y_f. Equations 6.9 and 6.21 look quite different from each other, but they describe the same thing. Now, just as we did earlier, it is convenient to talk about the potential energy relative to some reference point at which the potential energy is defined as zero. For this general form of gravitational potential energy, we shall take the condition of zero potential energy to be when the two masses are infinitely far apart.

In Figure 6.30 a mass m is shown a distance r from a mass M. What is the potential energy of this system? In asking that question, we are implying a more complete question: "What is the potential energy of this system *relative to* the potential energy when the two masses are infinitely far away from each other?" Equation 6.21 provides the change in potential energy of the final condition relative to the initial. In our present case, then, the initial separation distance is infinity and the final separation distance is r:

$$PE = GMm\left[\frac{1}{\infty} - \frac{1}{r}\right]$$

$$PE = -G\frac{Mm}{r} \qquad \textbf{(6.22)}$$

Notice that the gravitational potential energy is always negative as defined by this equation. We have already encountered negative potential energy in the simpler case of $PE = mgy$ near Earth's surface. In that case, a negative potential energy meant that work must be done to restore the system to the condition defined as having zero potential energy. We had to do work in lifting the book from the floor, where its potential energy was negative, to get it back to the table, where its potential energy was zero. We have now defined zero potential energy to be the condition when the two masses are infinitely far away. At a

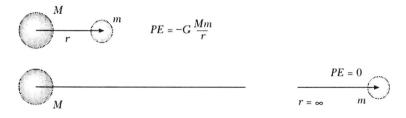

Figure 6.30
Gravitational potential energy can be measured by taking it to be zero when the masses are infinitely far away from each other (r_i = infinity).

distance r apart, there is an attractive force pulling the masses together, and we must do a positive amount of work on them to pull them farther and farther away until they are finally infinitely far apart.

Example 6.16

What is the gravitational potential energy of a 300-kg communications satellite in a geosynchronous orbit?

Reasoning

What we are after here is the potential energy of the satellite-Earth *system*. The satellite has potential energy only because of the gravitational attraction of Earth.

In Section 5.7 on satellite orbits, we found the radius of a geosynchronous orbit to be 4.22×10^7 m. We can use that distance in Equation 6.22 to calculate the potential energy.

Solution

$$PE = -G\frac{Mm}{r}$$

$$= -6.67 \times 10^{-11} \frac{\text{N·m}^2}{\text{kg}^2} \left[\frac{(5.98 \times 10^{24}\text{kg})(300 \text{ kg})}{4.22 \times 10^7 \text{m}} \right]$$

$$= -2.84 \times 10^9 \text{ J}$$

This result means that 2.84×10^9 J of work must be done on the satellite to move it infinitely far away from Earth.

A communication satellite is launched into a geosynchronous orbit from the Space Shuttle. *(NASA)*

A satellite is usually put into orbit by a multistage rocket that does work on the satellite over some large distance and reasonably large time. A space probe sent far, far away from Earth would likewise be sent by a rocket that exerts a force and does work over great distances and large times. Suppose, instead, that we wanted to launch a space probe from Earth's surface, with no rocket involved. What speed would it need to entirely escape from Earth, go infinitely far away, and never return? This speed is known as the **escape speed.**

With the gravitational force varying as the separation distance increases and the infinite final distance, this question would be impossible to handle with only $F = ma$ and our constant-acceleration kinematics equations. However, energy conservation is a useful tool we can apply to this situation. The initial and final conditions are sketched in Figure 6.31. Initially, the space probe, of mass m, is at Earth's surface ($r = R_E$), so its initial potential energy is

$$PE_i = -G\frac{Mm}{R_E}$$

Its initial velocity is v_i, so its initial kinetic energy is

$$KE_i = \tfrac{1}{2}mv_i^2$$

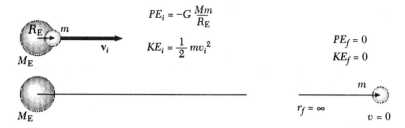

Figure 6.31
Escape speed. An object at Earth's surface with initial velocity $v_i = v_{esc}$ will escape from Earth's gravitational pull.

The initial total energy is the sum of these:

$$E_i = PE_i + KE_i = -G\frac{Mm}{R_E} + \tfrac{1}{2}mv_i^2$$

When the probe is infinitely far from Earth so that $r_f = \infty$, its final potential energy is zero. We can take its final speed to be zero also, so that $KE_f = 0$. Thus the final total energy is zero. This will provide us with the minimum value of v_i necessary just to get the probe infinitely far away from Earth. (Firing it with a greater velocity means its velocity infinitely far away will also be greater, and it will continue to drift forever.) We can apply energy conservation and solve for v_i, the escape speed from Earth's surface:

$$E_i = -G\frac{Mm}{R_E} + \tfrac{1}{2}mv_i^2 = 0 = E_f$$

$$\tfrac{1}{2}mv_i^2 = G\frac{Mm}{R_E}$$

$$v_i^2 = \frac{2GM}{R_E}$$

$$v_i = \sqrt{\frac{2GM}{R_E}}$$

$$= \sqrt{\frac{2(6.67 \times 10^{-11} \text{N·m}^2/\text{kg}^2)(5.98 \times 10^{24} \text{ kg})}{6.38 \times 10^6 \text{m}}}$$

$$= 11\,200 \text{ m/s} = 40\,300 \text{ km/h} = 25\,000 \text{ mi/h} \qquad \textbf{(6.23)}$$

Anything near Earth's surface moving with a speed of 40 300 km/h or greater will escape Earth's gravity. This idea has interesting consequences in areas quite different from satellites or space probes. Lightweight atoms in our upper atmosphere sometimes move with a speed greater than escape speed; these atoms leak from our atmosphere into space. Escape speed is much lower on the Moon or on Mars than on Earth. Most gaseous atoms around those two bodies have speeds greater than the local escape speed, and consequently neither the Moon nor Mars has any appreciable atmosphere.

Example 6.17

What is the escape speed for a deep-space probe "docked" in an orbit having a radius of 20 000 km around Earth? That is, what speed given to the probe while it is in a 20 000-km orbit will make it go infinitely far from Earth?

Reasoning

Equation 6.23 and the discussion preceding it assumed we were starting at Earth's surface, with $r = R_E$. The point of the discussion was that, at infinite separation from Earth, the probe would have zero total energy if it were drifting very slowly. Then it must have zero total energy initially as the probe leaves Earth's surface.

We apply that same reasoning here, but this time we are starting from a "docking orbit" with $r = 20\,000$ km. We can use Equation 6.23 directly if we use $r = 20\,000$ km instead of R_E.

Solution

$$v_i = \sqrt{\frac{2GM}{r}} = \sqrt{\frac{2(6.67 \times 10^{-11} \text{ N·m}^2/\text{kg}^2)(5.98 \times 10^{24} \text{ kg})}{2.00 \times 10^7 \text{m}}}$$

$$= 6320 \text{ m/s} = 22\,800 \text{ km/h} = 14\,100 \text{ mi/h}$$

SUMMARY

When a force **F** acts on an object, as the object moves through a displacement **s**, the work done on the object is

$$W = Fs \cos \theta \qquad \textbf{(6.1)}$$

where θ is the angle between the force vector and the displacement vector. Work is a scalar quantity and has units of joules (1 J = 1 N·m).

The force exerted by a spring is given by *Hooke's law*:

$$F_s = -kx \qquad \textbf{(6.3)}$$

where k is the *spring constant* and x is the spring's displacement from its equilibrium length. The work done in stretching or compressing a spring is

$$W_s = \tfrac{1}{2}kx^2 \qquad \textbf{(6.4)}$$

The rate at which work is done is *power*:

$$P = \frac{\Delta W}{\Delta t} \qquad \textbf{(6.5)}$$

Power is measured in watts (1 W = 1 J/s) or kilowatts in the SI and in horsepower in the British system.

Work must be done to cause an object to move, and any moving object can do work and therefore possesses *kinetic energy*, or energy of motion. An object of mass m moving with speed v possesses kinetic energy

$$KE = \tfrac{1}{2}mv^2 \qquad \textbf{(6.8)}$$

The work-energy theorem is a statement that the work done on an object equals the change in kinetic energy of the object:

$$W = KE_f - KE_i \qquad \textbf{(6.7')}$$

Lifting an object requires that work be done on the object. Once it has been lifted, the object can do work or increase its kinetic energy as it falls to a lower height and therefore possesses *gravitational potential energy*. An object of mass m that has been lifted through a height y has had its potential energy increased by

$$PE_g = mgy \qquad (6.11)$$

A compressed or stretched spring possesses *elastic potential energy*

$$PE_{el} = \tfrac{1}{2}kx^2 \qquad (6.14)$$

When only conservative forces act on a closed system, the total mechanical energy of the system remains constant. When nonconservative forces act on a system, some mechanical energy is converted to thermal energy.

Mass and energy are related according to Einstein's famous equation

$$E = mc^2 \qquad (6.20)$$

where $c = 3.00 \times 10^8$ m/s is the speed of light.

To escape from the pull of Earth's gravitational attraction, an object must be moving at a minimum speed called the *escape speed*:

$$v = \sqrt{2GM/r} \qquad (6.23)$$

where G is the gravitational constant, M is the mass of Earth, and r is the initial distance between the object and the center of Earth.

QUESTIONS

1. How does the meaning of the word *work* as used in physics differ from the common, ordinary usage?
2. How much work is done by a centripetal force as an object revolves once around a circular orbit?
3. Explain how you can get tired trying unsuccessfully to lift a baby grand piano and yet have a physicist say you have done no work on the piano.
4. How can a string supporting a pendulum exert a force over several complete cycles and yet do no work?
5. If an advertisement for a sports car states that the power is 120 kW, does that imply the car has an electric engine?
6. How can power be measured in both kilowatts and horsepower?
7. If someone quoted you the power of a light bulb in horsepower, how could you change that to watts?
8. Since power companies bill for usage in kilowatt-hours rather than kilowatts, would a power company ever have any concern about the total power used by its customers?
9. Work can be either positive or negative. Is that true for kinetic energy as well?
10. Can a bullet ever have as much kinetic energy as a semitrailer truck?
11. Why will increasing your car's speed by 10 percent require about 20 percent greater stopping distance?
12. Can gravitational potential energy ever be negative?
13. Describe the energy transformations that occur as someone bounces on a pogo stick.
14. If the height of a playground slide is kept constant, will it make any difference in the final speed of children playing on it, how long the slide is, or whether it is straight or has bumps (Fig. 6.32)? Answer this question twice—once for friction present and once for friction absent.

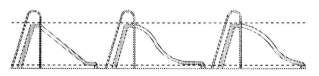

Figure 6.32   Question 14.
On which slide will a child have the fastest final speed?

15. You watch a television commercial for a "super-duperball" that bounces twice as high as the height from which it is dropped. Should you send your money to buy one?

16. The force between Earth and our Moon is obviously very real. How then can we say that the gravitational potential energy between the two is negative?

17. Explain the connection between escape speed and the atmosphere surrounding a planet.

18. In common usage, *energy conservation* means more efficient design of cars and home insulation to use less energy, but this term has a much different meaning in physics. What does it mean in physics?

19. Give an example of energy conservation using more than just mechanical energy—include chemical or electrical energy. Be careful with your definition of "the system" that conserves energy.

PROBLEMS

6.1 Work

6.1 A horizontal force of 80 N is applied to a 950-kg car for a distance of 28 m. How much work is done on the car?

6.2 A 2.0-kg block is pushed 2.3 m along a table by a horizontal force of 1.2 N. How much work has this force done on the block?

6.3 How much work must be done to push a thumbtack 0.6 cm into a bulletin board against an average opposing force of 3.5 N?

6.4 A wooden box with a weight of 350 N sits on a concrete floor. How much work is done if the box is moved at a slow and constant speed 2.5 m along the floor against a friction force of 100 N?

6.5 How much work is done by gravity on a 600-N skier who slides 500 m down a hill that makes an angle of 30° with the horizontal, as shown in Figure 6.33?

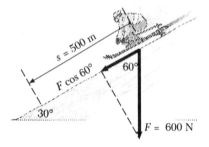

Figure 6.33
Problem 6.5. How much work is done by gravity as the skier goes down the hill?

6.6 A 35-kg shipping crate sits on a concrete floor. The coefficient of friction between crate and floor is 0.45. What horizontal force is necessary to move the crate at a constant speed? How much work is done if it is moved by this force for a distance of 3.5 m along the floor?

6.7 How much work is done by an engine that pulls a train of mass 5.0×10^6 kg for a distance of 2.0 km over a level track if the effective coefficient of friction is 0.015?

6.8 A girl moves a wagon 40 m along the street by pulling on a rope with a force of 60 N. The rope is directed 20° above the horizontal. How much work does she do?

6.9 A horizontal force of 10 N is used to push a box up a plane that is 8.0 m long and inclined at an angle of 12° from the horizontal. How much work is done?

6.10 A water skier is pulled along a canal by a rope attached to a jeep driving along the bank. The tension in the rope is 500 N, and the rope makes an angle of 15° with the skier's direction of motion. How much work is done as the skier travels 400 m?

6.11 A lawn mower is pushed with a 120-N force directed 30° below the horizontal. How much work is done as the mower cuts a strip of grass 30 m long?

6.12 Two people tow a barge 500 m at a constant speed of 1.52 m/s (Fig. 6.34). Each person exerts a force of 75 N on the barge from

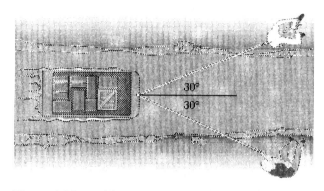

Figure 6.34   Problem 6.12.

opposite banks of a canal. The forces are directed at an angle of 30° on either side of the line of motion of the barge.   (a) How much work is done by each person?
(b) How much work is done by the barge on the water?

6.13   An 85-kg crate is pushed up a ramp inclined 10° above the horizontal by a force parallel to the ramp. The coefficient of friction between crate and ramp is 0.15. What force is necessary to keep the crate moving at constant speed? How much work is done in moving the crate 2.8 m?

○6.14   In playing shuffleboard, a long wooden stick is used to give a push to a 0.3-kg puck and cause it to move across the floor. A player pushes with a force of 25 N along the stick, inclined 40° to the ground. The coefficient of friction between puck and floor is 0.10.
(a) How much work is done on the puck by the stick as the puck moves 0.7 m?
(b) What will the puck's speed be after it has moved that 0.7 m, assuming it started at rest?   (c) After it has moved that 0.7 m the puck leaves the stick and slides across the floor. How far will it slide?

6.2 Power

6.15   A 20-kW motor is used to hoist a 1000-kg bucket of concrete to the eighth floor of a building, a height of 45 m. If the system is 50 percent efficient (that is, half the power is lost to friction), how much time is required to lift the bucket?

6.16   A 10-hp motor is used to hoist a 1000-kg bucket of concrete to the tenth floor of a building, a height of 55 m. If the system is 60 percent efficient (that is, 40 percent of the power is lost to friction), how much time is required to lift the bucket?

6.17   How fast can a 25-kW motor raise a 1500-kg load?

6.18   A physics student whose mass is 75.0 kg runs up the stairs of the physics laboratory building, a height of 28.5 m, in 1 min, 37 s. Determine his power in kilowatts and in horsepower.

6.19   An electric bill lists 1284 kW·h as the energy used for a month. If the cost is 11.5 ¢/kW·h, what is the total bill? How much energy, in joules, was used that month?

6.20   A 100-W light bulb in a closet is accidently left on for one month (24 h a day for 30 days). At a rate of 12.3¢/kW·h, what is the cost of this oversight?

6.21   A ventilating fan is operated by a 0.50-hp electric motor. How much does it cost to run the fan continuously for 30 days if the cost of electricity is 12.5¢/kW·h?

6.22   The output of a common student laboratory laser is about 250 mW. How much energy is delivered in 1.0 h?

6.23   What is the power output of a motor that raises a 2100-kg elevator car to the top of a 50-m building in 75 s?

6.24   A 1200-kg sports sedan accelerates from rest to 90 km/h in 8.25 s. What is the average force exerted? What distance is covered? What is the average power delivered to the wheels?

6.25   A 1500-kg sports car accelerates from rest to 100 km/h in 8.9 s. What is the average force exerted? What distance is covered? What is the average power delivered to the wheels?

6.26   A certain car needs an engine that provides 22 hp of power to the wheels in order to move along a level highway at 80 km/h. What total force of air resistance and rolling friction does the car experience?

6.27 Each of four jet engines on an airliner develops 7500 kW when cruising at 800 km/h. What is the thrust force of each engine?

6.28 Each of four jet engines on an airliner develops 6500 hp when cruising at 440 mi/h. What is the thrust force of each engine?

6.29 An anchor windlass on a ship has to lift a load of 10 000 N at a speed of 0.4 m/s. What power is necessary for the motor? Assume that 60 percent of the motor's power is available to lift the load.

6.30 What is the power of an engine that pulls a 500 000-kg train at a steady speed of 20 m/s along a horizontal track for which the effective coefficient of friction is 0.02?

○**6.31** An escalator is designed to carry shoppers from the basement to the ground floor 10 m above. It is to have a capacity of 80 shoppers per minute and assumes an average weight per shopper of 650 N. Find the necessary power, in kilowatts and in horsepower, of the motor if half its output is lost to friction.

○**6.32** A ski resort has a tow lift that pulls skiers 200 m up a 20° incline. The skiers are pulled at a speed of 2 m/s and at a rate of 300 skiers per hour. How far apart, on the average, are the skiers on the lift? How long is the lift? How many skiers are on it at one time? What motor power is necessary to run this tow system? Assume an average mass of 80 kg for each skier and 65 percent efficiency for the system.

○**6.33** A storage battery lights a 150-W bulb and also runs a motor that is 80 percent efficient. The useful power output of the motor is 200 W. How much electrical energy does the battery provide in 10 min?

○**6.34** What is the power of an engine that keeps a boat moving at a steady speed of 12 m/s if water resistance exerts on the boat a constant drag of 3200 N?

○**6.35** Energy is stored by pumping 2.00×10^6 kg of water into a reservoir that is a shallow lake 40.0 m above a river. Calculate the power of a pump that can fill this reservoir in 10.0 h.

●**6.36** The power of a boat's motor at a speed of 40.0 km/h is 22.0 kW. If the power is 15.0

percent greater while pulling a skier at the same speed, what is the tension in the tow rope?

6.3 Kinetic Energy

6.37 Determine the kinetic energy of a 3.8-kg cart traveling at a speed of 2.4 m/s.

6.38 What is the kinetic energy of a 1750-kg car traveling at 80 km/h?

6.39 What is the speed of a 0.2-kg air track glider if it has 0.8 J of kinetic energy?

6.40 What is the change in kinetic energy when a 1200-kg automobile accelerates from rest to 15 m/s? From 15 m/s to 30 m/s?

6.41 A young man pushes with a steady force of 60 N on a friend who is on (frictionless) roller blades. He pushes in the direction of his friend's motion. How far must he push in order to give the friend 300 J of kinetic energy?

6.42 How much work is required to accelerate a 1080-kg car from rest to 80 km/h? From rest to 100 km/h?

6.43 An 1100-kg sports sedan accelerates from rest to 90 km/h in 7.8 s. What is its final kinetic energy? What is the average power delivered to the wheels?

○**6.44** A 2.3-kg rabbit and a 10.0-kg hound have the same kinetic energy. If the hound has a speed of 3.50 m/s, what is the speed of the rabbit?

○❖**6.45** Below are data for eight air track gliders (Fig. 6.35) having different masses and speeds. Rank them in terms of their kinetic energies.

Glider	Mass (kg)	Speed (m/s)
A	0.225	0.85
B	0.225	0.90
C	0.250	0.35
D	0.250	0.50
E	0.267	0.65
F	0.315	0.82
G	0.450	0.45
H	0.450	0.52

Figure 6.35 Problem 6.45.

6.4 Gravitational Potential Energy

6.46 How much work is done in lifting a 2.0-kg book from the floor to a shelf 1.8 m high?

6.47 A 1200-kg elevator is lifted 50 m and then lowered 15 m. Calculate the work done in each move. Consider the elevator to be moving so slowly that kinetic energy can be neglected.

6.48 What is the change in gravitational potential energy as a 12-kg bucket is *(a)* raised 4 m? *(b)* lowered 4 m? *(c)* moved horizontally 4 m?

6.49 In the operation of a pile driver, a 400-kg mass is released from a height of 4 m above the pile to be driven. If the pile is driven 0.1 m into the ground with each blow of the mass, what is the average force exerted on the pile?

6.50 A skateboarder coasts up a hill that has a vertical rise of 0.75 m. If he starts with a speed of 5.9 m/s, with what speed does he arrive at the top?

6.51 A 6-m ladder weighs 240 N, and its center of mass is at its geometric center. What is the increase in its potential energy when it is raised from a horizontal position lying on the ground to a standing vertical position?

6.52 A student who weighs 700 N sleeps in a bed 42 cm above the floor. The student's center of mass is 90 cm from the bottoms of his feet. What is the increase in his potential energy when he gets up in the morning? Where does this energy come from?

6.53 When a girl who weighs 400 N walks up a hill that is 140 m long, her potential energy increases by 28 kJ. What is the vertical rise of the hill?

6.54 A 60-kg diver stands on a diving board 2 m above the water's surface over a diving well that is 3 m deep. Find the diver's potential energy with respect to the water's surface and with respect to the bottom of the well.

○**6.55** A ski slope drops at an angle of 20° and is 600 m long. Find the change in potential energy of a 70-kg skier who goes down this slope.

○**6.56** A 5000-kg cable car in San Francisco is pulled 400 m up a hill inclined 13° from the horizontal. Find the change in gravitational potential energy of the cable car.

○**6.57** A waterfall is 40 m high, and 3.5×10^4 kg of water flows over it each second. How much power does this waterfall produce? If 10 percent of this power can be converted to commercial electrical power, how many 100-W light bulbs could this light?

6.5 Spring Potential Energy

6.58 A spring with a spring constant of 280 N/m is compressed 0.30 m. How much work is required?

6.59 A spring with a spring constant of 480 N/m is compressed by doing 96 J of work on it. How far is it compressed?

6.60 A spring with a spring constant of 180 N/m is stretched by doing 36 J of work on it. How far is it stretched?

○**6.61** In loading a certain dart gun, the dart is pushed in 12 cm before the gun is cocked. A maximum force of 20 N is required to insert the dart. Find the spring constant of the spring in the gun and the total energy stored in the spring.

○ ❖ 6.62 Below are data for eight mass-spring systems (Fig. 6.36). For each case, find the speed of the mass as it passes through equilibrium and then rank the eight in order of increasing speed.

Case	Spring constant (N/m)	Mass (kg)	Initial position (m)	Initial velocity (m/s)
A	60	0.350	0.15	3.0
B	72	0.225	0.15	4.5
C	72	0.225	0.15	3.0
D	72	0.225	0.30	4.5
E	85	0.225	0.15	3.0
F	85	0.250	0.15	3.0
G	65	0.250	0.15	3.0
H	65	0.325	0.15	4.0

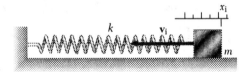

Figure 6.36 Problem 6.62.

6.6 Energy Conservation

6.63 A 2-kg ball is released from the edge of a 30-m canyon. What is its initial potential energy relative to the bottom of the canyon? What are its potential energy, kinetic energy, and speed (a) after it has fallen 10 m and (b) just before it strikes bottom?

6.64 A bowling ball is returned to the bowler by a mechanism that places the ball on a return ramp 100 cm above the alley level, as sketched in Figure 6.37. If the ball is given an initial speed of 0.6 m/s at point A, what is its speed when it reaches the bowler at point E? Ignore friction and the rotational kinetic energy of the ball.

Figure 6.37 Problem 6.64.

6.65 Running at 9.0 m/s, Tarzan grabs a vine hanging vertically from a tall tree. How high will he swing upward? Does the length of the vine enter into the calculation?

6.66 A practice machine for serving tennis balls is composed of a spring that has a spring constant of 800 N/m and is compressed 0.35 m. If no energy is lost, with what initial speed will 150-g tennis balls be launched by this device? If they are fired straight up, how high will they go?

6.67 A 0.3-kg pinecone falls from a branch 22 m above the ground. If air resistance is ignored, with what speed do you expect the cone to hit the ground?

6.68 A 0.5-kg rock is thrown straight upward. As it passes the eave of a house, 3 m after leaving your hand, its speed is 2 m/s. With what initial speed did it leave your hand?

○6.69 A Ping-Pong ball rolls off the edge of a table 1.0 m high. When the ball strikes the floor, its speed is 6 m/s. How fast was it moving when it left the table?

○6.70 A stone is thrown at 12 m/s from a cliff 20 m high. What is its speed just before it strikes the water below the cliff?

○6.71 A 30-kg sled is coasting 4 m/s over perfectly smooth, level ice. It enters a rough stretch of ice 20 m long in which the force of friction is 9 N. With what speed does the sled emerge from the rough stretch? (Notice that this is the same as Problem 4.55. With the tools of energy conservation, the solution is now far easier than it was then.)

○6.72 An arrow of mass 0.1 kg is fired directly upward by an archer who exerts an average force of 75 N on the bowstring in pulling it back 0.6 m. After it comes down, the arrow penetrates 20 cm vertically into the ground. Use the ideas of work and energy to determine (a) the arrow's speed as it leaves the bow, (b) how high it goes, and (c) the average upward force exerted by the ground on the arrow while stopping it.

○6.73 Indiana Jones swings on a hanging vine 10 m long that initially makes an angle of 37° with the vertical. He kicks off from his starting point with a speed of 12 m/s. What is his maximum speed at the bottom of his swing?

○6.74 The cart in Figure 6.38 approaches the hill with the minimal initial speed necessary to just get over the hill. Ignore friction and find (a) the initial speed of the cart and (b) its speed at the bottom of the hill on the far side.

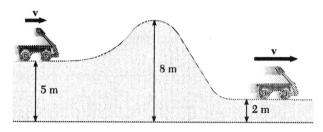

Figure 6.38 Problem 6.74.

●6.75 A ball is attached to a cord of length L that is initially held horizontal from a fixed support at P, as sketched in Figure 6.39. The ball is released from rest at point A. (a) Find its speed at point B, the bottom of the swing. A peg is fixed at point O, a distance y directly below point P. The cord hits the peg, and the ball swings upward in an arc as shown. (b) Find the ball's speed at C, where the cord between ball and peg is horizontal. (c) Find the ball's height (relative to its starting position) at D, the top of its swing after hitting the peg (assume $y > L/2$). (d) Find the ball's speed at D.

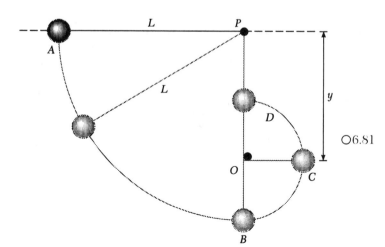

Figure 6.39 Problem 6.75.

●6.76 A car of mass m is moving with speed v along a level road. The coefficient of static friction between tires and road is μ_s. Use the ideas of work and energy to derive a formula for the minimum stopping distance without skidding in terms of v, μ_s, and g. Why is the coefficient of *static* friction appropriate here?

●6.77 A glider is on a frictionless air track that is inclined 10° from the horizontal. The glider is released 100 cm from a bumper at the end of the track and rebounds after losing 20 percent of its kinetic energy during the collision with the bumper. What is the glider's speed (a) just before hitting the bumper and (b) just after hitting the bumper? (c) How far from the bumper does the glider travel on the rebound?

6.7 Nonconservative Forces

○6.78 A skater weighing 500 N glides across the ice with an initial speed of 4.0 m/s. If there is a coefficient of friction of 0.115 between skates and ice, how far does she glide before coming to rest?

○6.79 A .22-caliber bullet of mass 0.0026 kg is fired into a wooden target at 320 m/s and penetrates 8 cm. What is the average force exerted on the bullet by the wood?

○6.80 In Example 6.12 we found that a person's speed at the bottom of a tall and thrilling water slide was about 20 m/s. After attaining this speed, the person is slowed by going into a long, shallow pool. The force exerted on the person by the water clearly depends upon the person's initial speed just as he or she enters the water. Assume the average of this force is one fourth the person's weight. What distance is required to bring the person to rest?

○6.81 A dolphin of mass 160 kg starts from rest and reaches a speed of 8 m/s by expending 8500 J of muscular energy. How much work was done to overcome the friction of the water?

*6.8 More on Energy

○6.82　The mass of a uranium nucleus is 3.952×10^{-25} kg. In a nuclear fission reaction, this nucleus breaks apart into many "fission fragments" and gives up 3.2×10^{-17} J of energy. What is the combined mass of all the fragments, expressed as a percentage of the original nuclear mass?

○6.83　The mass of a uranium nucleus is 3.952×10^{-25} kg. In a nuclear fission reaction, this nucleus breaks apart and gives up 3.2×10^{-17} J of energy. What mass of uranium is needed to fuel a 100-MW power station for one 30-day month?

*6.9 Escape Speed

6.84　What is the gravitational potential energy of the Moon-Earth system?

○6.85　What is the gravitational potential energy of Earth relative to the Sun?

○6.86　How much energy would have to be added to Mars to completely remove it from our solar system? Be sure to include its orbital kinetic energy in your calculation.

6.87　Consider a 1000-kg communications satellite in a geosynchronous orbit ($r = 42\,200$ km). (a) What is its gravitational potential energy relative to Earth? (b) What is its orbital kinetic energy? (c) What is its total energy? (d) How much energy must be added to it to remove it infinitely far from Earth? (e) What is its escape speed from its orbit?

○6.88　Calculate escape speed from the surface of Mars.

○6.89　Calculate escape speed from the surface of Jupiter.

○❖ 6.90　Calculate escape speed from the surface of the Moon and from the surface of each planet in our solar system. Rank them in order of increasing speed.

General Problems

○6.91　In playing shuffleboard, a long wooden stick is used to give a push to a 0.3-kg puck and cause it to move across the floor. The coefficient of friction between puck and floor is 0.15. A player pushes with a force along the stick, inclined at an angle θ to the ground. Find the minimum angle θ for which the player will not be able to move the puck at all.

○6.92　The pilot of the first successful human-powered heavier-than-air flight could generate an average of 0.32 kW during the flight. The aircraft covered 2.4 km at an average speed of 4.1 m/s. How much energy was expended during the flight?

○6.93　An attendant pushes a patient on a cart 20.7 m down a hall with a constant speed of 0.880 m/s, doing 2000 J of work on the cart. When the cart is released, it travels 0.480 m before coming to rest. Assuming the attendant exerts a horizontal force on the cart, determine the total weight of cart plus patient.

○6.94　A 5.00-g bullet travels at 400 m/s toward a wooden target that is 5.00 cm thick. The force of the wood on the bullet is 4.25 kN. What is the speed of the bullet as it emerges from the back of the target?

○6.95　A 2.0-kg ball is thrown from the top of the Science Building, 20 m above the parking lot below. The ball is thrown at an angle of 30° from the horizontal with an initial speed of 8 m/s. Use energy conservation to find the speed of the ball when it hits the parking lot. Does it make any difference in the answer if the initial velocity is 30° above or 30° below the horizontal?

○6.96　A 70-kg rock climber starting from rest slides down a 15° hill that is 50 m long. The climber arrives at the bottom with a speed of 3.0 m/s. How much thermal energy—shared between the surface of the hill and the climber's pants—has been generated in this slide?

PHYSICS AT WORK

This experiment is designed to measure, among other things, the force exerted on people in a car during a crash. When a car hits a solid object, different parts of the car (and its occupants) decelerate at different rates. The bumper is subjected to the greatest force, and the large forces can cause lots of damage. Modern automobiles are designed with "crumple zones" between the front end and the passenger compartment, so the seats decelerate more slowly than the bumper—but still quickly enough to launch you through the windshield in a severe crash. Seat belts help to reduce the average force on you, and they help prevent injury by spreading the force over a larger area of your body. An air bag reduces your rate of deceleration even more, and can keep you from striking the steering wheel or dashboard. (Never rely on an air bag alone. Always wear a seat belt.)

Does the picture show an elastic collision or an inelastic one? What are the benefits of seat belts and air bags when the car is hit from the side or rear?

(Photo courtesy of SAAB)

7

Linear Momentum and Collisions

Key Issues

- The momentum of an object—which Newton called the "quantity of motion" of an object—is the product of the object's mass and its velocity; momentum is a vector quantity, having direction as well as magnitude.

- The total momentum of a closed system—that is, in the absence of external forces—is constant; this is known as conservation of momentum. Even though objects undergo complex interactions—or violent collisions—the change in momentum for one object will be compensated by changes in the momentum of the rest of the system. This is a result of Newton's third law of motion.

- Energy, too, is always conserved, but in some collisions, some kinetic energy is converted to heat energy; this type of collision is known as an inelastic collision. Collisions in which kinetic energy is conserved are known as perfectly elastic collisions. Collisions between billiard balls and those between air track gliders with special bumpers come close to being perfectly elastic.

In the previous chapter we discussed conservation of energy, and now we discuss conservation of linear momentum. In later chapters we shall encounter additional conservation principles, such as conservation of angular momentum and conservation of electric charge. Because conservation is a fundamental idea that runs through all of physics, the various conservation principles are useful tools for understanding and solving real-world problems.

7.1 Linear Momentum

CONCEPT OVERVIEW

- Linear momentum is a description of the motion of an object.

- In a closed system, the total linear momentum is conserved, a consequence of Newton's third law of motion.

You already know that a large truck moving at 50 km/h is more difficult to stop than a little sports car also moving at 50 km/h. A car moving at 80 km/h is more difficult to stop than the same car moving at 40 km/h. In describing these and many other situations involving motion, it is useful to talk of linear momentum. The **linear momentum** of an object is defined as the mass m of the object multiplied by its velocity $\mathbf{v}$. We give this product the symbol $\mathbf{p}$:

$$\text{Linear momentum} = \mathbf{p} = m\mathbf{v} \qquad (7.1)$$

This definition of momentum[1] means that it is a vector. The magnitude of momentum is equal to the product of an object's mass m and its speed v. The direction of the momentum is the direction of the velocity $\mathbf{v}$. In SI units, momentum is measured in kilogram-meters per second (kg·m/s).

Momentum may seem similar to kinetic energy ($KE = \frac{1}{2} mv^2$) since both depend upon mass and speed, but momentum is a vector and kinetic energy is a scalar. Momentum and kinetic energy both describe the motion of an object, and we need both to fully understand and predict motion.

Momentum is a measure of how difficult it is to stop an object, a measure of "how much motion" an object has. Newton called it the "quantity of motion." Momentum agrees with our common ideas of motion. A football player may sustain greater injury if blocked by a heavy opponent running at full speed than if blocked by a lighter, slower opponent. A heavy lead shot, with its greater mass, is more difficult to stop than a baseball moving at the same velocity. It is also more difficult to accelerate the lead shot from rest to a given velocity than to do the same to a baseball. A few examples are illustrated in Figure 7.1.

7.2 Impulse

CONCEPT OVERVIEW

- The product $\mathbf{F} \Delta t$ is known as the impulse of the force $\mathbf{F}$. A force $\mathbf{F}$ applied for a time interval Δt causes a change in momentum $\Delta \mathbf{p} = \mathbf{F}\Delta t$.

- For a variable force, the change in momentum is the area under the curve of a force-time graph.

Newton originally worked out his second law of motion in its momentum form, $F = \Delta p/\Delta t$. We learned the simpler form $F = ma$ in Chapter 4.

[1] Later we shall encounter a quantity known as *angular* momentum, which describes rotational motion just as linear momentum describes linear motion. When there is no chance of confusing the two, we shall say "momentum" instead of "linear momentum."

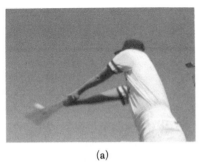
(a)

(b)

(c)

Figure 7.1

Momentum depends on both mass and velocity. (a) A player can swing the bat with considerable force, imparting to the less massive ball a momentum characterized by a high velocity. (© *Jeffrey Sylvester/FPG International Corporation*) (b) The explosive power of gun powder gives the low-mass bullet an extremely high velocity and therefore a substantial momentum. (© *Telegraph Colour Library/FPG International Corporation*) (c) The momentum of large football players results mostly from their mass. (© *Toby Rankin/The Image Bank*)

Accelerating something from rest requires applying a force **F** to it for some time Δt. This relationship between force and time is so strong and complete that we give a special name, **impulse,** to the product of force multiplied by time:

$$\text{Impulse} = \mathbf{F}\Delta t \tag{7.2}$$

Like momentum, impulse is a vector. It has units of force multiplied by time:

$$(\text{Impulse}) = \text{N·s} = \left(\frac{\text{kg·m}}{\text{s}^2}\right)(\text{s}) = \text{kg·m/s}$$

Note that the units for impulse are the same as the units for momentum, implying a connection between the two. To study this relationship between impulse and momentum, we begin with Newton's second law:

$$\mathbf{F} = m\mathbf{a} = m\,\frac{\Delta \mathbf{v}}{\Delta t} \tag{7.3}$$

$$\mathbf{F}\Delta t = m\Delta \mathbf{v} = \Delta(m\mathbf{v}) = \Delta\mathbf{p} \tag{7.4}$$

Thus, the impulse of a force $\mathbf{F}\Delta t$ is equal to the change in momentum of the body to which the force is applied:

$$\textbf{Impulse} = \text{change in } \textbf{momentum} = \Delta\mathbf{p} \tag{7.5}$$

(In deriving Equation 7.4, we assumed constant mass so that $m\Delta v = \Delta(mv)$. When we study special relativity in Chapter 27, we will find that this assumption must be modified for extremely high speeds.) Note from Equation 7.5 that *impulse* and *change in momentum* are two ways of saying the same thing. Throughout the remainder of this book, we will use the two terms interchangeably.

The force in Equations 7.2 and 7.4 is usually the *average* force because only in rare circumstances is the force constant. In most cases, the force varies with time, as indicated in Figure 7.2(a). We can find the impulse (also known as

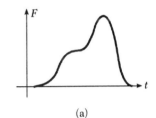
(a)

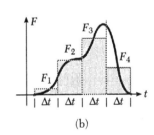
(b)

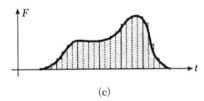
(c)

Figure 7.2

(a) Force often varies with time. (b) The impulse can be found by breaking the total time into small time intervals Δt and using the average force during each interval. In this graph, Impulse $\approx F_1\Delta t + F_2\Delta t + F_3\Delta t + F_4\Delta t$. (c) Accuracy is improved by using a smaller value for Δt. The impulse is the area under the force-time curve.

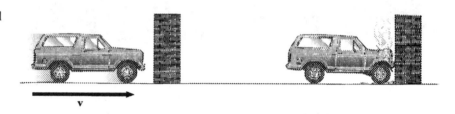

Figure 7.3

To stop the car, we can exert a small force over a long time, as the haystack might, or we can exert a large force over a short time, as the brick wall might. In both cases, the impulse—the change in momentum—is the same.

the change in momentum) by taking the average value of the force during many short time intervals, as shown in Figure 7.2(b). There the impulse (the change in momentum) is $F_1\Delta t + F_2\Delta t + F_3\Delta t + F_4\Delta t$, where each force is the average force over the corresponding short time interval Δt. Of course, we can calculate the change in momentum more accurately if we use a smaller value of Δt, as shown in Figure 7.2(c).

We can generalize from this figure and say that

For a variable force, the change in momentum is the area under the curve on a force-time graph.

However, finding the average force will usually suffice.

Note from Equation 7.2 that the same change in momentum can be accomplished by a small force acting for a long time or a large force acting for a short time, as illustrated in Figure 7.3. If your car runs into the brick wall and you come to rest along with the car, there is a significant change in your momentum. If you are wearing a seat belt or if the car is equipped with an air bag, your change in momentum occurs over a relatively long time interval. If you stop because you hit the dashboard, your change in momentum occurs over a very short time interval. The force-time curves for these two cases are sketched in Figure 7.4. If a seat belt or air bag brings you to a stop over a time interval that is, say, five times as long as the time required to stop when you strike the dashboard, then the forces involved are reduced to one fifth of the dashboard values. That is the purpose of seat belts, air bags, collapsible steering wheels, and padded dashboards. By extending the time during which you come to rest, these safety devices help reduce the forces exerted on you.

It is useful to rewrite Equations 7.3 and 7.4 as

$$\mathbf{F} = m\mathbf{a} = \frac{m\Delta\mathbf{v}}{\Delta t} = \frac{\Delta(m\mathbf{v})}{\Delta t} = \frac{\Delta\mathbf{p}}{\Delta t}$$

$$\mathbf{F} = \frac{\Delta\mathbf{p}}{\Delta t} \qquad\qquad\qquad \textbf{(7.6)}$$

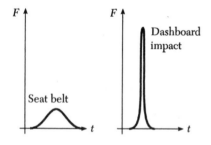

Figure 7.4

The area under the seat belt curve (on the left) is the same as the area under the dashboard curve. Using a seat belt means you come to a stop over a larger time interval, which means that the forces acting on you are smaller and therefore less damaging.

This form of the relationship tells us that the damage sustained in bringing a car—or person—to rest depends upon more than just the change in momentum. The force exerted on the object being stopped, and thus the damage, is equal to the *time rate of change of momentum*. In other words, if your speed must be decreased from 40 km/h to zero, damage will be slight if this change in momentum is accomplished in 1.0 s and extensive if the change is accomplished in 0.1 s.

An impulse applied to a body causes the body's momentum to change. The faster this change takes place, the more damage there is to the body because the force is greater.

Example 7.1

The head of a golf club is in contact with a 46-g golf ball for 0.40 ms. The ball leaves the club with a speed of 60 m/s. What is the *average* force acting on the ball while it is being hit?

Reasoning

The change in momentum of the golf ball is equal to the average force multiplied by the length of time the force acts on the ball; this is just Equation 7.4. Because the ball is initially at rest, the final momentum is equal to the change in momentum.

Solution

We begin by changing the mass to kilograms:

$$m = 46 \text{ g} = 0.046 \text{ kg}$$

After it leaves the club, the ball's momentum is

$$p_f = mv_f = (0.046 \text{ kg}) (60 \text{ m/s}) = 2.76 \text{ kg·m/s}$$

$$\Delta p = p_f - p_i = 2.76 \text{ kg·m/s}$$

From Equation 7.4,

$$\Delta p = F\Delta t$$
$$F = \frac{\Delta p}{\Delta t} = \frac{2.76 \text{ kg·m/s}}{0.40 \times 10^{-3} \text{ s}} = 6900 \frac{\text{kg·m/s}}{\text{s}} = 6900 \text{ N}$$

This is the *average* force. During part of the 0.40 ms that club and ball are in contact, the force is surely greater and during part of that time it is less.

Practice Exercise

A 145-g baseball, initially traveling to the left at 30.0 m/s, is in contact with a baseball bat for 0.800 s and leaves the bat traveling to the right at 40.0 m/s. What was the *average* force on the ball while it was being hit?

Answer 12.7 N

Example 7.2 ─────────────────────────────

Water shoots from a garden hose with a speed of 20 m/s and a mass flow rate of 1.25 kg/s. It leaves the hose horizontally and runs into a vertical wall, where it stops without splashing and gently runs down the wall (that is, its horizontal motion stops). What is the horizontal force exerted on the water by the wall?

Reasoning

The force exerted on a body is the time rate of change of the body's momentum, according to Equation 7.6. In 1.0 s, 1.25 kg of water, moving at 20 m/s, strikes the wall. This water has an initial horizontal momentum $p_i = mv_i$. After the water hits the wall, its horizontal momentum is zero.

Solution

First, we find the initial horizontal momentum of the 1.25 kg of water that strikes the wall in 1.0 s:

$$p_i = mv_i = (1.25 \text{ kg}) (20 \text{ m/s}) = 25 \text{ kg·m/s}$$

$$\Delta p = p_f - p_i = -25 \text{ kg·m/s}$$

We can now find the force exerted on the water from Equation 7.6:

$$F = \frac{\Delta p}{\Delta t} = \frac{-25 \text{ kg·m/s}}{1.0 \text{ s}} = -25 \text{ N}$$

The negative sign means that the direction of the force exerted by the wall on the water is opposite the direction of the water's velocity.

Practice Exercise

Water comes from a fire hose with a speed of 80 m/s and a mass flow rate of 20 kg/s. The water leaves the hose horizontally and strikes a vertical wall without splashing. What is the horizontal force exerted by the water on the wall?

Answer −1600 N

7.3 Conservation of Linear Momentum

CONCEPT OVERVIEW

■ In a closed system of objects, linear momentum is conserved as the objects interact (or collide).

Now we have two conservation laws to use as tools: conservation of mechanical energy and conservation of momentum.

We have described and calculated the linear momentum of a single object, but the importance of momentum is that it is conserved. The total momentum of a pair of objects—or of a system of many objects—remains constant if the objects are not acted upon by forces from outside the system.

Figure 7.5 illustrates two objects that interact with each other—perhaps two billiard balls that run into each other. According to Newton's third law, the force on object 1 exerted by object 2, $\mathbf{F}_{12}$, is equal in magnitude to the force on object 2 exerted by object 1, $\mathbf{F}_{21}$, but the two forces act in opposite directions. This is true for every instant of time, as indicated on the graph in Figure 7.5. This means the impulses are equal and opposite, too. Since impulse equals the momentum change of each ball, the momentum changes are also equal and opposite. Therefore the net change, the sum of the changes experienced by the objects, is zero. The value of the total momentum—the sum of the momenta of the two objects—remains constant; momentum is conserved.

The momentum of a system of particles is always conserved when only internal forces are present. Internal forces are the forces between the particles that make up the system. An *external* force is one from the outside; one example of an external force would be striking the billiard balls of Figure 7.5 with a cuestick.

Problem-Solving Tip

- Find the total momentum of the system before a collision and the total momentum after the collision (doing this will probably involve using some unknown quantity). By conservation of momentum, these two momenta must be equal to each other. Set them equal to each other and solve for the unknown.

Example 7.3

Figure 7.6 illustrates two gliders at rest on an air track with a compressed spring between them. They are held in place by a string. Once the string is cut, the gliders move away from each other. The gliders have masses of $m_1 = 672$ g and $m_2 = 336$ g. The lighter glider moves off with a speed of 28 cm/s. Find the speed of the more massive one.

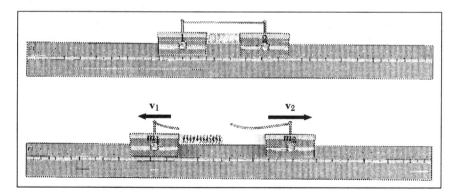

Figure 7.6
How do the two gliders move after undergoing an "explosion" caused by the expanding spring between them?

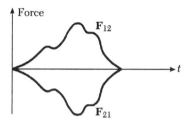

Figure 7.5
During a collision, the force varies greatly with time, but Newton's third law, $F_{12} = -F_{21}$, remains valid at every instant.

Reasoning

The only force on either glider is the internal force of one on the other because of the spring. The vertical gravitational force and normal force on each glider just cancel; there are no other external forces since the air track is frictionless. Therefore, momentum is conserved.

The initial total momentum is zero. The final total momentum is just the sum of the final momenta of the two gliders. Setting the final total momentum equal to the initial total momentum, we can solve for the unspecified velocity. Qualitatively, because momentum is conserved, we expect the more massive glider to have a speed less than the speed of the lighter glider.

Solution

We start with $p_i = 0$, and then find the final momentum and set it equal to the initial momentum:

$$p_f = p_{1f} + p_{2f} = m_1v_1 + m_2v_2 = 0 = p_i$$
$$m_1v_1 + m_2v_2 = 0$$
$$m_1v_1 = -m_2v_2$$

Remember that v_1 and v_2 are vectors, so this minus sign means the two gliders move in opposite directions. Solving for the velocity of m_1, we get

$$v_1 = -\frac{m_2}{m_1}v_2$$

This expression tells us that m_1 moves off with the lower speed:

$$v_1 = -\left(\frac{336 \text{ g}}{672 \text{ g}}\right)(28 \text{ cm/s}) = -14 \text{ cm/s}$$

Practice Exercise

Consider two air track gliders again held together with a compressed spring between them. The gliders have masses of 308 g and 602 g. When released, the 602-g glider moves with a speed of 25 cm/s to the right. What is the speed of the 308-g glider?

Answer   49 cm/s

7.4 Inelastic Collisions

CONCEPT OVERVIEW

- A perfectly inelastic collision is one in which the colliding objects stick together and move off with a common final velocity.

- Momentum is conserved in all collisions. Kinetic energy as well as momentum is conserved in a perfectly elastic collision; momentum is conserved in an inelastic collision but kinetic energy is converted to other forms of energy.

In any collision, the total momentum after collision is the same as the total momentum before collision; in other words, momentum is always conserved in

In the real world, the results of collisions can have serious consequences. (*Roger Viollet*)

any collision. While the total momentum remains unchanged, the collision can redistribute that momentum between the colliding objects.

There are two types of collision that are particularly interesting to look at because they can be completely solved. In this section we look at **perfectly inelastic collisions,** those in which the two colliding objects stick together and then move off with a common velocity. As we shall see later, kinetic energy is lost in inelastic collisions. In the next section, we shall look at **perfectly elastic collisions,** those in which both momentum and kinetic energy are conserved.

For perfectly inelastic collisions, there is only one unknown quantity—typically the common velocity of the two objects after they stick together—and conservation of momentum alone is sufficient to solve for that velocity.

Even though kinetic energy is not conserved in an inelastic collision, do not think that the law of conservation of energy is being violated. It is not, because the *total* energy is always conserved in a collision. Think about an automobile wreck. Much of the original kinetic energy is converted to heat energy and goes into deforming, tearing, and breaking metal. If we quickly and accurately measured the temperature of the bent sheet metal, we would find that it is warmer after being deformed. (Prove this to yourself by bending a paperclip back and forth and then touching the bending point—it will be noticeably warmer.) Also, some of the kinetic energy present before the wreck is converted to sound energy as we hear the crash.

Remember, no matter whether a collision is perfectly elastic, perfectly inelastic, or anything in between,[2] *the total momentum of the system is always conserved.* Thus, when we are given the initial velocities of two bodies involved

---

[2] We refer to any collision that is not perfectly elastic or perfectly inelastic as being simply *inelastic,* without the *perfectly.* We shall not discuss these in-between types in this book.

in a perfectly inelastic collision, we can easily solve for a single unknown, the common final velocity of the merged bodies.

In the examples that follow, collisions are often said to be *head-on*. By this we mean that all the motion, both before and after collision, is along one straight line. Collisions that are not head-on are called *glancing collisions*.

Problem-Solving Tip

■ After a perfectly inelastic collision, the two objects move off with a common velocity; there is only one final velocity to find.

Example 7.4

Two balls of clay suffer a head-on collision and stick together, as sketched in Figure 7.7. Ball 1 has a mass $m_1 = 200$ g and is initially moving with a velocity $v_{1i} = 45$ cm/s to the right. Ball 2 has a mass $m_2 = 600$ g and is initially moving with a velocity $v_{2i} = 25$ cm/s to the left. What is their final velocity?

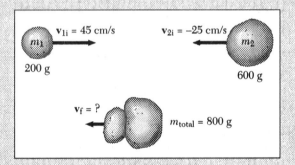

Figure 7.7
What is the common final velocity?

Reasoning

We shall find the momentum before the collision and set it equal to the momentum after the collision. Directions are important in calculating momentum because it is a vector. As usual we take motion to the right as positive and motion to the left as negative.

Solution

Before the collision, the total initial momentum of the two balls is

$$p_i = p_{1i} + p_{2i} = m_1 v_{1i} + m_2 v_{2i}$$
$$= (200 \text{ g})(45 \text{ cm/s}) + (600 \text{ g})(-25 \text{ cm/s}) = -6000 \text{ g·cm/s}$$

The negative sign means that the total initial momentum is a vector pointing to the left.

After the collision, both balls move with the same velocity $v_{1f} = v_{2f} = v_f$. Therefore the total final momentum of the two balls stuck together is

$$p_f = p_{1f} + p_{2f} = m_1 v_f + m_2 v_f = (m_1 + m_2)v_f$$
$$= (200 \text{ g} + 600 \text{ g})v_f = (800 \text{ g})v_f$$

Since momentum is conserved,

$$p_f = (800 \text{ g})v_f = -6000 \text{ g·cm/s} = p_i$$

$$v_f = \frac{-6000 \text{ g·cm/s}}{800 \text{ g}} = -7.5 \text{ cm/s}$$

Since the final velocity is negative, the two balls, now stuck together, move to the left.

Practice Exercise

Two air track gliders have a perfectly inelastic collision. Glider 1 has a mass of 308 g and before the collision is moving to the right at 35 cm/s. Glider 2 has a mass of 602 g and before the collision is moving to the left at 22 cm/s. What is their velocity after the collision?

Answer   2.7 cm/s, to the left

Problem-Solving Tip

■ Momentum is a vector. That means the component in the x direction is conserved *and* the component in the y direction is conserved.

Example 7.5 ───────────────────────────────

Figure 7.8 is a sketch of a traffic accident. Car 1 has a mass $m_1 = 1300$ kg and is moving with a velocity $v_{1i} = 30.0$ km/h due east as it enters the intersection. Car 2 has a mass $m_2 = 1100$ kg and is moving with a velocity $v_{2i} = 20.0$ km/h due north as it enters the intersection. The two cars collide perfectly inelastically, and their bumpers lock so that the two are stuck together as a single unit. Immediately after the collision, what is their velocity (both magnitude and direction)?

Reasoning

As before, we shall find the momentum before the collision and set it equal to the momentum after the collision. Because the motion is not confined to a single direction, we must treat momentum as a vector. We can always write a single vector equation like

$$\mathbf{p}_i = \mathbf{p}_{1i} + \mathbf{p}_{2i}$$

but we must replace this single expression with two corresponding scalar equations before we can proceed. We choose the positive x axis pointing east and the positive y axis pointing north.

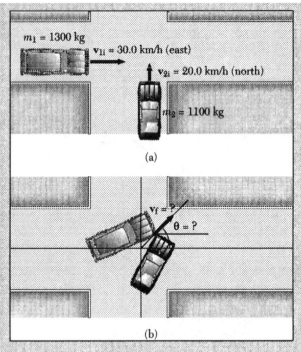

Figure 7.8
In asking about the final velocity after a perfectly inelastic collision, we must specify direction as well as magnitude.

Solution

For the initial momentum in the x and y directions, we can write, remembering that $v_{2ix} = 0$,

$$p_{ix} = p_{1ix} + p_{2ix} = m_1 v_{1ix} + m_2 v_{2ix}$$

$$= (1300 \text{ kg})(30.0 \text{ km/h}) + 0 = 39\ 000 \text{ kg·km/h}$$

Likewise, since $v_{1iy} = 0$, we have

$$p_{iy} = p_{1iy} + p_{2iy} = m_1 v_{1iy} + m_2 v_{2iy}$$

$$= 0 + (1100 \text{ kg})(20.0 \text{ km/h}) = 22\ 000 \text{ kg·km/h}$$

The total initial momentum is

$$p_i = \sqrt{p_{iy}^2 + p_{ix}^2} = 44\ 800 \text{ kg·km/h}$$

As in Example 7.4, and as in all perfectly inelastic collisions, both cars move with the same velocity after the collision: $\mathbf{v}_{1f} = \mathbf{v}_{2f} = \mathbf{v}_f$. We can therefore write the vector equation

$$\mathbf{P}_f = \mathbf{p}_{1f} + \mathbf{p}_{2f} = m_1 \mathbf{v}_{1f} + m_2 \mathbf{v}_{2f} = (m_1 + m_2)\mathbf{v}_f = (2400 \text{ kg})\mathbf{v}_f$$

$$p_f = (2400 \text{ kg})v_f = 44\ 800 \text{ kg·km/h} = p_i$$

$$v_f = \frac{44\,800 \text{ kg·km/h}}{2400 \text{ kg}} = 18.7 \text{ km/h}$$

We need the direction as well:

$$\tan \theta = \frac{p_y}{p_x}$$

Notice that we do not have to specify p_{iy}/p_{ix} or p_{fy}/p_{fx}. All we need is p_y/p_x because conservation of momentum means that $p_x = p_{ix} = p_{fx}$ and $p_y = p_{iy} = p_{fy}$. Thus,

$$\tan \theta = \frac{22\,000 \text{ kg·km/h}}{39\,000 \text{ kg·km/h}} = 0.564$$
$$\theta = 29.4°$$

We have found the velocity immediately after the collision. Of course, there are all sorts of external forces—due to friction between tires and pavement—affecting the motion of the wreckage.

Practice Exercise

A 1.0-kg ball of clay initially moving north at 2.4 m/s strikes a 1.0-kg ball of clay initially moving east at 3.2 m/s. The two stick together. What is their velocity after the collision?

Answer $v_x = 1.6$ m/s, $v_y = 1.2$ m/s; $\mathbf{v} = 2.0$ m/s, 37°N of E or 53°E of N

7.5 Elastic Collisions

CONCEPT OVERVIEW

■ A perfectly elastic collision is one in which kinetic energy as well as linear momentum is conserved.

As we mentioned in the preceding section, a **perfectly elastic collision** is one in which kinetic energy as well as linear momentum is conserved. Colliding billiard balls are one example of a perfectly elastic collision, and the collisions between elementary particles are often perfectly elastic; if a neutron collides with a proton, there is no loss in kinetic energy. Two gliders colliding on an air track with a spring between them undergo a perfectly elastic collision. At very low speeds, two automobiles might hit bumpers and bounce off each other in a (nearly) perfectly elastic collision.

In any collision, momentum is conserved, and so we can write

$$p_i = p_f$$

$$m_1 v_{1i} + m_2 v_{2i} = m_1 v_{1f} + m_2 v_{2f} \qquad \textbf{(7.7)}$$

Rearranging this equation so that the m_1 terms are on one side and the m_2

Billiard ball collisions give a good approximation of elastic collisions. *(David Rogers)*

Polished steel balls colliding on a low-friction grooved track provide a good approximation of elastic collisions in one dimension and make clear how momentum is conserved in such collisions. *(Courtesy of Central Scientific Company)*

terms on the other, we get

$$m_1(v_{1i} - v_{1f}) = m_2(v_{2f} - v_{2i}) \qquad (7.8)$$

For a perfectly elastic collision, kinetic energy is also conserved, so we can also write

$$KE_i = KE_f$$

$$\tfrac{1}{2}m_1v_{1i}^2 + \tfrac{1}{2}m_2v_{2i}^2 = \tfrac{1}{2}m_1v_{1f}^2 + \tfrac{1}{2}m_2v_{2f}^2 \qquad (7.9)$$

Multiplying this expression by 2 and rearranging so that the m_1 terms are on one side and the m_2 terms on the other, we get

$$m_1(v_{1i}^2 - v_{1f}^2) = m_2(v_{2f}^2 - v_{2i}^2)$$

which can be rewritten as

$$m_1(v_{1i} - v_{1f})(v_{1i} + v_{1f}) = m_2(v_{2f} - v_{2i})(v_{2f} + v_{2i})$$

Dividing this by Equation 7.8 gives

$$v_{1i} + v_{1f} = v_{2f} + v_{2i}$$

which can be rearranged to yield

$$v_{1i} - v_{2i} = -(v_{1f} - v_{2f}) \qquad (7.10)$$

which tells us that, in a perfectly elastic collision, the magnitude of Δv_i is equal to the magnitude of Δv_f. That is, the relative speed of one body with respect to the other is the same before the collision as after.

Example 7.6

Figure 7.9 shows two gliders approaching each other on a frictionless air track. Glider 1 has a mass $m_1 = 600$ g and is moving with a velocity $v_{1i} = 25$ cm/s to the right. Glider 2 has a mass $m_2 = 300$ g and is moving with a velocity $v_{2i} = 10$ cm/s to the left. Because each glider is equipped with a spring "bumper," the collision will be perfectly elastic. What is the final velocity of each?

Reasoning

The two gliders approach each other with a relative speed of 35 cm/s. According to Equation 7.10, which comes from conservation of momentum and conservation of kinetic energy, they will depart with a relative speed of 35 cm/s:

$$v_{2f} - v_{1f} = 35 \text{ cm/s}$$

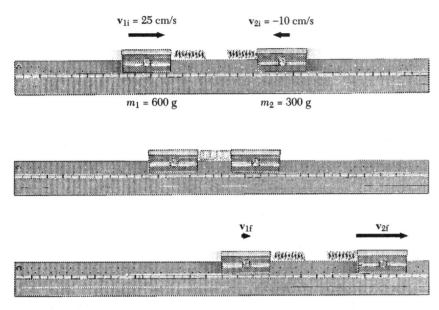

Figure 7.9
In a perfectly elastic collision, kinetic energy is conserved.

This is one equation with two unknowns; therefore an additional equation is needed. We can go back to conservation of momentum and Equation 7.7 or to conservation of kinetic energy and Equation 7.9 for the additional information. Because Equation 7.7 is linear while Equation 7.9 contains quadratic terms, it will surely be easier to use Equation 7.7.

Solution

We have two equations with two unknowns:

$$v_{2f} - v_{1f} = 35 \text{ cm/s} \qquad (1)$$

$$m_1 v_{1i} + m_2 v_{2i} = m_1 v_{1f} + m_2 v_{2f}$$

$$(600 \text{ g})(25 \text{ cm/s}) + (300 \text{ g})(-10 \text{ cm/s}) = (600 \text{ g})v_{1f} + (300 \text{ g})v_{2f} \quad (2)$$

Remember that signs are important; that glider 2 is initially moving to the left at 10 cm/s means $v_{2i} = -10$ cm/s. We can divide Equation 2 by 300 g to simplify:

$$50 \text{ cm/s} - 10 \text{ cm/s} = 2v_{1f} + v_{2f}$$

$$2v_{1f} + v_{2f} = 40 \text{ cm/s} \qquad (3)$$

We can then subtract Equation 1, the relative velocity equation, from Equation 3:

$$2v_{1f} + v_{2f} - (v_{2f} - v_{if}) = 40 \text{ cm/s} - 35 \text{ cm/s}$$
$$3v_{1f} = 5 \text{ cm/s}$$
$$v_{1f} = 1.7 \text{ cm/s}$$

From Equation 1 we have

$$v_{2f} = 35 \text{ cm/s} + v_{1f} = 35 \text{ cm/s} + 1.7 \text{ cm/s} = 37 \text{ cm/s}$$

Note that both final velocities are positive: both gliders move to the right after the collision. Initially, the larger glider has positive momentum (to the right) $p_{i1} = (600 \text{ g})(25 \text{ cm/s}) = 15\,000 \text{ g·cm/s}$ and the smaller glider has negative momentum (to the left) $p_{i2} = (300 \text{ g})(-10 \text{ cm/s}) = -3000 \text{ g·cm/s}$. Therefore the total momentum is positive:

$$p_{tot} = p_{i1} + p_{i2} = 15\,000 \text{ g·cm/s} - 3000 \text{ g·cm/s} = 12\,000 \text{ g cm/s}$$

This is the total momentum initially and finally. In order to conserve kinetic energy as well, both gliders move to the right after the collision.

Practice Exercise

Consider two gliders approaching each other. Glider 1 has a mass of 450 g and moves with an initial velocity of 20 cm/s to the right. Glider 2 has a mass of 320 g and moves with an initial velocity of 15 cm/s to the left. They undergo a perfectly elastic collision. What is the final velocity of each?

Answer $v_{f1} = 9.1$ cm/s to the left; $v_{f2} = 26$ cm/s to the right

In Example 7.6, involving a perfectly elastic collision, the kinetic energy after is equal to the kinetic energy before. We did not need to look at the details of what happens *during* the collision. That is the beauty and the usefulness of energy conservation. However, although not necessary, it is very *interesting* to look at these details. For a moment, at the very heart of the collision, both gliders stop and there is no kinetic energy! All of the initial kinetic energy is stored as elastic potential energy in the bumpers. That potential energy is then converted to the final kinetic energy of the gliders.

We shall now find equations for a head-on perfectly elastic collision in which one of the objects is initially at rest. This is sketched in Figure 7.10. A glider of mass m_1 comes in with speed v_i and collides with a glider of mass m_2 that is initially at rest. Because the gliders are on an air track, the motion is constrained to a line, so we know that the collision is head-on. If the gliders collide in a perfectly elastic collision, what is the final motion like? Three possibilities are sketched in Figure 7.10(c). What are the final velocities $\mathbf{v}_{1f}$ and $\mathbf{v}_{2f}$?

With glider 2 at rest ($v_{2i} = 0$), Equation 7.7 becomes

$$m_1 v_{1i} = m_1 v_{1f} + m_2 v_{2f} \tag{7.11}$$

and Equation 7.10 becomes

$$v_{1f} = v_{2f} - v_{1i} \tag{7.12}$$

We can solve Equation 7.11 for v_{1f}:

$$v_{1f} = v_{1i} - \frac{m_2}{m_1} v_{2f}$$

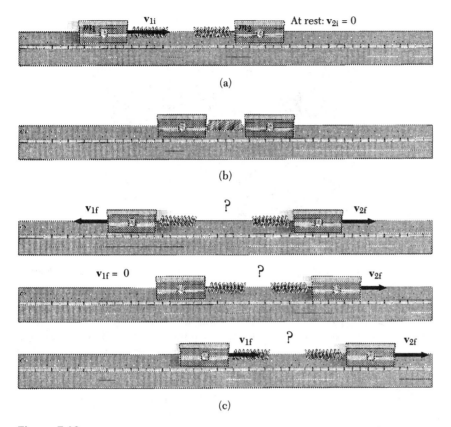

(a)

(b)

(c)

Figure 7.10

What happens to the two gliders after they collide? Assume an elastic collision and determine $\mathbf{v}_{1f}$ and $\mathbf{v}_{2f}$.

We can then set these two expressions for v_{1f} equal to each other and solve for the single unknown v_{2f}:

$$v_{2f} - v_{1i} = v_{1i} - \frac{m_2}{m_1} v_{2f}$$

$$v_{2f}\left(1 + \frac{m_2}{m_1}\right) = 2v_{1i}$$

$$v_{2f} = \frac{2m_1}{m_1 + m_2} v_{1i} \qquad (7.13)$$

Substituting this expression for v_{2f} into Equation 7.12, we obtain

$$v_{1f} = \frac{m_1 - m_2}{m_1 + m_2} v_{1i} \qquad (7.14)$$

Equation 7.13 tells us that nothing exciting happens to v_{2f}; the glider initially at rest always moves off to the right, no matter what the relative masses are. Look at Equation 7.14, however. The numerator can be positive, zero, or negative. If $m_1 > m_2$, the moving body continues on to the right after the

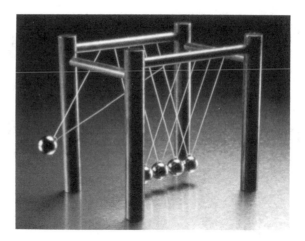

Figure 7.11
The Newtonian demonstrator. How does this thing work? *(Courtesy of Central Scientific Company)*

collision but is slowed down some; you might think of this as a Mack truck running into a parked sports car. If $m_1 < m_2$, the moving body rebounds and goes off to the left; you might think of this as a sports car running into a parked Mack truck and bouncing backwards. And if $m_1 = m_2$, then $v_{1f} = 0$ (the moving body stops) and $v_{2f} = v_{1i}$ (the body initially at rest moves off with the same speed as the incoming body had initially). You may have seen this $m_1 = m_2$ case demonstrated with the apparatus pictured in Figure 7.11.

(Equations 7.13 and 7.14 are interesting, but they apply only to this very special case where the initial conditions are that one body is moving, the other body is at rest, and the collision is head-on. Consequently, they should *not* be memorized.)

Perfectly elastic collisions do not have to be head-on. Conservation of momentum provides two equations (from its two components), and conservation of kinetic energy provides an additional equation. Thus a perfectly elastic collision that is glancing rather than head-on can be solved for up to three unknowns—perhaps two speeds and one direction or one speed and two directions. That is, if one of the four final conditions—two speeds and two directions—is known, the other three may be solved for in a perfectly elastic collision.

7.6 Rocket Propulsion

CONCEPT OVERVIEW

- Rocket engines demonstrate conservation of momentum. The momentum carried by exhaust gases in one direction is exactly balanced by the momentum change in the other direction due to an increase in the speed of the rocket.

We can describe how a car or airplane or boat moves along a road or through the air or through the water by using Newton's third law of action and reaction, $\mathbf{F}_{AB} = -\mathbf{F}_{BA}$ (Fig. 7.12), but how does a rocket operate in the emptiness of space, where there is nothing to push against? Rocket propulsion is a good

(a)

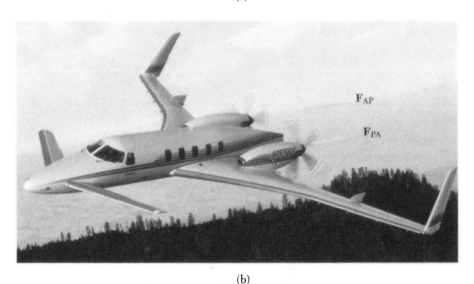

(b)

Figure 7.12
Pairs of action-reaction forces are involved in many common forms of motion.
(a) An automobile. *(© L. W. Black)* (b) The Beech Starship-1, a unique propeller
driven aircraft. *(Beech Aircraft Co.)*

application of conservation of momentum. As exhaust gases burn and leave the
rocket, momentum is carried with them. To conserve momentum, the rocket
must gain the same amount of momentum in the opposite direction.

Consider the rocket sketched in Figure 7.13, traveling far from Earth, and
take as our system only the rocket and its contents. When fuel is burned, it
carries momentum $\Delta\mathbf{p}_{\text{fuel}}$ to the left. In order for the total momentum of the
system to be conserved, this same amount of momentum $\Delta\mathbf{p}_{\text{rocket}}$ must be
given to the rocket moving to the right:

$$\Delta\mathbf{p}_{\text{rocket}} = -\Delta\mathbf{p}_{\text{fue}}$$

If we know that the exhaust gases leave the system with a speed u and a mass
flow rate $\Delta m/\Delta t$, we can find the force exerted on the rocket. In a short time

$\Delta\mathbf{P}_{\text{fuel}}$ $\Delta\mathbf{P}_{\text{rocket}}$

Figure 7.13
Conservation of momentum means
that a rocket is propelled forward as
exhaust gases are expelled backward.

The combined thrust of the Space Shuttle's engines and the booster rockets allows it to lift off more quickly than the manned rockets of the 1960s. *(NASA)*

interval Δt, an amount of fuel Δm leaves the rocket with speed u, as measured from the rocket, so that

$$\Delta p_{\text{fuel}} = -u\,\Delta m$$

where the minus sign indicates the velocity u and change in momentum Δp_{fuel} are to the left.

From Equation 7.6, we know

$$F = \frac{\Delta p}{\Delta t} = \frac{-u\,\Delta m}{\Delta t} = -u\frac{\Delta m}{\Delta t}$$

As mentioned in passing in earlier chapters, **thrust** is the magnitude of the force exerted by an engine. For our rocket, we can see that

$$F_{\text{thrust}} = u\frac{\Delta m}{\Delta t} \qquad\qquad \textbf{(7.15)}$$

where we have dropped the minus sign to define the magnitude of the thrust. That is, the thrust of a rocket is equal to the velocity u of the exhaust gases multiplied by the mass flow rate $\Delta m/\Delta t$. Of course, the direction of the thrust is opposite the direction of the exhaust gases coming from the rocket. To lift off the launch pad, a rocket's thrust must exceed the rocket's initial weight.

Pay attention to the change in acceleration of a large rocket the next time you watch a launch on television. A large rocket lumbers along at first, moving slowly with a very small acceleration. As the rocket continues to burn fuel, its mass decreases. Since the thrust of the engine remains constant, the acceleration of the rocket increases significantly.

Example 7.7

A particular rocket burns fuel at the rate of 500 kg/s. What must the exhaust velocity of the burned gases be in order to achieve liftoff of the 30 000-kg rocket?

Reasoning

Thrust is given by Equation 7.15. The thrust must be greater than the initial weight of the rocket.

Solution

$$F = u\frac{\Delta m}{\Delta t} = m_i g$$

$$u = \frac{m_i g}{\Delta m/\Delta t} = \frac{(30\ 000\ \text{kg})(9.80\ \text{m/s}^2)}{500\ \text{kg/s}} = 588\ \text{m/s}$$

Thus any exhaust-gas speed greater than this value will achieve liftoff.

Practice Exercise

What is the thrust of a rocket engine that burns fuel at the rate of 430 kg/s and has an exhaust velocity of 750 m/s?

Answer: 322 000 N

Example 7.8

A certain rocket burns fuel at the rate of 600 kg/s and exhausts it with a velocity of 700 m/s. The initial mass of the rocket is 30 000 kg, of which 22 000 kg is fuel. Find the initial and final accelerations of the rocket.

Reasoning

From the data given, we can calculate thrust. That and the weight of the rocket give the net force on the rocket. The net force and Newton's second law let us find the acceleration.

Solution

The thrust of the engine is, from Equation 7.15,

$$F_{thrust} = u\frac{\Delta m}{\Delta t} = (700\ \text{m/s})(600\ \text{kg/s}) = 420\ 000\ \text{N} = 4.20 \times 10^5\ \text{N}$$

The initial weight of the rocket is

$$w_i = m_i g = (30\ 000\ \text{kg})(9.80\ \text{m/s}^2) = 2.94 \times 10^5\ \text{N}$$

So the initial net force is

$$F_{net,i} = F_{thrust} - w_i = 1.26 \times 10^5\ \text{N}$$

which will cause an acceleration

$$a_i = \frac{F_{\text{net,i}}}{m_i} = \frac{1.26 \times 10^5 \text{ N}}{3.00 \times 10^4 \text{ kg}} = 4.20 \text{ m/s}^2$$

Once all the fuel is expended, the final weight of the rocket is

$$w_f = m_f g = (8000 \text{ kg})(9.80 \text{ m/s}^2) = 7.84 \times 10^4 \text{ N}$$

Now the net force is

$$F_{\text{net,f}} = F_{\text{thrust}} - w_f = 3.42 \times 10^5 \text{ N}$$

$$a_f = \frac{F_{\text{net,f}}}{m_f} = \frac{3.42 \times 10^5 \text{ N}}{8000 \text{ kg}} = 42.8 \text{ m/s}^2$$

This calculation assumes the final weight is based upon an acceleration of gravity equal to that at Earth's surface. As the rocket gets farther and farther from Earth, however, the acceleration of gravity will naturally decrease. That correction would provide an even higher final acceleration. Also, drag forces exerted by Earth's atmosphere have not been included.

SUMMARY

The linear momentum of an object is

$$\mathbf{p} = m\mathbf{v} \tag{7.1}$$

Momentum, a vector quantity, is a description of how much mass is moving and what its velocity is.

The product of a force multiplied by how long the force acts is defined as the impulse of the force:

$$\text{Impulse} = \mathbf{F} \, \Delta t \tag{7.2}$$

Impulse, also a vector quantity, is equal to the change in momentum:

$$\mathbf{F} \, \Delta t = \Delta \mathbf{p} \tag{7.5}$$

Force is equal to the time rate of change of momentum:

$$\mathbf{F} = m\mathbf{a} = \frac{\Delta \mathbf{p}}{\Delta t} \tag{7.6}$$

The real importance of momentum is that it is conserved: the total momentum of a closed system always remains constant.

In an inelastic collision, momentum is conserved but kinetic energy is not. When a collision is perfectly inelastic, the colliding bodies stick together and have a common final velocity. In a perfectly elastic collision, both momentum and kinetic energy are conserved.

Thrust is the force a rocket engine exerts and is given by

$$\mathbf{F} = -\mathbf{u}\frac{\Delta m}{\Delta t} \tag{7.17}$$

where $\mathbf{u}$ is the exhaust velocity of the gases and $\Delta m/\Delta t$ is their mass flow rate.

QUESTIONS

1. How can a lightweight bullet have more momentum than a heavy bowling ball?
2. How can a loaded cement truck have less momentum than a small sports car?
3. We are concerned with forest conservation and energy conservation because these are limited resources. Momentum conservation is very important. Does that imply that momentum is a limited resource?
4. Use conservation of momentum to discuss what happens to a passenger in an automobile crash. Discuss the use of seat belts and shoulder harnesses in bringing the passenger to rest. Discuss accidents that cause whiplash injuries.
5. Two objects initially traveling along the same straight line collide perfectly inelastically. After they collide, will their final motion necessarily be along the same straight line as their initial motion?
6. Two objects initially traveling along the same straight line collide perfectly elastically. After they collide, will their final motion necessarily be along the same straight line as their initial motion?
7. Is momentum conserved in an inelastic collision?
8. The momentum of a child swinging back and forth on a playground swing obviously changes during each cycle of the motion. Does this changing momentum mean that the momentum of the system is not conserved? Explain your answer.
9. If you jump from a table and land with your knees locked, why are you more likely to be hurt than if you land with your legs relaxed?
10. A basketball player jumps into the air with both arms extended vertically overhead. Just as he reaches maximum height, he quickly pulls one arm down. Do the fingertips of the still-extended hand go higher or lower as a result of his maneuver?
11. Construct a "balloon rocket" by releasing an inflated balloon with the opening untied. Explain how its propulsion is similar to that of a NASA rocket despite the fact that there is no flame belching from the balloon.
12. How can rockets eventually move so fast if they initially take off so slowly?
13. Why are rockets often built with several stages so that the fuel tanks and other assemblies can be discarded once they are no longer needed (Fig. 7.14)?

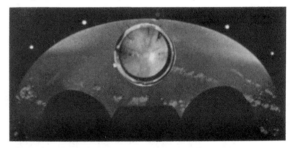

Figure 7.14
Question 7.13. A rocket stage separation. *(NASA)*

PROBLEMS

7.1 Linear Momentum

7.1 What is the momentum of a 25-g bullet traveling at 120 m/s?

7.2 Calculate the momentum of an 1100-kg sports sedan traveling at 40 km/h and at 80 km/h.

7.3 Calculate the momentum of a fullback who has a mass of 120 kg and is running at 6.2 m/s.

7.4 What is the momentum of a 1200-kg sedan traveling at 90 km/h? At what speed must a 3600-kg truck travel to have the same momentum?

7.5 A car with a mass of 1080 kg is initially traveling at 35 km/h. If its momentum increases by 12 000 kg·m/s, what is its final velocity? If its momentum decreases by 12 000 kg·m/s, what is its final velocity?

7.2 Impulse

7.6   What is the change in momentum as an
        1100-kg sports sedan accelerates from
        40 km/h to 100 km/h? What impulse is nec-
        essary to cause this change?

7.7   (a) What is the momentum of a golf ball that
        has a mass of 60 g and is moving with a
        speed of 80 m/s?   (b) If the impact between
        golf club and ball lasted for 2.0×10^{-4} s,
        what was the rate of change of momentum?
        (c) What average force acted on the ball
        while it was in contact with the club?
        (d) What average force acted on the club?

7.8   A 57-g tennis ball initially travels at 20 m/s
        before it is hit by a racket. After leaving the
        racket, its velocity is 30 m/s in the opposite
        direction.   (a) What is its change in momen-
        tum?   (b) Is the change in momentum di-
        rected along the initial velocity, the final ve-
        locity, or some other direction?   (c) What is
        the impulse delivered by the tennis racket?
        (d) In hitting a tennis ball, why is it impor-
        tant to "follow through," that is, keep the
        racket moving and make a full swing, rather
        than to stop right after the ball is hit?

7.9   A 150-g baseball initially traveling at 30 m/s
        is struck by a bat and leaves in the opposite
        direction at 35 m/s.   (a) What is its change
        in momentum?   (b) Is the change in mo-
        mentum directed along the initial velocity,
        the final velocity, or some other direction?
        (c) What is the impulse delivered by the
        bat?   (d) In hitting a baseball, why is it im-
        portant to "follow through," that is, keep the
        bat moving and make a full swing, rather
        than to stop right after the ball is hit?

7.10   A ball with a mass of 0.50 kg falls vertically,
        hits the floor with a speed of 5.0 m/s, and
        bounces back up, leaving the floor with a
        speed of 4.0 m/s. Calculate the ball's change
        in momentum (both magnitude and direc-
        tion). If the ball is in contact with the floor
        for 0.10 s, calculate the average force exerted
        on the ball by the floor.

7.11   A baseball bat is in contact with a 150-g
        baseball for 0.0015 s and exerts an average
        force of 8000 N during that time. If the ball

had an initial velocity of 25 m/s before being
hit and leaves along the same line it came in
on (its direction is exactly reversed), what is
its velocity after leaving the bat?

7.12   A baseball bat is in contact with a 150-g
        baseball for 0.0012 s. The ball had an initial
        velocity of 25 m/s before being hit and leaves
        along the same line it came in on (its direc-
        tion is exactly reversed). If its velocity after
        leaving the bat is 20 m/s, what average force
        was exerted by the bat?

7.13   Water from a hose strikes a wall with a hori-
        zontal speed of 20 m/s at a rate of 1.2 kg/s.
        The water runs down the wall with no hori-
        zontal speed after hitting it. What force does
        the water exert on the wall?

7.14   A fire hose sends 20 kg of water per second
        onto a burning building. If the water leaves
        the nozzle at 40 m/s and does not bounce
        back from the building, what force is exerted
        on the building by the water?

7.15   A ventilator fan moves 90 m^3 of air per min-
        ute, and the blast of air strikes a wall at a
        speed of 20 m/s. What force is exerted by
        the air on the wall? Each cubic meter of air
        has a mass of 1.3 kg.

O7.16   Momentum has been given in units of kilo-
        grams-meters per second. Show that this is
        equivalent to units of newtons-seconds.

O7.17   At 4:40 p.m. the momentum of a speedboat
        is 20 000 kg·m/s, and at 4:43 it is
        35 000 kg·m/s. The boat moves along a
        straight course, and its speed increases stead-
        ily (implying a constant force). What force
        was exerted on the boat? What exerted the
        force?

O7.18   What force, acting for 0.0010 s, will change
        the velocity of a 150-g baseball from 30 m/s
        eastward to 40 m/s westward?

O7.19   A fire hose sends 1000 kg of water per min-
        ute against a burning building. The water
        strikes the building at 20 m/s and does not
        bounce back.   (a) What is the rate of change
        of momentum of the water?   (b) What force
        does the building exert on the water?
        (c) What force does the water exert on the
        building?

○7.20 What force, acting for 3×10^{-7} s, will change the velocity of a proton ($m = 1.67 \times 10^{-27}$ kg) from zero to 1×10^7 m/s?

○7.21 A 60-kg woman jumps from a burning building and falls 10 m before making contact with a safety net, which stops her in 0.12 s. What is the average force exerted on her by the net?

●7.22 A fire hose sends 20 kg of water per second onto a burning building. The water strikes the roof horizontally at 40 m/s and is deflected 60° as shown in Figure 7.15. What are the magnitude and direction of the force exerted on the roof by the water? (*Hint:* Treat the horizontal and vertical components separately.)

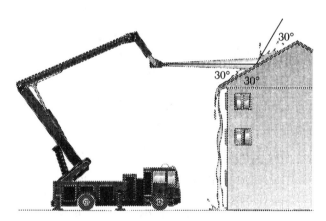

Figure 7.15
Problem 7.22.

7.3 Conservation of Linear Momentum

7.23 A 600-kg cannon fires a 5.0-kg cannonball with a horizontal muzzle velocity of 120 m/s. What is the recoil velocity of the cannon?

7.24 A 60-kg girl initially at rest on roller blades throws a 5.0-kg book bag due north with a speed of 3.0 m/s. Describe the girl's motion immediately afterwards.

7.25 Two gliders sit at rest on an air track with a spring compressed between them and held together by a string tied around them. Their masses are 650 g and 400 g. The string is burned in two, and the gliders move off in opposite directions. The 650-g glider moves off at 25 cm/s. What is the speed of the 400-g glider?

7.26 Two air track gliders, with masses of 350 g and 600 g, are tied together by a string with a compressed spring between them. They are moving to the right at 10 cm/s when the string is burned in two. The 350-g glider moves off to the right at 25 cm/s after they separate. What is the velocity (speed and direction) of the 600-g glider?

7.27 What is the recoil velocity of a 3-kg gun that fires a 0.060-kg bullet at a velocity of 200 m/s?

7.28 A young man holds a 1.50-kg air rifle loosely and fires a 3.00-g bullet. The muzzle velocity of the bullet is 300 m/s. (*a*) What is the recoil velocity of the gun? (*b*) If he holds the rifle tightly against his body, the recoil velocity will be less. Explain. (*c*) Find the new recoil velocity if his body mass is 37.5 kg.

7.29 An astronaut and spacesuit have a combined mass of 120 kg. While floating outside the space shuttle, the astronaut pushes on a 30-kg piece of equipment so that it drifts away at a speed of 4.0 m/s. What happens to the astronaut?

7.30 An astronaut of mass 105 kg carrying an empty oxygen tank of mass 40 kg is stationary relative to a nearby space shuttle. She throws the tank away from herself with a speed of 2.0 m/s (measured relative to the shuttle). With what velocity relative to the shuttle does she start to move through space?

7.31 The 120-kg astronaut of Problem 7.29 forgot to check her tether and finds herself drifting away from the shuttle at 0.20 m/s. She has no propulsion device with her. She has only a 1.5-kg camera. With what velocity must she throw the camera to give herself a speed of 0.10 m/s back toward the shuttle?

7.32 A 10 000-kg empty railroad grain car coasts under a grain storage bin at 2.0 m/s. As it goes by the bin, 2000 kg of grain is dropped into the car. What is the velocity of the loaded car as it leaves the bin? What has happened to its initial kinetic energy?

7.33 A 10 000-kg railroad grain car and its load of 3000 kg of grain coast along a level track at 3.0 m/s. A door is open slightly and lets the grain pour out at a rate of 100 kg/s. What is the speed of the car after the grain is all gone? What has happened to the initial kinetic energy of the car?

● **7.34** An inflated rubber raft of mass 30 kg carries two swimmers of masses 50 kg and 70 kg. The raft and swimmers are initially floating at rest when the swimmers simultaneously dive off from the midpoints of opposite ends of the raft, each with a horizontal velocity of 3.0 m/s. The 50-kg swimmer dives to the left; the 70-kg swimmer dives to the right. With what speed and in what direction does the raft start to move?

● **7.35** Two hockey players skate toward the puck (and each other) with equal speeds of 14 m/s in the directions shown in Figure 7.16. Player 1 has a mass of 85 kg, player 2, 70 kg. If the players collide and become entangled, find the magnitude and direction of their common velocity after the collision.

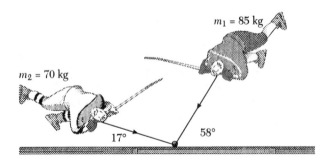

Figure 7.16
Problem 7.35.

7.4 Inelastic Collisions

7.36 A 600-g glider sitting at rest on an air track is struck by a 250-g glider traveling at 25 cm/s. The two stick together. What is their common final velocity?

7.37 A car of mass 600 kg, traveling at 20 m/s, collides with a stationary truck of mass 1400 kg. The two vehicles are locked together after the collision. What is their common velocity immediately after the collision?

7.38 A loaded 30 000-kg railroad car traveling at 8 km/h overtakes and runs into a 10 000-kg railroad car initially traveling at 5 km/h. The two couple and move off together. What is their speed?

7.39 A 600-g air track glider moves to the right with a speed of 20 cm/s while a 350-g glider moves to the left with a speed of 30 cm/s. The two collide and stick together. What is their velocity after the collision?

7.40 A train is being made up in a freight yard. An empty freight car, coasting at 6.0 m/s, strikes a loaded car that is stationary, and the cars couple together. Each car has a mass of 3000 kg when empty, and the loaded car contains 14 000 kg of bottled soft drinks. With what speed do the two coupled cars move?

7.41 A 23-g bullet traveling horizontally at 120 m/s hits a 4.0-kg block of wood sitting on a table and embeds itself in the wood. The coefficient of friction between block and table is 0.35. How far does the block slide across the table before coming to rest?

7.42 A 20-g bullet traveling horizontally at 120 m/s hits a 4.5-kg block of wood and embeds itself in the wood. The wood is sitting on a table and slides 0.65 m across the table before coming to rest. What is the coefficient of friction between block and table?

7.43 A 25-g bullet is fired horizontally with a speed of 100 m/s into a 5.0-kg block of wood, and the block and bullet move off together with speed v_f. What is this speed v_f? The block is suspended by cords so that it moves in an arc (but does not rotate), as illustrated in Figure 7.17 (this setup is called a ballistic pendulum). What is the height h to which the block rises?

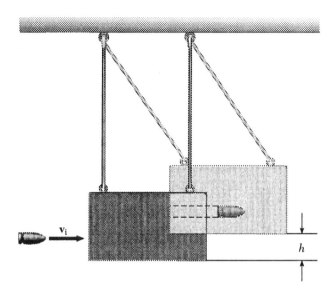

Figure 7.17
Problem 7.43. A ballistic pendulum.

O7.44   A 100-kg fullback is moving at 8.0 m/s as he is tackled head-on by a 110-kg linebacker moving at 3.0 m/s. With what speed and in what direction do the pair of players move after the tackle?

O7.45   An object of mass 1 kg, moving to the right at 8 m/s, catches up and collides with an object of mass 3 kg moving to the right at 4 m/s. The two objects stick together after impact. *(a)* What are the magnitude and direction of the velocity of the combined mass after the collision? *(b)* How much kinetic energy is lost in this collision?

O7.46   An object of mass 1 kg, moving to the right at 8 m/s, collides with an object of mass 3 kg moving to the left at 4 m/s. The two objects stick together after impact. *(a)* What are the magnitude and direction of the velocity of the combined mass after the collision? *(b)* How much kinetic energy is lost in this collision?

O7.47   Earth moves around the Sun at a speed of 29.9 km/s and has a mass of 5.99×10^{27} kg. If Earth is struck head-on by a meteorite of mass 2.00×10^9 kg moving at 60.0 km/s, by how much will Earth's velocity decrease?

O7.48   A stationary uranium nucleus with atomic mass $238 \times 1.67 \times 10^{-27}$ kg decays by emitting an alpha particle, which has atomic mass $4.00 \times 1.67 \times 10^{-27}$ kg and leaves the nucleus at a speed of 1.50×10^7 m/s. What is the recoil velocity of the remaining part of the original uranium nucleus?

O7.49   A 2.0-kg lump of clay is moving with a velocity of 2.0 m/s due north when it collides with a 1.0-kg lump of clay moving with a velocity of 1.0 m/s due east. What is the velocity (speed and direction) of the combined lump after the collision? How much kinetic energy is lost in this collision?

O7.50   A 1300-kg sedan moving east at 40 km/h collides with a 900-kg sports car moving north at 60 km/h, and the two get stuck together. What is their velocity (speed and direction) immediately after the collision?

●7.51   A 30-g bullet is fired horizontally into a 6.0-kg block of wood suspended as a ballistic pendulum (see Problem 7.43 and Fig. 7.17). The bullet remains in the block. The bullet and block move up in an arc to a height of 2.8 cm. What was the speed of the bullet as it entered the block?

7.5 Elastic Collisions

7.52   A 600-g air track glider initially at rest and a 300-g glider initially moving to the right at 25 cm/s collide elastically. Determine the velocity of each glider after the collision.

7.53   A 600-g glider moves on an air track with a velocity of 10 cm/s to the right while a 400-g glider moves to the left with a velocity of 20 cm/s. The two collide elastically. What is the velocity of each after the collision?

O7.54   A neutron having a mass of 1 mass unit and moving with a velocity of v_i makes a head-on elastic collision with a stationary oxygen nucleus that has a mass of 16 mass units. *(a)* What forward velocity does the oxygen nucleus receive? *(b)* What fraction of the kinetic energy of the neutron is transferred to the oxygen nucleus?

○ ❖ 7.55 Below are eight sets of initial conditions for perfectly elastic collisions on an air track. Calculate the final velocity of both gliders in each case, and rank the eight sets in order of increasing v_{2f}.

Set	Mass of glider 1 (g)	Initial velocity of glider 1 (cm/s)	Mass of glider 2 (g)	Initial velocity of glider 2 (cm/s)
A	600	10	600	0
B	600	10	300	0
C	600	10	200	0
D	300	15	200	0
E	300	20	250	0
F	300	25	350	0
G	450	28	375	0
H	375	28	450	0

○ ❖ 7.56 Below are eight sets of initial conditions for perfectly elastic collisions on an air track. Calculate the final velocity of both gliders in each case, and rank the eight sets in order of increasing v_{2f}.

Set	Mass of glider 1 (g)	Initial velocity of glider 1 (cm/s)	Mass of glider 2 (g)	Initial velocity of glider 2 (cm/s)
A	300	10	600	0
B	300	10	300	0
C	300	10	200	0
D	600	15	200	0
E	600	20	250	0
F	300	25	350	0
G	250	28	375	0
H	375	28	250	0

○7.57 A 3-kg ball makes a perfectly elastic, head-on collision with a stationary ball and rebounds with one quarter of its original speed. What is the mass of the initially stationary ball?

●7.58 A ball of mass m makes an elastic, head-on collision with a stationary ball of mass M and rebounds with $1/n$ of its original speed. Derive a formula for M in terms of m and n.

●7.59 Two identical billiard balls collide. Ball 1 is initially at rest, and ball 2 is initially moving at 1.0 m/s. Ball 2 leaves at an angle of 30° from its incoming direction. What is its final speed? What is the velocity (speed and direction) of ball 1 after the collision?

●7.60 Two identical billiard balls collide. Ball 1 is initially at rest, and ball 2 is initially moving at 1.5 m/s. Ball 1 leaves at an angle of 20° from the initial velocity direction of ball 2. What is the final speed of ball 1? What is the velocity (speed and direction) of ball 2 after the collision?

●7.61 A 2.0-kg ball is moving with a velocity of

2.0 m/s due north when it collides elastically with a 1.0-kg ball moving with a velocity of 1.0 m/s due east. The 2.0-kg ball leaves at an angle of 20° east of north. Calculate (a) the final speed of the 2.0-kg ball and (b) the final velocity (speed and direction) of the 1.0-kg ball.

*7.6 Rocket Propulsion

7.62   A certain rocket burns fuel at the rate of 200 kg/s, and gases leave the engine with a speed of 250 m/s. What thrust is produced?

7.63   Exhaust gases leave a certain rocket with a speed of 300 m/s. What must the mass flow rate be in order to lift the 50 000-kg rocket?

7.64   Rocket fuel is burned in a particular 40 000-kg rocket at a rate of 125 kg/s. What must the exhaust velocity of the gases be in order to lift this rocket?

General Problems

○7.65   A train of empty coal cars is being pulled along a straight, level section of track at 20 m/s. If rain starts to fall and all the cars together receive a combined total of 300 kg of water per minute, how much additional force must the engine supply to keep the trains moving at a constant speed of 20 m/s?

●7.66   An air table puck with mass $m_1 = 0.20$ kg has an initial velocity of 1.6 m/s at an angle of 45° to the x axis, as shown in Figure 7.18. It makes a glancing collision with an initially stationary puck of mass $m_2 = 0.40$ kg. After

the collision, puck 1 moves at 1.0 m/s at 20° to the x axis. What are the magnitude and direction of the final velocity of puck 2? The collision is elastic.

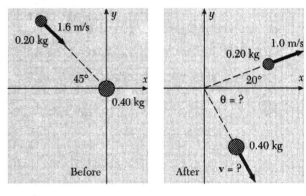

Figure 7.18
Problem 7.66.

●7.67   A 20-g bullet is fired horizontally into a 5.0-kg block of wood sitting on a horizontal surface. The bullet embeds in the wood and causes the block to slide 25 cm before coming to rest. The coefficient of friction between block and surface is 0.25. What was the speed of the bullet?

●7.68   An alpha particle of mass m collides elastically with a carbon-12 nucleus of mass $3\,m$. The carbon nucleus is initially at rest, and the alpha particle moves off at an angle of 20° measured from its initial direction. What is the angle between the direction of the outgoing carbon nucleus and the direction of the incoming alpha particle?

PHYSICS AT WORK

At 616 meters, the SkyBridge commuter rail bridge over the Fraser River in British Columbia is the longest cable-stayed bridge in the world devoted totally to transit use. In ancient times bridge builders relied mostly on empirical experience. With the development of first, iron, and later steel and concrete, bridge construction moved from guesswork to science. Based on Newton's laws, the science of statics and equilibrium grew quickly to provide the analytical tools needed by bridge designers and engineers.

In traveling over long suspension-type bridges, you may have noticed that the center spans are not perfectly level between the two supporting towers but curve slightly upward. Why do you think these bridges are built this way?

(Marie van Manen/B.C. Transit)

8

Static Equilibrium

Key Issues

- An object at rest which remains at rest is in static equilibrium.

- The first condition of equilibrium is that the net force on an object in equilibrium must be zero.

- Torques are the rotation equivalent of forces. Torque depends upon the force exerted and where that force is located as well as the axis about which rotation is being considered. Torques are always calculated with respect to some reference point.

- The second condition of equilibrium is that the sum of the torques on an object in equilibrium must be zero.

- An object may be in stable, neutral, or unstable equilibrium depending upon whether its center of gravity is raised, unaffected, or lowered if it is moved slightly from equilibrium.

Statics is the study of bodies at rest. If a body is to remain at rest, all the external forces acting on it must be balanced—that is, the net force on the body must be zero. In this chapter, we learn how the forces on a body must be arranged if they are to balance and so provide a net force of zero. In doing this, we once again use the ideas and techniques of vector addition and the resolution of vectors into components.

From Newton's second law of motion, $\mathbf{F} = m\mathbf{a}$, we know that if the net force on an object is zero, the acceleration of the object is also zero. Such an object remains at rest if it was initially at rest or remains moving with constant velocity if it was initially moving. An object that remains at rest is in a state of **static equilibrium.**

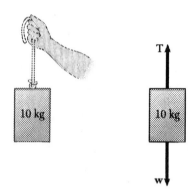

Figure 8.1
Free-body diagrams are as important in statics as they were in dynamics.

Statics is the branch of mechanics that deals with bodies that are standing still.

8.1 Balancing Forces

CONCEPT OVERVIEW
■ The forces acting on an object that remains at rest must balance or add to zero.

It is always best to begin any new topic with familiar situations or with ordinary occurrences whenever that is possible. Think of the forces that act on a 10-kg mass suspended as sketched in Figure 8.1. From the free-body diagram shown in the figure, you see that gravity pulls downward with a force $\mathbf{w}$, where

$$w = mg = (10 \text{ kg})(9.8 \text{ m/s}^2) = 98 \text{ N}$$

and the rope pulls upward with a force $\mathbf{T}$, the tension in the rope. These two forces must just balance if the mass is in equilibrium; therefore the tension must have a magnitude of 98 N.

What forces act on you when you sit in a chair? If the chair were not there, you would fall down; the chair supports you with an upward force. The force of gravity acts downward on you, as Figure 8.2 shows, and the chair exerts a normal force upward. If you are to remain at equilibrium seated in the chair, the upward force must equal the downward force. If your weight $\mathbf{w}$ is 600 N downward, then the chair pushes up on you with a normal force $\mathbf{F}_N$ of 600 N.

We generalize this discussion in the next section to take care of more complicated situations in which there are additional forces.

Figure 8.2
Normal forces and weight are present in most situations.

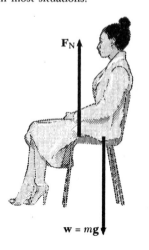

8.2 First Condition of Equilibrium

CONCEPT OVERVIEW
■ An object in static equilibrium has zero net force acting on it.

The **first condition of equilibrium** is that, in order for a body at rest to remain at rest, the *vector sum* of all the forces acting on it vanishes:

$$\Sigma\mathbf{F} = \mathbf{F}_1 + \mathbf{F}_2 + \mathbf{F}_3 + \mathbf{F}_4 + \mathbf{F}_5 + \cdots = 0 \qquad \textbf{(8.1)}$$

Consider the 10-kg mass of Figure 8.1. Because the weight of this mass is 98 N, the tension in the rope is just 98 N. Suppose that we now tie this rope with two other ropes, as shown in Figure 8.3. What forces are exerted by these two new ropes? To answer this question, look at the forces exerted on the knot

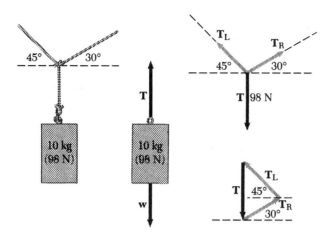

Figure 8.3
The net force on a knot due to the tensions in three ropes must be zero if the knot is stationary.

where the three ropes are joined. The knot is at rest; therefore the sum of the forces acting on it must be zero. Figure 8.3 contains a free-body diagram showing just those forces. The tension T is the magnitude of the force exerted on the knot by the rope attached to the weight, T_L is the magnitude of the force exerted on the knot by the rope on the left, which makes an angle of 45° with the horizontal, and T_R is the magnitude of the force exerted on the knot by the rope on the right, which makes an angle of 30° with the horizontal. These three forces must add to zero. Graphically, they can be added as shown in the figure; the force vectors form a triangle.

The graphical method of adding vectors shown in Figure 8.3 is certainly a valid procedure, but usually it is more accurate to add forces analytically using components, as in Figure 8.4. In terms of components, our first condition of equilibrium, Equation 8.1, is

$$\Sigma F_x = 0 \qquad\qquad \textbf{(8.2a)}$$

$$\Sigma F_y = 0 \qquad\qquad \textbf{(8.2b)}$$

Forces could have components in all three directions, in which case we would also require

$$\Sigma F_z = 0$$

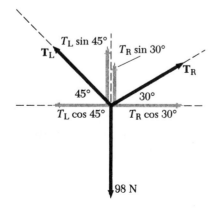

Figure 8.4
As always, resolve vectors into components and then add the components.

Here, however, we consider only *coplanar* forces, that is, forces that lie in the same plane. Therefore only two components are needed. The z component of all the forces we shall consider is zero.

Now apply Equations 8.2a and 8.2b to the forces acting on the knot in Figure 8.3. First, determine all the x components and set their sum equal to zero:

$$\Sigma F_x = T_x + T_{Rx} + T_{Lx} = 0$$

$$T_x = 0$$

$$T_{Rx} = T_R \cos 30° = 0.866 T_R$$

$$T_{Lx} = -T_L \cos 45° = -0.707 T_L$$

$$\Sigma F_x = 0 + 0.866 T_R - 0.707 T_L = 0$$

$$0.866 T_R = 0.707 T_L$$

$$T_R = 0.816 T_L$$

This is one equation with two unknowns. We need another equation, $\Sigma F_y = 0$:

$$\Sigma F_y = T_y + T_{Ry} + T_{Ly} = 0$$

$$T_y = -98 \text{ N}$$

$$T_{Ry} = T_R \sin 30° = 0.500 T_R$$

$$T_{Ly} = T_L \sin 45° = 0.707 T_L$$

$$\Sigma F_y = -98 \text{ N} + 0.500 T_R + 0.707 T_L = 0$$

$$98 \text{ N} = 0.500(0.816 T_L) + 0.707 T_L = 1.115 T_L$$

$$T_L = 87.9 \text{ N} \approx 88 \text{ N}$$

$$T_R = 71.7 \text{ N} \approx 72 \text{ N}$$

Problem-Solving Tips

- Free-body diagrams remain essential.
- A free-body diagram shows *all* the forces acting on a particular body and *only* those forces. It shows none of the forces acting on other bodies in the system.

Example 8.1

Figure 8.5 shows a rope fixed at the left end, supporting a 1000-N weight, passing over a pulley, and attached to a 4000-N weight. The 1000-N weight is attached by a second rope to a lightweight pulley that is free to move. This pulley remains at the center of the longer rope between the fixed support on the left and the pulley on the right. Therefore, the angles labeled θ in the figure are equal. Find the value of θ.

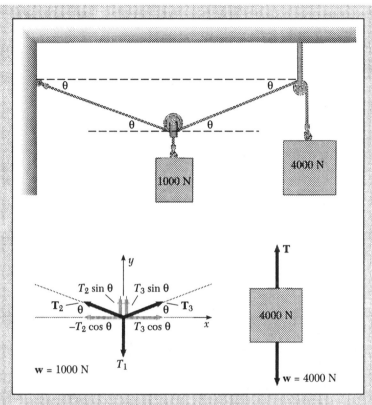

Figure 8.5
What angle θ does the rope make with the horizontal?

Reasoning

Figure 8.5 shows a free-body diagram of all the forces acting on the movable pulley: $\mathbf{T_1}$, $\mathbf{T_2}$, and $\mathbf{T_3}$. Because this pulley is at rest, the sum of these forces is zero:

$$\Sigma \mathbf{F} = \mathbf{T_1} + \mathbf{T_2} + \mathbf{T_3} = 0$$

We must do our calculations with the component form of this equation. Therefore we write the components of these three forces and set the sum of the x components equal to zero and the sum of the y components equal to zero. Then we can use all these components in the first condition of equilibrium.

Solution

$$T_{1x} = 0$$

$$T_{1y} = -1000 \text{ N}$$

$$T_{2x} = T_2 \cos \theta$$

$$T_{2y} = T_2 \sin \theta$$

$$T_{3x} = T_3 \cos \theta$$

$$T_{3y} = T_3 \sin \theta$$

The magnitude of the tension in the rope is the same everywhere along the entire length. In the vertical section of this rope supporting the 4000-N weight, the tension is clearly 4000 N. Therefore the tension *throughout* this rope is 4000 N:

$$T_2 = T_3 = T = 4000 \text{ N}$$

Therefore,

$$T_{2x} = -(4000 \text{ N}) \cos \theta$$

$$T_{2y} = (4000 \text{ N}) \sin \theta$$

$$T_{3x} = (4000 \text{ N}) \cos \theta$$

$$T_{3y} = (4000 \text{ N}) \sin \theta$$

Now all these components can be used in the first condition of equilibrium, Equations 8.2a and 8.2b. Beginning with the x components, we have

$$\Sigma F_x = (4000 \text{ N}) \cos \theta - (4000 \text{ N}) \cos \theta = 0$$

Although this is a true statement for any value of θ, it does not provide any real information. For the y components, we have

$$\Sigma F_y = (4000 \text{ N}) \sin \theta + (4000 \text{ N}) \sin \theta - 1000 \text{ N} = 0$$

$$(8000 \text{ N}) \sin \theta = 1000 \text{ N}$$

$$\sin \theta = 0.125$$

$$\theta = \sin^{-1} (0.125) = 7.2°$$

Practice Exercise Replace the hanging weight on the right with a 2000-N weight and find the angle θ.

Answer 14.5°

Problem-Solving Tip

■ When resolving forces into components, choose axes that correspond to directions in the situation at hand; x and y axes do not always have to be horizontal and vertical.

Example 8.2

Consider a 100-N sled being held at rest on a slick, icy (in other words, frictionless) hill, as sketched in Figure 8.6. The hill makes an angle of 30° with the horizontal, and the sled is held in place by a rope parallel to the hill. Find the tension in the rope.

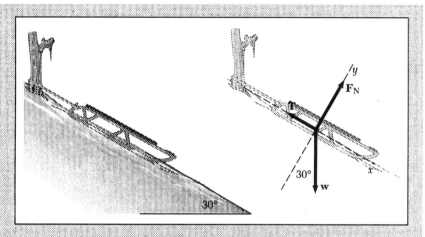

Figure 8.6
Find the tension in the rope holding the sled.

Reasoning

A free-body diagram showing all the forces on the sled has been drawn in the figure. Choose the x axis along the hill and the y axis perpendicular to that. Use the component form of the first condition of equilibrium and calculate x and y components of all the forces.

Solution

$$T_x = -T$$

$$T_y = 0$$

$$F_{Nx} = 0$$

$$F_{Ny} = F_N$$

$$w_x = w \sin 30° = (100 \text{ N})(0.500) = 50 \text{ N}$$

$$w_y = -w \cos 30° = -(100 \text{ N})(0.866) = -86.6 \text{ N}$$

Applying the x component of the first condition of equilibrium, we get

$$\Sigma F_x = T_x + F_{Nx} + w_x = 0$$

$$-T + 0 + 50 \text{ N} = 0$$

$$T = 50 \text{ N}$$

We can stop here since we have solved for the tension. The other part of the first condition of equilibrium gives us something we were not asked for, the value of the normal force:

$$\Sigma F_y = T_y + F_{Ny} + w_y = 0$$

$$0 + F_N + (-86.6 \text{ N}) = 0$$

$$F_N = 86.6 \text{ N} \approx 87 \text{ N}$$

A careful choice of x and y axes has greatly reduced the amount of work required to find the answer. Other choices would have increased our work and therefore our chance of error.

_____

Practice Exercise Find the tension in a rope that keeps a 150-N sled from sliding down a frictionless hillside inclined 20° from the horizontal. The rope is parallel to the hillside.

Answer 51 N

8.3 Static Friction

CONCEPT OVERVIEW

■ A friction force occurs whenever two surfaces can move relative to each other, with the friction force always opposing the relative motion.

■ The friction force is proportional to the force that pushes the two surfaces together.

We encountered friction in Chapter 4, but since that chapter dealt with moving objects, the emphasis was on sliding, or kinetic, friction. Now we concentrate on **static friction.**

Figure 8.7 shows a block sitting on a horizontal surface. An external force $\mathbf{F}_{ext}$ is being exerted, attempting to push the block to the right. This motion is opposed by the friction force $\mathbf{F}_f$ that the surface exerts on the block. Until the block slips and starts to move, the friction force just equals the external force. As the external force changes, the interaction between the two surfaces in contact changes and consequently the friction force changes. The friction force may have any value from zero to a maximum value $F_{f,max}$ given by Equation 4.7:

$$F_f \leq F_{f,max} = \mu_s F_N \tag{4.7}$$

where F_N is the normal force and μ_s is the coefficient of static friction; typical values for this coefficient are listed in Table 4.1. Because the friction force always opposes the motion that would occur if friction were not present, its direction is always opposite the direction of that motion. Often it is easy to see what these two opposing directions must be; sometimes it is not.

Figure 8.7
The force of static friction F_f is proportional to the normal force F_N between the two surfaces.

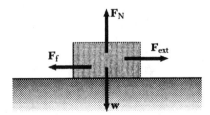

Example 8.3

A 100-N block sits on a plane inclined 25° from the horizontal, as shown in Figure 8.8. The block is held by a rope that is parallel to the plane and keeps the block from slipping down the plane. The coefficient of static friction between block and plane is 0.35. Find the tension in the rope.

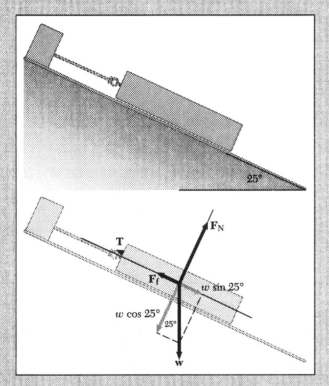

Figure 8.8
The weight of the block is 100 N. What is the tension in the restraining rope?

Reasoning

Because the block is not moving, the first condition of equilibrium must apply. Therefore we must resolve all the forces acting on the block into components and use Equations 8.2a and 8.2b.

From the fact that the rope is needed to keep the block from moving, we know that the friction force must be at its maximum value $\mu_s F_N$. Remember, the magnitude of F_f is always particular to a specific situation, with its maximum value in any situation determined by the normal force between the two surfaces in contact.

Solution

In the free-body diagram of Figure 8.8, the x and y axes have been chosen parallel and perpendicular to the plane. With this choice of axes, the normal force lies in the positive y direction, and the tension and friction forces both lie in the negative x direction. The weight we must resolve into components:

$$w_x = w \sin 25° = 42.3 \text{ N}$$

$$w_y = -w \cos 25° = -90.6 \text{ N}$$

The first condition of equilibrium in the x direction gives

$$\Sigma F_x = w_x - F_f - T = 0$$

$$w_x - F_{f,\text{max}} - T = 0$$

$$w_x - \mu F_N - T = 0$$

$$42.3 \text{ N} - 0.35 F_N - T = 0 \qquad (1)$$

The first condition of equilibrium in the y direction gives

$$\Sigma F_y = F_N - 90.6 \text{ N} = 0$$

$$F_N = 90.6 \text{ N}$$

This value of the normal force can be substituted in Equation 1 to find

$$42.3 \text{ N} - 0.35(90.6 \text{ N}) - T = 0$$

$$T = 11 \text{ N}$$

If we wanted the friction force, we would find that from

$$F_f = F_{f,\text{max}} = \mu F_N = (0.35)(90.6 \text{ N}) = 32 \text{ N}$$

Practice Exercise　A 50-N block on a plane inclined 30° from the horizontal is kept from sliding down the plane by a rope parallel to the plane. The coefficient of static friction between block and plane is 0.20. What is the tension in the rope?

Answer　16 N

Notice that the normal force is *not* equal to the weight in this case. Because there are numerous special cases where the normal force does happen to be equal to the weight, it is sometimes (and wrongly) assumed that the normal force and weight always have the same magnitude. Be aware that this assumption simply is not true!

Example 8.4

Consider a 30-N sled on which an 80-N child sits, as sketched in Figure 8.9. The sled is being pulled with a rope that makes an angle of 30° with the snowy, horizontal ground. The coefficient of static friction between snow and sled is 0.20. What is the tension in the rope just as the sled starts to move?

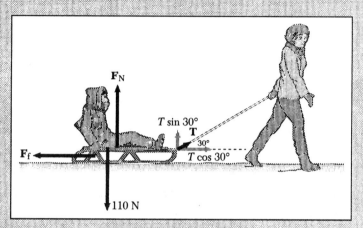

Figure 8.9
What is the tension in the rope just before the sled starts to move?

Reasoning

Just before the sled starts to move, it is in static equilibrium and the friction force has its maximum value, μF_N. Therefore we should apply the first condition of equilibrium.

Note that in the figure, rather than draw a free-body diagram, we have chosen to show all the forces acting on the sled right on the sketch. This shortcut is acceptable in very simple situations, but you should use it sparingly and carefully.

Solution

The components of **T** are

$$T_x = T \cos 30° = 0.866T$$

$$T_y = T \sin 30° = 0.500T$$

The total weight of child and sled is $\mathbf{w} = -110$ N, directed downward. The friction force $\mathbf{F}_f$ is toward the left, opposite the direction of impending motion, so it too is negative:

$$F_f = -\mu F_N = -0.20 F_N$$

Now we have all the pieces we need to use the first condition of equilibrium:

$$\Sigma F_x = T_x + F_f = 0$$

$$0.866T - 0.20 F_N = 0$$

$$T = 0.231 F_N \tag{1}$$

This result tells us the tension in terms of the normal force, but what we need is a numerical value for T. We can get a value for F_N from the other half of the first condition of equilibrium and then use that value to get our T value:

$$\Sigma F_y = F_N + T_y - 110 \text{ N} = 0$$

$$F_N = 110 \text{ N} - 0.500T$$

This is still just a relationship between F_N and T, but we can use it in our earlier equation,

$$T = 0.231 F_N = 0.231(110 \text{ N} - 0.500T)$$

$$T + 0.116T = 25.4 \text{ N}$$

$$T = \frac{25.4 \text{ N}}{1.116} = 22.8 \text{ N} \approx 23 \text{ N}$$

Even though we are not asked to, we can easily continue and find the normal force:

$$F_N = 110 \text{ N} - 0.500(23 \text{ N}) = 98 \text{ N}$$

Notice once again that the normal force is not the weight of the child or the sled.

Practice Exercise   A sled and child together have a weight of 200 N. A rope which makes an angle of 20° with the horizontal pulls on the sled. The coefficient of friction between the sled and the snow is 0.15. What will be the maximum tension in the rope before the sled starts to move?

Answer   30 N

Problem-Solving Tip

- The friction force always opposes the motion that would occur if friction were not present.

Example 8.5

In Figure 8.10 a heavy, rough crate sits on an inclined plane. The weight of the crate is 600 N. The plane is inclined 37° from the horizontal, and the coefficient of static friction between plane and crate is 0.83. Find the force, exerted parallel to the plane, needed to just get the crate moving up the plane. Then find the force, exerted parallel to the plane, needed to just get the crate moving down the plane. What happens to the friction force in each case?

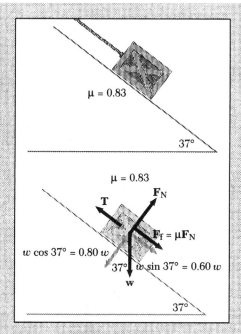

Figure 8.10
What tension will just get the crate moving up the plane?

Reasoning

Figure 8.10 also shows a free-body diagram for the case when the crate is on the verge of moving up the plane. The force of friction is directed down the plane since friction always opposes the motion that would occur if friction were not present. The rope exerts a force **T** up the plane, and the weight **w** acts vertically downward. The plane exerts two forces: the friction force F_f parallel to its surface and the normal force F_N perpendicular to its surface.

If the crate is on the verge of moving, the friction force is its maximum value, $F_f = \mu F_N$. As shown in the figure, choose the x axis parallel to the plane and the y axis perpendicular to it. Resolve the weight into its components and apply the first condition of equilibrium.

Solution

For the x components,

$$\Sigma F_x = -T + F_f + w \sin 37° = 0$$

$$-T + \mu F_N + (600 \text{ N})(0.60) = 0$$

$$T = 0.83F_N + 360 \text{ N}$$

For the y components,

$$\Sigma F_y = F_N - w \cos 37° = 0$$

$$F_N = (600 \text{ N})(0.80) = 480 \text{ N}$$

Now that the normal force is known, we also know the friction force and so can solve for the tension:

$$F_f = \mu F_N = (0.83)(480 \text{ N}) = 400 \text{ N}$$

$$T = 400 \text{ N} + 360 \text{ N} = 760 \text{ N}, \text{ directed up the plane}$$

Reasoning

Figure 8.11 is a free-body diagram for the case when the crate is on the verge of moving down the plane. The rope now exerts a force **T** down the plane, and the force of friction is now up the plane since friction always opposes the motion that would occur if friction were not present. Apply the first condition of equilibrium again.

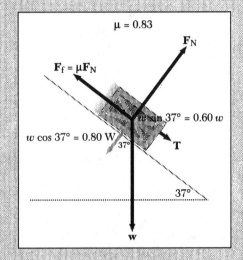

Figure 8.11
What tension will just get the crate moving down the plane?

Solution

$$\Sigma F_x = T - F_f + w \sin 37° = 0$$

$$T - \mu F_N + (600 \text{ N})(0.60) = 0$$

$$T = 0.83 F_N - 360 \text{ N}$$

$$\Sigma F_y = F_N - w \cos 37° = 0$$

$$F_N = (600 \text{ N})(0.80) = 480 \text{ N}$$

$$F_f = \mu F_N = (0.83)(480 \text{ N}) = 400 \text{ N}$$

$$T = 400 \text{ N} - 360 \text{ N} = 40 \text{ N}, \text{ directed down the plane}$$

It is instructive to work this example again with a different coefficient of static friction. This time, sand the bottom of the crate so that it is smoother and $\mu = 0.40$. The problem seems the same. The free-body diagrams are the same.

Proceed just as before. For the case of pulling the crate up the plane, we have

$$\Sigma F_x = -T + F_f + w \sin 37° = 0$$

$$-T + \mu F_N + (600 \text{ N})(0.60) = 0$$

$$T = 0.40\, F_N + 360 \text{ N}$$

$$\Sigma F_y = F_N - w \cos 37° = 0$$

$$F_N = (600 \text{ N})(0.80) = 480 \text{ N}$$

$$T = (0.40)(480 \text{ N}) + 360 \text{ N} = 550 \text{ N, directed up the plane}$$

So far, so good; this is exactly what we did in the example. Now, look at the case of sliding down the plane. The free-body diagram looks the same.

$$\Sigma F_x = T + w \sin 37° - F_f = 0$$

$$T = \mu F_N - (600 \text{ N})(0.60)$$

$$\Sigma F_y = F_N - w \cos 37° = 0$$

$$F_N = (600 \text{ N})(0.80) = 480 \text{ N}$$

$$T = (0.40)(480 \text{ N}) - (600 \text{ N})(0.60) = 190 \text{ N} - 360 \text{ N} = -170 \text{ N}$$

How do we interpret a negative value for the tension in a rope? You simply cannot push with a rope! The negative value means something went wrong! The maximum value of the friction force is 190 N. The component of the crate's weight directed down the plane is 360 N. Therefore if we turn the crate loose, it will slide down the plane. With this smoother bottom on the crate, we must pull up the plane with a tension force of 550 N to have the crate just ready to move up the plane. When it is on the verge of moving *down* the plane, however, we must also pull *up* the plane, this time with a tension force of 170 N.

Equations are useful and therefore important, but never simply plop numbers into an equation, solve for an answer, and leave without asking what the answer means. As already mentioned, equations are a *summary* of ideas and understanding. They must never become a *substitute* for them.

8.4 Rotational Forces: Torques

CONCEPT OVERVIEW

- Torque is the rotational analog of force.
- Torque depends upon the magnitude, point of application, and direction of the applied force.

As we have seen over and over again, a force is a push or a pull, and an unbalanced force causes an object to move (to accelerate). All the motion we have dealt with so far, however, has moved our objects from one point in space to another. This type of motion involving a displacement is called **translational motion.** We now consider another type of motion, **rotational motion,** where an object moves not from one place to another but rather around some axis of rotation.

Figure 8.12
Why does it make such a difference where you exert a force when you try to rotate a door?

In *moment arm*, *moment* has the sense of influence or importance rather than the sense of time, and *arm* conveys the idea of length or distance. Therefore moment arm tells us the *distance* between a center of rotation and the force *influencing* that rotation.

As you may have guessed, forces are also necessary for rotational motion, but how must a force act to cause something to rotate? Have you ever pushed on the edge of a door next to the hinges, as in Figure 8.12? What happens? The door is difficult to move; you must either exert a very large force if you insist on pushing on the hinged side or else move your hands to the other edge of the door, far from the hinges.

How can three children balance on a teeter-totter? In Figure 8.13, the two children on the left exert about twice as much downward force as the one child on the right. The support in the middle exerts an upward force on the board. How are all these forces balanced?

To analyze situations such as these two, we must look at *where* the forces are applied. The rotational effect of a force depends upon its point of application. A door opens easily if the force is applied far from the hinges. The two children sit closer to the teeter-totter's support than the single child does.

The rotational analog of force is **torque** and is given the symbol τ (lower-case Greek "*tau*"). A torque is not a force; it is something different. Consider the force **F** in Figure 8.14(a), which acts perpendicular to a wrench. As this force is exerted on the wrench, the wrench rotates clockwise. The point (or line) around which the rotation occurs, point O in the figure, is called either the **pivot** or the **fulcrum**. In a door, for instance, the pivot is along the line through the hinges; on a teeter-totter the pivot is the point where the support touches the board. The *perpendicular distance* from the pivot point to the line of action[1] of the force is $r_\perp$; this perpendicular distance is known as the **moment arm** of the force relative to the pivot. The torque—the rotational effect of the force—is defined by

$$\tau = r_\perp F \tag{8.3}$$

Notice that the force in Figure 8.14(b) exerts no torque relative to the pivot. The moment arm for this force is zero, and therefore the torque is zero.

Figure 8.14(c) shows a force **F** exerted at some angle θ other than the $\theta = 90°$ of (a) or the $\theta = 0$ of (b). This angle θ is the angle between the line of action of the force and a line drawn from the point of application of the force to the pivot. Equation 8.3 still defines the rotational effect of this force. The moment arm $r_\perp$ is once again the perpendicular distance from the pivot to the

---

[1] The **line of action** of a force passes through the point at which the force is applied and is parallel to the force. The dashed lines in Figure 8.14 indicate the lines of action of the forces.

Figure 8.13
Children on a teeter-totter soon learn about balancing torques.

line of action of the force. If the distance from the pivot to the point of application of the force is r, then the moment arm $r_\perp$ is one side of a right triangle in which r is the hypotenuse. Because $\sin\theta = \sin(180° - \theta)$, we can write

$$r_\perp = r\sin\theta \tag{8.4}$$

and the torque can be written as

$$\tau = r_\perp F = (r\sin\theta)F \tag{8.5}$$

In Figure 8.14(d) this same force **F** is resolved into its components. Just as in (b), there is no torque due to the parallel component $F_\parallel$. The distance r is the moment arm for the perpendicular component $F_\perp$. Therefore the torque can also be written as

$$\tau = rF_\perp = (r)(F\sin\theta) \tag{8.6}$$

Notice that Equations 8.3, 8.5, and 8.6 are all equivalent. Thus we can write the general expression for torque as

$$\tau = rF\sin\theta \tag{8.7}$$

where θ is the angle between the force and a line drawn from its point of application to the pivot point.

Calculating a torque *always requires* that you choose one point as your only pivot and calculate *all* torque relative to it. A torque that tends to rotate something counterclockwise is usually defined as positive; a clockwise torque, negative. The unit of torque is a unit of length times a unit of force, usually newton-meters (N·m).

Example 8.6

Consider again the children on the horizontal teeter-totter of Figure 8.13. The children on the left have a combined weight of $w_L = 800$ N and are located at a distance $r_L = 1.0$ m from the fulcrum (pivot). The child on the right has a weight of $w_R = 400$ N and is located a distance $r_R = 2.0$ m from the fulcrum. Find the torque exerted by the children on the left and the child on the right.

Reasoning

The distances from the fulcrum to the children are perpendicular to their weights; therefore the angles between these forces and the distances are 90°, and $\sin 90° = 1$. This happy result always makes calculating torques easier.

Solution

The torque due to the children on the left is

$$\tau_L = r_{\perp L}w_L = (1.0\text{ m})(800\text{ N}) = 800\text{ N·m}$$

This torque tends to rotate the teeter-totter counterclockwise and so is positive.

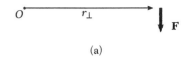

(a)

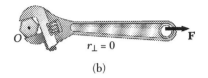

(b)

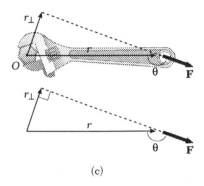

(c)

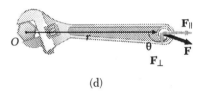

(d)

Figure 8.14
Any force on the wrench that causes it to rotate produces a torque. No torque is produced in (b).

By using a long-handled wrench, this plumber can exert considerable torque on the pipe, insuring a water-tight fit. (© *Richard Megna/Fundamental Photographs, New York*)

> The torque due to the child on the right is
>
> $$\tau_R = r_{\perp R} w_R = (2.0 \text{ m})(400 \text{ N}) = 800 \text{ N·m}$$
>
> This torque tends to rotate the teeter-totter clockwise and so is negative:
>
> $$\tau_R = -800 \text{ N·m}$$
>
> The teeter-totter remains at rest because the clockwise torques are balanced by the counterclockwise torques.
>
> The fulcrum in the center of the board exerts an upward force to support the board and children. Notice that this force cannot cause any torque about the fulcrum because its moment arm is zero.

8.5 Second Condition of Equilibrium

CONCEPT OVERVIEW

■ For an extended body to remain in static equilibrium, the sum of all the torques about any chosen axis must be zero.

Imagine you and a friend facing each other across a table on which a meter stick lies flat. You apply at the 0-cm end a force **F** directed toward your friend, who at the same time applies at the 100-cm end an equal force **F** directed toward you. That these two balanced forces cause the stick to rotate about the 50-cm point tells you that the first condition of equilibrium alone is not sufficient for describing static equilibrium.

We already know that the sum of all the forces acting on a body in equilibrium is zero; that is the first condition of equilibrium. Two equal but opposite forces applied at the ends of and perpendicular to a meter stick will cause the meter stick to rotate. These forces are balanced so the first condition of equilibrium alone is not sufficient to ensure equilibrium. The rotational effects of forces—in other words, the torques—on a body in equilibrium also must balance. This means the sum of the clockwise torques must equal the sum of the counterclockwise torques:

$$\Sigma \tau_\circlearrowleft = \Sigma \tau_\circlearrowright \tag{8.8}$$

Since we consider counterclockwise torques to be positive and clockwise torques to be negative, we can also write

$$\Sigma \tau = 0 \tag{8.9}$$

This is the **second condition of equilibrium.** When using it to solve problems, be careful that all the torques are calculated about the *same* point.

For a small body, like a knot on a rope, all the forces pass through a common point. Therefore their moment arms about that point are zero, their torques are zero, and the second condition of equilibrium is always satisfied. Forces that pass through a common point are known as **concurrent forces.**

Problem-Solving Tips

■ All torques must be calculated about the same pivot.

■ Calculating torques about a point through which one of the forces passes causes the torque due to that force to be zero; this usually makes the calculation easier.

Example 8.7

An ornamental brass ball with a weight of 100 N is supported at the end of a flagpole as shown in Figure 8.15. The flagpole is horizontal and has a length of 5.00 m. The flagpole's weight is so much less than that of the brass ball that we shall ignore it. The flagpole is attached to a building at one end and supported at the other end by a guy wire that makes an angle of 25.0° with the horizontal. Find the tension in the guy wire and the force exerted by the building on the pole.

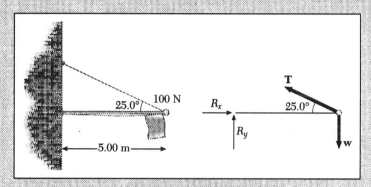

Figure 8.15
What are the forces acting on the horizontal flagpole when we neglect its weight?

Reasoning

The forces acting on the flagpole are shown in the free-body diagram. The components of the reaction force **R** that the building exerts on the pole are drawn as R_x and R_y. The direction of the force **T** is known, and both the magnitude and direction of the weight **w** are known.

Solution

The first condition of equilibrium must be met because the pole is in static equilibrium. Therefore

$$\Sigma F_x = R_x - T \cos 25.0° = 0$$

$$R_x = T \cos 25.0° = 0.906\,T$$

$$\Sigma F_y = R_y + T \sin 25.0° - 100\text{ N} = 0$$

$$R_y = 100\text{ N} - 0.423T$$

Notice that we have three unknowns—R_x, R_y, and T—but only two equations. There is not (yet) enough information to solve for these unknowns.

The second condition of equilibrium, which must apply because the flagpole is not rotating, provides the additional information we need. We must first choose a pivot point about which to calculate all the torques. If we choose the point where the flagpole is attached to the building, our calculating task will be minimized. The torques due to R_x and R_y vanish because these forces pass through this point and therefore have zero moment arms.

It is often useful to make a list of all the forces and the torque caused by each:

Force	Torque
$\mathbf{R}_x$	$rR \sin \theta = 0 \quad (r = 0)$
$\mathbf{R}_y$	$rR \sin \theta = 0 \quad (r = 0)$
$\mathbf{T}$	$rT \sin \theta = (5.00\ \text{m})(T)(\sin 25.0°) = (2.11\ \text{m})T \quad \text{ccw} \circlearrowleft$
$\mathbf{w}$	$rw \sin \theta = (5.00\ \text{m})(100\ \text{N})(\sin 90.0°) = 500\ \text{N·m} \quad \text{cw} \circlearrowright$

We can now use Equation 8.8:

$$\Sigma \tau_{\circlearrowright} = \Sigma \tau_{\circlearrowleft}$$

$$500\ \text{N·m} = (2.11\ \text{m})T$$

$$T = 237\ \text{N}$$

The tension in the wire is 237 N. This information can now go back into the earlier equations to find R_x and R_y:

$$R_x = 0.906T = 215\ \text{N}$$

$$R_y = 100\ \text{N} - 0.423T = 0$$

Thus the building exerts on the pole a 215-N force directed along the x axis (horizontal).

Practice Exercise   Another light, horizontal flagpole, 3.00 m long, with a 50-N ornament at its end, is attached to a building by a hinge, and supported by a guy wire, attached at the ornament end, that makes an angle of 30.0° with the horizontal. Find the tension in the guy wire and the force exerted by the building.

Answer $T = 100$ N; $F_{bldg} = 87$ N, horizontal

Practice Exercise   Consider another light flagpole, 6.00 m long, with a 50-N ornament at its end, attached to a building by a hinge. This flagpole makes an angle of 30.0° with the horizontal and is held in place by a horizontal wire attached at its end. Find the tension in the guy wire and the force exerted by the building.

Answer $T = 87$ N; $F_{bldg} = 100$ N, 30.0° from horizontal

Example 8.8

Two people lift an 800-N load as illustrated in Figure 8.16, exerting upward forces at the ends of a lightweight pole that is 4.0 m long. The load is placed 1.5 m from the right end. How much of the load is carried by each person?

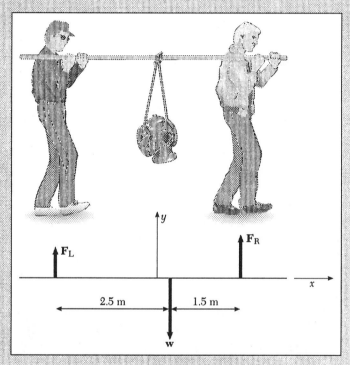

Figure 8.16
What force is exerted by each person?

Reasoning

Figure 8.16 includes a free-body diagram showing the forces on the pole and their points of application. The location of these forces is very important, and you should always position them carefully when you draw your own free-body diagrams. There are no forces in the x direction in this case, so the first condition of equilibrium provides only one equation.

To apply the second condition of equilibrium, we must choose a pivot point about which to calculate the torques. Any point will do, but the *same point* must be used for *all* the torques. Choosing the left end of the pole makes the moment arm zero for F_L.

Solution

The first condition gives

$$\Sigma F_y = F_L + F_R - w = 0$$

$$F_L + F_R = w = 800 \text{ N}$$

where F_L is the force exerted by the person on the left and F_R is the force exerted by the person on the right. The sum of their upward forces is 800 N.

The forces and their torques are

Force	Torque about left end
$\mathbf{F_L}$	$r_\perp F_L = 0$
$\mathbf{F_R}$	$r_\perp F_R = (4.0 \text{ m}) F_R \quad \text{ccw}$
$\mathbf{w}$	$r_\perp w = (2.5 \text{ m})(800 \text{ N}) = 2000 \text{ N·m} \quad \text{cw}$

The second condition of equilibrium gives

$$\Sigma \tau_\circlearrowright = \Sigma \tau_\circlearrowleft$$

$$2000 \text{ N·m} = (4.0 \text{ m}) F_R$$

$$F_R = 500 \text{ N}$$

This result provides the information necessary to solve for F_L:

$$F_L + F_R = 800 \text{ N}$$

$$F_L = 800 \text{ N} - 500 \text{ N} = 300 \text{ N}$$

It is left as a homework problem (Problem 8.31) for you to show that calculating torques about the right end of the pole in Example 8.8 gives exactly the same results.

Example 8.9

Figure 8.17 shows a 10-kg child sitting in a portable highchair that attaches to a table. The figure also gives approximate dimensions of the portable chair. Determine the forces exerted on it. Assume that the weight of the chair is negligible compared with the weight of the child.

Reasoning

The chair remains at rest and therefore is in static equilibrium, with both conditions of equilibrium fulfilled. The child's weight exerts a downward force **w** on the chair. Because the chair is touching the top and bottom surfaces of the table, both these surfaces exert a reaction force on the chair; let us call them F_{top} and F_{bottom}.

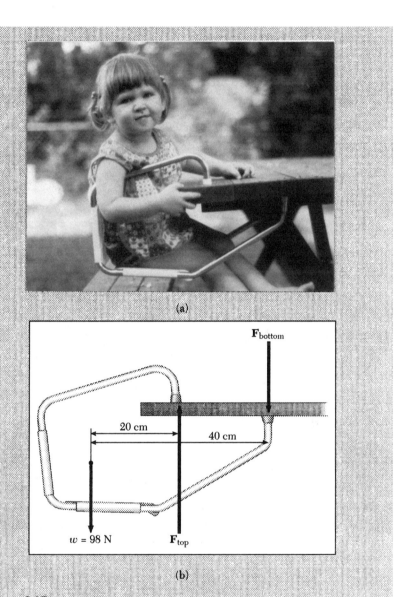

Figure 8.17
What forces are exerted on the highchair? The child's mass is 10 kg.

Let us choose the midpoint of the child's body as our pivot point because the force **w** passes through this point and therefore produces no torque about it. With this choice of pivot, F_{top} produces a counterclockwise torque and F_{bottom} produces a clockwise torque. The moment arms for these torques are as noted in Figure 8.17.

Solution

$$w = mg = (10 \text{ kg})(9.8 \text{ m/s}^2) = 98 \text{ N}$$

There are no forces exerted in the x direction, so the first condition gives only

$$\Sigma F_y = F_{top} - 98 \text{ N} - F_{bottom} = 0$$

$$F_{top} - F_{bottom} = 98 \text{ N}$$

As we know by now, this one equation with two unknowns tells us we need more information. The second condition of equilibrium gives

$$\Sigma \tau_\circlearrowleft = \Sigma \tau_\circlearrowright$$

$$F_{bottom}(40 \text{ cm}) = F_{top}(20 \text{ cm})$$

$$2 F_{bottom} = F_{top}$$

Now we can solve our simultaneous equations:

$$F_{top} - F_{bottom} = 98 \text{ N}$$

$$F_{top} = 2 F_{bottom}$$

$$2 F_{bottom} - F_{bottom} = 98 \text{ N}$$

$$F_{bottom} = 98 \text{ N}$$

$$F_{top} = 2(98 \text{ N}) = 196 \text{ N} \approx 200 \text{ N}$$

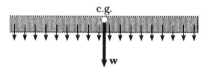

Figure 8.18
We can treat a body as if it were a rigid shell with all its weight concentrated at a single point—the center of gravity.

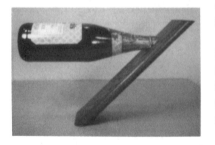

This one-bottle wine holder is an example of a balanced mechanical system. How does the holder—which cannot stand up on its own—support the bottle without collapsing? (© L. W. Black)

8.6 Center of Gravity

CONCEPT OVERVIEW

■ For many purposes, an extended body may be considered a rigid shell with all its weight or mass concentrated at the center of gravity or center of mass.

Think of the force of gravity acting on something like the long rod in Figure 8.18. Gravity pulls down on every atom of the rod, and we could therefore draw a multitude of little force vectors to represent the gravitational force distributed all along the rod. All those separate forces have the same effect as one single force: the weight **w.** We shall show that the force **w** acts as if all the weight of the rod were concentrated at one particular point. That point is called the **center of gravity.**

Center of gravity is a very useful concept. Because of it, we can, when looking at the force of gravity acting on an extended body,[2] treat the body as if it were a rigid shell with all its weight concentrated at the single point we call the center of gravity. In extended bodies that have a regular, symmetric shape and a uniform composition, the center of gravity is located at the geometric center of the body. A few simple cases are shown in Figure 8.19.

---

[2] An **extended body** simply means any real body of finite size rather than the idealized points we have been using as bodies. For instance, the child in Example 8.9, the load in Example 8.8, the brass ball in Example 8.7, and so on, are all extended bodies, but we treated each as if it were only a point through which the weight acted.

Figure 8.19
For regular, uniform, symmetric objects, the center of gravity is at the geometric center of the object.

The center-of-gravity concept allows us to look at extended objects whose weight cannot be ignored, such as the flagpole in the next example.

Example 8.10

A 100-N flagpole is held horizontal as shown in Figure 8.20. It is made of strong steel and has a length of 4.00 m. At one end is a 200-N ornamental brass ball, and the other end is attached to a building. The flagpole is supported by a guy wire that makes an angle of 20.0° with the horizontal. Find the tension T in the wire and the reaction force R exerted by the building on the pole.

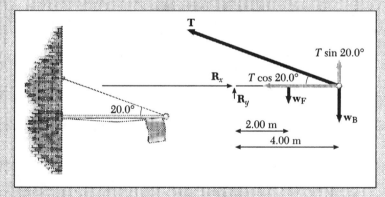

Figure 8.20
What are the forces acting on the horizontal flagpole when we do *not* neglect its weight?

Reasoning

This situation strongly resembles the one in Example 8.7, with one significant difference—here our flagpole is heavy and therefore we cannot ignore its weight. In the free-body diagram in Figure 8.20, $\mathbf{w_F}$ is the weight of the flagpole, and we locate it at the center of gravity, the geometric center of the symmetric flagpole. The weight of the ball $\mathbf{w_B}$ we locate at the geometric center of the ball.

We shall apply both conditions of equilibrium. From the three equations we get, we can solve for the three unknowns—R_x, R_y, and T.

Solution

From the first condition, we have

$$\Sigma F_x = R_x - T \cos 20.0° = 0$$

$$R_x = 0.940T \qquad\qquad (1)$$

$$\Sigma F_y = R_y + T \sin 20.0° - w_F - w_B = 0$$

$$R_y + 0.342T = 100 \text{ N} + 200 \text{ N} = 300 \text{ N} \qquad (2)$$

In using the second condition, we choose the point where the flagpole meets the building as our pivot because the reaction force **R** produces zero torque about this point.

Force	Torque
R_x	0
R_y	0
w_F	$(2.00 \text{ m})(100 \text{ N}) = 200 \text{ N·m}$ cw $\circlearrowright$
w_B	$(4.00 \text{ m})(200 \text{ N}) = 800 \text{ N·m}$ cw $\circlearrowright$
T	$(4.00 \text{ m})(T)(\sin 20.0°) = (1.37 \text{ m})T$ ccw $\circlearrowleft$

$$\Sigma \tau_{\circlearrowright} = \Sigma \tau_{\circlearrowleft}$$

$$1000 \text{ N·m} = (1.37 \text{ m})T$$

$$T = 730 \text{ N}$$

Once the tension is known, the reaction forces can be found from Equations 1 and 2:

$$R_y = 300 \text{ N} - (0.342)(730 \text{ N}) = 50.0 \text{ N}$$

$$R_x = (0.940)(730 \text{ N}) = 686 \text{ N}$$

Notice that here there is a vertical component of the reaction force that was not present in Example 8.7 when we ignored the weight of the flagpole.

Practice Exercise   Another horizontal flagpole, 30.0 N in weight, 3.00 m long, and with a 50-N ornament at its end, is attached to a building by a hinge, and supported by a guy wire, attached at the ornament end, makes an angle of 30.0° with the horizontal. Find the tension T in the wire and the force exerted by the building.

Answer $T = 130$ N; $F_{bldg} = 114$ N, 7.6° above horizontal

Practice Exercise   Consider another flagpole, 6.00 m long, with a weight of 30.0 N, and with a 50-N ornament at its end, attached to a building by a hinge. This flagpole makes an angle of 30° with the horizontal and is held in place by a horizontal wire attached at its end. Find the tension in the wire and the force exerted by the building.

Answer $T = 110$ N; $F_{bldg} = 140$ N at 35° above horizontal

Example 8.11

An 80.0-kg painter climbs a ladder as shown in Figure 8.21. The wooden ladder is 4.00 m long and has a mass of 30.0 kg. It leans against a wall at an angle of 37.0° with the vertical. The driveway is rough, and the coefficient of friction between it and the ladder is 0.450. The wall is smooth (that is, frictionless). How far up can the painter go before the foot of the ladder starts to slide away from the wall?

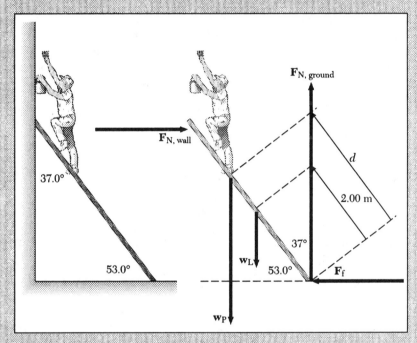

Figure 8.21
How far up can the painter climb before the ladder starts to slip?

Reasoning

Notice in the free-body diagram for the ladder that w_L, the weight of the ladder, is located at the ladder's center of gravity, 2.00 m from the foot. When the foot of the ladder is on the verge of slipping, the friction force has its maximum value $F_f = \mu F_N$. If the ladder slips, the foot will move to the right, and so the friction force is directed to the left, to oppose that possible motion. The force $F_{N,ground}$ is the normal force exerted by the ground on the ladder, and $F_{N,wall}$ is the normal force exerted by the wall on the ladder.

Solution

First, we convert the masses to weights:

$$w_L = m_L g = (30.0 \text{ kg})(9.80 \text{ m/s}^2) = 294 \text{ N}$$

$$w_P = m_P g = (80.0 \text{ kg})(9.80 \text{ m/s}^2) = 784 \text{ N}$$

Figure 8.22
The center of gravity is always directly below the point of support; this fact can be used to determine the location of the center of gravity for an irregular body. *(Photo © Charles D. Winters)*

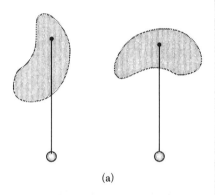

(a)

(b)

The first condition of equilibrium provides

$$\Sigma F_y = F_{N,\text{ground}} - w_P - w_L = 0$$

$$F_{N,\text{ground}} = w_P + w_L = 1078 \text{ N} \approx 1080 \text{ N}$$

$$\Sigma F_x = F_{N,\text{wall}} - F_f = 0$$

$$F_{N,\text{wall}} = F_f = \mu F_{N,\text{ground}} = (0.450)(1080 \text{ N}) = 486 \text{ N}$$

The distance up the ladder at which the force w_P is applied for these calculations involving $F_{f,\text{max}}$, labeled d in the figure, is the distance the painter can safely climb without having the ladder slip. This distance can be determined from the second condition of equilibrium. If we choose the foot of the ladder as our pivot, two torques are zero because forces $F_{N,\text{ground}}$ and F_f pass through this point and thus have zero moment arm relative to it.

Force	Torque
F_f	0
$F_{N,\text{ground}}$	0
w_L	$(294 \text{ N})(2.00 \text{ m})(\sin 37.0°) = 354 \text{ N·m}$ ccw
w_P	$(784 \text{ N})(d)(\sin 37.0°) = (472 \text{ N})d$ ccw
$F_{N,\text{wall}}$	$(486 \text{ N})(4.00 \text{ m})(\sin 53.0°) = 1550 \text{ N·m}$ cw

$$\Sigma \tau_\circlearrowright = \Sigma \tau_\circlearrowleft$$

$$1550 \text{ N·m} = 354 \text{ N·m} + (472 \text{ N})d$$

$$d = 2.53 \text{ m}$$

The painter can walk 2.53 m up the ladder, and it will remain in equilibrium. In order for him to walk farther up the ladder without his weight causing it to slip, there must be a larger friction force from the ground and a larger normal force from the wall.

Practice Exercise If the ladder in the previous example is moved so it makes an angle of 25.0° with the vertical wall and the coefficient of friction between the foot of the ladder and the floor is 0.30, how far up the ladder can the painter now go before it starts to slide?

Answer 2.8 m

For bodies that are not regular, uniform and symmetric, the center of gravity can be located by suspending the body as shown in Figure 8.22. When the body comes to rest, its center of gravity will be directly below the point of support because, for the second condition of equilibrium to be met, the upward force of the support and the downward weight of the body must lie along the same line. Hang a plumb line from the point of support and mark its location on the body. The center of gravity is somewhere on that line. Then, suspend the body from another point and drop another plumb line. The intersection of the two lines is the center of gravity. (In practice, it is a good idea to check with a third measurement.)

The center of gravity is not necessarily within a body. Figure 8.23 illustrates this point with a protractor. The center of gravity of a donut is in the middle of the hole!

It is useful to have a procedure for calculating the position of the center of gravity without resorting to hammer, nail, and plumb line. To do so, we think of the object as being composed of many separate objects that we can treat either as particles or as bodies with enough symmetry that we can easily find their individual centers of gravity. Consider a thin, lightweight rod with three weights w_1, w_2, and w_3 attached to it at positions x_1, x_2, and x_3, as sketched in Figure 8.24 along with the corresponding free-body diagram.

The center of gravity is the position $x_{\text{c.g.}}$ at which the total weight $w_{\text{tot}} = w_1 + w_2 + w_3$ can be located and have the same effect as the actual weight distribution. To determine this location, calculate the torques about one end of the rod. Let us choose the left end:

$$\Sigma\tau = x_1 w_1 + x_2 w_2 + x_3 w_3$$

The total weight located at center of gravity $x_{\text{c.g.}}$ must produce the same torque about the left end of the rod:

$$\Sigma\tau = x_{\text{c.g.}} w_{\text{tot}}$$

Setting these two expressions equal to each other gives

$$x_{\text{c.g.}} w_{\text{tot}} = x_1 w_1 + x_2 w_2 + x_3 w_3$$

$$x_{\text{c.g.}} = \frac{x_1 w_1 + x_2 w_2 + x_3 w_3}{w_{\text{tot}}}$$

This procedure can be generalized to more than three weights, and we can also generalize to two directions to handle objects more complicated than weights on a rod. Then the x and y coordinates of the center of gravity are

$$x_{\text{c.g.}} = \frac{x_1 w_1 + x_2 w_2 + x_3 w_3 + \cdots}{w_{\text{tot}}} = \frac{\Sigma_i x_i w_i}{w_{\text{tot}}} \tag{8.10}$$

$$y_{\text{c.g.}} = \frac{y_1 w_1 + y_2 w_2 + y_3 w_3 + \cdots}{w_{\text{tot}}} = \frac{\Sigma_i y_i w_i}{w_{\text{tot}}} \tag{8.11}$$

where $w_{\text{tot}} = w_1 + w_2 + w_3 + \cdots$.

For most common situations, mass and weight are proportional to each other. Therefore these definitions can just as well be stated in terms of the masses of the individual bodies. In this case, what we are locating is the **center of mass** (c.m.):

$$x_{\text{c.m.}} = \frac{x_1 m_1 + x_2 m_2 + x_3 m_3 + \cdots}{m_{\text{tot}}} = \frac{\Sigma_i x_i m_i}{m_{\text{tot}}} \tag{8.12}$$

$$y_{\text{c.m.}} = \frac{y_1 m_1 + y_2 m_2 + y_3 m_3 + \cdots}{m_{\text{tot}}} = \frac{\Sigma_i y_i m_i}{m_{\text{tot}}} \tag{8.13}$$

where $m_{\text{tot}} = m_1 + m_2 + m_3 + \cdots$.

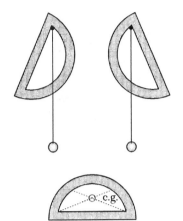

Figure 8.23
A protractor provides an example of an object whose center of gravity is located outside the object.

Figure 8.24
The total weight $w_{\text{tot}} = w_1 + w_2 + w_3$ located at the center of gravity has the same effect as the three weights at their individual locations; the rod is considered massless.

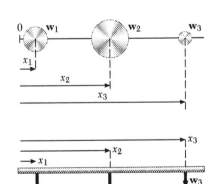

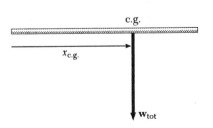

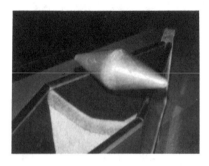

Can you explain how the cone mysteriously rolls "uphill" when released from a starting point (rear of photo) that is lower than its ending point? (© L. W. Black)

For an astrophysicist describing the motion of a galaxy in which the force of gravity may vary greatly from one part to another, the center of mass and center of gravity may be located at two different positions. For common objects near Earth's surface, however, the two positions coincide. In this book, we shall restrict our study to the latter, much simpler case.

Example 8.12

A metal part cut from a uniformly thick sheet of steel is shown in Figure 8.25. Locate its center of mass, measuring distances from the lower left corner of the part.

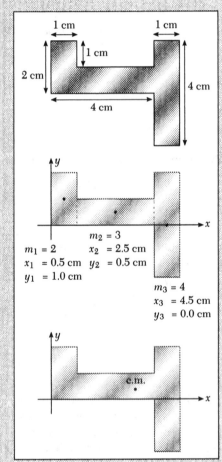

Figure 8.25
The center of mass of an odd-shaped object can be found by breaking the object down into a collection of regular-shaped objects.

Reasoning

First, think of the part as a composite of three regular components as shown in the middle drawing. The first part is 1.0 cm wide by 2.0 cm high. Its center of mass is 0.50 cm to the right of the left corner and

1.0 cm up. Its area is $A_1 = (1.0 \text{ cm})(2.0 \text{ cm}) = 2.0 \text{ cm}^2$. Its mass is proportional to the area, so we can write $m_1 = 2.0$ in some unspecified, arbitrary units that will later cancel. The center of mass for this rectangle can be determined by symmetry, and its coordinates are $x_1 = 0.50 \text{ cm}$, $y_1 = 1.0 \text{ cm}$.

The second part is 3.0 cm wide by 1.0 cm high; its area is therefore 3.0 cm^2, and its mass is $m_2 = 3.0$ arbitrary units. Its center of mass is 1.5 cm in from its left edge and 0.50 cm up. However, we must locate its center of mass from our initial coordinate origin, the lower left corner of the entire part. With that change accounted for, we have $x_2 = 2.5 \text{ cm}$, $y_2 = 0.50 \text{ cm}$.

The third piece is analyzed similarly, and we have $m_3 = 4.0$, $x_3 = 4.5 \text{ cm}$, $y_3 = 0$.

Solution

These values can be used in Equations 8.12 and 8.13:

$$m_{\text{tot}} = m_1 + m_2 + m_3 = 2.0 + 3.0 + 4.0 = 9.0$$

$$x_{\text{c.m.}} = \frac{(0.50 \text{ cm})(2.0) + (2.5 \text{ cm})(3.0) + (4.5 \text{ cm})(4.0)}{9.0}$$

$$x_{\text{c.m.}} = \frac{(1.0 + 7.5 + 18.0) \text{ cm}}{9.0}$$

$$x_{\text{c.m.}} = 2.94 \text{ cm}$$

$$y_{\text{c.m.}} = \frac{(1.0 \text{ cm})(2.0) + (0.50 \text{ cm})(3.0) + (0 \text{ cm})(4.0)}{9.0}$$

$$y_{\text{c.m.}} = \frac{(2.0 + 1.5) \text{ cm}}{9.0}$$

$$y_{\text{c.m.}} = 0.39 \text{ cm}$$

The center of mass is located at $x_{\text{c.m.}} = 2.9 \text{ cm}$, $y_{\text{c.m.}} = 0.39 \text{ cm}$.

*8.7 Forces in the Human Body

CONCEPT OVERVIEW

■ The human body offers many examples of an object in equilibrium under the influence of several forces.

We have used the conditions of equilibrium to find various forces acting on ladders and flagpoles and the like. It is interesting to use these same techniques to find the forces acting on or within our own bodies.

We begin by asking for the magnitude of the force exerted by the biceps muscle when a 15-N physics book is picked up. Figure 8.26 shows a sketch of an arm showing approximate dimensions plus a sketch of a simplified model of the forearm and forces. In the latter, $\mathbf{F}_E$ is the force exerted by the elbow joint and $\mathbf{F}_B$ is the force exerted by the biceps muscle, the force we are looking for.

Figure 8.26
What force F_B is exerted by the biceps when a 15-N book is lifted?

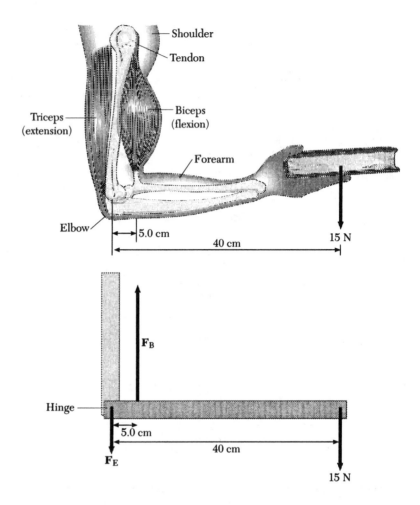

The laws of physics apply not only to blocks, planes, and other inanimate objects but to muscles and bones of a living body.

The second condition of equilibrium can be applied with torques calculated with respect to the elbow. In that case, F_E provides no torque. There is one counterclockwise torque, $(F_B)(5 \text{ cm})$, and one clockwise torque, $(15 \text{ N})(40 \text{ cm})$.

$$\Sigma\tau_\circlearrowleft = \Sigma\tau_\circlearrowright$$

$$(F_B)(5.0 \text{ cm}) = (15 \text{ N})(40 \text{ cm})$$

$$F_B = 120 \text{ N}$$

With the dimensions estimated here, the biceps muscle is required to exert a force equal to eight times the force being lifted by the hand!

Example 8.13

Actually, the biceps muscle does not pull straight up as indicated in Figure 8.26. A closer approximation is shown in Figure 8.27, where F_B makes an angle of 15° with the vertical. Using this model, calculate the force exerted by the biceps.

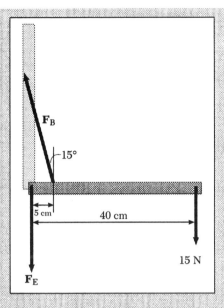

Figure 8.27
To improve our model, consider the biceps muscle attached to the forearm at an angle of 15°.

Reasoning

This calculation proceeds exactly as before except that now the counterclockwise torque is $(F_B)(5.0 \text{ cm})(\sin 75°)$. Notice that we need the sine of 75° rather than the sine of 15°.

Solution

$$\Sigma \tau_\circlearrowleft = \Sigma \tau_\circlearrowright$$

$$(F_B)(5.0 \text{ cm})(\sin 75°) = (15 \text{ N})(40 \text{ cm})$$

$$F_B = 124 \text{ N} \approx 120 \text{ N}$$

This refinement does not change the value very much, telling us that, even without detailed anatomical data, we can make interesting and useful estimates of some of the forces exerted by our bodies.

Example 8.14 _____

What force is exerted by each Achilles tendon when you lift both heels off the floor simultaneously to stand on tiptoe? The forces on one foot are sketched in Figure 8.28.

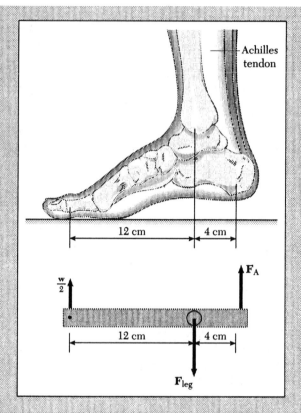

Figure 8.28
What force F_A must each Achilles tendon exert in order to lift your heels off the floor and allow you to stand on tiptoe?

Reasoning

The estimated dimensions of a foot are given in the figure. The free-body diagram is a simplified model showing the forces applied to one foot. It is assumed that you are standing on two feet so the force each foot exerts down on the floor is one-half your weight, $w/2$. By Newton's third law, the force exerted by the floor on each foot must be $w/2$. F_A is the upward force exerted by each Achilles tendon, and F_{leg} is the downward force exerted by your leg on your ankle.

Solution

We are not interested in F_{leg}; we calculate torques about the ankle so that F_{leg} provides no torque. Thus,

$$\Sigma \tau_{\curvearrowleft} = (F_A)(4 \text{ cm}) = (w/2)(12 \text{ cm}) = \Sigma \tau_{\curvearrowright}$$

$$F_A = 1.5w \approx 2w$$

Each Achilles tendon exerts a force of about twice your weight as you start to stand on tiptoe. Of course, all these dimensions are merely estimates; actual dimensions and, consequently, actual forces vary considerably from person to person.

Example 8.15

Figure 8.29 shows a person using both hands to lift a 30-kg mass. When the lifter's back is horizontal, what is the tension T in the back muscles and the compressive force C in the spinal disks?

Reasoning

Figure 8.29 shows dimensions and weights appropriate for an 820-N person. The weight $w_1 = 380$ N is the weight of the upper torso, $w_2 = 60$ N is the weight of the head, and $w_3 = 394$ N is the weight of the arms (100 N) plus the weight being lifted (294 N). The tension $\mathbf{T}$ in the back muscles is directed 12° above the horizontal.

We shall calculate torques about the hip joint and apply both conditions of equilibrium.

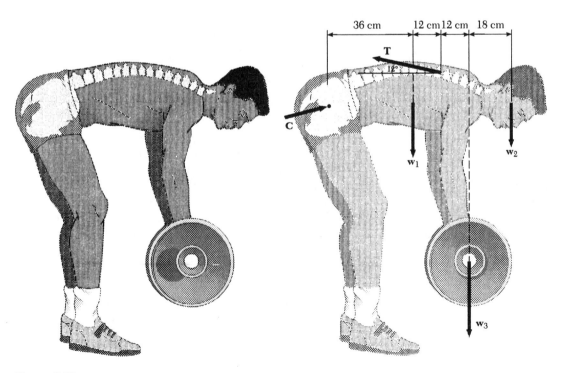

Figure 8.29
What forces are exerted on the back when someone lifts a 30-kg mass?

Solution

Force	Torque
C	0
T	$(T)(48 \text{ cm})(\sin 12°) = T(9.98 \text{ cm})$ ccw↺
w_1	$(380 \text{ N})(36 \text{ cm}) = 13\,700 \text{ N·cm}$ cw↻
w_2	$(60 \text{ N})(72 \text{ cm}) = 4320 \text{ N·cm}$ cw↻
w_3	$(394 \text{ N})(60 \text{ cm}) = 23\,600 \text{ N·cm}$ cw↻

$$\Sigma \tau_↺ = (9.90 \text{ cm})T = 41\,600 \text{ N·cm} = \Sigma \tau_↻$$

$$T = 4170 \text{ N} \approx 4200 \text{ N}$$

To calculate the compressive force C, we apply the first condition of equilibrium:

$$\Sigma F_x = C_x - T \cos 12° = 0$$

$$C_x = (4200 \text{ N})(\cos 12°) = 4110 \text{ N}$$

$$\Sigma F_y = T \sin 12° - C_y - w_1 - w_2 - w_3 = 0$$

$$C_y = (4200 \text{ N})(\sin 12°) - 380 \text{ N} - 60 \text{ N} - 394 \text{ N} = 39.2 \text{ N} \approx 40 \text{ N}$$

$$C = \sqrt{C_x^2 + C_y^2} = 4110 \text{ N} \approx 4100 \text{ N}$$

In British units, we have

$$4100 \text{ N} \left(\frac{0.2248 \text{ lb}}{1 \text{ N}} \right) = 920 \text{ lb}$$

almost half a ton!

Lifting with your legs while keeping your back vertical greatly reduces the forces exerted on your back muscles and spinal disks.

Vertebrae in the spine are separated by disks. These disks are flexible, fluid-filled sacks that cushion the vertebrae and allow them to move, as sketched in Figure 8.30. Just straightening up with nothing in your hands puts a compressive force of a quarter of a ton on the disks. Lifting a 30-kg weight requires a force of nearly *half a ton* on the disks! That is like resting a grand piano on a disk! If the compressive force on a disk is great enough, the disk ruptures. It is no wonder that so many people experience lower back pain.

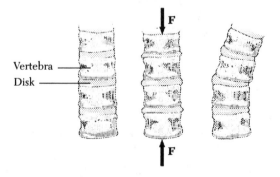

Figure 8.30
Vertebrae in the spine are separated by disks.

*8.8 Types of Equilibrium

CONCEPT OVERVIEW

■ An object in equilibrium may or may not remain in equilibrium if moved slightly.

■ The type of equilibrium an object has—stable, neutral, or unstable—is determined by what happens to its center of gravity when it is moved.

Why is it easier to balance an umbrella by hanging it from a doorknob than by balancing it on its tip as illustrated in Figure 8.31? Free-body diagrams showing the forces in the two cases are shown in Figure 8.32. The force and torque equations for equilibrium are both satisfied in both cases. The normal force $\mathbf{F}_N$ exerted upward by the knob or the floor is just equal to the umbrella's weight $\mathbf{w}$ but acts in the opposite direction. These two forces lie along the same line, so there is no unbalanced torque. The umbrella hanging from the knob is in **stable equilibrium,** because its center of gravity is raised for any movement of the umbrella. If it is moved slightly and released, the torques caused by this movement away from its equilibrium will cause this umbrella to move back to equilibrium. The umbrella balanced on its tip is in **unstable equilibrium**

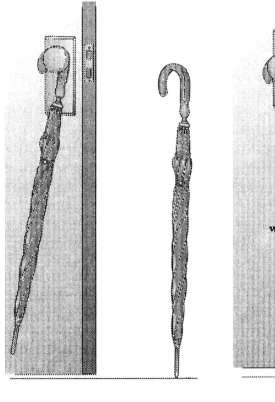

Figure 8.31

Why is an umbrella hanging from a doorknob stable and an umbrella standing on the floor unstable?

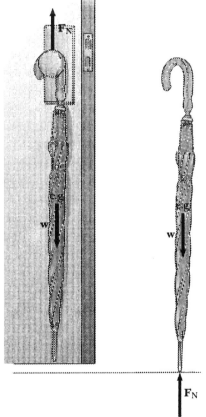

Figure 8.32

The forces and torques on the umbrella are balanced in both cases. If there is a slight movement from equilibrium, however, the results are quite different in the two cases.

Figure 8.33
This toy is stable because any shift in its position raises the center of gravity. A restoring force will then act to return the center of gravity to its lowest possible point. *(© Paul Silverman/Fundamental Photographs, New York)*

because its center of gravity is lowered for any movements of the umbrella. If it is moved slightly, the torques caused by this movement away from equilibrium will cause this umbrella to move still farther from equilibrium until it finally rests horizontally on the floor, in stable equilibrium.

When an object is in stable equilibrium, any movement away from equilibrium *raises* its center of gravity. This fact is the basis of toys like the one shown in Figure 8.33. The position of the masses causes the center of gravity to be below the point of support. (Remember that the center of gravity does not need to lie within an object.)

To understand why the two umbrellas of Figure 8.31 behave in different ways, we must pay attention to what happens to the center of gravity during the motion of each. When we move the hanging umbrella away from its equilibrium position, we are raising its center of gravity, and the resulting motion of the umbrella is to return its center of gravity to the lowest point possible. When we tip the balanced umbrella away from its (precarious) equilibrium position, we are lowering its center of gravity, and there is no energy source available to allow the umbrella to do work against gravity and raise its center of gravity to the original position. Instead, the umbrella keeps falling until its center of gravity is at the lowest possible point—the floor level.

Therefore, we say that equilibrium is **stable** if a slight movement causes the center of gravity to move higher and **unstable** if a slight movement causes the center of gravity to move lower.

A third type of equilibrium—**neutral equilibrium**—exists when you can move an object without changing the height of its center of gravity. The sphere in Figure 8.34 provides an example of neutral equilibrium. If we gently roll it to another position on the same surface, the sphere will stay put at the new position.

The cube resting on one face on the surface in Figure 8.34 is in stable equilibrium (its center of gravity is at the lowest point possible). The pencil resting on its tip is in unstable equilibrium (its center of gravity will move lower if it is disturbed).

Figure 8.34
The sphere is in neutral equilibrium, and the cube is in stable equilibrium. What happens to the objects if each is moved slightly?

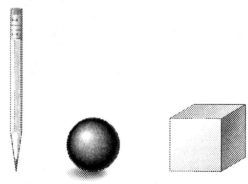

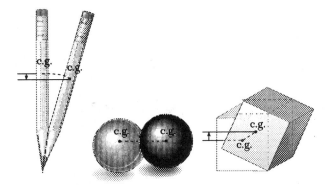

Figure 8.35
When the pencil moves from equilibrium, its center of gravity is lowered; the pencil is in unstable equilibrium. When the sphere moves from equilibrium, the position of its center of gravity is unchanged; the sphere is in neutral equilibrium. When the cube moves from equilibrium, its center of mass is raised; the cube is in stable equilibrium.

The gravitational potential energy of an extended body is defined by mgy, where y is the vertical position of the center of gravity. If an object, like a pencil initially balanced on its point, lowers its center of gravity when it falls, its gravitational potential energy is decreased. Work from the outside must be done to restore it to its initial position. When we roll a sphere along a table, with its center of gravity always at the same level, there is no change in its potential energy. Therefore no external work is required to restore it to its initial position. When a cube is tilted up on an edge, with its center of gravity being raised, its potential energy has increased. The cube can do work as it returns to its initial position. The three types of equilibrium are defined by what happens to the gravitational potential energy of the object. This is illustrated in Figure 8.35.

SUMMARY

An object at rest that remains at rest is in static equilibrium. For any object to be in static equilibrium, the sum of the forces acting on it must be zero:

$$\Sigma \mathbf{F} = 0 \qquad (8.1)$$

This is the first condition of equilibrium. In terms of components, it means that all the force components in the x direction must add to zero and all the force components in the y direction must add to zero:

$$\Sigma F_x = 0 \qquad (8.2a)$$
$$\Sigma F_y = 0 \qquad (8.2b)$$

Even for objects in static equilibrium, friction forces always oppose the motion that would result if friction were not present. The maximum friction force is proportional to the normal force between the surfaces in contact;

$$F_{f,\max} = \mu_s F_N \qquad (4.7)$$

The rotational analog of force is called torque τ. Torque must be defined with respect to some pivot point. The perpendicular distance from that pivot point to the line of action of the force is known as the moment arm $r_\perp$. The torque due to a force F with moment arm $r_\perp$ about a particular pivot point is

$$\tau = r_\perp F \qquad (8.3)$$

An equivalent definition of torque is

$$\tau = rF_\perp \qquad (8.6)$$

where $F_\perp$ is the component of force perpendicular to the line drawn from the pivot point to the point of application of the force. In general, torque is defined by

$$\tau = rF \sin \theta \qquad (8.7)$$

where r is the displacement from the pivot point to the point of application of the force and θ is the angle between that displacement and the direction of the force.

For an extended body to be in equilibrium—and that state includes lack of rotation—the sum of all the torques must be zero. Therefore the sum of the clockwise torques must equal the sum of the counterclockwise torques:

$$\Sigma\tau_\circlearrowleft = \Sigma\tau_\circlearrowright \qquad (8.8)$$

Counterclockwise torques are defined to be positive, and clockwise torques are defined to be negative.

The coordinates of the center of gravity are

$$x_{c.g.} = \frac{x_1 w_1 + x_2 w_2 + x_3 w_3 + \cdots}{w_{tot}} \qquad (8.10)$$

$$y_{c.g.} = \frac{y_1 w_1 + y_2 w_2 + y_3 w_3 + \cdots}{w_{tot}} \qquad (8.11)$$

The coordinates of the center of mass are

$$x_{c.m.} = \frac{x_1 m_1 + x_2 m_2 + x_3 m_3 + \cdots}{m_{tot}} \qquad (8.12)$$

$$y_{c.m.} = \frac{y_1 m_1 + y_2 m_2 + y_3 m_3 + \cdots}{m_{tot}} \qquad (8.13)$$

There are three types of equilibrium—stable, unstable, and neutral. For an object in stable equilibrium, any small movement raises its center of gravity and causes the object to move back toward its original equilibrium position. For an object in unstable equilibrium, any small movement lowers its center of gravity and causes the object to move still farther from its original equilibrium position. For an object in neutral equilibrium, any small movement neither raises nor lowers its center of gravity and does not cause the object to move back toward or away from its initial position.

QUESTIONS

1. Explain the statement "statics is just a special case of dynamics."
2. Can an object be in equilibrium when only one force acts upon it? Give examples.
3. Explain the meaning of torque so that someone outside of physics understands what the word means.
4. Can two forces of identical magnitude and direction exert anything but identical torques on an object? Give examples.
5. Can two forces of different magnitude and direction exert identical torques on an object? Give examples.
6. Can the sum of the forces on an object be zero and the sum of the torques be nonzero? Give examples.
7. Can the sum of the torques on an object be zero and the sum of the forces be nonzero? Give examples.
8. In discussing friction, the forces exerted by a surface have been treated as though there are two separate forces, a normal force $\mathbf{F}_N$ and a friction force $\mathbf{F}_f$. Is this really the case? Explain.
9. Why is a ladder more likely to slip when a painter is near the top than when the painter is near the bottom?

10. Why are sit-ups done with your hands behind your head more difficult than those done with your arms at your sides?
11. Why will a picture hang crooked if the hook is not centered on the frame?
12. The back muscles normally support the spine in a straight position. What would happen if the muscles on one side were to weaken?
13. How many different ways must you suspend a flat, essentially two-dimensional object to locate its center of mass? How many ways to locate its center of gravity?
14. During a high jump or pole vault, a good athlete's center of mass passes *under* the bar. Explain how that can be.
15. Try the experiment shown in Figure 8.36. Stand with your back to the wall and your heels touching the wall, then try to bend over and pick up a chair by grasping the seat. Explain any differences between female and male abilities, to accomplish this task.
16. Why do health and safety experts urge people to lift heavy objects by bending their knees and keeping their back straight and vertical rather than by simply bending from the waist?

Figure 8.36　Question 15.

17. The force exerted by body muscles lifting something is often many times the weight of the object lifted. How can this be?
18. Why do you tend to lean to the left when you carry a heavy suitcase with your right hand?

PROBLEMS

8.1 Balancing Forces

8.2 First Condition of Equilibrium

8.1　A 10-kg mass hangs suspended from the center of a light rope suspended from two supports 2.0 m apart. The center of the rope is 0.20 m below a horizontal line through the two supports. What is the tension in the rope?

8.2　A light, horizontal wire is stretched between two posts 10 m apart. A bird weighing 3.0 N alights at the center of the wire, and the wire sags 6.0 cm. What is the tension in the wire when the bird is sitting on it?

8.3　A 2-kg ball is held in position by a horizontal string and a string that makes an angle of 30° with the vertical, as shown in Figure 8.37. Find the tension T in the horizontal string.

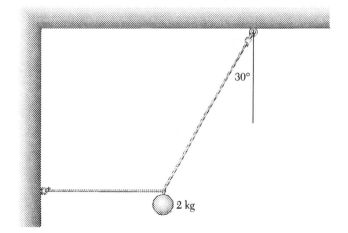

Figure 8.37　Problem 8.3.

8.4　A 250-N object is suspended by cables as shown in Figure 8.38. Find the tension in each cable.

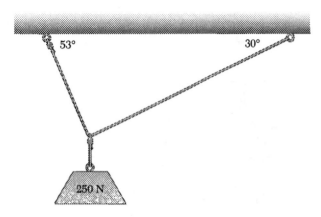

Figure 8.38 Problem 8.4.

8.5 Two 55-kg cheerleaders stand on the shoulders of a 80-kg cheerleader during a halftime performance. What is the normal force exerted by the ground? What is the vertical force exerted by the shoulders of the single cheerleader?

8.6 A 2.5-kN automobile engine is suspended by a chain (Fig. 8.39). What horizontal force must be applied to move the chain until it makes an angle of 10° with the vertical?

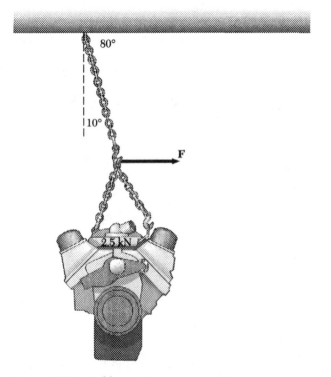

Figure 8.39 Problem 8.6.

8.7 A 500-N ball is suspended as shown in Figure 8.40. What is the tension in the horizontal cord?

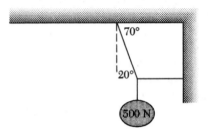

Figure 8.40 Problem 8.7.

8.8 To lift a patient, six nurses—three on each side—grip the sheet on which the patient is lying and lift, each with a force of 280 N at an angle of 20° from vertical. What is the weight of the patient?

8.9 A lightweight flagpole is attached at one end to a building and makes an angle of 20° with the horizontal (Fig. 8.41). At its other end is a guy wire that makes an angle of 40° with the horizontal and a massive bronze eagle whose weight is 80 N. Find the tension in the guy wire.

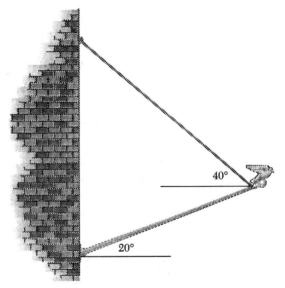

Figure 8.41 Problem 8.9.

8.10 The pulleys in Figure 8.42 are frictionless but have a mass of 4 kg each. What is the tension in the cord and in the pulley supports?

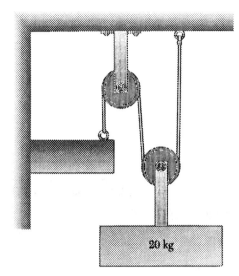

Figure 8.42   Problem 8.10.

○8.11   To pull an automobile out of the mud, the driver ties a rope to the car and to a tree 40 m away. The driver then pulls on the midpoint of the rope with a force of 400 N. The midpoint of the rope is displaced 2.0 m by the pull. Find the force exerted on the car.

○8.12   A light, horizontal wire is stretched between two posts 30 m apart. A bird, weighing 5.0 N, sits at the center of the wire and causes it to sag 8.0 cm. What is the tension in the wire when the bird sits on it?

○8.13   A 150-N object is suspended from two spring scales as in Figure 8.43. What does the scale on the right read and at what angle is it placed?

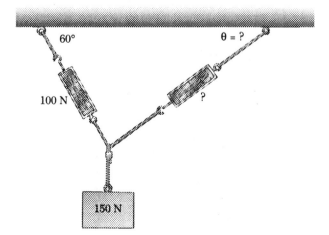

Figure 8.43   Problem 8.13.

○8.14   A person standing on crutches as shown in Figure 8.44 has a mass of 80 kg. Each crutch makes an angle of 22° with the vertical (as seen from the front). Half the person's weight is supported by the crutches. The other half is supported by the normal force exerted by the ground on the feet. The person is at rest, and the force exerted by the ground on the crutches acts *along the crutches.* Determine (*a*) the minimum coefficient of friction between crutches and ground and (*b*) the magnitude of the compressive force on each crutch.

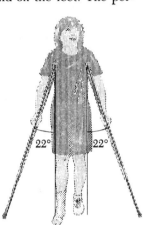

Figure 8.44   Problem 8.14.

●8.15   Determine the magnitude and direction of the net force on the leg in Figure 8.45.

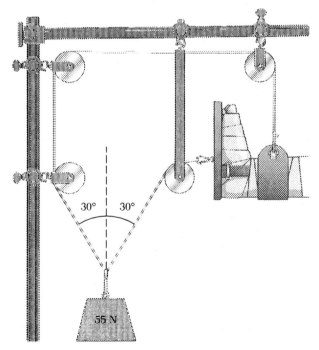

Figure 8.45   Problem 8.15.

8.3 Static Friction

8.16   A 250-N child sits on a 30-N sled. A horizontal force of 35 N can be exerted on the sled before it will move. What is the coefficient of static friction between snow and sled runners?

8.17   A 1-kN crate sits on a level floor. The coefficient of static friction between crate and floor is 0.4. What force directed at 30° above the horizontal can be exerted before the crate will start to move?

8.18   The coefficient of static friction between the 400-N block and the table in Figure 8.46 is 0.55. How much weight must be placed on top of this block to keep it from sliding?

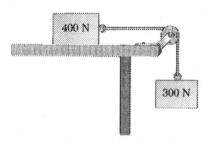

Figure 8.46   Problem 8.18.

8.19   A sled weighs 50.0 N. To just get it moving over a horizontal, snow-covered path, a horizontal force of 5.00 N is required. What is the minimum horizontal force required to move the sled if a person weighing 300 N sits on it? What force exerted at 30° above the horizontal is necessary to just get the sled and person moving?

8.20   A 700-N crate sits on a plane inclined 30° from the horizontal. The coefficient of static friction between crate and plane is 0.80. What force directed parallel to the plane can be exerted before the crate starts to slide *down* the plane? Before it starts to slide *up* the plane?

8.21   A 1-kg block is firmly attached to a wall by a rope. It sits atop a 5-kg block as shown in Figure 8.47. The coefficient of static friction between the two blocks and the coefficient of static friction between the 5-kg block and the floor are both 0.2. When the system is just on the verge of slipping, what is the magnitude of the external horizontal force **F** exerted on the 5-kg block? What is the tension in the rope attached to the 1-kg block?

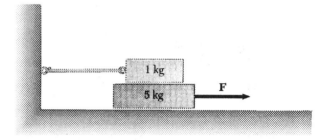

Figure 8.47   Problem 8.21.

8.22   A 2-kg block sits atop a 7-kg block as shown in Figure 8.48, attached by a cable and pulley arrangement. The coefficient of static friction between the two blocks is 0.25. The coefficient of static friction between the 7-kg block and the floor is 0.4. What is the magnitude of the maximum external horizontal force **F** that can be exerted on the lower block before it will move? What is the tension in the cable?

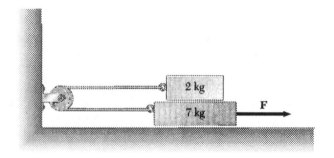

Figure 8.48   Problem 8.22.

8.23   A tractor pulls on two logs in tandem (Fig. 8.49); the logs are on the verge of slipping and are fastened together by a chain. Each log weighs 1800 N. The tractor exerts a force of 1600 N on the front log, and this log is acted on by a friction force of 900 N. Find the tension in the chain between the logs and the force of friction on the back log.

The coefficients of static friction between logs and ground are different for the two logs. What are these two coefficients?

Figure 8.49 Problem 8.23.

○8.24 A 20-N block sits on a plane inclined 30° from the horizontal. The coefficient of static friction between block and plane is 0.30. A cord runs from the block, over a pulley at the top of the plane, to a second block hanging vertically downward from the pulley. What minimum weight must the suspended block have to prevent the 20-N block from sliding down the plane? What maximum weight can the suspended block have before the 20-N block starts to slip up the plane?

○8.25 The *angle of repose* is the maximum angle of an inclined plane for which a block on the plane remains at rest. Show that the tangent of the angle of repose equals the coefficient of static friction.

●8.26 The bottom portion of a hill forms an arc having a radius of curvature of 17.0 m (Fig. 8.50). An object placed on this hill will begin to slide down if it is more than 2.15 m above the bottom. What is the coefficient of static friction between hill and object?

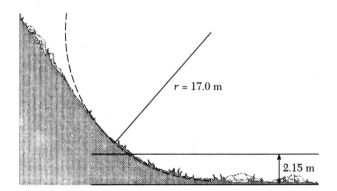

Figure 8.50 Problem 8.26.

●8.27 The coefficient of static friction between a mop and the floor is μ. Show that if the tangent of the angle between the handle of the mop and the floor is greater than μ, the mop cannot be moved, regardless of the magnitude of the force directed along the handle. Neglect the weight of the mop and handle.

8.4 Rotational Forces: Torques

8.28 A 20-cm wrench is used to loosen a bolt. What torque is exerted on the bolt when a force of 65 N is applied at the end of and perpendicular to the wrench?

8.29 A piece of pipe can sometimes be used as a "cheater" to effectively lengthen the moment arm of a wrench. If a pipe is used as shown in Figure 8.51 to allow a 150-N force to be exerted 30 cm from the bolt, what torque is exerted on the bolt?

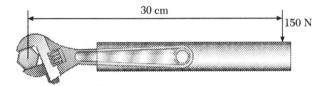

Figure 8.51 Problem 8.29.

8.30 A person grasps the rim of a bicycle wheel having a radius of 33.0 cm (Fig. 8.52). The hand exerts a force of 65.0 N at an angle of 75.0° with respect to an extension of the radius. Determine the torque on the wheel about the axle.

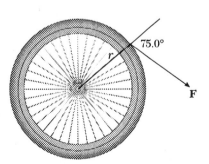

Figure 8.52 Problem 8.30.

8.5 Second Condition of Equilibrium

8.31 Re-do Example 8.9 but calculate the torques about the right-most support (that is, the point at which the force $\mathbf{F}_{bottom}$ is applied in Figure 8.17) and show that your answers are the same as calculated in the example.

8.32 Re-do Example 8.9 but calculate the torques about a point passing through $\mathbf{F}_{top}$ (that is, about the support that sits on top of the table) and show that your answers remain the same as those calculated in the example.

8.33 A person lifts a 750-N barbell overhead. The barbell is 1.8 m long, and the weight is loaded uniformly. If the hands grip the bar 30 cm and 40 cm from the center, how much weight is supported by each hand?

8.34 A light bamboo fishing pole 3.6 m long is supported by a horizontal string as shown in Figure 8.53. A 2.0-kg fish hangs from the end of the pole, and the pole is pivoted at the bottom. What is the tension in the horizontal string? What are the components of the force exerted by the pivot on the pole?

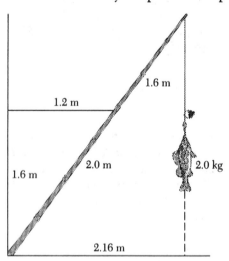

Figure 8.53 Problem 8.34.

8.35 A uniform pole 6.0 m long weighs 300 N and is attached at one end to a wall (Fig. 8.54). The pole is held at an angle of 30° above the horizontal by a horizontal guy wire attached to the pole 4.0 m from the end attached to the wall. A load of 600 N hangs from the free end of the pole. Find the tension in the guy wire and the components of the force exerted by the wall on the pole.

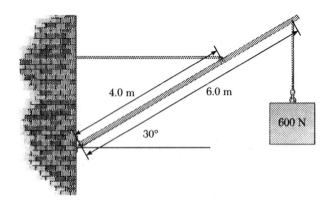

Figure 8.54 Problem 8.35.

8.36 A person holds a 2-kg carton of milk at arm's length, as shown in Figure 8.55. What force **B** must be exerted by the biceps muscle? Ignore the weight of the forearm.

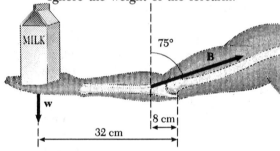

Figure 8.55 Problem 8.36.

○8.37 As part of a physical therapy program following a knee operation, a 10-kg weight is attached to an ankle and leg lifts are done as sketched in Figure 8.56. Calculate the torque about the knee due to this weight for the four positions shown.

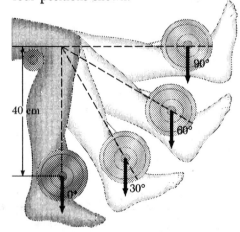

Figure 8.56 Problem 8.37.

○8.38 A lightweight rod is attached at one end to a vertical wall. The rod is inclined upward at an angle of 15° to the horizontal, held in this position by a horizontal wire fastened to its other end. A 500-N sign hangs from the center of the rod. Determine the tension in the wire and the force exerted by the wall on the rod.

○8.39 The sail in Figure 8.57 weighs 750 N. The force $\mathbf{F}_1$ exerted by the wind on the sail is horizontal and acts through point B. The weight of the whole boat (not including the weight of the sail or that of the sailor) is 1250 N and acts through point O, which is 0.800 m from point A along the line OA. The force $\mathbf{F}_2$ exerted by the water acts through point A. The person sailing the boat has a weight of 600 N. Determine the magnitude of $\mathbf{F}_1$.

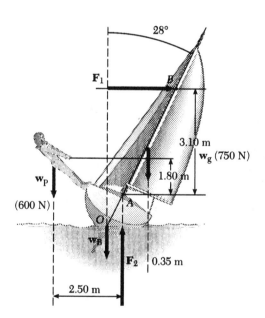

Figure 8.57   Problem 8.39.

8.6 Center of Gravity

8.40 Five uniform boards are all the same width and 4.0 cm thick. The boards are 2.9 m, 1.6 m, 0.80 m, 0.40 m, and 1.2 m long, and both mass and weight are proportional to board length. If the boards are stacked as shown in Figure 8.58, what are the vertical and horizontal coordinates of the center of gravity?

Figure 8.58   Problem 8.40.

8.41 A carpenter's square made of uniformly thick metal has the dimensions shown in Figure 8.59. Locate its center of gravity.

Figure 8.59   Problem 8.41.

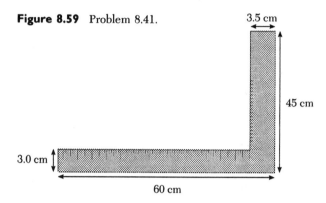

8.42 Two masses are attached by a lightweight, semicircular wire, as shown in Figure 8.60. Locate the center of gravity of the system.

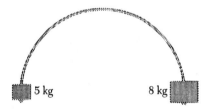

Figure 8.60   Problem 8.42.

8.43 Weights of 10 N, 20 N, and 30 N are placed at the corners of an equilateral triangle with sides of 1 m. Locate the center of gravity of the system.

8.44 A 70-kg painter stands 2.0 m from one end of a uniform, 80-kg, 3.0 m horizontal platform that is supported by two cables, one at each end. Find the force exerted by each cable.

8.45 A 60-kg diver stands at the end of a diving board that is 6.0 m long. The board is attached to the diving platform at the other end and is supported 4.0 m behind the diver. Find the forces on the board.

8.46 A 1.0-m uniform boom makes an angle of 35° with the vertical and is supported by a guy wire that is perpendicular to the boom and attached 60 cm from the supported end (Fig. 8.61). The boom has a weight of 80 N, and a 200-N weight is attached to its upper end. Find the tension in the guy wire.

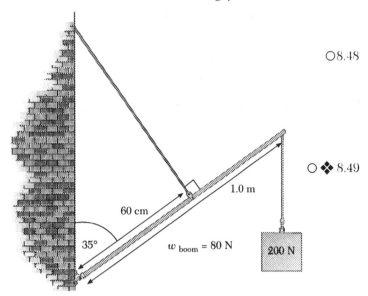

Figure 8.61   Problem 8.46.

○8.47   A 40-N flagpole 2 m long is mounted on a building and held horizontal by a guy wire that makes an angle of 30° with the horizontal. Find the tension in the guy wire and the force exerted by the building on the pole.

○8.48   A 450-N flagpole 3.5 m long is mounted on a building. A horizontal guy wire attached to the unmounted end holds the flagpole in place so that it makes an angle of 60° with the horizontal. Find the tension in the wire and the force exerted by the building on the pole.

○ ❖ 8.49   Figure 8.62 shows eight identical flagpoles, each 2.0 m long, of mass 50 kg, and with a 100-kg ornamental brass ball at the end. Calculate the tension in the guy wire for each case and rank the eight in order of increasing tension.

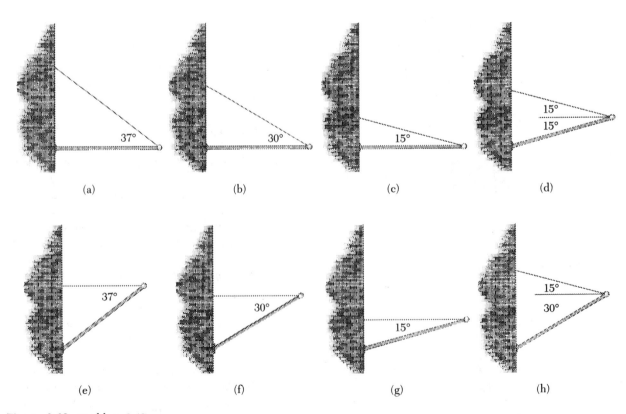

Figure 8.62   Problem 8.49.

O8.50 A 500-N stuntwoman walks out to the end of a uniform, horizontal, unattached plank that is lying on a flat roof and extends a certain distance over the edge of the roof. The plank is 6.0 m long and weighs 450 N. How far beyond the roof can it extend and still support the woman?

O8.51 A uniform rectangle of plywood weighing 200 N is in a vertical plane, with a 1.2 m side resting on the floor and a 0.90 m side vertical. What force **F** applied to an upper corner at 30° to the horizontal will cause the plywood to just start to raise off the floor and be pivoted about the diagonally opposite corner?

O8.52 An 800-N man climbs to the top of a 6.0-m ladder that is leaning against a smooth (frictionless) wall at an angle of 60° with the horizontal, as sketched in Figure 8.63. The nonuniform ladder weighs 400 N, and its center of gravity is 2.0 m up from the foot. What minimum coefficient of static friction between ground and ladder will keep the ladder from slipping?

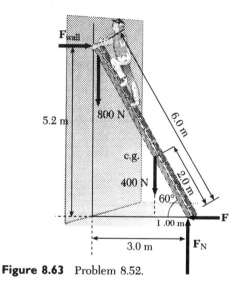

Figure 8.63 Problem 8.52.

O8.53 A ladder 5.0 m long weighs 200 N, with its center of gravity 2.0 m from the lower end. The ladder leans against a smooth wall and makes an angle of 53° with the horizontal. What minimum coefficient of static friction is needed to keep the ladder from slipping?

●❖8.54 Figure 8.64 shows eight ladders—each uniform, 2.0 m long, and with a mass of 30 kg.

Neglecting the friction between ladder and wall, calculate the minimum coefficient of static friction needed between ground and ladder to allow a 75-kg painter to climb 1.5 m up each ladder. Rank the eight by increasing value of this coefficient.

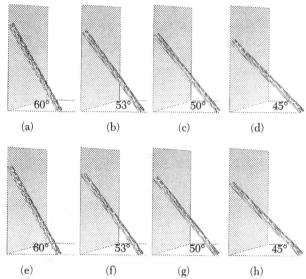

Figure 8.64 Problem 8.54.

O8.55 A uniform, rectangular, 600-N sign 0.80 m tall and 2.0 m wide is held in a vertical plane, perpendicular to a wall, by a horizontal pin through the top inside corner and by a guy wire that runs from the outer top corner of the sign to a point on the wall 1.5 m above the pin (Fig. 8.65). Calculate the tension in the wire and the force exerted by the pin on the sign.

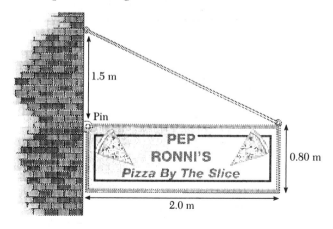

Figure 8.65 Problem 8.55.

8.56 A 100-N child sits in a portable highchair as shown in Figure 8.66. The center of the child's torso is 18 cm away from the right-hand end of the top supports and 45 cm away from the right-hand end of the bottom supports. Determine the forces $\mathbf{F}_{top}$ and $\mathbf{F}_{bottom}$ that hold the chair in place.

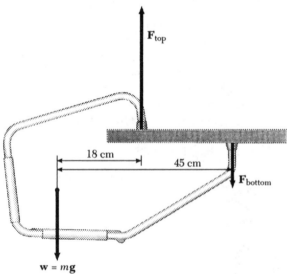

Figure 8.66   Problem 8.56.   *(Photo by the author)*

8.57 A 73-kg diver stands at the end of a 35-kg board that is 4.0 m long and supported at two places: at the opposite end and 1.5 m from that end. Determine the forces acting on the board.

●8.58 A designer plans to construct a mobile as shown in Figure 8.67. Each horizontal support has a weight of 150 N. What weights w_2 and w_3 must be used to balance the system?

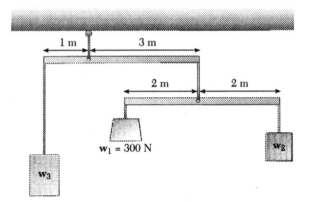

Figure 8.67   Problem 8.58.

●8.59 Locate the center of gravity of the piece of uniform plywood with a hole cut out shown in Figure 8.68.

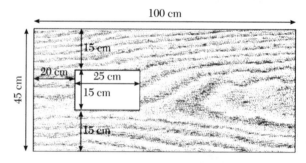

Figure 8.68   Problem 8.59.

●8.60 Locate the center of mass of the piece of uniform plywood with a hole cut out shown in Figure 8.69.

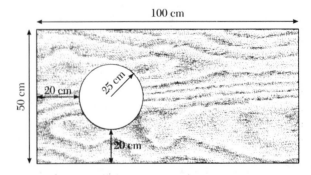

Figure 8.69   Problem 8.60.

●8.61   Locate the center of gravity of the piece of uniform plywood with a hole cut out shown in Figure 8.70.

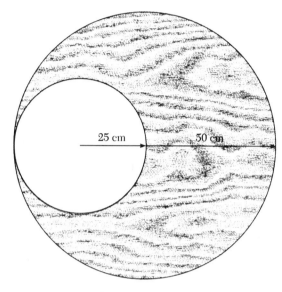

Figure 8.70   Problem 8.61.

*8.7 Forces in the Human Body

8.62   The wire in a set of dental braces carries a tension of 3.0 N. If the wire bends around the teeth so that the forces are directed as shown in Figure 8.71, what force must be exerted by the gums and root to keep the tooth in place?

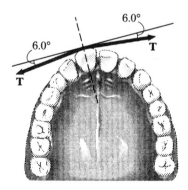

Figure 8.71   Problem 8.62.

○8.63   In Figure 8.72, a hand exerts a downward force of 100 N on a table top. What must the tension in the triceps muscle be? Ignore the weight of the forearm and hand and ignore the biceps muscle.

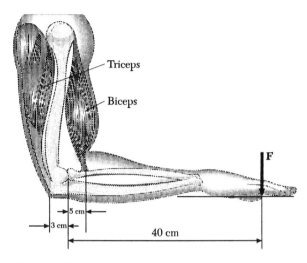

Figure 8.72   Problem 8.63.

○8.64   What is the tension T exerted by the hamstring muscles in the back of the thigh and the compressive force C in the knee joint when a horizontal force of 100 N is applied to the ankle as shown in Figure 8.73?

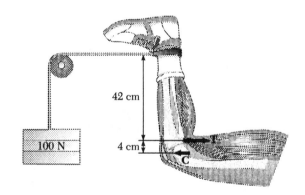

Figure 8.73   Problem 8.64.

○8.65   The biceps runs approximately from the shoulder to the elbow. Measure the length of your arm and estimate the length of your biceps. Then determine the tension in your biceps when you lift your physics book with your forearm horizontal and your upper arm vertical. What is the tension if you lift a 10-kg weight in this same way?

○8.66   What force is needed in the deltoid muscle to keep an outstretched arm horizontal, as sketched in Figure 8.74? Assume that the weight of the arm plus hand is 25 N and that the hand is empty. (The ball is for Problem 8.67.)

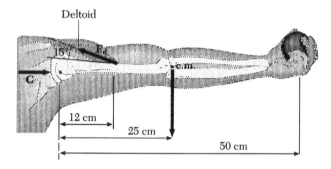

Figure 8.74   Problems 8.66 and 8.67.

○8.67   If a 100-N ball is held 50 cm from the shoulder in an outstretched hand as sketched in Figure 8.74, what must the tension in the deltoid muscle be to keep the arm horizontal?

○8.68   Calculate the tension necessary in the back to pick up the 250-N weight in the two positions shown in Figure 8.75. Assume that the spine is rigidly attached to the pelvis and that the compressive force in the spine acts along the spine.

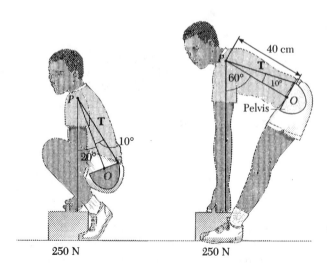

Figure 8.75   Problem 8.68.

○8.69   A 1000-N man stands on tiptoe on one foot so that all his weight is borne by the ground beneath the ball of the foot, as shown in Figure 8.76. If the foot and ankle are considered as a free body, the three forces that keep that body in equilibrium are the reaction force $\mathbf{w}'$ exerted by the ground, the tension $\mathbf{T}$ exerted by the Achilles tendon, and the compressive force $\mathbf{C}$ exerted by the tibia. Calculate the magnitude of $\mathbf{C}$ and $\mathbf{T}$.

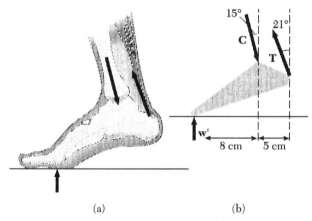

(a)   (b)

Figure 8.76   Problem 8.69.

○8.70   Determine the magnitude of $\mathbf{F}_1$ and $\mathbf{F}_2$ in Figure 8.77. The arm is at rest.

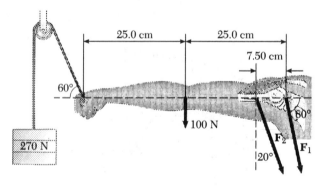

Figure 8.77   Problem 8.70.

○8.71   The principle forces acting on a foot when a person is squatting are shown in Figure 8.78. Determine the magnitude of the force $\mathbf{F}_H$ exerted by the Achilles tendon on the heel at point H and the magnitude of the force $\mathbf{F}_J$ exerted on the ankle joint at point J.

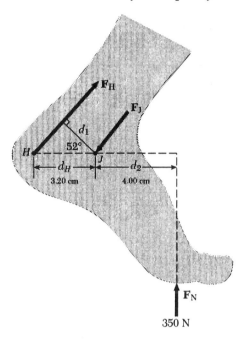

Figure 8.78 Problem 8.71.

● 8.72 An upraised forearm pulls down on a grip with a force of 100 N as shown in Figure 8.79. For simplicity, ignore the radius and ulna parts of the arm. Assume the force exerted by the biceps acts horizontally. Calculate this force.

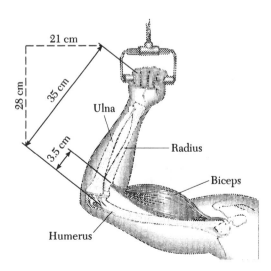

Figure 8.79 Problem 8.72.

● 8.73 Figure 8.80 (see page 310) shows a person rising to an upright position from being bent over. The erector spine muscles in the lower back, attached between hips and spine, exert a tension and cause the trunk to rotate about the hips to an upright position. The figure shows a greatly simplified mechanical model, with the upper torso considered entirely rigid, and the rotation about the hips replaced by a pivot. The force $\mathbf{T}$ is the tension exerted by the erector spinae muscles, which are attached halfway between hips and head. The force $\mathbf{w}_1$ is the weight of the trunk and arms when the hands are on the hips; the center of gravity of trunk plus arms is assumed to be 36 cm from the hips. The force $\mathbf{w}_2$ is the weight of the head; the center of gravity of the head is assumed to be 72 cm from the hips. These values—and that $w_1 = 480$ N and $w_2 = 60$ N—are reasonable for an 820-N person. The compressive force $\mathbf{C}$, applied to the base of the spine, is the force on the disks between the vertebrae. Determine the magnitude of $\mathbf{C}$.

General Problems

● 8.74 Figure 8.81 (see page 310) shows a person lifting a 30-kg mass. When the lifter's back is horizontal, as in the sketch, what is the tension T in the back muscles and the compressive force C in the spinal disks?

● 8.75 Consider a table with four legs of negligible mass equally spaced around the circumference of a uniform circular top weighing 80 N. Find the smallest weight that will upset the table when placed right at the edge.

○ 8.76 A uniform boom 6 m long weighs 300 N and is attached by a pivot at one end to a vertical wall. The pole is held at an angle of 35° above the horizontal by a horizontal guy wire between the wall and the upper end of the pole. A load of 600 N hangs from the upper end of the boom. Find the tension in the wire. What are the horizontal and vertical components of the force of the wall on the boom?

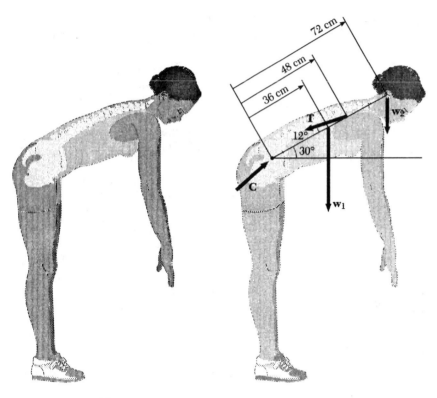

Figure 8.80   Problem 8.73.

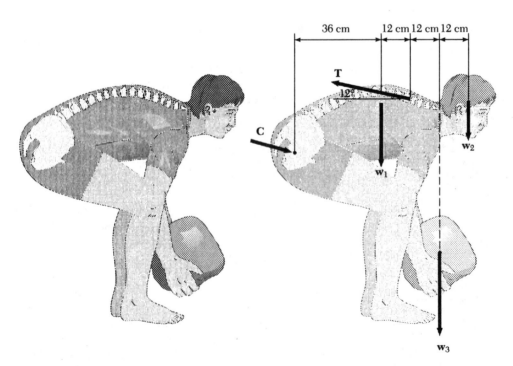

Figure 8.81   Problem 8.74.

○8.77 Tom and Dick are carrying their friend Harry on a uniform, horizontal plank 8 m long that weighs 140 N. Harry, who weighs 300 N, sits 5 m from Tom and 3 m from Dick. How much weight does each man support?

●8.78 A uniform ruler of length L and weight **w** leans without slipping against a smooth vertical wall at an angle θ with the vertical, with its foot on a rough table. Show that the coefficient of friction between the table top and their ruler must be at least $\frac{1}{2}$ tan θ. Show that your result makes sense in the limiting cases as $\theta \rightarrow 0°$ and as $\theta \rightarrow 90°$.

PHYSICS AT WORK

On the *Moonraker* ride, thrillseekers sit along the rim of the wheel, which spins rapidly. When the wheel reaches sufficient angular speed, it is tilted up to a vertical position. What keeps the "gravity-defying" passengers from falling as they reach the top of the wheel? Can you calculate the minimum angular speed that will keep them in their seats? How is this related to roller coasters that make loops and to satellites that orbit the Earth?

(© Marco Corsetti/FPG International Corp.)

9

Rotation

- Rotational motion is analogous to translational motion. Rotational motion about a fixed axis is similar to one-dimensional, linear motion. Angular displacement is defined in terms of rotation relative to a reference line. Angular velocity can then be defined in terms of angular displacement and time; likewise, angular acceleration can be defined in terms of angular velocity and time. These definitions lead to mathematical statements which are shorthand ways of describing these aspects of rotational motion.

- Equations which look very much like those for a one-dimensional, linear motion describe rotation about a fixed axis.

- The moment of inertia is equivalent to the "rotational mass" of an object. The moment of inertia describes how difficult it is to change the rotational motion of an object. The rotational equivalent of Newton's second law is that the torque on an object equals the product of its moment of inertia and its angular acceleration, $\tau = I\alpha$.

- Still analogous to translational motion, a rotating object has angular momentum and rotational kinetic energy. Angular momentum is conserved during interactions within a closed system. Rotational kinetic energy is another form of energy that must be included when energy conservation is considered.

Most of our study of motion so far has concerned *translation;* objects have moved either along a straight line or along a curved trajectory. Our introduction to rotation in Chapter 8 was concerned with the possible motion about a pivot resulting from unbalanced torques. Now we continue our study of rotational motion, preparing to answer many questions about rotating bodies. In this chapter we refine our description of motion and examine this additional way objects can move.

9.1 Angular Quantities

CONCEPT OVERVIEW

- Rotational motion, like translational motion, is described relative to an origin.

- Angular position can be measured in radians, degrees, or revolutions.

Our study of motion goes back to Chapter 2, where we began with motion in one dimension. There we described the position of an object by giving its displacement *s* from some reference position that we called the origin as shown in Figure 9.1. How can we give the equivalent information for rotational motion? How can we describe where something is on a wheel or a stereo turntable or a merry-go-round?

Figure 9.2 shows a rotating rigid body, a phonograph turntable. A radius has been drawn from the center to a point × on the edge and then extended beyond the circle as a reference. We call this the *x* direction and take counterclockwise rotations as positive and clockwise rotations as negative.[1]

There are three common units used in describing rotation. You are most familiar with **degrees** (°); there are 360° in one rotation or cycle. **Revolutions** (rev) are another common rotation unit. As any given point on the label on a phonograph record goes around three times, we say the record has turned through three revolutions. The record is probably turning at a rate of 33.3 revolutions per minute, which we abbreviate 33.3 rpm.

For our needs, there is a third unit of angular measure that is more useful than either of these, the **radian.** In Figure 9.3 the rim point marked × , originally located along the *x* axis, has traveled through a distance *s* along the

A circle is divided into 360° because ancient astronomers measured the Sun to take 360 days to complete one revolution through the collection of stars we call the zodiac.

---

[1] Other conventions are possible. Airplane and ship navigators measure angles from north, which this text usually calls the *y* direction, and considers clockwise rotation to be positive while this text usually considers counterclockwise as positive.

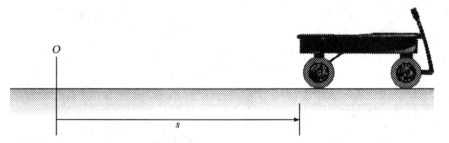

Figure 9.1

Linear displacement *s* is measured from an origin *O*.

circumference. We define the angle θ through which the turntable has rotated as the ratio of the arc length s to the radius r:

$$\theta = \frac{s}{r} \qquad (9.1)$$

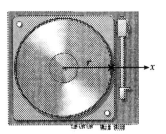

Figure 9.2
The x axis serves as a convenient reference line for rotational motion.

Notice the units; because both s and r are measured in length units (meters), this method of specifying the angle gives a dimensionless number. To remind us that we are measuring an angle, however, we quote it in **radians** (rad).

How do radians compare with degrees and revolutions? In one revolution, the × on the rim of our turntable moves through an arc length equal to the circumference, $s = C = 2\pi r$. Therefore, $\theta = s/r = 2\pi r/r = 2\pi = 2\pi$ rad (Fig. 9.4). That is,

$$360° = 1 \text{ rev} = 2\pi \text{ rad} \qquad (9.2)$$

A right angle is one fourth of a revolution, which is $\pi/2$ rad = 1.571 rad. From Equation 9.2,

$$1 \text{ rad} = \frac{360°}{2\pi} = 57.3°$$

In describing linear motion, we defined speed and velocity as the rate at which something changed its position as it moved. In an identical manner, we define the **angular velocity** ω (lower-case Greek *"omega"*) to be

$$\omega = \frac{\Delta\theta}{\Delta t} \qquad (9.3)$$

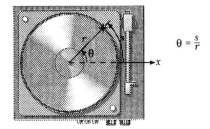

Figure 9.3
Angle θ measures the rotation from a reference direction that here is taken to be the x axis. A point on the circumference that is initially aligned with the reference direction is a distance r from the center and moves along an arc of length s. Angle θ measured in radians is defined by $\theta = s/r$.

As in the linear case, this is the *average* angular velocity over the time interval Δt. In the limit as Δt becomes very small, this becomes the *instantaneous* angular velocity. Angular velocity ω is measured in radians per second (rad/s = $1/s = s^{-1}$).

For most of our purposes, we can use angular velocity and angular speed interchangeably, but there is a difference. Just as with linear motion, angular speed is the magnitude of the vector quantity angular velocity. We postpone discussing the vector nature of angular velocity until Section 9.6, however. For now, when we are dealing only with rotation about a fixed axis, we keep track of the vector nature of angular velocity by calling it positive when the rotation is in the counterclockwise direction and calling it negative when the rotation is in the clockwise direction. This sign convention is analogous to our convention for translational velocity **v** along a straight line, where we took movement to the right as positive and movement to the left as negative.

For rotation, as with linear motion, the rate with which the angular velocity changes is important. We call this rate the **angular acceleration** α (lower-case Greek *"alpha"*) and define it identically to its linear counterpart:

$$\alpha = \frac{\Delta\omega}{\Delta t} \qquad (9.4)$$

$s = C = 2\pi r$

$\theta = \frac{s}{r} = \frac{2\pi r}{r} = 2\pi$

$\theta = 360°$

Figure 9.4
A complete circle is $360° = 1$ rev $= 2\pi$ rad.

This is the *average* angular acceleration for any time interval Δt and the *instantaneous* angular acceleration in the limit as Δt becomes very small. Angular acceleration has units of radians per second per second ($\text{rad/s}^2 = 1/\text{s}^2 = \text{s}^{-2}$).

Example 9.1

A unicyclist rides through a recently painted area on a sidewalk and then continues on 4.3 m in 3.2 s before realizing what has happened and stops. A 4.3-m trail of paint has been left by the wheel, which has a radius of 33 cm. Through what angle, in radians and in revolutions, did the wheel turn once it left the wet-paint area? What was its average angular velocity?

Reasoning

From the definition of angular measure, Equation 9.1, we can find the angle of rotation. The average angular velocity is given by Equation 9.3.

Solution

A point on the edge of the tire has traveled a distance $s = 4.3\text{ m} = 430$ cm. Thus,

$$\theta = \frac{s}{r} = \frac{430\text{ cm}}{33\text{ cm}} = 13 = 13\text{ rad}$$

$$\theta = 13\text{ rad}\left(\frac{1\text{ rev}}{2\pi\text{ rad}}\right) = 2.1\text{ rev}$$

The average angular velocity is

$$\omega = \frac{\Delta\theta}{\Delta t} = \frac{13\text{ rad}}{3.2\text{ s}} = 4.1\text{ rad/s}$$

Practice Exercise   Through how many radians does the wheel turn in laying down a 1.0-m trail of paint?

Answer   3.1 rad

9.2 Rotational Kinematics

CONCEPT OVERVIEW

■ Rotational motion about a fixed axis can be described by concepts and equations very similar to those for one-dimensional translational motion.

The equations for angular velocity ω and angular acceleration α are identical to Equations 2.1 and 2.3 for linear velocity v and linear acceleration a. From those equations for v and a, we derived three very useful kinematics equations for straight-line motion. We can immediately write the corresponding equations for rotational motion; these are listed in Table 9.1.

TABLE 9.1 Comparison between the definitions and equations for linear motion and those for rotation about a fixed axis

Linear motion	Rotation	Equation numbers
	Definitions	
Displacement s	Angular displacement θ	
Velocity $v = \frac{\Delta s}{\Delta t}$	Angular velocity $\omega = \frac{\Delta \theta}{\Delta t}$	2.1, 9.3
Acceleration $a = \frac{\Delta v}{\Delta t}$	Angular acceleration $\alpha = \frac{\Delta \omega}{\Delta t}$	2.3, 9.4
	Equations	
$s = s_i + v_i t + \frac{1}{2}at^2$	$\theta = \theta_i + \omega_i t + \frac{1}{2}\alpha t^2$	2.6, 9.5
$v = v_i + at$	$\omega = \omega_i + \alpha t$	2.5, 9.6
$v^2 = v_i^2 + 2a(s - s_i)$	$\omega^2 = \omega_i^2 + 2\alpha(\theta - \theta_i)$	2.7, 9.7

*These equations apply only when the linear or angular acceleration is constant.

With these equations for rotational kinematics, we now have the tools to solve rotation problems similar to the problems tackled in Chapter 2 involving straight-line motion.

Example 9.2

A turntable has an angular acceleration of 1.5 rad/s^2. If it starts from rest, what is its angular velocity after 4.0 s? Through what angle has it rotated in that time?

Reasoning

As with linear motion, we have three rotational kinematics equations that involve angular acceleration, angular velocity, angular displacement, and time. Equation 9.6, $\omega = \omega_i + \alpha t$, describes the final angular velocity. Equation 9.5, $\theta = \theta_i + \omega_i t + \frac{1}{2}\alpha t^2$, describes the final angular position.

Solution

$$\omega = \omega_i + \alpha t = 0 + (1.5 \text{ rad/s}^2)(4.0 \text{ s}) = 6.0 \text{ rad/s}$$

$$\theta = \theta_i + \omega_i t + \frac{1}{2}\alpha t^2$$
$$\theta = 0 + 0 + \frac{1}{2}(1.5 \text{ rad/s}^2)(4.0 \text{ s})^2 = 12 \text{ rad}$$

In 4.0 s, the turntable has acquired an angular velocity of 6.0 rad/s and has rotated through 12 rad (1.9 rev).

Example 9.3

A turntable starts from rest and undergoes uniform angular acceleration such that it is rotating at 5.0 rad/s in 8.5 s. Through what angle has it rotated during this time? Find the turntable's angular velocity when it has rotated through 100 rad.

Reasoning

We assume that the angular acceleration is uniform and begin by calculating it. We can immediately write the angular acceleration as the change in angular velocity divided by the time required. We can calculate the angle rotated through during 8.5 s from Equation 9.5.

Note that in asking for the angular velocity when θ is 100 rad, there is no mention of time. The third of our "big three" rotational kinematics equations, Equation 9.7, states a relationship between α, ω, and θ that does not involve time and, therefore, that is the one we need.

Solution

$$\alpha = \frac{\Delta\omega}{\Delta t} = \frac{5.0 \text{ rad/s} - 0 \text{ rad/s}}{8.5 \text{ s}} = 0.59 \text{ rad/s}^2$$

Equation 9.5 relates angle, time, and angular acceleration:

$$\theta = \theta_i + \omega_i t + \tfrac{1}{2}\alpha t^2$$

$$= 0 + 0 + \tfrac{1}{2}(0.59 \text{ rad/s}^2)(8.5 \text{ s})^2 = 21 \text{ rad}$$

From Equation 9.7 we get

$$\omega^2 = \omega_i^2 + 2\alpha(\theta - \theta_i)$$

$$= 0 + 2(0.59 \text{ rad/s}^2)(100 \text{ rad}) = 118 \text{ rad}^2/\text{s}^2$$

$$\omega = 11 \text{ rad/s}$$

Look back and you will see that Example 9.2 is the rotational version of Example 2.2, and Example 9.3 is the rotational version of Examples 2.3 and 2.4.

The definition of radian measure, Equation 9.1, can be solved for the arc length traveled by a point on a rotating body:

$$s = r\theta \qquad\qquad\qquad \textbf{(9.1')}$$

where r is the distance from the point to the axis of rotation. This expression can be used with Equations 9.3 and 9.4 to find the **tangential velocity** v_t and tangential acceleration a_t of a point on a rotating object:

$$v_t = \frac{\Delta s}{\Delta t} = \frac{r\Delta\theta}{\Delta t} = r\omega \qquad\qquad \textbf{(9.8)}$$

$$a_t = \frac{\Delta v_t}{\Delta t} = \frac{r\Delta\omega}{\Delta t} = r\alpha \qquad\qquad \textbf{(9.9)}$$

These two expressions tell us that the outside horse on a merry-go-round travels faster than the inside horse. The tangential velocity of a rotating object is the component of linear velocity in the direction tangential to the circle in which the object is traveling; this direction is perpendicular to the radius. Likewise, the tangential acceleration is the component of acceleration in the direction tangential to the circle in which a rotating object is moving.

We already know from Equation 5.1 that anything moving in a circle has a centripetal acceleration directed radially inward:

$$a_c = \frac{v^2}{r} = \frac{(r\omega)^2}{r}$$

$$a_c = r\omega^2 \qquad (9.10)$$

Outside horses on the merry-go-round cover a larger linear distance during each revolution, and therefore their translational speed is greater than that of inside horses. *(© D'Lynn Waldron/The Image Bank)*

The tangential component of acceleration and the centripetal component are mutually perpendicular, as you can note in Figure 9.5. Therefore the total linear acceleration is

$$a = \sqrt{a_t^2 + a_c^2} \qquad (9.11)$$

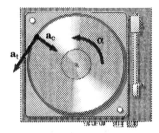

Example 9.4

A bicycle wheel rotates with an initial angular velocity $\omega_i = 5.3$ rad/s. In 2.8 s, it uniformly accelerates to $\omega_f = 9.4$ rad/s. For the valve stem ($r = 33$ cm), find *(a)* the initial tangential velocity, *(b)* the final tangential velocity, and *(c)* the total linear acceleration.

Reasoning

The first two parts are just direct applications of Equation 9.8, which relates tangential velocity to angular velocity and radius.

To find the total linear acceleration, we must first find its components, which are the centripetal acceleration and the tangential acceleration. To use Equation 9.9 to find the tangential acceleration, we need the angular acceleration from Equation 9.4; for the centripetal acceleration, we can use Equation 9.10 directly. Remember that the unit of radian is dimensionless; it came from $\theta = s/r$ or [rad] = [m]/[m]. That means

$$\frac{\text{cm·rad}}{\text{s}} = \frac{\text{cm}}{\text{s}}$$

Solution

$$v_{ti} = r\omega_i = (33 \text{ cm})(5.3 \text{ rad/s}) = 170 \text{ cm/s}$$

$$v_{tf} = r\omega_f = (33 \text{ cm})(9.4 \text{ rad/s}) = 310 \text{ cm/s}$$

Figure 9.5
Tangential acceleration a_t is tangent to a circular path, which means it is also perpendicular to the radius. Centripetal acceleration a_c is directed toward the center of the circle, along a radius.

The tangential acceleration is proportional to the angular acceleration, which is the change in angular velocity divided by the time:

$$\alpha = \frac{\Delta\omega}{\Delta t} = \frac{(9.4 - 5.3) \text{ rad/s}}{2.8 \text{ s}} = 1.5 \text{ rad/s}^2$$

$$a_t = r\alpha = (33 \text{ cm})(1.5 \text{ rad/s}^2) = 50 \text{ cm/s}^2$$

$$a_c = r\omega^2 = (33 \text{ cm})(5.3 \text{ rad/s})^2 = 930 \text{ cm/s}^2$$

Because the tangential and centripetal components of the acceleration are perpendicular to each other, the magnitude of their vector sum is given by the Pythagorean theorem:

$$a = \sqrt{50^2 + 930^2} \text{ cm/s}^2 = 931 \text{ cm/s}^2$$

9.3 Rotational Dynamics

CONCEPT OVERVIEW

■ Rotational dynamics, the study of the causes of rotational motion, is very similar to one-dimensional dynamics.

■ The rotational analog of mass is moment of inertia $I = \Sigma mr^2$.

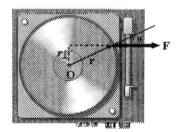

Figure 9.6
A force **F** acts on an object, with the angle between the force and the radial vector **r** being θ. The moment arm is $r_\perp$. "Angular force," or torque, $\tau = rF \sin\theta$ is produced.

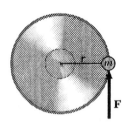

Figure 9.7
The force **F** applied to mass m is perpendicular to the radius.

Now we are ready to explain the causes of rotational motion—that is, to look for the rotational equivalent of Newton's second law. From our study of statics in Chapter 8, we know that the rotational equivalent of a force is a torque:

$$\tau = r_\perp F \tag{8.3}$$

$$\tau = rF \sin\theta \tag{8.7}$$

where, as shown in Figure 9.6, $r_\perp$ is the moment arm, F is the applied force causing the torque, r is the distance between the axis of rotation and the point of application of the force, and θ is the angle between **F** and the line drawn along the vector **r**. You should remember that torque is always defined relative to some chosen point or axis.

Figure 9.7 shows a small object of mass m moving in a circle of radius r. A force F is applied to the mass, tangential to the circle (which is the same thing as perpendicular to the radius). The acceleration in the direction of this force is the tangential acceleration a_t. Thus Newton's second law tells us that

$$F = ma_t$$

Multiply both sides by r to get the torque:

$$\tau = rF = mra_t$$

Since $a_t = r\alpha$ (Equation 9.9), we can say

$$\tau = (mr)(r\alpha) = mr^2 \alpha \tag{9.12}$$

That is, the torque applied to an object is equal to the object's mass times the square of its distance from the axis of rotation times the angular acceleration.

Note the similarity between Equation 9.12 and our old friend $F = ma$. In Equation 9.12 F is replaced by "angular force" τ, the torque, m is replaced by "angular mass" mr^2, and a is replaced by angular acceleration α. This relationship has been derived for the special case of a single particle moving in a circle, but the result is general. The "angular mass" mr^2 is given a special name, **moment of inertia,** I, and therefore we can write

$$\tau = I\alpha \qquad\qquad (9.13)$$

which describes rotation about a fixed axis in exactly the same manner that $F = ma$ describes straight-line motion.

As just noted, $I = mr^2$ for a single particle. For several particles or for an extended body, we take the sum of all the masses or imagine the body as being made up of many small masses:

$$I = \Sigma mr_{\perp}^{2} \qquad\qquad (9.14)$$

where we have written $r_{\perp}$ to remind us that this is the perpendicular distance from the axis of rotation. From Equation 9.14, we see that the units of moment of inertia are kilograms-meters squared: $kg \cdot m^2$.

Using Equation 9.14 to determine moments of inertia for extended bodies requires integral calculus. Because we do not use calculus in this course, the moments of inertia for several common shapes are given in Table 9.2, and we shall use these results in working problems.

As shown in Table 9.2, shape makes quite a difference in moments of inertia. To begin to develop a feeling for why these differences exist, let us compare some of the shapes. Compare a hollow cylinder or ring with a solid cylinder of the same mass M and radius R. All the mass in the hollow cylinder or the ring is located a distance r from the axis of rotation, but only a small percentage of the mass in the solid cylinder is at its edge, a distance R from the axis of rotation. Most of the mass in the solid cylinder has some $r_{\perp} < R$. Indeed, some of the mass of the solid cylinder is right at the axis of rotation and does not contribute to the moment of inertia at all. Next, compare the two moments of inertia given for a rod. When the rod is rotated about its center of mass, all of the mass lies within $L/2$ of the axis of rotation. When the rod is rotated about an end, only half of the mass has $r_{\perp} < l/2$. For the remaining half of the mass, $r_{\perp} > L/2$. These changes in $r_{\perp}$ are quite significant, as can be seen from the different values of the moments of inertia.

In Chapter 4 we applied Newton's second law to various systems. We now do the same with $\tau = I\alpha$.

Problem-Solving Tip

■ As always, free-body diagrams are essential.

TABLE 9.2 Selected Moments of Inertia

Shape	Moment of Inertia
Ring	MR^2
Hollow cylinder	MR^2
Thick-walled cylinder	$\dfrac{M(R_1{}^2 + R_2{}^2)}{2}$
Solid cylinder	$\dfrac{MR^2}{2}$
Hollow sphere	$\dfrac{2MR^2}{3}$
Solid sphere	$\dfrac{2MR^2}{5}$
Rod, axis through center	$\dfrac{ML^2}{12}$
Rod, axis through one end	$\dfrac{ML^2}{3}$
Rectangular plate	$\dfrac{M(A^2 + B^2)}{12}$

Example 9.5

Figure 9.8 shows a drum with a rope wound around it. The drum has a radius of 1.0 m and a moment of inertia of 360 kg·m². A constant tension of 120 N is kept in the rope, and there is no friction in the bearings that support the drum. Find its angular acceleration.

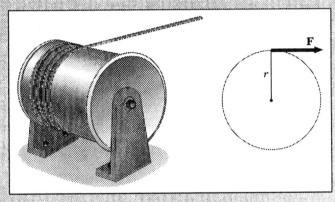

Figure 9.8
Tension in the rope creates a torque that gives the cylinder an angular acceleration.

Reasoning

Just as $F = ma$ describes linear dynamics, this rotational dynamics problem is described by Equation 9.13, $\tau = I\alpha$. We can solve this equation for the angular acceleration.

Solution

From Equation 8.5 we know that the torque created in this situation is

$$\tau = r_\perp F = (1.0\ \text{m})(120\ \text{N}) = 120\ \text{N·m}$$

Then

$$\alpha = \frac{\tau}{I} = \frac{120\ \text{N·m}}{360\ \text{kg·m}^2} = 0.33\ \frac{\text{N}}{\text{kg·m}}\left(\frac{1\ \text{kg·m/s}^2}{1\ \text{N}}\right) = 0.33/\text{s}^2 = 0.33\ \text{rad/s}^2$$

Practice Exercise What is the angular acceleration if the constant tension is 180 N?

Answer 0.50 rad/s²

Problem-Solving Tips

- In situations involving both translation and rotation, look for the connection between the two. Often this connection will be something like a cord that unwinds from a rotating body. If that is the case the relationships $s = r\theta$, $v_t = r\omega$, and $a_t = r\alpha$ will prove useful.

- Ensure that your free-body diagrams make clear which variable connects the motion of the different objects (often this will be a common acceleration).

Example 9.6 _____

Figure 9.9 shows a solid cylinder with a mass of 100 kg and a radius of 0.100 m. According to Table 9.2, this shape has a moment of inertia of $MR^2/2 = 0.500$ kg·m². A cord wrapped around the cylinder runs over a lightweight, frictionless pulley and then is attached to a 12.0-kg hanging mass. Find the linear acceleration of the hanging mass and the angular acceleration of the cylinder.

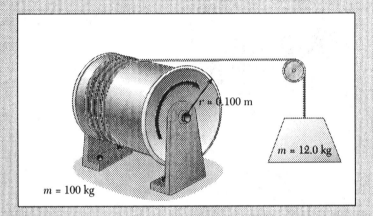

Figure 9.9
The mass falls with linear acceleration while the cylinder rotates with angular acceleration.

Reasoning

Individual free-body diagrams are necessary for a detailed solution and are shown in Figure 9.10. The unknown tension in the cord is labeled T. For the linear motion of the hanging mass, we apply $F_{net} = ma$, and for the rotational motion of the cylinder, we apply $\tau = I\alpha$.

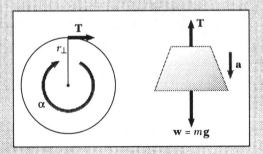

Figure 9.10
Free-body diagrams remain essential to solving problems.

Solution

First let us apply $F_{net} = ma$ to the hanging mass. Instead of our usual choice of up as positive, however, let us take down as positive because we expect the linear acceleration to be downward:

$$F_{net} = mg - T = ma$$

$$(12.0 \text{ kg})(9.80 \text{ m/s}^2) - T = (12.0 \text{ kg})a$$

$$118 \text{ kg·m/s}^2 - T = (12.0 \text{ kg})a \tag{1}$$

Because this equation has two unknowns, we know we must find additional information. The free-body diagram of the cylinder will supply some of that. For such a rotating body, we use $\tau = I\alpha$, the rotational analog of $F = ma$. From Chapter 8, we know that τ is also equal to $r_\perp T$. Combining these two expressions for τ, we get

$$\tau = r_\perp T = (0.100 \text{ m})T = (0.500 \text{ kg·m}^2)\alpha = I\alpha$$

$$(0.100 \text{ m})T = (0.500 \text{ kg·m}^2)\alpha \tag{2}$$

The linear acceleration a of the hanging mass is the tangential acceleration of the edge of the cylinder, and that quantity is related to the angular acceleration α of the cylinder by Equation 9.9:

$$a_t = r\alpha = (0.100 \text{ m})\alpha = a \tag{3}$$

We now substitute this value for a in Equation 1 and solve for T:

$$118 \text{ kg·m/s}^2 - T = (12.0 \text{ kg})(0.100 \text{ m})\alpha$$

$$T = 118 \text{ kg·m/s}^2 - (1.20 \text{ kg·m})\alpha \tag{4}$$

This equation has the same two unknowns as Equation 2, and so we can solve them:

$$(0.100 \text{ m})[118 \text{ kg·m/s}^2 - (1.20 \text{ kg·m})\alpha] = (0.500 \text{ kg·m}^2)\alpha$$

$$11.8 \text{ kg·m}^2/\text{s}^2 - (0.120 \text{ kg·m}^2)\alpha = (0.500 \text{ kg·m}^2)\alpha$$

$$11.8 \text{ kg·m}^2/\text{s}^2 = (0.120 \text{ kg·m}^2)\alpha + (0.500 \text{ kg·m}^2)\alpha$$

$$\alpha = \frac{11.8 \text{ kg·m}^2/\text{s}^2}{0.620 \text{ kg·m}^2} = 19.0/\text{s}^2 = 19.0 \text{ rad/s}^2$$

Now that we have the angular acceleration of the cylinder, we use it in Equation 3 to determine a, the linear acceleration of the hanging mass:

$$a = (0.100 \text{ m})(19.0 \text{ rad/s}^2) = 1.90 \text{ m/s}^2$$

It is also interesting to return to Equation 4 and determine the tension in the cord:

$$T = 118 \text{ kg·m/s}^2 - (1.20 \text{ kg·m})(19.0 \text{ rad/s}^2) = 95 \text{ N}$$

Notice that the tension is *not* equal to the weight $(12.0 \text{ kg})(9.80 \text{ m/s}^2) = 118 \text{ N}$ of the hanging mass.

Practice Exercise   What is the angular acceleration if the hanging mass is 25 kg?

Answer $\alpha = 33$ rad/s^2

Example 9.7

Figure 9.11 shows an Atwood's machine made with a frictionless but massive pulley of radius 0.150 m and moment of inertia 0.900 kg·m^2. A cord runs over this pulley without slipping or stretching and has attached to its ends masses $m_1 = 5.00$ kg and $m_2 = 8.00$ kg as shown. Find the linear acceleration of the masses, the angular acceleration of the pulley, and the tension on the left and right sides of the cord.

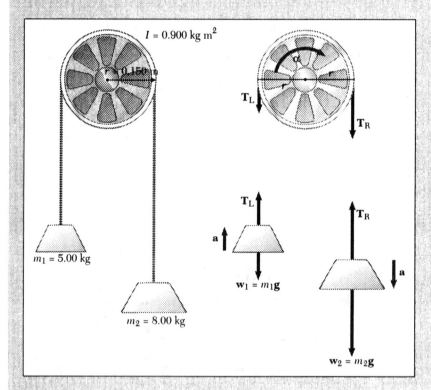

Figure 9.11
The tension is different on the two sides of a massive pulley. In the free-body diagrams, this difference is indicated by $\mathbf{T}_L$ and $\mathbf{T}_R$.

Reasoning

When a cord passes over a frictionless pulley of negligible mass, there is a single value of the tension throughout. If the pulley is heavy, however—

as this one is—or if friction is present in the pulley's bearing, the tension on one side of the pulley is different from the tension on the other side. In the free-body diagram, this difference has been indicated by T_L for the tension on the left side and T_R for the tension on the right side.

We can write $F = ma$ or $\tau = I\alpha$ for each of the three free-body diagrams.

Solution

We begin with m_1 and up as the positive direction:

$$F_{net,1} = m_1 a$$

$$T_L - m_1 g = m_1 a$$

$$T_L - (5.00 \text{ kg})(9.80 \text{ m/s}^2) = (5.00 \text{ kg})a$$

$$T_L - 49.0 \text{ N} = (5.00 \text{ kg})a$$

If we are careful that all forces (meaning tensions) are measured in newtons, all masses in kilograms, and accelerations in meters per second squared, we can omit units and write this equation as

$$T_L - 49.0 = 5.00\, a$$

Since we expect the angular acceleration of the pulley to be clockwise, we take clockwise as positive in calculating the torques:

$$\tau = I\alpha$$

$$T_R r - T_L r = I\alpha$$

$$T_R(0.150 \text{ m}) - T_L(0.150 \text{ m}) = (0.900 \text{ kg·m}^2)\alpha$$

$$T_R - T_L = (6.00 \text{ kg·m})\alpha$$

Again, if we are careful to have consistent units, we can omit them and write

$$T_R - T_L = 6.00\, \alpha$$

For m_2 and this time taking down as the positive direction, we have

$$F_{net,2} = m_2 a$$

$$m_2 g - T_R = m_2 a$$

$$(8.00 \text{ kg})(9.80 \text{ m/s}^2) - T_R = (8.00 \text{ kg})a$$

$$78.4 \text{ N} - T_R = (8.00 \text{ kg})a$$

As before, we can drop units and write

$$78.4 - T_R = 8.00\, a$$

Notice that we took up as positive for m_1 and down as positive for m_2. Since we expect the acceleration to be up for one and down for the other, it is convenient to make that choice; it is not necessary.

We now have three equations in four unknowns—a, α, T_L, and T_R. We can relate a and α by Equation 9.9 (a is the *tangential* acceleration of the *rim* of the pulley as well):

$$a = r\alpha = (0.150 \text{ m})\alpha$$

or, dropping units as before, this becomes

$$a = 0.150\,\alpha$$

With four equations and four unknowns, we have enough information to solve. Of course, we can substitute $a = 0.150\,\alpha$ to immediately reduce this to three equations in three unknowns—T_L, T_R, and α:

$$T_L - 49.0 = (5.00)(0.150\,\alpha) = 0.750\,\alpha \tag{1}$$

$$T_R - T_L = 6.00\,\alpha \tag{2}$$

$$78.4 - T_R = (8.00)(0.150\,\alpha) = 1.20\,\alpha \tag{3}$$

Equations 1 and 3 may be rewritten as

$$\alpha = \frac{T_L - 49.0}{0.750} = 1.33\,T_L - 65.3$$

$$\alpha = \frac{78.4 - T_R}{1.20} = 65.3 - 0.833\,T_R$$

These two expressions for α may then be set equal to each other:

$$1.33\,T_L - 65.3 = 65.3 - 0.833\,T_R$$

$$1.33\,T_L = 130.6 - 0.833\,T_R$$

Dividing through by 1.33 gives

$$T_L = 98.2 - 0.626\,T_R \tag{4}$$

Substituting T_L from Equation 4 into Equation 2 gives

$$T_R - (98.2 - 0.626\,T_R) = 6.00\,\alpha$$

$$1.626\,T_R - 98.2 = 6.00\,\alpha$$

$$1.626\,T_R = 98.2 + 6.00\,\alpha$$

Dividing through by 1.626 gives

$$T_R = 60.4 + 3.69\,\alpha \tag{5}$$

Substituting T_R from Equation 5 into Equation 3 gives

$$78.4 - (60.4 + 3.69\,\alpha) = 1.20\,\alpha$$

$$4.89\,\alpha = 18.0$$

$$\alpha = 3.68$$

We now know the numerical value for α is 3.68. To be consistent with the units we started with, this means that

$$\alpha = 3.68 \text{ rad/s}^2$$

$$a = r\alpha = (0.150 \text{ m})(3.68 \text{ rad/s}^2) = 0.552 \text{ m/s}^2$$

From Equation 3, the tension in the right side of the cord is

$$T_R = 78.4 - 1.20 \, \alpha$$

$$T_R = 78.4 - (1.20)(3.68) = 74.0$$

This is only the numerical part; we must add the proper units to fully specify T_R. Consistent units require that we measure force (or tension) in newtons. Therefore $T_R = 74.0$ N.

For the left side of the cord, Equation 4 gives

$$T_L = 98.2 - 0.626 T_R$$

$$T_L = 98.2 - (0.626)(74.0) = 51.9 \text{ N}$$

Practice Exercise What is the acceleration if the masses are 10 kg and 15 kg?

Answer 0.77 m/s^2

9.4 Angular Momentum

CONCEPT OVERVIEW

- Angular momentum is conserved.

- Changing the moment of inertia of a body changes the angular velocity in order to conserve angular momentum.

We have already discussed the conservation of linear momentum and the conservation of energy, and now we meet yet another conservation law: conservation of angular momentum. By analogy with the definition of linear momentum as $p = mv$ (Equation 7.1), we define angular momentum L for rotation about a fixed axis as the moment of inertia I multiplied by the angular velocity ω:

As you will see time and time again in this course, analogy is one of the most powerful tools a physicist has.

$$L = I\omega \tag{9.15}$$

Angular momentum has units of kg·m^2/s, which has no special name.

For straight-line motion, we found it useful to use both $F = ma$ and $F = \Delta p / \Delta t$. We have already used the rotational equivalent of the former, $\tau = I\alpha$. The rotational equivalent of the latter,

$$\tau = \frac{\Delta L}{\Delta t} \tag{9.16}$$

tells us that the net torque is equal to the time rate of change of the angular momentum. For a closed system, the net torque is zero so the angular momentum does not change. In other words, *angular momentum is conserved in any closed system.* We may shift the masses in a system around so that the moment

(a)

(b)

Figure 9.12
(a) When the student's arms are extended, the moment of inertia is larger and the angular speed of the stool is slow. (b) When the arms and weights are brought inward, the moment of inertia decreases and the angular speed increases. What role do the weights play?   *(© Charles D. Winters)*

of inertia changes, or we may change the rate at which part of the system rotates. Any such change in one part of the system causes in another part of the system a change that keeps the total angular momentum constant (Fig. 9.12).

Problem-Solving Tip

- For a closed system, find the angular momentum before the interaction and set this equal to the angular momentum after the interaction. This is what conservation of angular momentum means.

Example 9.8

An ice skater begins to spin with her arms and one leg extended, as illustrated in Figure 9.13, with an initial angular velocity $\omega_i = 3.0$ rad/s. With arms and leg extended, her moment of inertia is $I_i = 4.5$ kg·m². As she continues to spin, she draws her arms and leg in close to the axis of rotation, causing her moment of inertia to decrease to $I_f = 0.90$ kg·m². Neglecting friction, what is her final angular velocity ω_f?

Figure 9.13
Similar to the spinning student of Figure 9.12, Olympic gold medalist Kristi Yamaguchi decreases her moment of inertia by pulling her arms and leg in close to her body. Conservation of angular momentum requires that her angular velocity increase. *(Mike Powell/Allsport USA)*

Reasoning
We shall apply conservation of angular momentum.

Solution

$$L_f = L_i$$
$$I_f\omega_f = I_i\omega_i$$
$$(0.90 \text{ kg·m}^2)\omega_f = (4.5 \text{ kg·m}^2)(3.0 \text{ rad/s})$$
$$\omega_f = 15 \text{ rad/s}$$

Once off the board, the diver in Figure 9.14 provides another example of conservation of angular momentum. Because gravity exerts a force on her body, the diver does not constitute a closed system. However, gravity exerts no torque about her center of mass. Therefore her angular momentum is conserved just as a skater's is. She leaves the board with arms and legs outstretched to provide maximum moment of inertia. Then she pulls herself into a tuck position, with arms and legs pulled in close to the axis of rotation. In this position, her moment of inertia is reduced and her angular velocity increases significantly. Just before entering the water, she will again stretch out and thus slow her rotation so that she enters the water with very little rotation.

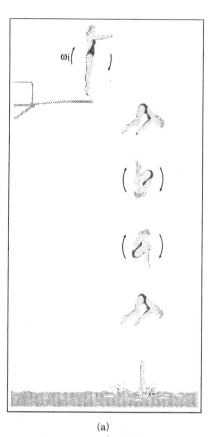

(a)

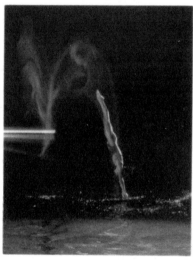

(b)

Figure 9.14
Just as the skater does, a diver increases her angular velocity by decreasing her moment of inertia. *(Gerard Planchenault/Agence Vandystadt–Allsport)*

Example 9.9

Figure 9.15 shows a physics demonstration in which a disk with moment of inertia $I_1 = 0.0080$ kg·m^2 is initially rotating with angular velocity $\omega_i = 3.6$ rad/s. Another disk with moment of inertia $I_2 = 0.0040$ kg·m^2, initially at rest, is dropped on top of the first disk. The two disks eventually rotate together with the same final angular velocity ω_f. Find that final angular velocity.

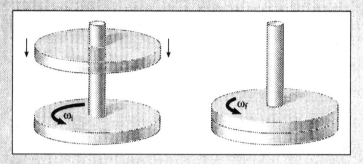

Figure 9.15
How is angular momentum conserved when the top disk lands on the bottom disk, which is already spinning, and the two-disk system continues to spin?

Reasoning

This is another example of conservation of angular momentum. The two disks exert torques on each other, but there are no torques exerted on them from the outside. Therefore, angular momentum is conserved. The initial angular momentum of the single rotating disk equals the final angular momentum of the two disks rotating together.

Solution

First find the initial angular momentum of the two-disk system:

$$L_i = L_{1i} + L_{2i} = I_1\omega_i + 0 = 0.029 \text{ kg·m}^2/\text{s}$$

Now find the final angular momentum after the two disks reach a common angular velocity ω_f:

$$
\begin{aligned}
L_f &= L_{1f} + L_{2f} \\
&= I_1\omega_f + I_2\omega_f \\
&= (I_1 + I_2)\omega_f \\
&= (0.0080 \text{ kg·m}^2 + 0.0040 \text{ kg·m}^2)\omega_f \\
&= (0.012 \text{ kg·m}^2)\omega_f
\end{aligned}
$$

$$L_f = L_i$$

$$(0.012 \text{ kg·m}^2)\ \omega_f = 0.029 \text{ kg·m}^2/\text{s}$$

$$\omega_f = \frac{0.029 \text{ kg·m}^2/\text{s}}{0.012 \text{ kg·m}^2} = 2.4 \text{ rad/s}$$

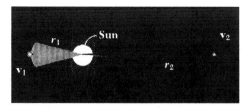

Figure 9.16
As a planet revolves around the Sun in an elliptical orbit, a radius of the orbit sweeps out equal areas during equal time intervals. This is Kepler's second law.

Angular Momentum and Kepler's Laws

From Kepler's second law of planetary motion (Section 5.6), we know that the radius from our Sun to any planet sweeps out equal areas in equal times. Therefore a planet in an elliptical orbit must move slower when it is farther from the Sun and faster when it is nearer the Sun, as sketched in Figure 9.16.

Gravity acts along the line from the Sun to the planet and therefore exerts no torque on the planet. With no torque exerted, the angular momentum of the planet remains constant. For a single object of mass m moving with speed v perpendicular to a radius r, Equation 9.15 becomes

$$L = (mr^2)\left(\frac{v}{r}\right) = mrv \tag{9.17}$$

where we have used the relationships $I = \Sigma mr_\perp^2$ (Equation 9.14) and $v_t = r\omega$ (Equation 9.8). Equation 9.17 describes the situation of a planet at the extremes of its orbit, as sketched in Figure 9.17, where the velocity $\mathbf{v}$ is perpendicular to the radius $\mathbf{r}$ at those extremes. The area ΔA of the little triangle swept out in time Δt is

$$\Delta A = \frac{rv\Delta t}{2} = \frac{(L/m)\Delta t}{2}$$

$$\frac{\Delta A}{\Delta t} = \frac{L}{2m}$$

If a tether ball is given a good swing, it will go faster and faster as its rope wraps around the pole and the ball is pulled in closer to the pole. This is necessary to conserve angular momentum. Likewise, as the end person in an ice-skating whipline is pulled in toward the center, conservation of angular momentum causes that person to speed up. $L = mrv$, and a decrease in r means an increase in v.

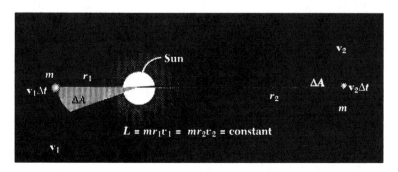

Figure 9.17
Angular momentum remains constant for planets orbiting the Sun: $L = mr_1v_1 = mr_2v_2$.

Equal areas are swept out in equal times because angular momentum is conserved (that is, L = constant means $\Delta A/\Delta t$ = constant).

To be complete, we can make Equation 9.17 general by writing it

$$L = mvr_\perp \tag{9.18}$$

where $r_\perp$ is the perpendicular distance from the origin to the velocity of the moving object. This perpendicular distance $r_\perp$ is very much like the moment arm we encountered in our study of torques.

9.5 Rotational Kinetic Energy

CONCEPT OVERVIEW

■ Any rotating body has rotational kinetic energy $KE_r = \frac{1}{2}I\omega^2$.

A particle of mass m moves in a circle of radius r with tangential velocity v, as sketched in Figure 9.18. It has translational kinetic energy given by Equation 6.8:

$$KE_t = \tfrac{1}{2}mv^2$$

Now we would like to describe this same kinetic energy in terms of rotational quantities. We know that angular velocity and tangential velocity are related by Equation 9.8:

$$v_t = r\omega$$

and therefore we can write the rotational kinetic energy as

$$KE_r = \tfrac{1}{2}m(r\omega)^2 = \tfrac{1}{2}(mr^2)\,\omega^2$$

$$KE_r = \tfrac{1}{2}I\omega^2 \tag{9.19}$$

Notice that this expression for rotational kinetic energy has the same form as its linear counterpart.

For an object that is rotating and translating simultaneously, the total kinetic energy is the sum of its translational kinetic energy ($\frac{1}{2}mv^2$) and its rotational kinetic energy ($\frac{1}{2}I\omega^2$). Rotational kinetic energy is another form of energy to consider when applying the law of conservation of energy. The "rotational collision" of Example 9.9 is entirely analogous to the perfectly inelastic collisions we considered in Chapter 7. In both cases, kinetic energy is lost as it is turned into heat due to friction.

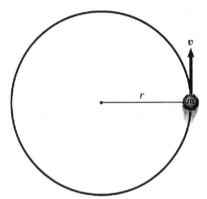

Figure 9.18
The particle has kinetic energy because it is moving.

Example 9.10

A bicycle wheel starts from rest and rolls down an inclined plane from a vertical height h, as illustrated in Figure 9.19. Find its linear speed at the bottom of the plane.

Figure 9.19
Initial potential energy equals final translational kinetic energy plus rotational kinetic energy.

Reasoning

Because a bicycle wheel has almost its entire mass M at the same radius R, we can use $I = MR^2$ from Table 9.2. Since numerical values for M and R are not given, we expect these variables to drop out of the calculations before the final result.

Initially, the wheel is at rest so there is no kinetic energy. There is potential energy $PE_i = Mgh$, however. When the wheel reaches the position $y = 0$, it has no potential energy, but now it has kinetic energy—both translational kinetic energy and rotational kinetic energy. We apply conservation of energy to find its final velocity.

Solution

We can write the initial total energy as

$$E_i = PE_i + KE_{ti} + KE_{ri} = Mgh$$

At the bottom, just as the wheel leaves the plane,

$$E_f = PE_f + KE_{tf} + KE_{rf} = 0 + \tfrac{1}{2}Mv^2 + \tfrac{1}{2}I\omega^2$$

The tangential speed v_t of the rim is the same as the translational speed v of the axle so v and ω are connected by

$$v_t = v = R\omega$$
$$\omega = v/R$$

We know the moment of inertia to be $I = MR^2$, and therefore we write the final total energy as

$$E_f = \tfrac{1}{2}Mv^2 + \tfrac{1}{2}(MR^2)(v/R)^2 = Mv^2$$

Now we apply conservation of energy:

$$E_f = E_i$$

$$Mv^2 = Mgh$$

$$v^2 = gh$$

$$v = \sqrt{gh}$$

Practice Exercise   What is the final linear speed of a solid disk that rolls down the same inclined plane?

Answer $\sqrt{\tfrac{4}{3}gh}$

As an extension of Example 9.10, consider a ring of mass m_{ring} and radius r_{ring} and a solid disk of mass m_{disk} and radius r_{disk} rolling from rest down an inclined plane. If started from the same vertical height h, which body will win the race? From Table 9.2 we know that the moment of inertia of the ring is $I_{\text{ring}} = m_{\text{ring}} r_{\text{ring}}^2$ and that of the solid disk is $I_{\text{disk}} = \frac{1}{2}m_{\text{disk}} r_{\text{disk}}^2$. Just as in the example, the initial total energies are $E_{\text{i,ring}} = m_{\text{ring}} gh$ and $E_{\text{i,disk}} = m_{\text{disk}} gh$. At the bottom, the same algebra we used in Example 9.10 gives us $E_{\text{f,ring}} = m_{\text{ring}}v^2$, but for the cylinder we get

$$E_{\text{f,disk}} = \frac{1}{2}m_{\text{disk}}\, v^2 + \frac{1}{2}I_{\text{disk}}\omega^2$$

$$= \frac{1}{2}m_{\text{disk}}\, v^2 + \frac{1}{2}(\frac{1}{2}m_{\text{disk}}\, r_{\text{disk}}^2)(v/r_{\text{disk}})^2 = \frac{3}{4}m_{\text{disk}}v^2$$

Applying energy conservation to both bodies, we have

$$m_{\text{ring}}\, v_{\text{ring}}^2 = m_{\text{ring}}\, gh$$

$$v_{\text{ring}} = \sqrt{gh}$$

$$\tfrac{3}{4}m_{\text{disk}}v_{\text{disk}}^2 = m_{\text{disk}}\, gh$$

$$v_{\text{disk}} = \sqrt{\tfrac{4}{3}gh} = 1.2\sqrt{gh}$$

The cylinder wins the race. This is true regardless of the masses and radii of the two bodies. The result is the same when $m_{\text{ring}} = m_{\text{disk}}$ and when $m_{\text{ring}} \neq m_{\text{disk}}$. You may want to try this several times from different heights and with rings and disks of different masses and radii. Try it with a sphere, too.

*9.6 Vector Nature of Angular Quantities

CONCEPT OVERVIEW

- For rotation that is not about a fixed axis, we must use the full vector nature of rotational quantities.

- The precession of a gyroscope is a result of the rotational nature of Newton's second law, $\tau = \Delta \mathbf{L}/\Delta t$.

When we considered only straight-line motion, the vector nature of displacement, velocity, and acceleration showed up only in whether the values were positive or negative. Rotation about a fixed axis is very much like straight-line motion, and consequently we have not needed to discuss the vector nature of the quantities we have introduced in this chapter. But many of them are indeed vectors quantities. A physics professor stands on a nearly frictionless turntable holding a weighted bicycle wheel in a horizontal plane. ("Nearly frictionless" means there will be no external torques.) The professor now starts the wheel rotating counterclockwise (as viewed from the top), and she rotates in the opposite direction, as sketched in Figure 9.20. We shall come back to this demonstration shortly after we have the tools to fully explain it.

To see that angular velocity $\boldsymbol{\omega}$ is a vector quantity, consider the rotating turntable in Figure 9.21. Every point on the turntable has a different linear velocity, as shown in the figure. The only direction that is constant for the rotation of all these points is the axis of rotation. But even that leaves two possibilities for a vector: pointing up along the axis or down along the axis. By

Figure 9.20
As the disk is rotated in one direction, the person rotates in the opposite direction.

Figure 9.21
The axis of rotation offers a unique direction to describe rotation. The right-hand rule says that the direction of $\boldsymbol{\omega}$ is the direction in which the thumb points when the fingers are curved in the direction of rotation.

definition, we assign a direction to the angular velocity by applying the **right-hand rule** shown in Figure 9.21. Curl the fingers of your right hand in the direction of rotation and your thumb points along the axis of rotation in the direction of the angular velocity. Notice that nothing moves through space in the direction of $\boldsymbol{\omega}$! This vector gives the direction of the axis of rotation. Its magnitude, given by Equation 9.3, describes how fast the object is rotating about this axis.

Since $\boldsymbol{\omega}$ is a vector quantity, angular acceleration $\boldsymbol{\alpha}$ is also. If an object rotating about a fixed axis is accelerated such that the magnitude of $\boldsymbol{\omega}$ increases, the direction of $\boldsymbol{\alpha}$ is the same as the direction of $\boldsymbol{\omega}$. If an object rotating about a fixed axis is accelerated such that the magnitude of $\boldsymbol{\omega}$ decreases, the direction of $\boldsymbol{\alpha}$ is opposite the direction of $\boldsymbol{\omega}$. These sign conventions are illustrated in Figure 9.22. This is just like the case of linear velocity and acceleration for straight-line motion. If the magnitude of $\mathbf{v}$ increases, then $\mathbf{a}$ lies in the same direction as $\mathbf{v}$. If the magnitude of $\mathbf{v}$ decreases, then $\mathbf{a}$ lies in the direction opposite $\mathbf{v}$.

Just as in translational motion, a negative angular acceleration, which slows a rotating body down, is sometimes called a deceleration.

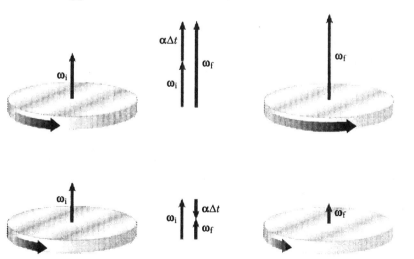

Figure 9.22
The angular acceleration vector points along the axis of rotation for rotation about a fixed axis. When $\boldsymbol{\omega}$ is increasing, $\boldsymbol{\omega}$ and $\boldsymbol{\alpha}$ point in the same direction; when $\boldsymbol{\omega}$ is decreasing, the direction of $\boldsymbol{\alpha}$ is opposite the direction of $\boldsymbol{\omega}$.

Just as $F = ma$ and $F = \Delta p/\Delta t$ are the scalar versions of the vector equations $\mathbf{F} = m\mathbf{a}$ and $\mathbf{F} = \Delta \mathbf{p}/\Delta t$, Equation 9.13 is the scalar version of the vector equation

$$\tau = I\alpha \tag{9.20}$$

and Equation 9.16 is the scalar version of the vector equation

$$\tau = \frac{\Delta \mathbf{L}}{\Delta t} \tag{9.21}$$

We have seen that $\boldsymbol{\omega}$ and $\boldsymbol{\alpha}$ are vectors, and we can easily see that angular momentum $\mathbf{L}$ is a vector given by

$$\mathbf{L} = I\boldsymbol{\omega} \tag{9.22}$$

Let us now find how to assign a direction to the torque without having to rely on knowing the direction of $\boldsymbol{\alpha}$.

Figure 9.23 shows a force $\mathbf{F}$ applied to a wheel at position $\mathbf{r}$. We know from Equation 8.6 that the magnitude of this torque is

$$\tau = rF \sin \theta$$

If the wheel is initially at rest, this torque will cause it to begin rotating clockwise. If the wheel is initially rotating clockwise, this torque will cause it to rotate clockwise faster. From our discussion of the directions of $\boldsymbol{\omega}$ and $\boldsymbol{\alpha}$, this means the angular acceleration $\boldsymbol{\alpha}$ must here be a vector that points into the page. Since $\boldsymbol{\tau} = I\boldsymbol{\alpha}$, the torque vector also points into the page.

There is an operation, known as either the *vector product* or the *cross product*, that takes two vectors $\mathbf{A}$ and $\mathbf{B}$ and produces a third vector $\mathbf{C} = \mathbf{A} \times \mathbf{B}$. The magnitude of $\mathbf{C}$ is

$$C = |\mathbf{C}| = |\mathbf{A} \times \mathbf{B}| = AB \sin \theta \tag{9.23}$$

where θ is the angle between $\mathbf{A}$ and $\mathbf{B}$ when they are drawn from a common origin. The direction of $\mathbf{C}$ is *perpendicular* to both $\mathbf{A}$ and $\mathbf{B}$, with its sense being given by the right-hand rule as $\mathbf{A}$ is rotated into $\mathbf{B}$; this is sketched in Figure 9.24. Notice that $\mathbf{A} \times \mathbf{B}$ is not the same as $\mathbf{B} \times \mathbf{A}$. In fact,

$$\mathbf{A} \times \mathbf{B} = -\mathbf{B} \times \mathbf{A}$$

Figure 9.25 shows the vectors $\mathbf{r}$ and $\mathbf{F}$ from Figure 9.23 redrawn with a common origin. If you line up the fingers of your right hand along $\mathbf{r}$ with the palm facing the smaller angle between $\mathbf{r}$ and $\mathbf{F}$, your thumb will point out of the page. This is the direction the torque $\boldsymbol{\tau}$ must point. Or, use the fingers of your right hand to rotate $\mathbf{r}$ into $\mathbf{F}$ (through the smaller angle) and your thumb points in the direction of the torque $\boldsymbol{\tau}$. Consistent with this, we define torque as

$$\boldsymbol{\tau} = \mathbf{r} \times \mathbf{F} \tag{9.24}$$

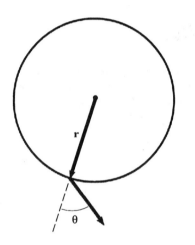

Figure 9.23
Torque $\boldsymbol{\tau}$ is produced when a force $\mathbf{F}$ is applied at a distance r from the center of rotation; $\tau = rF \sin \theta$.

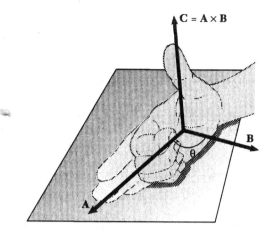

Figure 9.24
The vector $\mathbf{C} = \mathbf{A} \times \mathbf{B}$ is known as a vector product or cross product. The direction of $\mathbf{C}$ is obtained from the right-hand rule. Point the fingers of your right hand along the direction of $\mathbf{A}$ with your palm facing the smaller angle between $\mathbf{A}$ and $\mathbf{B}$. Your thumb now points in the direction of $\mathbf{C}$.

Figure 9.25
The vectors of Figure 9.23 redrawn: $\boldsymbol{\tau} = \mathbf{r} \times \mathbf{F}$.

We now have the background to look at some interesting—and even surprising—situations. Figure 9.26 shows a person standing on a nearly frictionless turntable and holding a weighted bicycle wheel. Because the turntable is nearly frictionless, there are no external torques acting on the person. With both person and wheel at rest, the total angular momentum is clearly zero. If the person now starts the wheel rotating counterclockwise (as viewed from above), the wheel gains angular momentum that can be described as a vector pointing up. To keep the total angular momentum zero, the person must acquire angular momentum that can be described as a vector pointing down; thus, she will rotate in the clockwise direction.

This classroom demonstration helps explain the design of helicopters (Fig. 9.27). The rotor blades atop a helicopter act like the bicycle wheel in Figure 9.26.

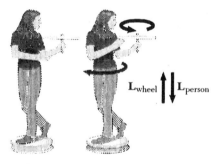

Figure 9.26
An interesting demonstration (with a dizzy physics professor) is explained by conservation of angular momentum and the vector nature of angular momentum.

Figure 9.27
Helicopter designers must be very familiar with angular momentum. (a) A single rotor helicopter. *(© J. Smith/The Image Bank)* (b) A double rotor helicopter. *(© Bob Graham/FPG International Corporation)*

The gyroscope gets its name from the verb "to gyrate," which means to move in either a circular or a spiral motion.

A single rotor rotating in one direction would cause the helicopter body to rotate in the opposite direction. To avoid this, some helicopters have two counter-rotating blades on top and others have a single rotor on top and a small propeller set far away. That propeller exerts a torque on the air comparable to the person in Figure 9.26 putting a foot down on the floor to exert a torque and hold herself steady.

Figure 9.28 shows a toy gyroscope. Exuberant ads for these delightful toys sometimes claim they "defy the law of gravity!" Try to balance one at rest by placing the tip of its axis on the support piece and the gyroscope topples over every time, just as you expect. Get the wheel rotating rapidly, though, and when placed on the support, the device does not topple over but rather revolves slowly around the support point in a motion we call *precession*. Gravity still applies, of course, so just what *is* keeping the gyroscope hovering above the tabletop rather than crashing into it? In answering this question, we assume that the gyroscope is spinning so rapidly that its angular momentum vector $\mathbf{L}$ always lies along the axis of rotation. As the rotation slows down, this approximation becomes less and less accurate.

Consider a gyroscope rotating in the direction indicated by the arrow in Figure 9.28 or 9.29. In addition to this rotation about its central axis, the gyroscope is also precessing about the support point, as indicated by the dashed circles. The weight of the gyroscope exerts a torque $\mathbf{r} \times m\mathbf{g}$ that has the support point as its pivot. The direction of this torque, determined by the right-hand rule illustrated in Figures 9.24, 9.28 and 9.29, is out of the plane of the page. (Do not mistakenly think that the direction of $\boldsymbol{\tau}$ is the same as the direction of $m\mathbf{g}$; it is not.)

From Equation 9.16, we know that the direction of $\boldsymbol{\tau}$ is also the direction of $\Delta\mathbf{L}$. Thus the angular momentum changes from $\mathbf{L}_0$ to $\mathbf{L}_f$ $\mathbf{L}_0 + \Delta\mathbf{L}$. Because the rotation axis must align with $\mathbf{L}$ (remember our starting assumption), the gyroscope precesses in a circle that has its plane perpendicular to the plane of the page. In this precession, which is counterclockwise as viewed from above, the tip of the gyroscope goes into the page on the right (at "three o'clock") and comes back out on the left (at "nine o'clock"). Thus the rotation axis cannot

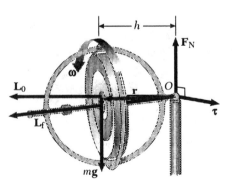

Figure 9.28
A toy gyroscope reveals the basics of all gyroscopic motion. *(Courtesy of Central Scientific Company)*

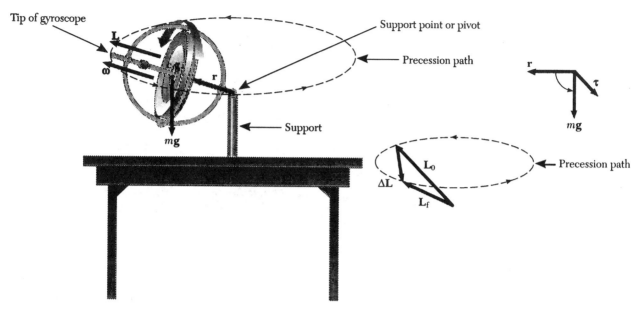

Figure 9.29
Counterclockwise precession of a gyroscope.

swing down and hit the tabletop because it has to follow the $\mathbf{L} + \Delta\mathbf{L}$ path. It is not "defying gravity" but rather the changing angular momentum that governs the motion of the rotation axis, and the gyroscope keeps it from toppling over.

The direction of the precession depends on the direction of rotation about the gyroscope's axis. The right-hand rule tells us that $\boldsymbol{\omega}$ is directed away from the support point in Figure 9.29 so this is the direction of $\mathbf{L}$. When we reverse the direction of central-axis rotation, as indicated by the arrow in Figure 9.30, the direction of $\boldsymbol{\omega}$ and consequently that of $\mathbf{L}$ is reversed, but the direction of

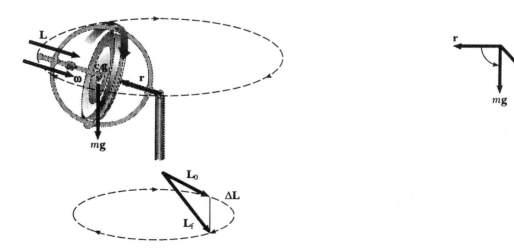

Figure 9.30
What happens to the precession direction when the direction of central-axis rotation is reversed?

Figure 9.31
Gimbal-mounted navigational gyroscopes are necessities for safe high-speed air and sea travel. A demonstration model is shown. *(Courtesy of Central Scientific Company)*

If you were to try leaning far out over your feet the way a gyroscope leans far out over its support point, you would succumb to gravity and fall over.

τ and consequently that of $\Delta \mathbf{L}$ remains the same—out of the page. Therefore the changing angular momentum vector $\mathbf{L} + \Delta \mathbf{L}$ now enters the page on the left and comes back out on the right. Because the rotation axis stays aligned with $\mathbf{L}$, the precession is now clockwise when viewed from the top.

Once the central-axis spinning stops, the gyroscope topples over. There is no "defying gravity" or "violating the laws of physics." In fact, the precession that keeps the gyroscope from toppling over is *predicted* by the laws of physics.

Equation 9.21, $\tau = \Delta \mathbf{L}/\Delta t$, tells us that, in the absence of a net torque, $\Delta \mathbf{L}$ is zero so $\mathbf{L}$ remains constant, in direction as well as in magnitude. This constant direction of $\mathbf{L}$ has an important application. Figure 9.31 shows a gyroscope mounted on gimbals that exert zero torque. Such a gimbal-mounted gyroscope can be used as a "gyrocompass" to always point in the same direction even as the ship or airplane holding it moves in various directions.

SUMMARY

There are three common units used in describing rotation—degrees, revolutions, and radians. The angle θ, in radians, through which an object rotates is defined as the ratio of the arc length s through which a point on the object moves to the distance r between this point and the axis of rotation:

$$\theta = \frac{s}{r} \qquad (9.1)$$

Degrees, revolutions, and radians are related by

$$360° = 1 \text{ rev} = 2\pi \text{ rad} \qquad (9.2)$$

The angular velocity ω is the rate at which an object rotates:

$$\omega = \frac{\Delta \theta}{\Delta t} \qquad (9.3)$$

The angular acceleration α is the time rate of change of the angular velocity ω:

$$\alpha = \frac{\Delta\omega}{\Delta t} \qquad (9.4)$$

The arc length traveled by a point on a rotating object is

$$s = r\theta \qquad (9.1')$$

which can be extended to find the tangential velocity,

$$v_t = r\omega \qquad (9.8)$$

and the tangential acceleration,

$$a_t = r\alpha \qquad (9.9)$$

The moment of inertia I of a rotating body is defined as

$$I = mr_\perp^2$$

for a single particle and as

$$I = \Sigma mr_\perp^2 \qquad (9.14)$$

for an extended body, where $r_\perp$ is the perpendicular distance from the axis of rotation to the individual mass m.

The angular momentum L for rotation about a fixed axis is defined as "angular mass" or moment of inertia I multiplied by the angular velocity ω,

$$L = I\omega \qquad (9.15)$$

Much as its straight-line motion counterpart, rotation about a fixed axis can be described by

$$\tau = I\alpha \qquad (9.13)$$

and

$$\tau = \frac{\Delta L}{\Delta t} \qquad (9.16)$$

For an isolated system, the total angular momentum L is conserved.

Rotational kinetic energy is

$$KE_r = \tfrac{1}{2}I\omega^2 \qquad (9.19)$$

and is another form of energy to include when considering energy conservation.

Angular velocity $\boldsymbol{\omega}$, angular acceleration $\boldsymbol{\alpha}$, torque $\boldsymbol{\tau}$, and angular momentum $\mathbf{L}$ are all vector quantities. The direction of $\boldsymbol{\omega}$ is the direction in which the thumb points when the fingers of the right hand are curled in the direction of rotation. The direction of $\boldsymbol{\alpha}$ is the same as the direction of $\boldsymbol{\omega}$ when a rotating body being accelerated gains angular velocity and opposite to the direction of $\boldsymbol{\omega}$ when a rotating body being accelerated loses angular velocity.

The direction of $\boldsymbol{\tau}$ is the same as the direction of the angular acceleration caused by $\boldsymbol{\tau}$.

QUESTIONS

1. Explain how, standing a known distance away, you could determine the height of one of the carvings on Mt. Rushmore using only a meter stick.
2. As you drive along the highway, why do road signs beside your car seem to go by faster than those far down the road?
3. If two bicyclers ride side by side at the same velocity and one rides a bicycle with 24-in. wheels while the other rides a bicycle with 26-in. wheels, which bicycle wheel has the greater angular velocity?
4. Consider a compact disc (CD) rotating at constant angular velocity. Does a scratch on the rim have a radial component of velocity? A tangential component? A radial component of acceleration? A tangential component?
5. Answer the four parts of Question 4 for a phonograph record rotating with constant angular acceleration as it changes from 33 rpm to 45 rpm.

6. Circus highwire performers (Fig. 9.32) will often use very long poles as balancing aids. Explain how this works.

Figure 9.32
Question 9.6   (© *Barbara Adams/FPG International Corporation*)

7. Why would reducing the mass of a bicycle's wheels be more advantageous to a racer than the same mass reduction in the frame?

8. Why are sit-ups more difficult to do if your hands are placed behind your head instead of at your sides? This is the same question asked in Chapter 8 when we could answer only in terms of forces and torques. In addition, how does the change in your own moment of inertia affect doing sit-ups?

9. Explain the difference between centripetal acceleration and angular acceleration.

10. In riding up a steep hill, a bicycle rider will probably shift down to a lower gear. That means shifting the chain to a rear-wheel gear of a different diameter. Will the diameter of the climbing gear be smaller or larger than the diameter of the level-road gear?

11. A solid disk and a solid sphere of the same mass and radius are released from rest at the top of an inclined plane. Which reaches the bottom first?

12. Two spheres look identical and have the same mass. One sphere is hollow; the other, solid. How might you determine which is which?

13. A hollow sphere and a solid sphere of the same mass and radius are released from rest at the top of an inclined plane. Which reaches the bottom first?

14. Two solid disks are released from rest at the top of an inclined plane. One has twice the mass and twice the radius of the other. Which reaches the bottom first?

15. Why does a yo-yo come back up its string?

16. Automatic record changers drop a new phonograph record on top of the record that has just been played. Explain why this may leave scratches on the records.

17. What is the direction of Earth's angular momentum vector?

18. You stand on a frictionless turntable and hold a weighted, spinning bicycle wheel by handles extending from its axis. You want to rotate in one direction, stop, and then rotate in the opposite direction. Should you start with the axis of the wheel horizontal or vertical, or will both starting positions give the same result?

PROBLEMS

9.1 Angular Quantities

9.1   Convert to radians: (*a*) 30°,   (*b*) 45°, (*c*) 60°,   (*d*) 90°,   (*e*) 180°,   (*f*) 225°.

9.2   Convert to degrees: (*a*) $\pi/2$ rad, (*b*) $3\pi/4$ rad,   (*c*) 1.20 rad,   (*d*) 2.5 rad, (*e*) 3.5 rad.

9.3   What is the angular velocity of the minute hand of a clock in radians per second?

9.4   At 553.33 m, the CN Tower in Toronto is the world's tallest free-standing structure. How far would you be from the tower if it subtended an angle of 0.1 rad (Fig. 9.33)?

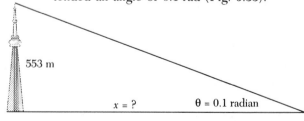

553 m x = ? θ = 0.1 radian

Figure 9.33   Problem 9.4.

9.5   The Statue of Liberty is 46 m tall. How far away would you be if it subtended an angle of 0.25 rad (Fig. 9.34)?

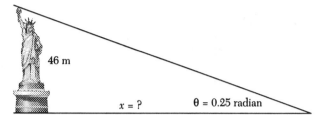

46 m

$x = ?$ $\theta = 0.25$ radian

Figure 9.34   Problem 9.5.

9.6   A compact disc turns at 500 rpm. What is its angular velocity in radians per second?

9.7   A circulating fan rotates at 960 rev/min. What is its angular speed in radians per second?

9.8   A Ferris wheel of radius r_1 is powered by a wheel of radius r_2 in contact with the circumference of the Ferris wheel. What is the ratio of the angular velocities of the two wheels?

9.9   Earth rotates at the rate of 1 rev/day. What is that in radians per hour? In degrees per hour? How large—in radians and in degrees—would you expect a one-hour time zone to be? At 45° latitude, how wide—in kilometers—would you expect a one-hour time zone to be? (The radius of Earth is 6500 km.)

9.2 Rotational Kinematics

9.10   It requires 6 s for a phonograph turntable initially rotating at 33 rpm to reach 45 rpm. If the angular acceleration is uniform, how many revolutions are made during this time?

9.11   A ship's windlass starts from rest and requires 9.3 s to reach a final angular velocity of 0.8 rev/s. What is that angular velocity in radians per second? If the windlass has a radius of 15 cm, how much line will be hauled in during that 9.3 s?

9.12   In coming back to idle, a gasoline engine slows down from 5000 to 1200 rev/min in 5.3 s. How many revolutions does the engine make during this time?

9.13   A hamster running inside a wheel turns the wheel at a rate of 37.2 rev/min. The wheel

has a radius of 0.125 m. If the hamster were running on the ground, how fast would it be running?

9.14   What is the centripetal acceleration experienced by a rider 4.5 m from the axis of rotation on a carousel that rotates five times each minute?

9.15   What is the centripetal acceleration of a scratch 6.1 cm from the center of a compact disc rotating at 500 rpm?

9.16   In civil aviation, a "standard turn" for level flight of a propeller-driven airplane is one in which the plane makes a complete circular turn in 2.0 min. If the speed of the plane is 180 m/s, what is the radius of the circle? What is the centripetal acceleration of the plane?

9.17   What is the centripetal acceleration of the tip of a second hand if the tip is 10.0 cm away from the rotation axis?

○**9.18**   An ultracentrifuge used to separate uranium isotopes has a rotating chamber that is 7.5 cm in diameter and rotates at 90 000 rev/min. What is the acceleration of a point on the circumference of this chamber?

Problem 9.18. An ultracentrifuge.   (*Charles D. Winters*)

○**9.19**   In a particular ultracentrifuge used to separate uranium isotopes, the acceleration of a point on the edge is 300 000*g*. If the centrifuge is 10 cm in radius, what is its rotational speed?

O9.20 A bicyclist travels at 10.0 m/s. What is the angular speed of a point on the tire if the tire's radius is 33.5 cm?

O9.21 A ventilator fan is turning at 600 rev/min when the power is cut off, and it turns through 1000 rev while coasting to a stop. Find the angular acceleration and the time required to stop.

9.3 Rotational Dynamics

9.22 A thin, light, wooden meterstick has attached to it a 3.0-kg mass at the 20-cm mark and a 5.0-kg mass at the 70-cm mark. What is the moment of inertia about an axis through the zero end? About an axis through the 50-cm mark?

O9.23 A 0.20-kg meter stick has a 0.30-kg mass at the 20-cm mark and a 0.50-kg mass at the 70-cm mark. What is the moment of inertia about an axis through the zero end? About an axis through the 50-cm mark?

9.24 What is the moment of inertia about the central axis of a bicycle wheel of mass 2.5 kg and radius 0.30 m?

O9.25 What is the moment of inertia of a pole vaulter's pole that is pivoted about an axis through one end? The pole is uniform, 5.0 m long, and has a mass of 3.5 kg.

9.26 A wheel that is 0.80 m in diameter and has a moment of inertia of 4.8 kg·m² is free to rotate on its central axis. A constant tension of 12 N is kept in a cord wound around the circumference. Find the angular acceleration of the wheel. The wheel starts from rest. What is the angular velocity at the end of 5 s? How much cord is unwound in 5 s?

9.27 A grindstone is initially rotating with an angular velocity of 10.5 rad/s. A lawn mower blade is pressed tightly against the grindstone and brings it to rest in 6.5 s. The grindstone is a uniform disk of mass 10 kg and radius 0.30 m. (a) What is its angular acceleration (assumed constant)? (b) What is its moment of inertia? (c) What force (assumed constant) is exerted by the blade?

9.28 A horizontal, 6.5-N force is exerted tangentially on a Frisbee of mass 0.032 kg and ra-

Problem 9.28. Throwing a Frisbee. (© *Lawrence Migdale/Stock, Boston*)

dius 0.143 m. Assuming the Frisbee is initially at rest and the force is exerted for 0.12 s, determine the angular velocity about its center of mass when the Frisbee is released.

O9.29 A solid, uniform cylinder of radius 12 cm and mass 5.0 kg is free to rotate about its symmetry axis. A cord is wound around the cylinder, and a 1.2-kg mass is attached to the end of the cord. Find the linear acceleration of the hanging mass, the angular acceleration of the cylinder, and the tension in the cord.

O9.30 A playground merry-go-round is a solid, uniform, horizontal disk with a radius of 1.5 m and a mass of 100 kg. If a child applies a force of 80 N for 3.0 s tangent to the rim, what is the final angular velocity of the merry-go-round if it starts from rest?

O9.31 An Atwood's machine is composed of a 1.0-kg mass and a 1.5-kg mass attached to a string hanging over a wheel that has a radius of 8 cm and a moment of inertia of 12.5 kg·m². Find the linear acceleration of the masses, the angular acceleration of the wheel, and the tension in the string on each side.

O9.32 A 10-kg block sits on a horizontal surface, and the coefficient of friction between block and surface is 0.20. A string runs from this block over a wheel of radius 10 cm and moment of inertia 10 kg·m² and is attached to a hanging 2.0-kg mass. Find the acceleration of the masses, the angular acceleration of the wheel, and the tension in each side of the string.

9.4 Angular Momentum

9.33 What is the angular momentum of a 0.200-kg meter stick when rotated about its center at 6.28 rad/s?

9.34 What is the angular momentum of a uniform, solid disk having a radius of 0.500 m and a mass of 2.00 kg when rotated about its center at 12.5 rad/s?

9.35 What is Earth's angular momentum due to its daily rotation about its axis?

9.36 What is Earth's angular momentum due to its revolution about the Sun?

9.37 A playground merry-go-round that has a moment of inertia of 500 kg·m² rotates at 0.25 rev/min. A 50-kg child jumps onto the edge, 1.6 m from the axis of rotation. What is the angular velocity after the child jumps on? After riding for two revolutions, the child jumps off as gently as possible. What is the angular velocity of the merry-go-round after the child leaves?

9.38 Assume a playground merry-go-round to be a uniform cylinder of mass 150 kg and radius 1.8 m. What is its moment of inertia? It is at rest when a 50-kg child, running at 4.0 m/s in a direction tangential to the edge of the merry-go-round, jumps on. What is its angular velocity after the child sits down on the edge?

9.39 A solid, uniform disk of mass 2.0 kg and radius 20 cm rotates initially at 5.0 rad/s. A ring of mass 1.0 kg and radius of 20 cm, initially at rest, is carefully dropped on top of the rotating disk so that the ring circumference is exactly superimposed on the circumference of the disk. The two finally rotate together at a common angular velocity. Find this final angular velocity.

9.40 A particular communications satellite has a moment of inertia of 250 kg·m² for rotation about its symmetry axis. When initially placed in orbit, it rotates at 11.5 rad/s. This rotation is too fast. To decrease its rotation, two 2.3-kg masses are released on long lines symmetrically on either side of the satellite. What is the angular velocity of the satellite and masses system when the masses are each 50 m from the satellite?

9.5 Rotational Kinetic Energy

9.41 What is the kinetic energy of a tire that has a moment of inertia of 60 kg·m² and rotates at 150 rev/min?

9.42 What is the kinetic energy of a 0.200-kg meter stick when rotated about one end at 6.28 rad/s?

9.43 What is the kinetic energy of a uniform, solid disk having a radius of 0.500 m and a mass of 2.00 kg when rotated about its center at 12.5 rad/s?

9.44 A centrifuge has a moment of inertia of 0.05 kg·m². How much work must be done to bring it from rest to 12 000 rev/min?

9.45 Calculate the initial and final kinetic energy of the rotating disks of Example 9.9.

9.46 How much energy is lost in the process described in Problem 9.40?

◯9.47 A solid disk rolls without slipping down a hill of vertical height 10.0 m. If the disk starts from rest at the top of the hill, what is its translational speed at the bottom?

◯9.48 What is the rotational kinetic energy of a drum majorette's baton that is rotating about an axis through its center of mass at an angular speed of 3.0 rev/s? Assume the baton to be a uniform rod 0.70 m long with a mass of 0.50 kg. What is the rotational kinetic energy when 0.020-kg rubber tips at each end are included in your calculations?

Problem 9.48. Twirling a baton. (© *Thomas Freedman*)

○9.49   A solid cylinder starts from rest at height H, rolls down a hill and into a loop-the-loop of radius R. What minimum height H is necessary for the cylinder to remain in contact with the loop-the-loop at the top. The radius r of the cylinder is much, much less than the radius of the loop.

○9.50   A uniform, solid sphere of mass 1.8 kg and radius 8.5 cm is rolling along a horizontal plane with speed $v_i = 2.8$ m/s when it encounters an inclined ramp. To what vertical height will the sphere roll before coming to rest?

○9.51   An Atwood's machine is composed of a 2-kg mass and a 2.5-kg mass attached to a string hanging over a wheel that has a radius of 10 cm and a moment of inertia of 12.5 kg·m². The 2.5-kg mass is initially 1.0 m above the floor. Use conservation of energy to find the speed of this mass just before it hits the floor.

●9.52   A basket of tomatoes of mass 20.0 kg is being hoisted by a windlass. The rope is wrapped around an axle that is a solid cylinder of wood having a radius of 0.100 m and a mass of 10.0 kg. The mass of the crank handle is negligible. The operator lets go of the handle when the basket is 6.00 m above the ground. With what linear speed does the basket strike the ground?

●9.53   The frictionless puck in Figure 9.35 has a mass of 0.120 kg. Its distance from the axis of rotation is initially 40.0 cm, and the puck is moving with an initial speed of 80.0 cm/s. The string is pulled downward 15.0 cm through the hole. Determine the work done on the puck.

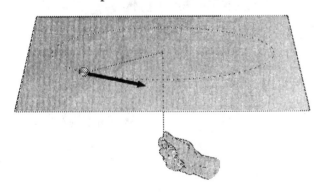

Figure 9.35   Problem 9.53.

*9.6 Vector Nature of Angular Quantities

○9.54   To demonstrate conservation of angular momentum, a physics professor stands on a frictionless turntable with a 2.0-kg mass in each outstretched hand. An assistant gives her a small initial angular velocity of 2.0 rad/s. She then drops her hands to her sides, and her angular velocity increases dramatically. As a rough estimate, consider her arms to have mass of 5.0 kg each and to be rods 1.0 m long that are hinged at the axis of rotation. The rest of her body has an approximate moment of inertia of 0.55 kg·m². Find her final angular velocity when the masses are 0.25 m from the axis of rotation. Calculate the initial and final values of the rotational kinetic energy and explain the cause of the difference in these values.

●9.55   A 0.35-kg gyroscope is composed of a 0.20-kg solid disk rotor that has a radius of 4.0 cm (Fig. 9.36). The remaining 0.15 kg of the gyroscope's mass is in the shaft and frame. The gyroscope rotates at 300 rev/min and is supported in a horizontal position by a pivot 5.0 cm from its center of mass (the center of the rotor). Find the angular velocity of its precession ω_p.

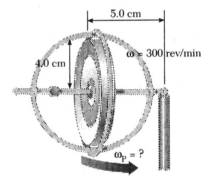

Figure 9.36   Problem 9.55.

General Problems

○9.56   A 20-kg block sits on a plane inclined 30° from horizontal. The coefficient of friction between block and plane is 0.15. A string runs from this block over a wheel of radius 12 cm and moment of inertia 16 kg·m² and is attached to a hanging 10-kg mass. Find the acceleration of the masses, the angular acceleration of the wheel, and the tension in each side of the string.

O9.57   A uniform turntable is a solid disk of mass 0.90 kg and radius 0.20 m. It is brought to a rotational velocity of 3.0 rad/s by having its drive motor turned on for 4.5 s. The drive motor is then turned off.
(a) What torque did the motor supply while bringing the turntable up to speed?
(b) What is the turntable's angular momentum while it is coasting frictionlessly at a constant 3.0 rad/s?   (c) While the turntable is coasting at 3.0 rad/s, a disk, whose moment of inertia is 0.0025 kg·m², is dropped onto the turntable and starts to rotate along with it. What is the new angular velocity of the turntable plus disk?   (d) How much rotational kinetic energy is lost during the inelastic collision of the record with the turntable?

●9.58   A man on a rotating platform holds a 2.0-kg mass in each hand. Initially his arms are extended, with each mass 1.0 m from the axis of rotation, and he rotates at 2.0 rad/s. The platform and man's body together (but not including the masses) may be assumed to have a constant moment of inertia of 1.00 kg·m².   (a) Calculate the total angular momentum of the system (man plus platform plus masses).   (b) What is the rotational kinetic energy of the system?   The man now draws his arms in until each mass is 0.25 m from the axis of rotation.   (c) What is the new angular velocity of the system?

(d) What is the final rotational kinetic energy of the system?   (e) What produces this change in rotational kinetic energy?

●9.59   When a truck is brought to rest by conventional brakes, its translational kinetic energy is transformed partly to heat energy in the brake system and partly to the energy of tearing loose fibers from the brake linings and forming steel dust from the brake shoes. It is proposed to save some of this energy by storing it in a flywheel on board the truck, to be reused as the truck starts up again. Suppose that a truck of mass 1.0×10^4 kg moving at 30 m/s stores 50 percent of its translational kinetic energy in a flywheel that is a solid disk of mass 16 kg and radius 0.50 m.
(a) What is the flywheel's moment of inertia?
(b) What is its final speed, in revolutions per minute?   (c) What is the tangential speed of a point on its edge?   (d) The flywheel will fly apart due to internal stresses if it stores more than 14×10^6 J of rotational kinetic energy. Should we be concerned with the safety of this situation?

●9.60   A pole vaulter's pole is released from a vertical position and topples over and falls to the ground. The pole has length L and mass M. Show that the linear velocity with which the end of the pole strikes the ground is $\sqrt{3\ gL}$.

●9.61   A solid sphere of mass M and radius R rolls without slipping from height H down an inclined plane of length L. What is the linear velocity of the sphere's center of mass when it reaches the bottom of the plane?

Fluids on Earth (due to gravity) exhibit interesting properties when under pressure. Biological organisms normally live under the influence of these fluid properties. Fluid flow in organisms, studied by the branch of biology known as physiology, presents interesting applications of fluid mechanics.

Because of their long necks, giraffes must pump blood a relatively great distance to their brains. Special vascular adaptations prevent the dangerous rise in cerebral blood pressure that would otherwise develop when the giraffe lowers its neck to drink. Paleontologists are applying these principles of physiology based on the physics of fluids to analyze how long-necked dinosaurs once moved and lived. Can you speculate as to how the largest dinosaurs—much longer, taller, and more massive than any giraffe—could maintain their equilibrium when raising or lowering their necks and heads?

(© M.P. Kahl/Photo Researchers)

10

Solids and Fluids at Rest

Key Issues

- There are three common states of matter—solid, liquid, gas. A solid has a definite shape and volume. A liquid has a definite volume but can flow to take the shape of its container. A gas expands or contracts to take the volume as well as the shape of its container.

- Solids can be deformed by exerting forces on them. To account for the initial size of the solid, we describe these deforming forces in terms of stress, the ratio of the deforming force to a characteristic dimension. Likewise, it is important to describe the resulting deformation in terms of strain, the ratio of the deformation to an initial dimension.

- Density is the ratio of mass to volume. For fluids—both liquids and gases—density is often more useful than mass.

- Pressure is the ratio of the force exerted by a fluid to the area over which that force acts. For fluids, pressure is often more useful than force. Pressure exerted by a fluid at some point is equal in all directions.

- Pressure increases with depth into a fluid. Thus pressure on a diver—or a submarine— increases the deeper she goes.

- For a closed system of fluid, an increase in the pressure at one point causes an equal increase in pressure throughout the system; this is known as Pascal's principle and is the principle behind the operation of hydraulic brakes or hoists.

So far we have studied the behavior of particular objects—whether the movement of an isolated particle or the rotation of an extended body. In this chapter and the next, we look at the behavior of matter in bulk. How does a pencil eraser deform when you push sideways on it? Why does a bag of potato chips inflate as you go up a mountain? How are you able to float in water—and why do you float higher in salt water than in fresh?

10.1 States of Matter

CONCEPT OVERVIEW

▄ Matter can be in the form of solid, liquid, gas, or plasma, and the form taken is a function of temperature.

There are three common states of matter: solid, liquid, and gas. A **solid** is matter that has a particular volume and shape that do not change when the matter is moved from one container to another. A **liquid** is matter that has an unchangeable volume but changes its shape to fit its container. A **gas** is matter that has no particular volume and no particular shape; expanding to fill any container it is put in. These ideas are illustrated in Figure 10.1.

(a)

(b)   (c)

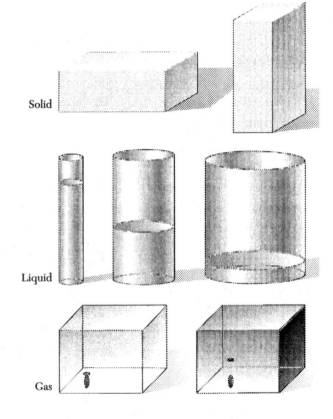

Figure 10.1
(a) A solid, such as sulfur, has a definite shape and volume. (© 1993 L. W. Black)
(b) A liquid, such as bromine, has no shape but does have a definite volume. (© Charles Steele) (c) A gas, such as chlorine, has no set shape or volume. (© Charles Steele)

There is also a fourth state of matter, known as **plasma.** A plasma is like a gas in that it has no particular volume and shape, but the electrons have been torn from the atoms, giving a plasma distinct electrical properties that are quite different from those of a gas. Because plasmas are formed only under extreme conditions, they are not part of our everyday world. We shall not treat them further in this chapter.

Solids may be classified as crystalline or amorphous. In a **crystalline solid,** the atoms are arranged in an orderly, repetitive manner; table salt—sodium chloride (NaCl)—is a crystalline solid. In an **amorphous solid,** the atoms are not in any orderly pattern; glass, butter, and wax are amorphous solids. Crystalline solids melt at a particular temperature. Amorphous solids melt gradually over a range of temperatures.

The atoms of a solid are held together very tightly. In Figure 10.2 the atoms of a solid are represented by dark spheres and the forces due to neighboring atoms are represented by springs. These atom-to-atom forces are called **atomic bonds.** Even in the presence of these bonds, the atoms can vibrate about a fixed position. As the temperature of the material increases, some of these bonds holding the atoms together are broken because the added heat makes the atoms vibrate more vigorously. The material becomes a liquid. For the liquid, enough atomic bonds remain to keep the volume fixed. However, since some of the bonds present in the solid are broken now, the atoms are free to move and "flow" as the liquid is poured from one container to another. As temperature is increased further, the remaining bonds are broken and the liquid becomes a gas. The atoms of a gas act as tiny particles that do not interact except when they collide with each other. They move at enormous speeds (to be discussed later, in thermal physics). At even higher temperatures, the electrons are stripped from the gas atoms and a plasma is formed.

Glass is an amorphous solid.
(© *L. W. Black*)

Since a **fluid** is any substance that flows, both liquids and gases are fluids. We shall use *fluid* when we are talking about either liquids or gases.

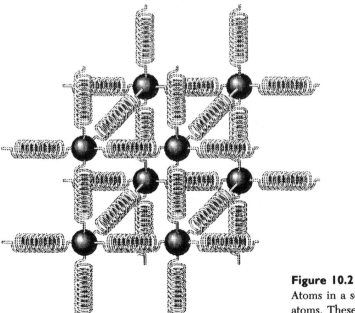

Figure 10.2
Atoms in a solid are held in place by forces from neighboring atoms. These forces can be pictured as springs attached to hard spheres, where the spheres represent the atoms.

10.2 Elastic Properties of Solids

CONCEPT OVERVIEW

- Objects are deformed when acted on by external forces.

- Stress and strain describe force and deformation in ways that account for the size of the object being deformed.

In everyday language, *elastic* means "able to bounce back after some interaction." Thus in the collisions we studied in Chapter 7, the colliding objects bounce back off each other in an elastic collision.

The deformation of a spring when it is either stretched or compressed is a good illustration of **elastic deformation.** Figure 10.3 shows a skate cart in a physics laboratory attached to a spring. When no force is exerted on the spring, it remains at its equilibrium position. As an external force is exerted to the right in the figure, the spring stretches. Experimentally, we find that the stretch is directly proportional to the applied force. If a force of magnitude F_0 causes a stretch of x_0, then a force of magnitude $3F_0$ causes a stretch of $3x_0$. As we learned in Section 6.1, this proportionality is known as **Hooke's law:**

$$F_{ext} = kx$$

where the proportionality constant k is the spring constant. Stretch being proportional to force also means that, when the external force is reduced, the stretch will also get smaller. When the external force is taken away entirely, the spring returns to its equilibrium position.

What we have just described for the specific case of a spring is true of any other elastic deformation. We can generalize and say all elastic deformations have these features in common:

1. There is a cause of the deformation.
2. The deformation can be measured.
3. There is a direct proportionality between the cause and the deformation.
4. As the cause is reduced, the deformation gets smaller, and the object returns to its original condition when the cause vanishes.

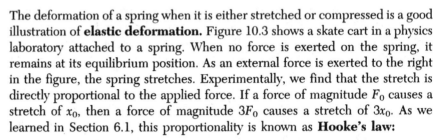
Equilibrium position ($x = 0$)

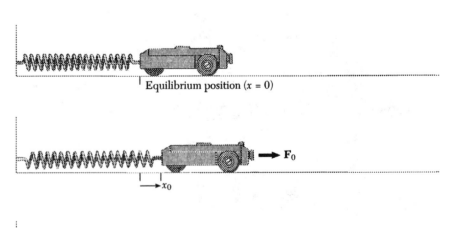

Figure 10.3
Elastic deformation of a spring.

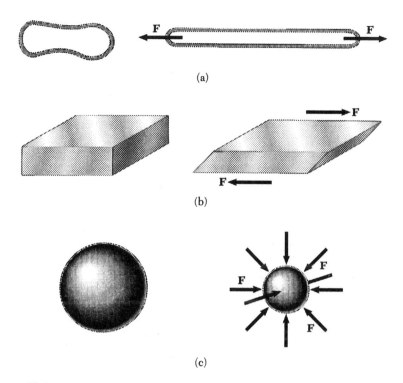

Figure 10.4
Elastic deformation in solid bodies: (a) length deformation, (b) shear deformation, (c) bulk deformation.

All solids deform somewhat when external forces act on them. If forces are applied to a rubber band, it stretches in what is called a **length deformation.** Push your hand across the cover of your physics book, and the book distorts in a **shear deformation.** Squeeze an inflated balloon, and it becomes smaller in a **bulk deformation.** Figure 10.4 shows these three basic types of deformations. All other deformations are either special cases or combinations of these. Figure 10.5 shows a board being bent, for instance. The fibers on the top surface are being stretched by *tensile* forces while the fibers on the bottom surface are being squeezed by *compressive* forces; both are length deformations. Tensile forces tend to pull something apart; compressive forces compress an object.

It is easy to see length deformation in a rubber band, but even a steel bar deforms if a force is applied to it, as sketched in Figure 10.6. Because of the force **F** exerted by the attached weight, the bar becomes longer by an amount

Figure 10.5
A bent board undergoes two types of length deformation: tensile deformation at the top and compressive deformation at the bottom.

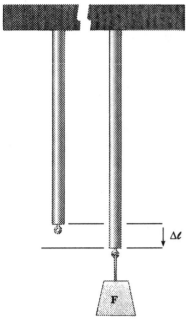

Figure 10.6
A tensile force causes even a steel
bar to deform. Here the length of
the bar increases by an amount Δl as
a result of the action of the force **F**.

Δl. Of course, this elongation is extremely small for most steel bars and therefore not very noticeable, but it *is* present and is readily measured in even an
introductory college physics laboratory.

We expect different elongations when the same force **F** is applied to bars
having different cross-sectional areas. To handle such different situations, we
use a quantity called **stress,** defined as the ratio of the deforming force to the
area over which the force acts:

$$\text{Stress} = \frac{\text{force}}{\text{area}} \tag{10.1}$$

The original size of the object must also be taken into account when we are
describing deformation. An elongation of 1 mm is negligible in a specimen that
is initially 10 m long but very significant in a specimen that is initially 10 mm
long. To handle this, we introduce the concept of **strain,** the ratio of the
amount of deformation to the original size of the object being deformed:

$$\text{Strain} = \frac{\text{amount of deformation}}{\text{original size}} \tag{10.2}$$

For each type of deformation we encounter, we shall describe a material's
ability to withstand being deformed by a **modulus,** which is the ratio of stress
to strain:

$$\text{Modulus} = \frac{\text{stress}}{\text{strain}} \tag{10.3}$$

To see how these three terms are used, let us consider again the steel bar
in Figure 10.6. The stress is the ratio of the applied tensile force F to the
cross-sectional area A of the bar; the tensile force is applied perpendicular to
the cross-sectional area. The strain is the ratio of the elongation Δl to the
original length l_O. The modulus for length deformation is known as **Young's
modulus** Y:

$$Y = \frac{\text{stress}}{\text{strain}} = \frac{F/A}{\Delta l/l_O} \tag{10.4}$$

Stress has SI units of newtons per square meter, also known as **pascals** (Pa),
and strain is dimensionless; therefore Young's modulus also has units of pascals.

Young's modulus is defined only over a certain stretching range called the
elastic region. If a steel bar is stretched as shown in Figure 10.6, a graph of
stress versus strain will look something like the one shown in Figure 10.7. As
the stress increases from zero, the strain increases in direct proportion. This
region of direct proportionality is known as the **elastic region.** We can solve

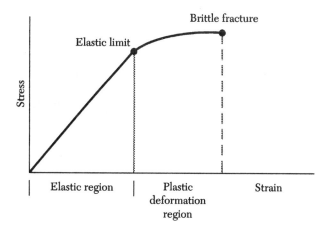

Figure 10.7
A typical stress-strain curve.

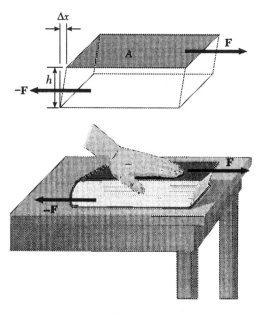

Figure 10.8
A shear deformation; the force **F** is applied parallel to the book's cover.

Equation 10.4 for the elongation Δl to find $\Delta l = (l_O/AY)F$; that is, the elongation is proportional to the force applied, as you might expect. Furthermore, in the elastic region the deformation is reversible. If the tensile force is reduced, so is the elongation. If the force is removed, the bar will return to its original length.

The elastic region extends only so far, however, and there is an **elastic limit** or **yield point.** The region beyond the elastic limit is called the **plastic deformation** region. Here the strain is no longer directly proportional to the stress; instead, each small change in stress produces a larger change in strain. Also, the deformation is not reversible in the plastic region. When the stress is removed, the bar will remain stretched—it has been permanently deformed. Young's modulus does not apply in the plastic region. A material stretches much more easily beyond the elastic limit, and when it reaches the point labeled **brittle fracture** in Figure 10.7, it breaks.

Figure 10.8 illustrates a shear deformation. If you place your closed physics book on your desk and exert a force parallel to the cover as shown, the book will deform. The initially rectangular cross section of the edge will deform into a parallelogram. The stress of interest in this case is the ratio of the force exerted to the area over which it is exerted (notice that the force is parallel to the area in this case). The strain is the ratio of Δx to h in Figure 10.8. Thus we can write the **shear modulus** S as

Elastic means "able to return to original shape," and *plastic* means "easily formed into a new shape." Squeezing a foam-rubber ball is an elastic deformation, because the ball returns to its original shape the instant you open your hand. Squeezing a ball of modeling clay is a plastic deformation. Even after you open your hand, the clay retains the shape formed by the squeeze.

$$S = \frac{F/A}{\Delta x/h}$$

(10.5)

The shear modulus is also measured in pascals.

If you squeeze uniformly on an inflated balloon, you will make its volume smaller as the balloon undergoes bulk deformation. Even though bulk defor-

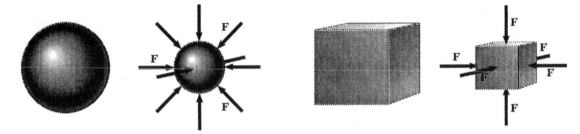

Figure 10.9
A force applied uniformly over the whole surface of an object results in a bulk deformation. The object's volume decreases, but its shape remains the same.

mation is easiest to see in objects like balloons and sponges, it also happens in other objects, as sketched in Figure 10.9. The stress is once again the ratio of applied force to area. The strain is the ratio of the change in volume $-\Delta V$ (negative because the volume is decreasing) to the original volume V. Thus, we can define the **bulk modulus** B as

$$B = \frac{F/A}{-\Delta V/V} \qquad (10.6)$$

The bulk modulus, too, is measured in pascals.

Some typical values for all three types of modulus are given in Table 10.1.

TABLE 10.1 Typical Values of Elastic Moduli			
Substance	Young's Modulus ($\times 10^{10}$ N/m^2)	Shear Modulus ($\times 10^{10}$ N/m^2)	Bulk Modulus ($\times 10^{10}$ N/m^2)
Aluminum	7.0	2.5	7.0
Brass	9.1	3.5	6.1
Copper	11	4.2	14
Iron	19	7.0	10
Steel	20	8.4	16
Tungsten	35	14	20
Glass	7	3	5
Silk	1.1	—	—
Nylon	0.27	—	—
Water	—	—	0.21
Mercury	—	—	2.8

Problem-Solving Tip

- To determine a modulus, which is characteristic of a material rather than of a particular object, information about force and deformation must be used in terms of stress and strain.

Example 10.1 ─────────────────────────

A load of 36.7 kg is suspended from the end of a wire that is initially 1.15 m long and has a circular cross section with a diameter of 0.850 mm. The wire stretches 2.13 mm. Find the tensile stress, tensile strain, and Young's modulus. Based upon the value you calculate for Young's modulus, what material do you think the wire is made of?

Reasoning

Using the dimensions of the wire, we find the stress and strain. From that we find Young's modulus and compare that to values in Table 10.1.

Solution

First, we need to find the cross-sectional area:

$$A = \pi r^2 = (3.14)(0.425 \text{ mm})^2 \left(\frac{1 \text{ m}^2}{10^6 \text{ mm}^2} \right) = 5.67 \times 10^{-7} \text{ m}^2$$

Once we convert the load to newtons, we are ready to calculate the tensile stress:

$$\text{Stress} = \frac{\text{force}}{\text{area}} = \frac{(36.7 \text{ kg})(9.80 \text{ m/s}^2)}{5.67 \times 10^{-7} \text{ m}^2} = 6.34 \times 10^8 \text{ N/m}^2$$

$$= 6.34 \times 10^8 \text{ Pa}$$

The strain is the ratio of the stretch to the original length:

$$\text{Strain} = \frac{\Delta l}{l_O} = \frac{2.13 \text{ mm}}{1.15 \text{ m}} \left(\frac{1 \text{ m}}{1000 \text{ mm}} \right) = 1.85 \times 10^{-3}$$

Now we are ready to use Equation 10.4 to calculate Young's modulus:

$$Y = \frac{F/A}{\Delta l/l_O} = \frac{6.34 \times 10^8 \text{ N/m}^2}{1.85 \times 10^{-3}} = 3.43 \times 10^{11} \text{ N/m}^2 = 3.43 \times 10^{11} \text{ Pa}$$

By comparing this to values of Young's modulus in Table 10.1, we can see that this wire must be made of tungsten.

Practice Exercise Another wire, with the same dimensions and same load, stretches 1.0 cm. Find the tensile stress, tensile strain, and Young's modulus. Based upon your answer, what material do you think the wire is made of ?

Answer 6.34×10^8 N/m^2, 8.7×10^{-3}, 7.3×10^{10} Pa, aluminum

10.3 Density

CONCEPT OVERVIEW

- Density is the ratio of mass to volume.
- Specific gravity is the density of a material divided by the density of water.

It is common to hear a statement such as "steel is heavier than wood," but what does the speaker mean? Ten kilograms of steel is no heavier than 10 kg of wood—or 10 kg of feathers! And 10 kg of steel is lighter than 25 kg of wood. The speaker surely means that a given piece of steel is heavier than a piece of wood of the same size. To avoid such ambiguous language, we define the density of materials. The **density** of a uniform material is the ratio of its mass to its volume—the mass per unit volume. For a nonuniform material, average density is the ratio of mass to volume. Thus, if a block of uniform material has a mass m and a volume V, we define the density ρ (Greek lower-case *rho*) of the material as

$$\rho = \frac{m}{V} \qquad (10.7)$$

In SI units, density is measured in kilograms per cubic meter. Other units—such as grams per cubic centimeter—are also commonly used.

The glass cylinder contains four fluids of different densities. Beginning at the top, the fluids are oil (orange), water (yellow), saltwater (green), and mercury (silver). The cylinder also contains four objects, which, starting at the top, are a Ping-Pong ball, a piece of wood, an egg, and a steel ball. Which of the four fluids in the cylinder has the lowest density, and which the greatest? What can you say about the relative densities of the four objects? *(Courtesy Henry Leap and Jim Lehman)*

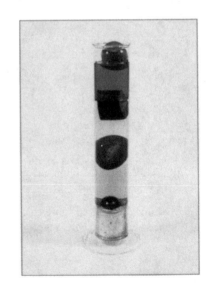

Table 10.2 lists densities for some common materials. Notice that most solids are approximately a thousand times more dense than most gases. On the atomic level, this difference in density means the atoms in a gas are about ten times as far apart as the atoms in a solid. Notice, too, that the density of water is 1.000 g/cm^3. This nice round value is not a mere coincidence. Rather, SI units were devised and the mass standard chosen so that 1.000 cm^3 of water would have a mass of exactly 1.000 g.

Instead of density, the same information is sometimes given in terms of the specific gravity of a material. **Specific gravity** is the ratio of the density of a

TABLE 10.2 Typical Densities

Substance	Density[a]	
	kg/m^3	g/cm^3
	Solids	
Ice	917	0.917
Concrete	2300	2.3
Glass	2600	2.6
Aluminum	2700	2.7
Iron, steel	7860	7.86
Copper	8920	8.92
Silver	10500	10.5
Lead	11300	11.3
Gold	19300	19.3
Platinum	21400	21.4
	Liquids	
Ethyl alcohol	806	0.806
Benzene	879	0.879
Water (4°C)	1000	1.000
Blood	1040	1.04
Seawater	1025	1.025
Glycerin	1260	1.26
Mercury	13600	13.6
	Gases	
Carbon dioxide	1.98	0.00198
Air	1.29	0.00129
Oxygen	1.43	0.00143
Hydrogen	0.0899	0.0000899
Helium	0.179	0.000179

[a]Except for water, all values are for a temperature of 0°C and a pressure of 1 atm.

material to the density of water; thus it is a *relative density* and is unitless. If densities are given in grams per cubic centimeter, the specific gravity is just the density without units (since the density of water is 1.000 g/cm^3). If densities are given in kilograms per cubic meter, specific gravity is the density divided by 10^3 without units.

Example 10.2

What is the radius of a 1.00-kg copper sphere?

Reasoning

This is an application of the definition of density. We know the density of copper from Table 10.2. We also need to know that for a sphere $V = 4\pi r^3/3$.

Solution

$$\rho_{copper} = 8.92 \times 10^3 \, \frac{kg}{m^3} = \frac{m}{V}$$

$$V = \frac{m}{\rho} = \frac{1.00 \, kg}{8.92 \times 10^3 \, kg/m^3} = 1.12 \times 10^{-4} \, m^3$$

$$r = \left(\frac{3}{4\pi} V\right)^{1/3} = 0.0299 \, m = 2.99 \, cm$$

Practice Exercise What is the edge length of a 1.00-kg aluminum cube?

Answer 7.2 cm

Example 10.3

A waterbed mattress is 1.85 m by 2.08 m by 20.4 cm deep. Find the mass and weight of the water in it.

Reasoning

From the definition of density, we can solve for $m = \rho V$. Therefore we need to find the volume. (Be sure to convert 20.4 cm to meters.)

Solution

$$V = (1.85 \, m)(2.08 \, m)(0.204 \, m) = 0.784 \, m^3$$

$$m = (1000 \, kg/m^3)(0.784 \, m^3) = 784 \, kg$$

$$w = mg = (784 \, kg)(9.80 \, m/s^2) = 7680 \, N = 1730 \, lb$$

The water weighs more than three quarters of a ton!

10.4 Pressure

CONCEPT OVERVIEW

- Pressure increases with depth in a fluid.

- Pressure depends only upon depth and is independent of the amount of fluid present and independent of any additional depth below the point of interest.

In understanding the behavior of individual solid objects, the idea of force is useful. In understanding the behavior of fluids, the idea of pressure is just as useful. **Pressure** P is the force F exerted on an area A divided by that area, or force per unit area:

$$P = \frac{F}{A} \qquad\qquad \textbf{(10.8)}$$

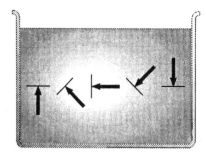

Figure 10.10
A fluid at rest exerts pressure equally in all directions.

Pressure is also measured in pascals.

Notice that pressure and stress are defined in exactly the same way: each is force per unit area. By convention, we use the term *stress* when we are discussing solids and the term *pressure* when we are discussing liquids and gases.

Pressure, even though it involves force, is not a vector quantity. Experimentally, it is observed that the magnitude of the force exerted on a small area is the same for any orientation of that area, as sketched in Figure 10.10. You experience this evenness of pressure every time you dive into a pool of water: when swimming underwater with your body horizontal, you feel the increased pressure equally on all parts of your body. We describe this by saying that a fluid at rest exerts pressure equally in all directions. The study of fluids at rest is called **hydrostatics.**

The force exerted by a fluid at rest is always perpendicular to the area on which the force is exerted. If this were not true, the fluid could exert both a perpendicular component of force and a parallel component on some area A of its container. By Newton's third law, the container would then exert an equal and opposite force on the fluid. Since fluids—by their very definition—can flow, the parallel component of this force exerted by the container would cause the fluid to flow and consequently it would not be at rest. Thus, a fluid at rest can exert only perpendicular forces.

Swimmers and divers also know that pressures at equal depths are the same. The pressure 2 m below the surface of a swimming pool is the same whether a diver is on the bottom of a 2-m-deep pool or 2 m deep in a 5-m-deep diving well or 2 m deep in one of the Great Lakes. The pressure increases with depth and depends only upon distance from the atmosphere-water interface. It is independent of the distance to the bottom and independent of the size of the body of water, as indicated in Figure 10.11.

Even nonswimmers have experienced changes in pressure with changes in depth (or height). Your ears have probably "popped" as you climbed a mountain or on an airplane trip. There is air inside your middle ear. At the bottom of the mountain, your middle ear contains air at sea-level atmospheric pressure. As you ascend the mountain—and thus rise higher in this "ocean of air" in which we live—the atmospheric pressure decreases, and the air inside your ear

Deep sea submersibles are built to withstand the extreme pressures encountered in the ocean depths.
(© *Franklin Viola/Comstock, Inc.*)

We Earth inhabitants live at the bottom of an ocean of air, where the fluid is not water, but air. Because pressure depends on depth in the fluid, air pressure is greatest at Earth's surface and decreases with altitude.

Figure 10.11
The Page Dam on the Colorado
River. The water pressure on any
point on the vertical face of the dam
depends only on how far below the
water surface the point is.   *(© Paul
Bodine)*

tries to expand because of the lower pressure. Your ear may fill up and everything will sound funny until some of the air escapes as your ear pops.

Having described the pressure-depth relationship qualitatively, we can
now calculate it quantitatively. Consider a column of liquid with cross-sectional
area A and height h, as sketched in Figure 10.12. By symmetry, the horizontal
forces on the sides of the column cancel each other. Therefore we need look at
only the vertical forces. At the top, the atmosphere pushes down with a force of
magnitude $F_{atm} = P_{atm}A$, where P_{atm} is atmospheric pressure. Gravity also exerts a downward force on the column of liquid equal to its weight, $w = mg =$
$\rho Vg = \rho Ahg$, where we have used the fact that the volume of the column is
$V = Ah$. At the bottom of the column, the surrounding liquid pushes up on the
column with a force of magnitude $F = P_hA$, where P_h is the pressure at depth
h. Since the column, along with the rest of the liquid, is at rest, the upward
force must balance the downward forces:

$$P_hA = P_{atm}A + \rho Ahg$$

$$P_h = P_{atm} + \rho gh \qquad (10.9)$$

That is, the pressure at depth h is greater than atmospheric pressure by an
amount ρgh that depends on the density of the fluid, the depth, and, of course,
gravity. It is useful to write this relationship as

$$\Delta P = P_h - P_{atm} = \rho gh \qquad (10.10)$$

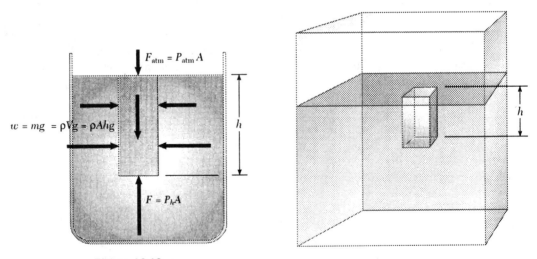

Figure 10.12
Pressure increases with depth. At depth h below the surface, the pressure is greater
than atmospheric pressure by an amount ρgh.

Example 10.4

What is the additional pressure 2 m under the surface of a swimming pool?

Reasoning

Pressure increases with depth below the surface according to Equation 10.9 or 10.10. Because we want only the change in pressure, we can apply Equation 10.10 directly.

Solution

$$\Delta P = \rho g h$$
$$= (1.0 \times 10^3 \text{ kg/m}^3)(9.8 \text{ m/s}^2)(2 \text{ m}) = 20\,000 \text{ N/m}^2 = 20\,000 \text{ Pa}$$
$$= 20 \text{ kPa}$$

Practice Exercise   What is the additional pressure 3 m under the surface of a swimming pool?

Answer   30 kPa

10.5 Atmospheric Pressure and Gauge Pressure

CONCEPT OVERVIEW

➡ Gauge pressure is the difference between absolute pressure and atmospheric pressure.

In Example 10.4 we calculated a pressure increase of 20 kPa. Is that a small pressure difference or a large one? Of course, the answer to this question depends upon the two pressures we want to compare. All of the air above us pushes down and creates pressure just as any other fluid. Because air density varies greatly with altitude and because the top of the atmosphere is difficult to define, we cannot simply calculate sea-level air pressure by directly using the equations of the previous section.

At any location, the pressure exerted by Earth's atmosphere will vary with the weather. At sea level, the average value of the air pressure is 1.013×10^5 Pa. For convenience in discussing pressures exerted by Earth's atmosphere, we define the unit **atmosphere** (atm), where

$$1 \text{ atm} = 1.013 \times 10^5 \text{ Pa} = 1.013 \times 10^5 \text{ N/m}^2$$

Besides this unit "atmosphere," there are several other units of pressure commonly used in calculating atmospheric pressure. One unit is the **bar** (from barometric):

$$1 \text{ bar} = 100 \text{ kPa} = 1.00 \times 10^5 \text{ N/m}^2$$

Thus, standard atmospheric pressure, 1 atm, is just a little more than 1 bar.

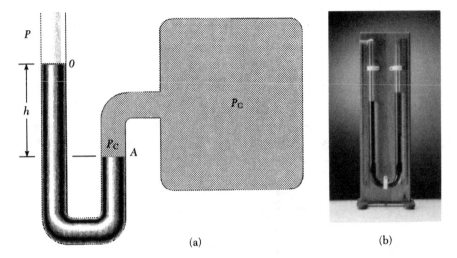

(a)   (b)

Figure 10.13
(a) A U-tube manometer measures gauge pressure, indicated by the height h of the column of liquid.
(b) A demonstration U-tube manometer. *(Courtesy of Central Scientific Company)*

We can now answer our question about whether 20 kPa is a large or a small pressure change. At a depth of 2 m, the pressure has increased by about 20 percent over its value at the water's surface, which is 1 atm = 100 kPa.

The force exerted by the atmosphere on a single page of your physics book is about 1000 N (about 200 lbs)! Why do such forces not crush our bodies? The answer is because the force on the other side of the page—and inside our bodies—is also about 1000 N, so the net force is zero. Any nonzero net force usually is the result of a difference in pressure, which can be measured with a manometer (Fig. 10.13).

A U-tube manometer, like that in Figure 10.13, consists of a U-shaped tube filled with a liquid. One end of the tube is open to the air, and the other end is attached to the container whose pressure we want to measure. The liquid in the open end (position O in the figure) has atmospheric pressure P_{atm} pressing down on it. The pressure P_C in the container is also the pressure at the interface between the fluid in the container and the liquid in the manometer (position A in the figure). From Equation 10.9, we know these two pressures are related by

$$P_C = P_{atm} + \rho g h$$

where ρ is the density of the liquid in the manometer tube. The pressure difference $\Delta P = \rho g h$ is known as **gauge pressure:**

$$P_{gauge} = \rho g h \qquad (10.11)$$

It is the pressure measured by a gauge such as this manometer. **Absolute pressure** is gauge pressure plus atmospheric pressure.

Because manometers are such common pressure gauges, pressure is sometimes given not in pascals or some other pressure unit but as the difference in the heights of the liquid in the manometer tube. Because mercury is often used in manometers, pressure is often given in mm Hg, which stands for "millimeters of mercury." While this unit sounds more like a distance than a pressure, it is really just shorthand for the value of h, which still needs to be multiplied by

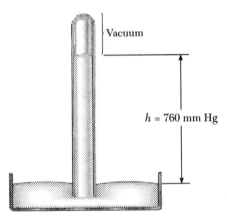

Figure 10.14
A mercury barometer measures atmospheric pressure in terms of the height of the column of mercury.

ρ (13.6 × 10^3 kg/m^3 for mercury) and by g to get a true pressure. When these numbers are multiplied, we find

$$1 \text{ mm Hg} = 133 \text{ Pa}$$

That means that

$$1 \text{ atm} = 101.3 \text{ kPa} = 760 \text{ mm Hg}$$

We have already talked about manometers—general instruments for measuring pressure. A barometer is a specialized instrument used to measure atmospheric pressure. In one type, a long tube sealed at one end is filled with mercury, inverted, and put into a dish of mercury, as in Figure 10.14. The tube is long enough that there is no mercury at the very top, leaving a vacuum (which means the pressure at the top is essentially zero). Under standard atmospheric conditions, the height of the mercury column is 760 mm. What keeps the mercury column 760 mm high is the pressure exerted by the atmosphere on the mercury in the dish. A column of air having a cross-sectional area equal to the surface area of the dish of mercury and a height that reaches to the top of the atmosphere is pressing down on the mercury in the dish and thereby forcing the mercury in the tube to stay at a height of 760 mm. When the air pressure increases, it presses down harder and forces the mercury column to a height greater than 760 mm (the weatherperson says the barometer is rising). When the air pressure decreases, it presses down less on the mercury in the dish, and some mercury flows out of the tube and into the dish (the barometer is falling).

As you can imagine, it would be extremely inconvenient to carry around an open U tube or open dish filled with mercury every time a manometer or barometer was needed. Therefore, other types of gauges are often used to measure pressure; two of these are illustrated in Figures 10.15 and 10.16. An *aneroid gauge* consists of a bellows of flexible metal and a lever arrangement moving a pointer. Most commonly used barometers use this device and are thus known as aneroid barometers. Gas—often nitrogen—is contained inside the bellows. When the outside air pressure increases, the bellows is compressed. When the outside air pressure decreases, the bellows expands. Either motion is transmitted to the pointer, which acts as the hand on a dial. A *Bourdon gauge* replaces the delicate bellows with a flexible tube. The pressure to be

Another name for mm Hg is the *torr*, in honor of the Italian scientist Evangelista Torricelli, inventor of the mercury barometer.

Figure 10.15
Aneroid barometers, because they are more compact, rugged, and transportable than mercury barometers, are the type most commonly found in homes. *(Taylor Scientific Instruments)*

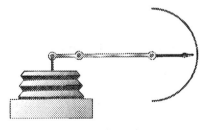

Aneroid gauge
(a)

Bourdon gauge
(b)

Figure 10.16
(a) An aneroid gauge is often used in barometers. (b) A Bourdon gauge is very sturdy and can be used in stressful situations that might break a more delicate instrument.

The inflatable band-device used to measure blood pressure is called a sphygmomanometer. The prefix *sphygmo-* is from a Greek word that means "pulsating."

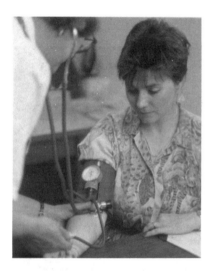

Figure 10.17
Measuring blood pressure with a sphygmomanometer. Why is the band inflated before the first measurement is taken? (© *Charles D. Winters*)

measured is inside the tube and tends to straighten the tube. Thus this type of gauge acts in much the same way as an aneroid gauge but is much sturdier and therefore used in industrial situations where an aneroid gauge would not survive.

Blood pressure is important, even vital(!). It is measured by placing an inflatable band around a person's arm, as in Figure 10.17. A small hand pump inflates the band while a gauge indicates the pressure in the band. In a physician's office, the gauge may be a wall-mounted mercury manometer. An aneroid gauge is also common, and electronic pressure-sensing devices are used in automatic blood-pressure measuring machines. In any case, while a stethoscope is used to listen to the blood flow in the veins below the band, the band pressure is increased until blood flow stops. Then the band pressure is gradually decreased until blood flow is again heard. This is the maximum pressure exerted by the heart when it is pumping and is called *systolic* pressure. Band pressure is further decreased until characteristic changes in the sound of the blood flow are heard, indicating that *diastolic* pressure has been reached; this is the pressure when the heart is resting. These values are given in millimeters of mercury and are written as a fraction—140/80, say, or 160/90—with the systolic pressure as the numerator and the diastolic pressure as the denominator.

Notice that when the band is placed around the upper arm, it is at about the same height as the heart; therefore what is measured is the blood pressure of the heart. A similar measurement on a leg would read a higher value because a leg is "deeper," or below the heart. The blood pressure at the brain is somewhat less than at the heart.

Example 10.5

What is the blood pressure in the feet of someone whose (heart) blood pressure is 140/80? When the person is standing upright, the feet are 1.5 m below the heart.

Reasoning

Just as diving deeper in a swimming pool increases the pressure, going "deeper" (or lower) in a person's blood (or circulatory system) increases the pressure. We use Equation 10.10 and get our value of ρ from Table 10.2.

Solution

$$\Delta P = \rho g h = (1.04 \times 10^3 \, \text{kg/m}^3)(9.8 \, \text{m/s}^2)(1.5 \, \text{m}) = 1.53 \times 10^4 \, \text{Pa}$$

Now we need to convert to millimeters of mercury:

$$1 \, \text{atm} = 101.3 \, \text{kPa} = 760 \, \text{mm Hg}$$

$$\Delta P = 15.3 \, \text{kPa}\left(\frac{760 \, \text{mm Hg}}{101.3 \, \text{kPa}}\right) = 115 \, \text{mm Hg} \approx 120 \, \text{mm Hg}$$

This value must be added to each heart blood pressure reading:

$$P_{\text{feet}} = \frac{140 + 120}{80 + 120} = \frac{260}{200}$$

The pressure in the feet is essentially double that at the level of the heart!

10.6 Pascal's Principle

CONCEPT OVERVIEW

■ In a liquid at rest, an increase in pressure at one point results in the same increase at every other point in the liquid.

Equation 10.9, $P_h = P_{\text{atm}} + \rho g h$, and the fact of equal pressures at equal depths have some useful applications. If P_{atm} increases, then the pressure everywhere else increases by the same amount. This is known as **Pascal's principle** and is attributed to the 17th-century French scientist and philosopher Blaise Pascal (for whom the pressure unit and the computing language are named).

Figure 10.18 illustrates two applications of Pascal's principle. The first is a hydraulic brake system used in cars and trucks. As the brake pedal is depressed, a force F_1 is applied to a piston in the master cylinder. The resulting increase in pressure in the hydraulic fluid in the master cylinder is transmitted through the line to the fluid in all the wheel cylinders. At each wheel, a force $F_2 > F_1$ is applied to the brake pads that stop the car either by clamping onto the brake disk as shown here or by pressing brake shoes against the inside of a drum. Such a system provides a convenient way to transmit a single force produced by the driver inside the car to four other locations. In addition, careful choice of the relative sizes of the master cylinder and wheel cylinders also allows F_2 to be much greater than F_1. This second aspect is the basis of the hydraulic lift or jack illustrated in Figure 10.18, best explained by means of an example.

A hydraulic system is one that involves water or some other liquid. Thus hydraulic cement is cement that hardens under water, a hydraulic elevator runs on fluid pressure, a hydraulic jack works by forcing liquid into a cylinder, and the hydraulic brakes on your car use brake fluid to transfer pressure from your foot on the brake pedal to the brakes at your car's wheels.

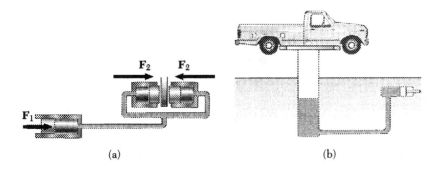

(a)   (b)

Figure 10.18
Pascal's principle has many useful applications, such as (a) a hydraulic brake system and (b) a hydraulic car lift.

Example 10.6

A hydraulic jack uses Pascal's principle. Consider the truck sketched in Figure 10.18. It has a mass of 1500 kg and is sitting on a mechanic's hoist supported by a hoist piston that has a radius $r_H = 15$ cm. Hydraulic fluid supports the hoist piston. Hydraulic lines connect it to a smaller piston—the pumping piston—connected to a motor and a valve-and-tank arrangement (not shown). The motor can exert a force of only 100 N on the pumping piston. What must the radius r_P of this piston be in order to make the pressure at its face the same as the pressure at the face of the hoist piston?

Reasoning

Pressure is force divided by area. We set the pressure in the pumping piston equal to the pressure in the hoist piston. The surface area of the hoist piston we can calculate easily, and we know the force on it. We also know the force on the pumping piston and therefore are able to calculate its area and its radius.

Solution

By Pascal's principle, the two pressures must be equal if the two pistons are at the same height (we assume they are). We have, therefore,

$$P_H = \frac{(1500 \text{ kg})(9.8 \text{ m/s}^2)}{\pi (0.15 \text{ m})^2} = \frac{100 \text{ N}}{\pi r_P^2} = P_P$$

$$r_P = 0.012 \text{ m} = 12 \text{ mm}$$

The pumping piston is quite small. What does all this mean? With these two pistons, a force of only 100 N is supporting the 15 000-N weight of the automobile! Is this a case of sleight of hand or are we using hydraulics to get out more than we put in? We clearly are getting out more force, but we cannot get out more work. To lift the hoist a distance d_H, the pumping piston has to travel a distance d_P. Because the fluid is not compressible, the volume of fluid that leaves the pumping piston equals the volume that enters the hoist piston:

$$d_H(\pi r_H^2) = d_P(\pi r_P^2)$$

We know the pressures are the same:

$$P_H = \frac{F_H}{\pi r_H^2} = \frac{F_P}{\pi r_P^2} = P_P$$

Solving these two equations for r_H^2/r_P^2 and setting the two expressions equal, we have

$$d_H F_H = d_P F_P$$

where we have used F_H for the output force (equal to the car's weight) and F_P (= 100 N) for the input force. The left side of this equation is the output work and is indeed just equal to the input work. Thus, while F_H is much larger than F_P, the distance d_H that the hoist cylinder moves is much smaller than the distance d_P the pumping cylinder moves. This is yet another example of energy conservation.

Practice Exercise If the hoist's piston has a radius of 20 cm, what must be the radius of the pump's piston?

Answer 16 mm

Example 10.7

Systolic blood pressure at heart level on a giraffe (Fig. 10.19) has been measured to be about 270 mm Hg. What is the blood pressure at the giraffe's brain, 2.17 m above the heart?

Reasoning

This is another application of Equation 10.10, which shows that pressure varies with vertical distance. Just as the water pressure at the bottom of a swimming pool is greater than at the surface, the blood pressure at the giraffe's feet is greater than at its heart, and the blood pressure at its brain is less than at its heart.

Solution

$$\Delta P = \rho g h$$

$$= (1.04 \times 10^3 \text{ kg/m}^3)(9.80 \text{ m/s}^2)(-2.17 \text{ m}) = -22.1 \text{ kPa}$$

$$= -166 \text{ mm Hg}$$

where the minus sign ($h = -2.17$ m) is needed because the reference pressure P_{heart} has been taken at heart level and h is above the heart. Since h in Equation 10.10 was initially measured positive downward, we end up with a minus sign here. The blood pressure at the brain is less than at the heart:

$$P_{\text{brain}} = P_{\text{heart}} + \Delta P$$
$$= 270 \text{ mm Hg} - 166 \text{ mm Hg}$$
$$= 104 \text{ mm Hg}$$

Figure 10.19
Giraffe physiology is an interesting application of fluid mechanics. (© M. P. Kahl/Photo Researchers)

A giraffe's heart exerts far more pressure than that of other mammals in order to have the blood reach the brain with about the same pressure as in most other mammals. This means an increase in pressure throughout the giraffe's body. Increased blood pressure can sometimes lead to swelling in legs and feet, but giraffes do not suffer from this problem because their thick, tight skin supports the increased pressure. The giraffe has been described as having been designed with its own "g-suit" like the ones pilots wear to compensate for drastic pressure changes as an airplane dives and rolls.

10.7 Buoyancy

CONCEPT OVERVIEW

▬ The buoyant force exerted by a liquid on an immersed object is equal to the weight of the liquid displaced by the object.

Objects submerged in a liquid seem to weigh less than in air because, when submerged, they are supported by an upward force exerted by the liquid. You may have noticed this effect when climbing out of a swimming pool. When you first start to climb the pool ladder—with most of your body still submerged—you feel lighter and it is easy to pull yourself up. As you continue to climb, your body begins to feel heavier and you must exert more force to pull yourself up. That is because less and less of your body is submerged and thus the water exerts less and less of an upward force on you. Heavy steel ships float because the water pushes upward on the hull, which is filled largely with air.

To understand this upward force, think of a submerged cylinder as sketched in Figure 10.20. The cylinder has cross-sectional area A and height h. Because of pressure, forces are exerted perpendicular to its entire surface. The horizontal forces on its sides cancel because of symmetry. There is a downward force $F_{top} = AP_{top}$ on the top surface and an upward force $F_{bottom} = AP_{bottom}$ on the bottom surface. Therefore the vector sum of the forces exerted by the liquid is $F_{net,liquid} = A(P_{bottom} - P_{top})$. We know from Equation 10.9 that $P_{bottom} = P_{top} + \rho g h$. Thus,

$$F_{net,liquid} = (P_{bottom} - P_{top})A = \rho g h A = \rho g V = (\rho V)(g) = mg \quad \textbf{(10.12)}$$

where $V = hA$ is the volume of liquid displaced by the cylinder. This net upward force exerted by the liquid is known as the **buoyant force.** Remember that here ρ is the density of the liquid, not of the cylinder. Thus, m is the mass and w is the weight of the liquid displaced by the cylinder, and we have

$$F_{net,liquid} = F_{buoyant} = w_{liquid} \quad \textbf{(10.13)}$$

The buoyant force exerted by the liquid on the submerged cylinder is equal to the weight of the liquid displaced by the cylinder.

A buoyant force acts on any object submerged in a liquid. Thinking first of a cylinder with its symmetry allows us to see that this buoyant force is directly

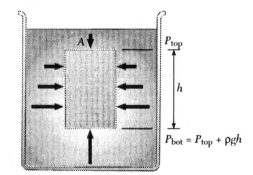

Figure 10.20
The buoyant force on a submerged object is a result of the fact that pressure increases with depth.

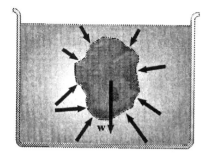

Figure 10.21
The buoyant force on a submerged object is equal to the weight of the liquid displaced by the object.

connected to the increase in pressure as depth increases. Figure 10.21 shows an irregularly shaped object submerged in a liquid. The liquid exerts forces all over the surface of this object, and gravity pulls down on it with a force w_{object}, the weight of the object. To understand the buoyant force exerted by the liquid, think of removing the object and replacing its volume with liquid. The liquid outside that volume remains as it was; the forces exerted by the liquid outside the volume on the liquid inside the volume are the same as the forces exerted on the original object. The net force is the buoyant force. Gravity now pulls down on the liquid inside the volume with a force w_{liquid}, the weight of the liquid inside the volume. Because this liquid is at rest, the vector sum of the upward buoyant force and the downward force w_{liquid} must be zero. In other words, the buoyant force is equal to the weight of the liquid displaced by a submerged object.

This fact was discovered two millennia ago by Archimedes and therefore is known as Archimedes' principle. According to history or legend, he discovered buoyancy when he climbed into a bath while thinking of a way to determine if the king's crown was made of pure gold.

Problem-Solving Tip

- The buoyant force is equal to the weight of the liquid displaced. This fact can be used to find the volume and thus the density of an object of known mass.

You may have heard someone say "Eureka!" upon figuring out a problem or finding a lost object. The word is Greek for "I've found it." It is what Archimedes is supposed to have shouted as he ran dripping through the streets after the rising water level, as he stepped into his bath, gave him his idea about buoyancy.

Example 10.8

A crown weighs 20.0 N in air and 17.8 N when submerged in water as in Figure 10.22. Is it made of solid gold?

Reasoning

The first question that usually comes to mind in this classic problem is "what is the volume of the crown?" Because it is a crown, it is probably inadvisable to melt or crush it to determine its volume. Can we find another way?

The crown experiences a buoyant force of 2.2 N (20.0 N − 17.8 N) when submerged in water. This buoyant force is equal to the weight of a

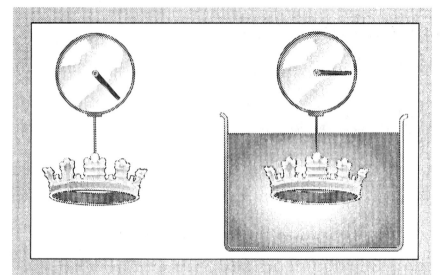

Figure 10.22
An object submerged in water (or any other liquid) appears to weigh less because of the buoyant force acting on it. Archimedes used this principle to determine the gold content of the king's crown.

volume of water equal to the volume of the crown and is given by Equations 10.12 and 10.13:

$$F_{buoyant} = F_{net.liquid} = \rho_w gV$$

where V is the volume of the crown as well as the volume of the water displaced by the crown and ρ_w is the density of the water.

Solution

We know the buoyant force is 2.2 N, so we can write

$$\rho_w gV = 2.2 \text{ N}$$

$$V = \frac{2.2 \text{ N}}{(1000 \text{ kg/m}^3)(9.8 \text{ m/s}^2)} = 2.24 \times 10^{-4} \text{ m}^3$$

The mass of the crown can be found from its weight in air:

$$w = mg$$

$$m = \frac{20.0 \text{ N}}{9.8 \text{ m/s}^2} = 2.04 \text{ kg}$$

Now we can find the density of the crown ρ_C:

$$\rho_C = \frac{m}{V} = \frac{2.04 \text{ kg}}{2.24 \times 10^{-4} \text{ m}^3} = 9.1 \times 10^3 \text{ kg/m}^3$$

Table 10.2 tells us that the density of gold is about two times this (19.3×10^3 kg/m^3), so we know the crown is definitely not solid gold. Looking at the densities of iron (7.86×10^3 kg/m^3) and copper (8.92×10^3 kg/m^3), we might guess that the crown is probably one of those two metals overlaid with a thin coat of gold.

So far we have talked only about the buoyant force on a *submerged* object. If an object floats in a liquid, some portion of it is under the surface and the liquid again pushes up with a buoyant force equal to the weight of the liquid that is displaced. For this reason, a ship's weight is often referred to as its "displacement."

Example 10.9

A cargo ship has a horizontal cross-sectional area of 750 m^2 at water level. It takes on a load of 620 automobiles, each having a mass of 1 metric ton. How much lower in the water does the ship float after taking on this load? The ship is floating in fresh water.

Reasoning
The water must now provide an additional buoyant force equal to the weight of the automobiles. This buoyant force is also equal to the weight of the water displaced as the ship sinks lower into the water. We know the cross-sectional area of the volume of water displaced, so can find the height of this volume. This height is the vertical distance the ship sinks.

Solution
The weight of the cars is

$$w_{cars} = (620 \times 1000 \text{ kg})(9.80 \text{ m/s}^2) = 6.08 \times 10^6 \text{ N}$$

Since this weight is equivalent to the additional buoyant force, we can write, using Equations 10.12 and 10.13,

$$F_{buoyant} = \rho g V = 6.08 \times 10^6 \text{ N}$$

$$V = \frac{6.08 \times 10^6 \text{ N}}{(1000 \text{ kg/m}^3)(9.80 \text{ m/s}^2)} = 620 \text{ m}^3$$

Now we determine the height of a "block" of water having this volume:

$$V = Ah = 620 \text{ m}^3$$

$$h = \frac{620 \text{ m}^3}{750 \text{ m}^2} = 0.827 \text{ m} \approx 0.83 \text{ m}$$

The ship will ride 0.83 m lower in the water.

Practice Exercise If the ship is in seawater ($\rho = 1025 \text{ kg/m}^3$), how much lower will the ship float?
Answer 0.81 m

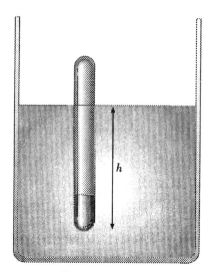

Figure 10.23
A hydrometer measures specific gravity.

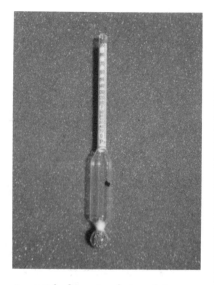

A spirit hydrometer designed for testing the proof and trille (percent) of alcohol in liquors and wines.
(© L. W. Black)

Example 10.10

A hydrometer is a simple and useful device used to measure specific gravity (and thus density). The hydrometer floats in a liquid, and the value of the specific gravity of the liquid is read from a scale imprinted on the hydrometer tube. The hydrometer in Figure 10.23 is a graduated glass tube with lead shot fixed at one end. The tube is 18.0 cm long and has a cross-sectional area of 1.50 cm². The tube-plus-lead combination has a mass of 20.0 g. How far from the weighted end should the "1.00" mark be placed, indicating a specific gravity of 1.00?

Reasoning
What we are being asked is "how much of the hydrometer will be submerged when it floats in a liquid having a specific gravity of 1.00 (water)?" We need to find the height of the column of displaced water that has the same weight as the weight of the hydrometer. In other words, we set the bouyant force equal to the weight of the hydrometer.

Solution
Let the length of the hydrometer that is under the water be h. Then the bouyant force is, from Equations 10.12 and 10.13,

$$F_{\text{buoyant}} = \rho g V = \rho g A h = mg = w_{\text{hydrometer}}$$

$$\rho g A h = mg$$

$$\rho A h = m$$

$$h = \frac{20.0 \text{ g}}{(1.00 \text{ g/cm}^3)(1.50 \text{ cm}^2)} = 13.3 \text{ cm}$$

The "1.00" mark should be 13.3 cm from the weighted end.

Practice Exercise   How high will this hydrometer float when placed in a container of ethyl alcohol ($\rho = 0.806$ g/cm³)?
Answer   16.5 cm of the tube length will be submerged.

If the liquid in your car's cooling system is part water and part ethyl alcohol (the main ingredient in most commercial antifreeze products), measuring its specific gravity is a quick and accurate way of measuring the concentration of antifreeze in the system. In fact, the hydrometer an auto mechanic uses may even be calibrated to show the freezing temperature of the liquid instead of its specific gravity. In the same manner, a hydrometer can be used to check the condition of a battery because the specific gravity of the liquid in the battery is a quick and accurate check of the electrolyte concentration in the battery.

*10.8 Surface Tension

CONCEPT OVERVIEW

■ Forces between molecules cause the interface between two fluids to act like a thin membrane coating the surface; this is known as surface tension.

In studying pressure and buoyancy, we examined the behavior of liquids below the surface, but the surface itself is quite interesting. If you look at the droplets of water sparkling on a spider's web as in Figure 10.24 (or a more mundane drop of water dripping from a faucet), you will see that the drop is nearly spherical—a tiny ball of water. What causes a drop to take such a shape?

If you are very careful, you can float a razor blade on the surface of some water contained in a glass. Insects can walk across a pond surface without sinking, with their feet making tiny indentations in the surface. In all these cases, the surface of the liquid acts as if it were covered with a thin layer of plastic wrap under tension. The liquid surface exerts a force called **surface tension** that causes this.

Think of a bag of grain, as illustrated in Figure 10.25. A slit of length l is cut in the bag. To keep the bag together, the slit is sewn, and then there is a total force **F** on each side of the slit due to the stitches that hold the edges together. Surface tension is this force per unit length:

$$\text{Surface tension} = \gamma = \frac{F}{l} \qquad \textbf{(10.14)}$$

Surface tension is represented by the symbol γ (Greek lower-case *gamma*). The SI unit for surface tension is newtons per meter (N/m).

Figure 10.24
Small droplets of water are spherical in shape because of surface tension. *(Hans Pfetschinger/Peter Arnold)*

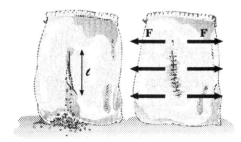

Figure 10.25
A bag of grain has a slit of length l. After the slit is sewn, there is a force **F** on each side of the cut. Surface tension is the force per unit length.

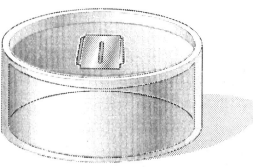

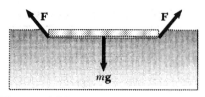

Figure 10.26
A razor blade floats on a water surface because of surface tension. *(Photo Courtesy Henry Leap and Jim Lehman)*

TABLE 10.3 Surface Tensions for Various Liquid-Air Interfaces

TABLE 10.3 Surface Tensions for Various Liquid-Air Interfaces

Liquid	Surface tension $(\times 10^{-3}$ N/m$)^a$
Benzene	28.9
Carbon tetrachloride	26.8
Ethyl alcohol	22.3
Glycerin	63.1
Mercury	465.0
Olive oil	32.0
Soap solution	25.0
Water	72.8
Oxygen ($-193°$C)	15.7
Helium ($-269°$C)	0.12

^{a}Except for oxygen and helium, all values at 20°C.

Table 10.3 lists a few values of surface tension. These particular examples are all for the surface between the given liquid and air. Surface tension depends upon both of the substances involved. The surface tension at a water-oil interface is quite different from that at a water-air interface.

Figure 10.26 shows a razor blade floating on water. The buoyant force is far too small to be responsible. Instead, the force needed to keep the blade from sinking comes from surface tension. The surface exerts a force tangent to itself. Therefore as the blade deforms the surface downward, the force exerted by the surface has an upward component, as shown in the figure. In a similar manner, as illustrated in Figure 10.27, surface tension provides the necessary upward force to balance an insect's weight and thereby allow the insect to walk on water.

A common demonstration of surface tension uses the U-shaped wire frame and slider shown in Figure 10.28. If the frame and slider are dipped into a soap solution and lifted out, the surface tension in the soap film within the frame pulls the slider up to the top of the frame. A weight w can be attached to the

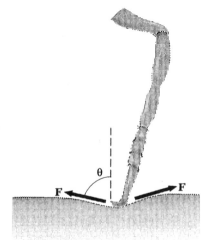

Figure 10.27
An insect is able to walk on a water surface because its foot deforms the water's surface without penetrating it and the resultant surface tension force balances the insect's weight.

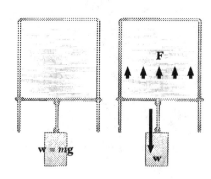

Figure 10.28
The sliding wire is held in equilibrium by an upward force of magnitude $F = 2\gamma l$ and a downward force of magnitude $w_{tot} = w + w_S$.

slider to hold it in equilibrium (the total downward force w_{tot} is this weight w plus the weight of the slider w_s). Once w_{tot} is large enough to hold the slider in equilibrium, it is in equilibrium for *any* position, telling us that the upward force of surface tension does not depend upon the surface area. As the surface area increases, molecules that were down inside the bulk of the film move up to the surface, and more surface is created. Thus the surface is not being stretched as a rubber membrane would be stretched.

The upward force exerted by the surface tension, which just balances the downward force w_{tot}, depends upon the length of the slider. That is why surface tension is measured in force per length. Call the length of the slider s. Since there are two surface layers (front and back) to the soap film, the total length of surface film exerting an upward force F on the slider is $2s$. Thus, from Equation 10.14,

$$F = \gamma l = 2\gamma s$$

and this force must equal the total weight pulling down:

$$2\gamma s = w_{\text{tot}}$$

$$\gamma = \frac{w_{\text{tot}}}{2s}$$

Because of surface tension, any liquid surface will contract as much as possible, contracting until the internal pressure just balances the effect of the surface tension. Drops of liquid falling freely in a vacuum are spherical because that shape has minimal area consistent with the volume.

Surface tension has the same effect on a soap bubble. Soap bubbles are spherical because that shape provides the smallest possible area. The size of the bubble is that which just balances the surface tension with the force due to the difference in pressure between inside and outside. Imagine cutting a soap bubble in half and looking at the forces on one hemisphere, as sketched in Figure 10.29. For a spherical bubble of radius R, the circumference is $2\pi R$. Along this circumference, there are two surfaces (inside and outside) that provide surface tension. So the upward force due to surface tension is

$$F_{\text{st}} = \gamma l = 2(2\pi R)\gamma$$

The downward force in Figure 10.29 is due to the air pressure inside the soap bubble. That force is just (gauge) pressure times area:

$$F_{\text{pressure}} = P(\pi R^2)$$

For a bubble in equilibrium, these two forces must be equal:

$$P\pi R^2 = 4\pi R\gamma$$

$$P = \frac{4\gamma}{R} \qquad (10.15)$$

This relationship means that the pressure inside small bubbles is greater than the pressure inside large bubbles. In Figure 10.30, the smaller bubble has

A very thin soap film is supported by a metal frame. The many beautiful colors observed are the result of the interference of white light striking the thin soap film. Note that a thin stream of water dropping from the tap is not able to break the soap film. Further note that the water stream passes *through* the soap film without breaking it. Can you explain this unusual phenomenon? *(Courtesy of Vittorio Zanetti, Università Degli Studi di Trento)*

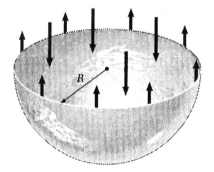

Figure 10.29
The pressure inside a soap bubble is greater than atmospheric pressure because of the surface tension in the soap film.

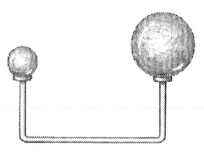

Figure 10.30
Which bubble will get larger and
which will get smaller?

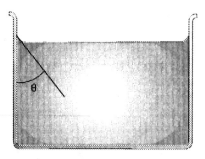

Figure 10.31
Water wets a glass surface and rises
higher where it touches the glass,
giving a concave liquid surface.

Water (top) has a concave meniscus,
and mercury (bottom) has a convex
meniscus. (© 1993 Charles D. Winters)

greater interior pressure and so will push air through the connection into the
larger bubble; the small bubble will get smaller as the large one gets larger!

Example 10.11

If the surface tension of a solution used to make soap bubbles is 0.040
N/m, what is the gauge pressure inside a soap bubble 5.0 cm in diameter?

Reasoning
As shown in Figure 10.29 and described by Equation 10.15, the surface
tension holding two halves of a bubble together is just balanced by the
force due to the internal pressure. We can use Equation 10.15 directly.

Solution

$$P = \frac{4\gamma}{R} = \frac{4(0.040 \text{ N/m})}{0.025 \text{ m}} = 6.4 \text{ N/m}^2 = 6.4 \text{ Pa}$$

Practice Exercise What is the gauge pressure when the diameter is
8.0 cm?

Answer 4.0 Pa

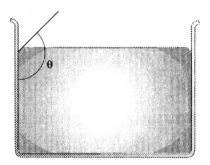

Figure 10.32
Because the contact angle between
mercury and glass is greater than
90°, liquid surface is convex.

So far our discussion of surface tension has assumed a liquid-gas (air)
interface. Different surface tension effects are present where a liquid and a
solid come in contact. Figure 10.31 shows a beaker containing water. Careful
inspection of the edge where water and glass meet shows that the water climbs
a small distance up the glass surface. This is because the force between a glass
molecule and a water molecule is greater than the force between two water
molecules. We can describe this quantitatively by determining the angle θ,
called the **contact angle** between the solid and the liquid. If θ is less than 90°,
as it is in a water-glass interaction, we say the liquid "wets" the solid, and the

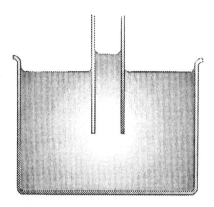 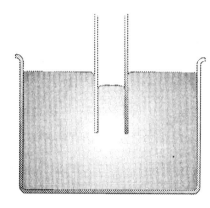

Figure 10.33
A liquid that wets the glass of the capillary tube rises in the tube (left), while a liquid that does not wet the glass is lower in the tube (right).

surface of the liquid is concave as in Figure 10.31. For materials like mercury and glass, the contact angle is greater than 90°, as illustrated in Figure 10.32. In this case, the liquid does not wet the solid, and the surface of the liquid is convex.

An important consequence of surface tension is the rising or falling of liquid in a narrow tube. Such a tube is called a **capillary tube,** and this phenomena is known as **capillarity.** For liquids such as water, which wet the capillary tube, the liquid rises in the tube. For liquids such as mercury, which do not wet the tube, the liquid falls in the tube. Both cases are illustrated in Figure 10.33; the curved liquid edge inside the tube is called a *meniscus.*

Figure 10.34 shows more detail. For the liquid rising in the tube, the upward component of the surface tension forces just balances the weight of the fluid that has risen up the tube. The surface tension forces F_{st} are proportional to the circumference, which means they are also proportional to r, the radius of the tube. The weight w of the fluid is proportional to the cross-sectional area and thus proportional to r^2:

$$F_{st} = \gamma (2\pi r) \cos \theta$$
$$w = \rho g V = \rho g h A = \rho g h (\pi r^2)$$
$$w = F_{st}$$

$$\rho g h (\pi r^2) = \gamma (2\pi r)\cos \theta$$

$$h = \frac{2\gamma \cos \theta}{\rho g r}$$

where h is the height the liquid rises in the tube. Notice that h is inversely proportional to the radius of the tube. Notice, too, that it depends upon $\cos \theta$. When θ, the angle of contact, is less than 90°, then $\cos \theta$ is positive and the liquid rises in the tube. When θ is greater than 90°, $\cos \theta$ is negative and the liquid is depressed in the tube.

Capillary action causes kerosene to rise up the wick of a lamp and melted wax to rise up a candle wick and into the flame. Capillarity explains water being drawn up a cloth in the same manner. That is why cotton is more comfortable in the summer than many synthetics—capillary action is more effective with cotton fibers and so perspiration can be carried away from your skin better. Capillary action also aids the rise of water and nutrients in plants.

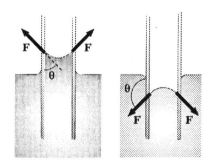

Figure 10.34
Surface tension forces are responsible for capillarity.

Example 10.12

A razor blade having a mass of 0.75 g and dimensions 2.2 cm by 4.5 cm is carefully placed on the surface of some water in a cup. What is the angle of contact between the edges of the razor blade and the water?

Reasoning

The upward component of the force exerted by the surface tension just balances the downward force of the razor blade's weight. The sine of the angle we seek is part of the upward component.

Solution

The force exerted by the surface tension is equal to the perimeter of the razor blade times the surface tension. The perimeter length is

$$L = 2(2.2 \text{ cm} + 4.5 \text{ cm})$$
$$= 13.4 \text{ cm} = 0.13 \text{ m}$$

From Table 10.3

$$\gamma = 72.8 \times 10^{-3} \text{ N/m}$$
$$F = \gamma L = (72.8 \times 10^{-3} \text{ N/m})(0.13 \text{ m}) = 9.5 \times 10^{-3} \text{ N}$$

The upward component of this force is $F_y = F \sin \theta$, where θ is the angle between the tangent to the water's surface and the horizontal direction. This upward force just balances the downward weight:

$$w = mg = (0.00075 \text{ kg})(9.8 \text{ m/s}^2) = 0.0074 \text{ N}$$

Setting these two forces equal, we have

$$F_y = w$$
$$(9.5 \times 10^{-3} \text{ N}) \sin \theta = 0.0074 \text{ N}$$
$$\sin \theta = 0.78$$
$$\theta = 51°$$

SUMMARY

Matter can exist as a solid, a liquid, a gas, or a plasma. Atoms in a solid or liquid are held together by forces called atomic bonds.

External forces applied to a solid can result in length deformation, shear deformation, or bulk deformation. Which type of deformation occurs depends upon the directions of the applied forces.

Stress is defined as

$$\text{Stress} = \frac{\text{force}}{\text{area}} \qquad (10.1)$$

Strain is the ratio

$$\text{Strain} = \frac{\text{amount of deformation}}{\text{original size}} \qquad (10.2)$$

The general definition for a modulus is

$$\text{Modulus} = \frac{\text{stress}}{\text{strain}} \qquad (10.3)$$

The modulus for length deformation is called Young's modulus:

$$Y = \frac{F/A}{\Delta l/l_0} \qquad (10.4)$$

The modulus for shear deformation is

$$S = \frac{F/A}{\Delta x/h} \qquad (10.5)$$

and the modulus for bulk deformation is

$$B = \frac{F/A}{-\Delta V/V} \qquad (10.6)$$

The density ρ of a material is

$$\rho = \frac{m}{V} \qquad (10.7)$$

where m is a given mass of the material and V is the volume occupied by that mass. Specific gravity is the ratio of the density of a material to the density of water.

Pressure is the force exerted on a given surface area divided by that area:

$$P = \frac{F}{A} \qquad (10.8)$$

A fluid at rest exerts pressure equally in all directions. The force exerted by a fluid at rest is always perpendicular to the area on which the force is exerted. Pressures at equal depths are the same. The pressure P_h at depth h is increased by an amount $\rho g h$ over atmospheric pressure:

$$P_h = P_{\text{atm}} + \rho g h \qquad (10.9)$$

At sea level, the average value of the air pressure, which we call one atmosphere (1 atm) of pressure, is 1.013×10^5 Pa.

The difference in pressure $\Delta P = \rho g h$ measured by a gauge is known as gauge pressure. Absolute pressure is gauge pressure plus atmospheric pressure.

Pressure applied to the surface of a confined liquid increases the pressure throughout the liquid by the same amount; this is known as Pascal's principle.

The buoyant force on an object submerged in a liquid is equal to the weight of the liquid displaced by the object; this is known as Archimedes' principle.

QUESTIONS
----

1. Why are gases generally much more compressible than liquids and solids?
2. If you were rock climbing, would you want Young's modulus for your safety ropes to be smaller or greater than that for an ordinary rope?
3. Is Young's modulus for a bungy cord smaller or larger than for a typical hemp rope?
4. When a cube of Jell-o jiggles on your plate, what kind of deformation are you watching?
5. When a spring is stretched, what sort of deformation is taking place in a very small section of the spring?
6. How is the force constant of a spring affected if the spring is cut in half?
7. What would Young's modulus be for a perfectly rigid material? What would the shear modulus be? The bulk modulus?
8. In comparing stress and strain, which is cause and which is effect?
9. Why does a paper clip become warm when you bend it back and forth?
10. A common physics demonstration is illustrated in Figure 10.35. Containers of various shapes are connected to a common tube at the bottom. Water is poured into one of the containers, flows into the bottom tube, and rises in the other con-

tainers. When the pouring has ceased and the water has come to rest, the vertical height of the water is exactly the same in each container. People sometimes describe this by saying "water seeks its own level." Explain why it happens.

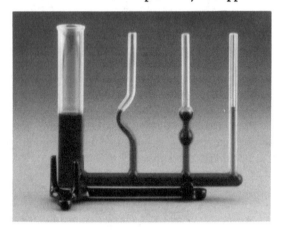

Figure 10.35
Question 10.   *(Courtesy of Central Scientific Company)*

11. Screw the cap tightly on an empty plastic milk container at sea level; although empty of milk, it still contains air. What will happen if you take the container to the top of a high mountain where atmospheric pressure is 80 percent of its sea-level value? What will happen if you take it scuba diving to a depth of 10 m, where the absolute pressure is twice sea-level atmospheric pressure?

12. If the air inside a can is removed by attaching the can to a vacuum pump, the can will collapse. Why? What force is exerted by air pressure on one side of a can measuring 20 cm by 25 cm?

13. In 1657, Otto von Guericke, inventor of the vacuum pump, joined two brass hemispheres with a gasket and then pumped the air from inside the sphere, creating a partial vacuum. A full-size replica of the two hemispheres is shown in Figure 10.36. Numerous attempts by two teams of eight horses each to pull the hemispheres apart failed (though some attempts were successful). *(a)* Explain why the sphere could withstand being pulled apart. *(b)* How is the force needed to separate the two hemispheres dependent on the radius of the sphere?

Figure 10.36
Question 13. Brass hemispheres similar to those used by von Guericke. *(© L. W. Black)*

14. Consider two lakes. The first is formed by a dam 75 m wide and 20 m deep and extends for 12 km behind the dam. The second is formed by a dam 75 m wide and 40 m deep and extends for only 1.2 km behind the dam. Which dam experiences the greater force?

15. If your blood pressure were measured with your arm held up over your head, how would the readings compare with those measured in the usual manner, with your arm held so that the band is at the level of your heart?

16. Why is it impractical to use a water barometer?

17. Why is a manometer commonly filled with mercury?

18. Explain how you can drink through a soda straw.

19. A large rock sits in a rowboat floating in a backyard pool. If the rock is thrown overboard, does the level of the water in the pool rise, lower, or stay the same?

20. An aluminum canoe floats in a swimming pool until a physics class comes along and sinks it. Does the water level in the pool rise, lower, or stay the same?

21. As a cargo ship leaves Toronto Harbor on Lake Ontario and travels through the St. Lawrence Seaway into the Atlantic Ocean, it travels from fresh water to salt water. Will the level at which it floats in the Atlantic be higher than, lower than, or the same as the level at which it floats in Lake Ontario?

22. A barge loaded with sand comes to a low-hanging obstruction. In order for the barge to pass under the obstruction, should sand be taken out or added? Explain your answer.
23. It is easy to understand how wooden ships can float. Explain how steel ships can.
24. What causes a hot-air balloon to rise? How can the balloonist control the balloon's altitude?
25. Ice cubes float in a glass of water filled to the brim. As they melt, does water overflow the glass or does the water level decrease?
26. A helium-filled balloon will rise only until its density is the same as that of the surrounding atmosphere, but a submarine will sink until it reaches bottom. Explain this difference in behavior.
27. Can you think of a mechanism that would explain why one kind of paper towel really might be the "quicker picker-upper"?

PROBLEMS

10.1 States of Matter

10.2 Elastic Properties of Solids

10.1 A spring with a force constant of 650 N/m stretches 2.5 cm when an external force acts on it. What is the magnitude of that force?

10.2 A telephone wire 125.00 m long and 1.00 mm in radius is stretched to a length of 125.25 m when a force of 800 N is applied. What is Young's modulus for the material of the wire?

10.3 A telephone wire 1.50 m long and 1.00 mm in diameter is stretched 7.83 mm when a force of 200 N is applied. What is Young's modulus for the material of this wire?

10.4 A guitar string made of steel is tightened to produce a strain of 1.5×10^{-4}. If this causes an elongation of 0.1 mm, what is the unstretched length of the string?

10.5 A steel wire has a length of 1.2 m and a cross-sectional area of 1.5 mm^2. What load suspended from the wire will cause it to stretch 2.0 mm?

10.6 A steel wire has a length of 0.80 m and elongates 0.60 mm when loaded with a 300-N weight. What is the diameter of the wire?

10.7 A silkworm hangs from a silk thread 15.0 cm long. The thread has a radius of 2.50 μm. If the thread is stretched 6.30 mm, how much does the worm weigh?

10.8 An upright pole, 2.0 m long, carries a load of 4.5 kN at its top. This load compresses the pole by 0.35 cm. Young's modulus for the pole is 1.2×10^{10} N/m^2. What is the cross-sectional area of the pole?

10.9 When supporting a 70-kg mountain climber, a nylon safety rope initially 20 m long stretches 1.1 m. The rope has a diameter of 0.75 cm. Find Young's modulus for this rope. How far will it stretch when supporting an 80-kg climber?

10.10 One face of a copper cube 10 cm on each edge is acted on by a 1000-N force directed parallel to the face while the opposite face is clamped in a vice. What are the shear stress and shear strain in the cube? By what distance is the unclamped face displaced?

10.11 Steel will rupture if subjected to a shear stress of more than about 4.2×10^8 N/m^2. What sideward force is necessary to shear a steel bolt 1 cm in diameter?

10.12 An aluminum block is 20 cm long, 10 cm wide, and 25 cm high. One of the faces that measures 20 cm × 10 cm is glued to a workbench, and then a horizontal force of 350 kN is applied to the opposite face. By what distance is the top edge displaced?

10.13 A uniform pressure of 6×10^4 N/m^2 is applied to a tungsten sphere that has a volume of 2×10^{-3} m^3. What is the change in volume of the sphere?

10.14 A uniform pressure of 8×10^4 N/m^2 is applied to a brass ball that has a volume of 0.5×10^{-3} m^3. What is the change in volume of the ball?

10.15 The pressure on a 2.2-m^3 container of water in the cargo bay of an airplane changes from 100×10^3 N/m^2 to 80×10^3 N/m^2 as the airplane gains altitude. Calculate the change in volume of the water.

○❖ **10.16** Listed below are dimensions for seven wires and the loads applied to them. Determine the amount of elongation in each wire, and then rank the seven in order of increasing elongation.

Wire	Material	Original Length (m)	Diameter (mm)	Load (N)
1	Steel	1.0	1.0	150
2	Steel	1.0	1.5	150
3	Steel	1.0	2.0	180
4	Steel	1.2	2.5	200
5	Aluminum	1.2	1.8	180
6	Copper	1.2	1.8	200
7	Copper	0.8	2.0	220

○ **10.17** A rubber eraser 3.0 cm × 1.5 cm × 8.0 cm is clamped at one end, with the 8.0-cm edge vertical. A horizontal force of 1.2 N is applied at the free end, perpendicular to the 1.5 cm edge, and the top of the eraser is displaced 2.4 mm horizontally. What are the shear stress, shear strain, and shear modulus?

○ **10.18** A bottle contains 4000.0 cm^3 of kerosene at atmospheric pressure. When the cork is pushed in slightly, the gauge pressure on the liquid increases to 800 kPa and its volume is reduced to 3998.0 cm^3. Calculate the bulk modulus of the kerosene.

○ **10.19** A telephone book contains 1000 pages, each 50.0 μm thick. A member of the PowerTeam rips the book by applying equal and opposite 600-N forces as shown in Figure 10.37. Assuming the forces are distributed over the shaded rectangular surfaces, determine the maximum shear deformation the paper can withstand before it begins to tear if paper can withstand a maximum shear stress of 1.0×10^7 N/m^2.

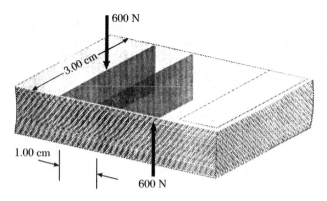

Figure 10.37 Problem 10.19.

○ **10.20** A crutch consists of two 1.50-m lengths of aluminum tubing. The tubing has an outer radius of 9.22 mm and walls that are 1.51 mm thick. If the crutch supports a weight of 430 N, determine the decrease in its length.

○ **10.21** A steel piano wire (Fig. 10.38) has a radius $r_1 = 0.800$ mm and a length l_0 of 83.0 cm.

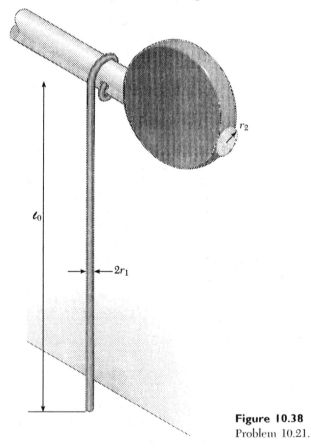

Figure 10.38
Problem 10.21.

One end of the wire is wound over a tuning peg in a circle that has a radius of $r_2 = 3.20$ mm. The other end is fastened rigidly to the soundboard. Assuming the tension in the wire is negligible initially, determine the tension after the peg has been turned through one complete turn. The vertical section of the wire, labeled l_O in the drawing, is 83.0 cm long.

O10.22 What is the elongation of the rod in Figure 10.39 if it is under a tension of 5.8×10^3 N?

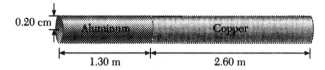

Figure 10.39 Problem 10.22.

O10.23 Without the weight hanging from its center, the wire in Figure 10.40 is essentially straight and horizontal (dashed line). If the wire is made of aluminum, what is its diameter?

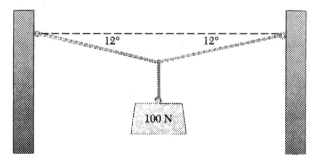

Figure 10.40 Problem 10.23.

O10.24 The scissors in Figure 10.41 cut a piece of paper 0.0100 mm thick. The gap between the blades is 0.0100 mm at the point of contact with the paper, and the blades contact the paper along a line 0.200 mm long. The paper can withstand a shear stress of 1.0×10^7 N/m^2 before it is cut. What is the minimum force necessary on the scissors handles to cut through the paper?

●10.25 A uniform pole that weighs 1600 N lies horizontally on the ground, pivoted about one

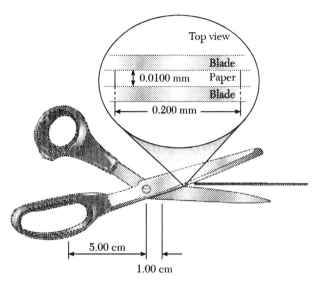

Figure 10.41 Problem 10.24.

end (Fig. 10.42). A nylon rope of length 20 m and diameter 1.5 cm is used to apply a vertical force at the free end of the pole. How much does the rope stretch before the pole starts to move?

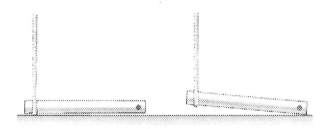

Figure 10.42 Problem 10.25.

●10.26 A person weighing 800 N stands on the ball of one foot. The tibia is 36 cm long; other dimensions are given in Figure 10.43. Find (a) the stress in the tibia, (b) the strain in the tibia, and (c) the change in length of the tibia. Young's modulus for the tibia is 2.0×10^{10} N/m^2.

●10.27 Consider a wire made from material with the following characteristics: Young's modulus = 10×10^{10} N/m^2, elastic limit = 3×10^8 N/m^2,

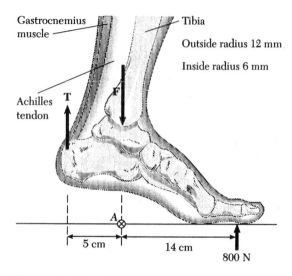

Gastrocnemius muscle

Tibia

Outside radius 12 mm

Inside radius 6 mm

Achilles tendon

T   F

A

5 cm   14 cm

800 N

Figure 10.43 Problem 10.26.

and brittle fracture point $= 5 \times 10^8$ N/m^2. Show that it is impossible to stretch this wire by 1 percent.

10.3 Density

10.28 What is the volume of a 2.5-kg sphere that is made of pure iron?

10.29 What is the volume of a 1.25-kg cube that is made of pure tungsten? ($\rho = 19.35$ g/cm^3 for tungsten)

10.30 A 35.4-kg cube is 15 cm on a side. Of what material is it made?

10.31 A sphere has a radius of 10.0 cm and a mass of 47.3 kg. Of what material is it made?

10.32 Show that 1 g/cm$^3 = 10^3$ kg/m^3.

○ ❖ 10.33 Listed below are characteristics of seven objects. Find the density of each and rank the seven in terms of increasing density.

Object	Mass (kg)	Shape	Dimension (cm)
1	1.2	Sphere	$r = 10$
2	1.4	Sphere	$r = 11$
3	1.6	Sphere	$r = 12$
4	1.6	Cube	side $= 10$
5	1.8	Cube	side $= 15$
6	2.0	Cylinder	$r = 20$, $h = 10$
7	2.5	Cylinder	$r = 10$, $h = 20$

10.4 Pressure

10.34 A 50-kg woman wears high-heel shoes with heels that have a cross-sectional area of 1 cm^2. When all her weight is supported on one heel, what pressure is exerted by the heel on the floor?

10.35 For a person 1.8 m tall, what is the difference in blood pressure between the top of the head and the bottom of the feet?

10.36 To drink from a waterhole, a giraffe lowers its head 4.8 m from the upright position. What is the change in blood pressure at the brain when the giraffe raises its head back to the upright position?

10.37 The collapsible plastic bag in Figure 10.44 contains a glucose solution (specific gravity =

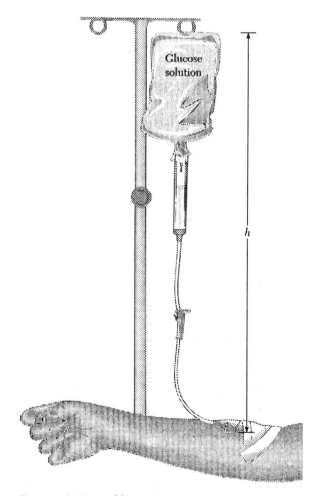

Glucose solution

h

Figure 10.44 Problem 10.37.

1.02). If the average gauge pressure in the artery is 100 mm Hg, what must the minimum height h of the bag be in order to infuse glucose into the artery?

10.38   If Earth's atmosphere had a uniform density of 1.290 kg/m³, how high would the atmosphere have to extend in order to give rise to the observed sea-level atmospheric pressure of 1.013×10^5 N/m²?

10.39   What total force is exerted on a dam that is 30.0 m deep and 80.0 m wide and has a lake behind it that is 4.83 km long?

10.40   A homeowner wishes to calibrate a new barometer. The home is on a hill 350 m above an airport where the barometric pressure is reported to be 99.1 kPa. What is the pressure difference, in kilopascals, between the home and the airport? To what kilopascal value should the homeowner set the barometer? What is this setting in millimeters of mercury?

10.41   Estimate the difference in air pressure between the base and the top of Toronto's 553-m CN Tower.

10.42   Crew members attempt to escape from a sunken submarine lying on the ocean floor 100 m below the surface. What force must they apply to the escape hatch, which has a surface area of 1.2 m × 0.5 m? The pressure inside the submarine is 1 atm.

10.5 Atmospheric Pressure and Gauge Pressure

10.43   A mercury U-tube manometer shows a height difference of 18.5 cm from one arm to the other. What gauge pressure does this height difference correspond to, in pascals and in atmospheres?

10.44   A U-tube manometer filled with ethyl alcohol is attached to a container of nitrogen. The difference in height of the two columns of alcohol is 28.3 cm. What is the absolute pressure of the nitrogen, in pascals and in atmospheres?

10.45   A household vacuum cleaner produces a partial vacuum that is measured by a U-tube water manometer to be 55 cm H₂O below

atmospheric pressure. What is the gauge pressure in pascals? The absolute pressure in pascals? What total force is available if the nozzle is 2.5 cm in diameter? (Treat the pressure unit "cm H₂O" exactly as you would treat the unit "mm Hg.")

○10.46   Mercury is poured into a U-tube open to the atmosphere at both ends. Then water is poured in one side until there is a column of water 20 cm long on top of the mercury, as shown in Figure 10.45. What is the height h of the water column above the mercury surface in the other arm of the U-tube?

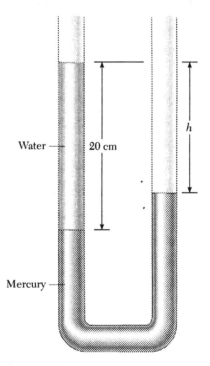

Figure 10.45
Problem 10.46.

○10.47   Oil is poured into a U-tube open to the atmosphere at both ends. Then a little mercury is poured into one side. The oil and mercury come to equilibrium as shown in Figure 10.46. Calculate the density of the oil.

○❖ 10.48   A pressurized container is connected to a U-tube manometer as shown in Figure 10.47. The manometer contains *two* liquids whose heights h_1 and h_2 are given below. For each

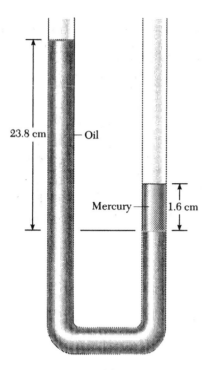

Figure 10.46 Problem 10.47.

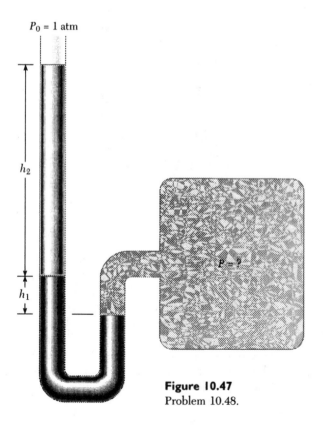

Figure 10.47
Problem 10.48.

set of data, determine the pressure in the container and then rank the seven liquid combinations in order of increasing pressure.

Combination	Material 1	h_1 (mm)	Material 2	h_2 (mm)
1	Mercury	15	Water	5.0
2	Mercury	15	Water	10
3	Mercury	18	Water	20
4	Mercury	20	Water	25
5	Mercury	20	Water	50
6	Water	10	Glycerin	20
7	Water	15	Glycerin	30

○10.49 The tank in Figure 10.48 contains oil and water. A partial vertical barrier positioned as shown separates the two liquids. Determine the difference x in the heights of the two surfaces. The oil has a density of 930 kg/m³.

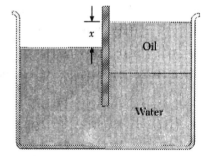

Figure 10.48 Problem 10.49.

10.6 Pascal's Principle

10.50 A hydraulic lift in a service station is used to raise a truck of 6 metric tons (m = 6 000 kg) by compressed air (Fig. 10.49). The cylinder

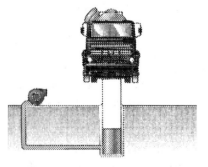

Figure 10.49 Problem 10.50.

of the lift has a radius of 13.5 cm. What air pressure is needed?

10.51 A hydraulic lift is used on a 1 200-kg automobile. A force of 40 N is applied to a small input piston whose cross-sectional area is 3.5 cm². What is the cross-sectional area of the large piston that is lifting the automobile?

○ 10.52 The hydraulic lift in Figure 10.50 supports a car weighing 1.2×10^4 N. The radius of piston 1 is 2.0 cm, and the radius of piston 2 is 10 cm. What is the magnitude of the force **F** that must be applied to piston 1 to support the car?

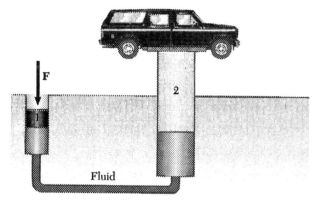

Figure 10.50 Problem 10.52.
A hydraulic lift.

○ 10.53 In the hydraulic press sketched in Figure 10.51, the small piston has a radius of 1.25 cm and the large piston has a radius of 20 cm. The worker operates the press by means of a lever as shown. If the worker

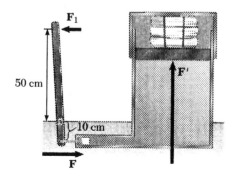

Figure 10.51 Problem 10.53.
A hydraulic press.

applies a force of 100 N on the lever, what is the magnitude of the force exerted on the bale of recycled scrap paper?

● 10.54 Figure 10.52 shows the essential parts of a hydraulic brake system. The area of the piston in the master cylinder is 6.40 cm², and the area of the piston in the brake cylinder is 1.75 cm². The coefficient of friction between brake shoe and drum is 0.500. If the wheel has a radius of 34.0 cm, determine the frictional torque about the axle of the wheel when a force of 44.0 N is exerted on the brake pedal.

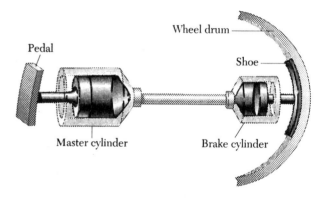

Figure 10.52 Problem 10.54.
Hydraulic brakes.

10.7 Buoyancy

10.55 What percentage of an ice cube sticks out above the water?

10.56 What percentage of a copper sphere floats above the surface of a container of mercury?

10.57 A block that weighs 40.0 N in air is suspended from a scale and submerged in water. The scale then indicates a weight of 33.5 N. What is the density of the block?

10.58 A crown that weighs 42.8 N in air has an apparent weight of 40.0 N when immersed in water. What is its density?

○ 10.59 An object weighs 2.0 N in air, 1.6 N when immersed in water, and 1.7 N when immersed in oil. What is its density? What is the density of the oil?

○10.60 The float in Figure 10.53 is constructed as follows. Piece A has a volume of 3.00 cm^3 and is totally submerged. The stem connected to it is 10.0 cm long, has a cross section of 0.200 cm^2, and is partially submerged. The combined mass of piece A plus the entire stem is 4.80 g. When floating in a liquid that has a specific gravity of 1.20, how much of the stem projects above the liquid surface?

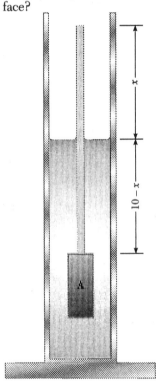

Figure 10.53   Problem 10.60.

○10.61 A cube of aluminum 3.0 cm on an edge floats in a beaker of mercury. What volume of aluminum is immersed in the mercury? What volume of lead placed on top of the cube would cause the cube to be just immersed in the mercury?

○10.62 A piece of moon rock weighs 32.0 N in air and 24.3 N when immersed in ethyl alcohol. What is the rock's volume? What is its density?

○10.63 A weather balloon made of light, flexible plastic weighs 8.0 N when empty. It carries a load of 80 N of apparatus when filled with helium near Earth's surface. What is the volume of the balloon?

●10.64 A 10-cm cube of wood floats in a container of oil and water as shown in Figure 10.54. The density of the oil is 650 kg/m^3. Determine the mass of the wood.

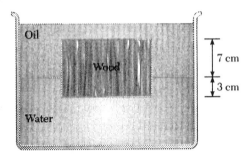

Figure 10.54   Problem 10.64.

*10.8 Surface Tension

10.65 The slider in the surface tension apparatus of Figure 10.28 has a mass of 0.350 g, and 1.25-g mass is suspended from it to establish equilibrium. If the slider is 5.0 cm long, what is the surface tension of the liquid?

10.66 What is the gauge pressure inside a soap bubble 3.0 cm in diameter? The surface tension of the soap solution is 0.030 N/m.

○10.67 An insect stands on a water surface. Each leg makes a spherical depression that has a radius of 1.5 mm. The angle of contact is 53°. Find the weight of the insect. (Recall that insects have six legs.)

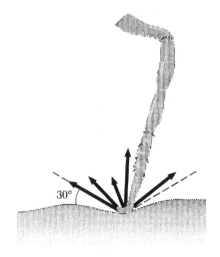

Figure 10.55   Problem 10.68.

○10.68 Figure 10.55 shows the foot of an insect on the surface of a pond. The arrows indicate the direction of the tension forces in the surface. If the radius of the foot is 0.05 mm and the insect has six legs, how much does the insect weigh?

○10.69 The surface tension of a liquid is measured with the apparatus shown in Figure 10.56. The film breaks when an upward force of 0.500 N is applied. The average radius of the ring is 1.20 cm. What is the surface tension of the liquid?

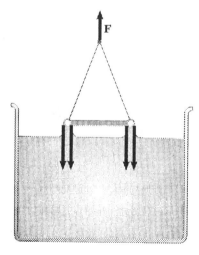

Figure 10.56 Problem 10.69.
(Photo Courtesy of Thomas B. Greenslade)

General Problems

○10.70 A picture is hung on a horizontal steel nail 0.805 mm in diameter. The effective point of support on the nail is 0.530 cm out from the wall. If the nail shears 0.0500 mm because of the picture hanging from it, what is the weight of the picture?

○10.71 Consider a wire made from material with the following characteristics: Young's modulus = 15×10^{10} N/m^2, elastic limit = 6×10^8 N/m^2, and brittle fracture point = 6×10^8 N/m^2. What is the strain in the wire just before it breaks?

○10.72 A particular house has 150 m^2 of living area with ceilings that are 2.4 m high. What is the mass of the air contained in this house?

○10.73 The viewing port of a deep-diving research vessel has an area of 0.45 m^2 and can withstand a net force of 1.8×10^6 N before it ruptures. The air inside the vessel is maintained at 1.0 atm. How deep can the vessel dive?

○10.74 Pascal dramatically demonstrated the effects and applications of his principle by placing a long hose into a tightly sealed wooden wine barrel filled with water. When water was poured into the hose to a height of 12 m, the barrel burst. For a hose radius of 0.30 cm and an average barrel radius of 20 cm, calculate the weight of the water in the hose. Calculate the force on the top of the barrel.

○10.75 A chemist wants to measure the density of olive oil using Archimedes' principle. An irregularly shaped piece of metal of unknown composition is available. The following data are taken: weight of metal in air, 10.8 N; weight of metal in water, 6.80 N; weight of metal in oil 7.20 N. Find the density of the metal and of the oil.

○10.76 A balloon and the helium that fills it together have a mass of 0.18 kg and a volume of 0.20 m^3. What buoyant force does the balloon feel from the air when atmospheric pressure is 1.0 atm?

○10.77 A "water skimmer" stands on the surface of a pond. Each of its six legs makes a spherical depression that has a radius of 0.80 mm. The angle of contact is 30°. What is the weight of the insect?

PHYSICS AT WORK

As hot gases rise from the lit cigarette, we see the upward movement of particulate matter—smoke—begin as an orderly column. This tight column of smoke soon expands into the swirls and disconnected puffs we call *turbulence*. When the hot gases and smoke from the burning tobacco and paper mix with the cooler air, the process appears chaotic. This study of turbulent flow is very complex and often requires the use of supercomputers. Yet we can study with some precision the orderly flow of fluids and then use this knowledge to make conjectures about turbulence.

The study of turbulence in the flow of fluids is of great interest to physicists and engineers. Knowledge about turbulent flow over the fuselage and wings of an airplane is critical for a safe and efficient design. Similarly, understanding turbulence and its causes can lead to the design of better heart valves and possibly artificial hearts that can function for years rather than months. Where else might the study of turbulence be beneficial?

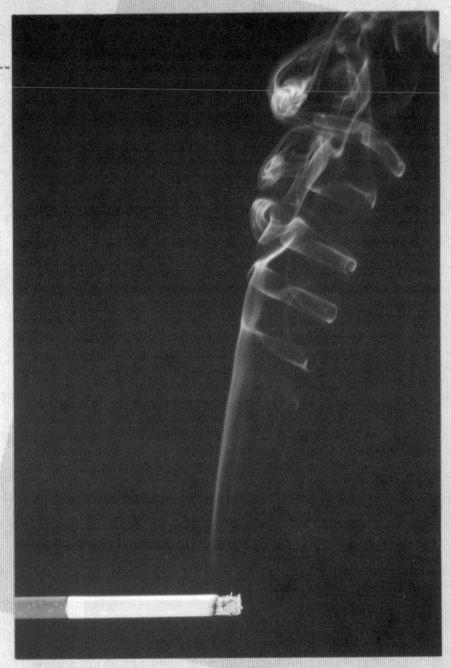

(© Diane Schiumo/Fundamental Photographs, New York)

Fluids in Motion

- Streamline flow describes the flow of a fluid—a liquid or a gas—in which each particle of the fluid which flows past a particular point follows the same path as each preceeding particle which has flowed past that point. We will direct our attention to situations that produce streamline flow. In another type of flow, known as turbulent flow, the motion of each particle may be quite different from the motion of a nearby particle. Turbulent flow is more difficult to predict, describe, and explain.

- The equation of continuity describes the relationship between density, flow velocity, and cross-sectional area and is the result of the conservation of mass in fluid flow. "Still waters run deep" is a colloquial form of the equation of continuity.

- Bernoulli's equation is a result of the conservation of energy applied to a moving fluid. A major result of Bernoulli's equation is that the pressure exerted by a fluid decreases as the flow speed increases. This can be used to explain the lift developed by airplane wings, why a "curve ball" does, indeed, curve, how a Pitot tube measures the air speed of an airplane, and why high winds can lift the roofs off of houses.

The previous chapter dealt with fluids at rest—hydrostatics. This chapter deals with fluids in motion—**hydrodynamics.** Just as Newton's second law of motion is of central importance in understanding the motion of a solid object, one equation is of central importance in understanding the motion of fluids: Bernoulli's equation. In this chapter, we develop this fundamental expression using the ideas of conservation of energy and the equivalence of work and energy (two concepts that are consequences of Newton's laws, as we have already seen). Bernoulli's equation is especially interesting because it was developed about a century before these two ideas were well understood.

Bernoulli's equation helps us understand such diverse situations as the lift of an aircraft wing, the bulging of a convertible top, and the circulation of air through the tunnels dug by prairie dogs. The equation does not take into account the frictional effect of moving fluids, a topic we consider briefly in the section on viscosity.

11.1 Fluid Flow

CONCEPT OVERVIEW

➡ In streamline flow, the motion of all particles which pass any given point will be exactly the same.

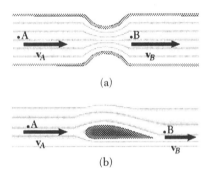

Figure 11.1
Streamline flow of fluid through a pipe and across an airplane wing. All particles of the fluid pass point A with velocity $\mathbf{v}_A$ and pass point B with velocity $\mathbf{v}_B$.

A difference in pressure produces a net force on a fluid and causes it to accelerate. Watch tonight's weather map on television; you will notice that air flows from areas of high pressure to areas of low pressure. As another example, blood flows from the arteries, where the pressure is about 100 mm Hg, through the capillaries and into the veins, where the pressure is only about 10 mm Hg.

Fluid flow can be described in terms of two main types—streamline flow and turbulent flow. In **streamline flow,** also known as **laminar flow** (Fig. 11.1), the motion of all the particles that pass any given point will be identical. In other words, all the fluid particles passing point A in Figure 11.1 have the velocity $\mathbf{v}_A$ at that point. When they get to B, all these particles have the velocity $\mathbf{v}_B$. Streamline flow does not mean the flow is identical along the entire pathway. It means that the flow at any point does not change over time.

The path a particle of flowing fluid takes is called a **streamline.** Every particle that passes a particular point follows the streamline that goes through

Streamline flow across the front of a modern automobile. What benefits can result from such aerodynamic design?
(*Courtesy of SAAB*)

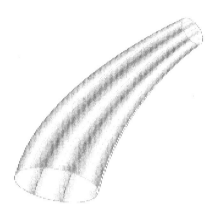

Figure 11.2
A bundle of streamlines composes a stream tube; fluid never flows across the surface of a stream tube.

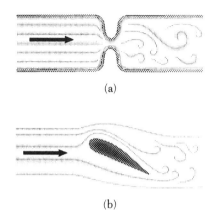

(a)

(b)

Figure 11.3
Turbulent flow through a constriction in a pipe and around an airplane wing.

that point. Streamline flow is always along a streamline, and consequently a flowing particle can never move from one streamline to another. A bundle of streamlines, like the ones sketched in Figure 11.2, is known as a **stream tube.** Fluid in a stream tube never intersects the surface of the tube; instead, the fluid always moves parallel to the surface. A water pipe serves as a model of a stream tube.

In **turbulent flow** (Fig. 11.3), the motion at any particular point may be quite different from one particle to the next. Turbulent flow is characterized by the randomness, the irreproducibility of the motion of individual particles. It usually occurs in fluids moving at high speeds. As you might expect, friction is far greater in turbulent flow than in streamline flow. Because of the randomness, analysis of turbulent flow is extremely complex. Therefore we shall concentrate our attention on streamline flow.

Streamlines and stream tubes are mental constructs that help us understand fluid flow.

11.2 The Equation of Continuity

CONCEPT OVERVIEW

- Flow rate is determined by the speed, cross-sectional area, and density of a fluid.

- For an incompressible fluid, an increase in cross-sectional area is accompanied by a decrease in speed (and vice versa).

The **mass flow rate** of a fluid is the mass of fluid Δm that flows through a given cross section of a stream tube in a time interval Δt:

$$\text{Mass flow rate} = \frac{\Delta m}{\Delta t}$$

The exhaust gases of a jet aircraft on takeoff exhibit turbulent flow. (© *Peter Gridley/FPG International Corporation*)

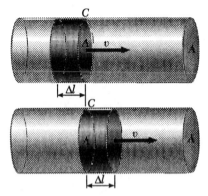

Figure 11.4
The shaded volume of fluid that passes the cross section at C in time Δt is $\Delta V = A\Delta l$. For fluid moving with speed v, the distance traveled is $\Delta l = v\Delta t$, so that $\Delta V = Av\Delta t$ and $\Delta m = \rho Av\Delta t$.

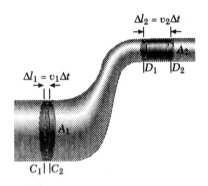

Figure 11.5
As the area of the cross section changes, the speed of the fluid also changes. The mass flow rate past cross section A_1 is the same as that past cross section A_2.

Figure 11.4 shows fluid of density ρ moving with speed v through a pipe (or stream tube) that has a cross-sectional area A. The shaded portion of the fluid moves past the cross section at C in time interval Δt; this portion of the fluid is of length Δl. The volume ΔV that passes the cross section at C in time Δt is therefore

$$\Delta V = A\Delta l$$

From Equation 2.2, we know that the length and speed are related by

$$\Delta l = v\Delta t$$

so that

$$\Delta V = A\Delta l = Av\Delta t$$

and the mass that passes C in time Δt is just this volume multiplied by the density ρ:

$$\Delta m = \rho\Delta V = \rho Av\Delta t$$

This means the mass flow rate can be written

$$\text{Mass flow rate} = \frac{\Delta m}{\Delta t} = \rho Av \qquad \textbf{(11.1)}$$

Of course, if the speed of the fluid is not uniform, then the v in Equation 11.1 is the average speed.

What if the cross section changes? Figure 11.5 shows fluid flowing through a stream tube that does change cross section. We shall concentrate on the two positions shown, where the cross-sectional areas are A_1 and A_2. The fluid is initially bounded by cross-sectional areas at C_1 and D_1; it then moves until it is bounded by C_2 and D_2. In time Δt, the mass Δm_1 that passes A_1 is

$$\Delta m_1 = \rho_1 A_1 v_1 \Delta t$$

and the mass Δm_2 that passes A_2 is

$$\Delta m_2 = \rho_2 A_2 v_2 \Delta t$$

For steady-state flow, mass is not stored in the pipe anywhere, and in any situation, mass is conserved. Therefore the mass that flows into the volume must equal the mass that flows out:

$$\Delta m_1 = \Delta m_2$$

or

$$\rho_1 A_1 v_1 = \rho_2 A_2 v_2 \qquad \textbf{(11.2)}$$

This is known as the **equation of continuity.** For an incompressible fluid (in other words, a liquid but not a gas), it reduces to

$$A_1 v_1 = A_2 v_2 \qquad \textbf{(11.3)}$$

As a consequence of the equation of continuity, a fluid speeds up as the pipe it is flowing through gets smaller and slows down as the pipe gets larger. In a river of constant width, for example, the water runs slowly where the riverbed is deep and the cross section large, and rapidly where the riverbed is shallow and the cross section small. River "rapids" occur in shallow water, not deep.

The equation of continuity is a mathematical expression of the adage "still water runs deep."

Problem-Solving Tip

■ The equation of continuity connects cross-sectional area and flow speed.

Example 11.1

The aorta, the large artery leading from the heart, has a radius of about 1.0 cm. Blood flows through it with a speed of about 30 cm/s. This blood eventually flows through numerous capillaries, each having a radius of about 4.0×10^{-4} cm. However, because there are so many capillaries, their total cross-sectional area is about 2.0×10^3 cm^2. Find the speed of blood as it flows through the capillaries.

Reasoning
Because the density of the blood does not change, we can use the Equation 11.3 form of the equation of continuity, which explains the relationship between cross-sectional area and flow speed.

Solution
First determine the cross-sectional area of the aorta:

$$A_A = \pi r^2 = 3.14 \text{ cm}^2$$

Then use Equation 11.3:

$$A_C v_C = A_A v_A$$

$$v_C = \frac{(3.14 \text{ cm}^2)(30 \text{ cm/s})}{2.0 \times 10^3 \text{ cm}^2} = 0.047 \text{ cm/s}$$

Example 11.2

Air flows from a 25 cm by 10 cm air duct into a room that is 2.6 m high, 3.5 m wide, and 4.2 m long. What must the speed of the air be as it exits the duct if the air in the room is to be changed twice an hour?

Reasoning
We consider the room as being an extension of the duct with a much larger cross-sectional area. After the air enters the room, its velocity is v_R.

It may seem as if the information we have is not sufficient since we do not know A_R. Where is the duct with respect to the rest of the room? Which dimensions make up the cross section of the room? Before we worry about these details, let us see whether we even need the cross section of the room, A_R.

Solution

Begin with the equation of continuity:

$$A_D v_D = A_R v_R$$

Replace the speed v_R with quantities we know:

$$A_D v_D = A_R(\Delta l_R/\Delta t) = \frac{A_R \Delta l_R}{\Delta t} = \frac{\Delta V_R}{\Delta t}$$

where ΔV_R is the room volume and Δt is the time each element of air takes to move through the room. Because we know these two values, we can determine v_D:

$$v_D = \frac{\Delta V_R}{A_D \Delta t} = \frac{38 \text{ m}^3}{(0.025 \text{ m}^2)(30 \text{ min})} = 51 \text{ m/min} = 0.85 \text{ m/s}$$

Practice Exercise   If the size of the duct is changed to 35 cm by 15 cm, what must the speed of the air be to still change the air twice an hour?

Answer   24 m/min = 0.40 m/s

When the fluid is incompressible and also in many common situations in which the density remains *nearly* constant, it is useful to consider volume flow rate instead of mass flow rate. This is always true for water and is often true for air. The **volume flow rate,** labeled Q, as you should expect, is the volume of fluid that passes through a cross section divided by the time required:

$$Q \text{ Volume flow rate} = \frac{\Delta V}{\Delta t} \tag{11.4}$$

where ΔV is the volume that passes the cross section in time Δt. Volume flow rate is normally measured in cubic meters per second in SI units. A heating system could be rated as moving the air at a rate of so many cubic meters per second; in the United States it is still more commonly rated in cubic feet per minute.

11.3 Bernoulli's Equation

CONCEPT OVERVIEW

■ Bernoulli's equation relates pressure, flow speed, and height: at a given height, as the flow speed of a fluid increases, its pressure decreases (and vice versa).

Bernoulli's principle helps us understand how an airplane wing lifts an airplane off the ground; that will probably be the main application you associate with

Daniel Bernoulli, the 18th-century mathematician/physicist who worked out the equation that bears his name, was a member of a famous family of mathematicians. Both his uncle James, and his father John, were professors at the university in Basel, Switzerland.

this principle. However, it also explains many other situations involving a moving fluid. For instance, how can air circulate through the underground homes of burrowing animals? Why is there often a draft up a chimney even with no fire in the fireplace? Can a baseball pitcher really throw a "curve ball"?

Bernoulli's principle, which helps us understand these and other situations, tells us that

> *the pressure exerted by a moving fluid is greater where the speed of the fluid is smaller and smaller where the speed of the fluid is greater.*

Figure 11.6
Bernoulli's principle explains the motion of a fluid through a stream tube.

Consider an incompressible fluid flowing through the stream tube of Figure 11.6; the cross section is greater at position 1 and smaller at position 2. From the equation of continuity, we know that the fluid must be flowing faster at position 2 than at position 1. In other words, the fluid accelerates as it moves from 1 to 2. Since a force is needed to cause this acceleration, the pressure at 1 must be greater than the pressure at 2; this is consistent with Bernoulli's principle.

Every time we write Bernoulli's equation, we include the density of the fluid. However, in discussing and applying it here, we use the approximation that we are dealing with incompressible fluids—that ρ is a constant. We can even gain a great understanding of the behavior of air by using the approximation that even it behaves as an incompressible fluid. We also assume steady-state, streamline flow, and no friction.

In Figure 11.7, we have a fluid flowing through a stream tube of varying cross section and varying height. Let us consider the shaded portion of the fluid, looking at the forces on it and the work done on it as it moves from the position shown in (a) to the position shown in (b). To accomplish this move, the fluid at position 1 moves a distance l_1 while the fluid at position 2 moves a distance l_2. At position 1, the fluid to the left of our shaded portion exerts pressure P_1 on the fluid. Therefore a force $F_1 = P_1 A_1$ pushes the fluid to the right through the distance l_1, and work $W_1 = F_1 l_1 = P_1 A_1 l_1$ is done on the shaded fluid by the fluid to its left. At the same time, the fluid to the right of our shaded portion at position 2 exerts pressure P_2 on the fluid. Therefore a force $F_2 = P_2 A_2$ pushes to the left on the fluid as it moves through the distance l_2, and work $W_2 = -F_2 l_2 = -P_2 A_2 l_2$ is done on the shaded fluid by the fluid to its right; this work is negative because the force and the displacement point in opposite directions.

Since the height changes, the force of gravity also does work on the fluid. The work done by gravity can be described as the negative of the change in the potential energy: $W_g = -mg(y_2 - y_1)$, where m is the mass of the fluid in the volume $A_1 l_1$ or $A_2 l_2$ (by the equation of continuity, these are the same volume. That is, $m = \rho_1 A_1 l_1 = \rho_2 A_2 l_2$). Thus the net work done on the shaded fluid is

$$W_{net} = W_1 + W_2 + W_g$$

From Equation 6.7', we know that the net work on an object is equal to its change in kinetic energy. Thus

$$W_{net} = \Delta KE = \tfrac{1}{2}mv_2{}^2 - \tfrac{1}{2}mv_1{}^2$$

$$W_1 + W_2 + W_g = \tfrac{1}{2}mv_2{}^2 - \tfrac{1}{2}mv_1{}^2$$

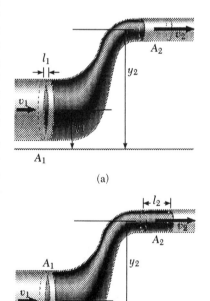

(a)

(b)

Figure 11.7
Details for developing Bernoulli's equation.

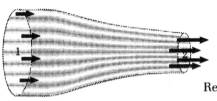

Figure 11.8
The pressure is lower at 2 than at 1 because fluid speed is greater at 2.

$$P_1A_1l_1 - P_2A_2l_2 - mg(y_2 - y_1) = \tfrac{1}{2}mv_2{}^2 - \tfrac{1}{2}mv_1{}^2$$

$$\tfrac{1}{2}mv_1{}^2 + P_1A_1l_1 + mgy_1 = \tfrac{1}{2}mv_2{}^2 + P_2A_2l_2 + mgy_2$$

$$\tfrac{1}{2}\rho_1A_1l_1v_1{}^2 + P_1A_1l_1 + \rho_1A_1l_1gy_1 = \tfrac{1}{2}\rho_2A_2l_2v_2{}^2 + P_2A_2l_2 + \rho_2A_2l_2gy_2$$

Recall that $A_1l_1 = A_2l_2$; call this volume V. Then we have

$$\tfrac{1}{2}\rho_1Vv_1{}^2 + P_1V + \rho_1Vgy_1 = \tfrac{1}{2}\rho_2Vv_2{}^2 + P_2V + \rho_2Vgy_2$$

$$\tfrac{1}{2}\rho_1v_1{}^2 + P_1 + \rho_1gy_1 = \tfrac{1}{2}\rho_2v_2{}^2 + P_2 + \rho_2gy_2 \qquad (11.5)$$

This is Bernoulli's equation. All the terms on the left refer to quantities at position 1 in Figure 11.7; all those on the right, to position 2. Thus, each side must be a constant, and we can also write Bernoulli's equation as

$$\tfrac{1}{2}\rho v^2 + P + \rho gy = \text{constant} \qquad (11.6)$$

It is easier to follow the changes in the terms of Bernoulli's equation if there are only two of them instead of three. Therefore let us look at special cases. Figure 11.8 shows flow along a stream tube at constant height but varying cross section. From the equation of continuity, we know the speed is less at position 1 and greater at position 2. The value of y in Bernoulli's equation remains constant, however, and for this special case the equation is

$$\tfrac{1}{2}\rho v^2 + P = \text{constant} \qquad (11.7)$$

or

$$\tfrac{1}{2}\rho v_1{}^2 + P_1 = \tfrac{1}{2}\rho v_2{}^2 + P_2 \qquad (11.8)$$

Now it is easy to see that the increased speed at 2 means a reduced pressure at 2.

This idea can be used to measure the speed of a fluid. Figure 11.9 shows an arrangement known as a *Venturi flowmeter*. The flow speed through the restriction is faster than the speed in the unrestricted sections of the tube. This

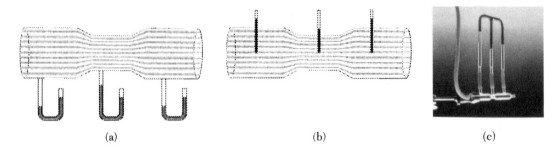

(a)   (b)   (c)

Figure 11.9
(a–b) These Venturi flowmeters show lower pressure in the narrow section, which means the fluid speed is greater there. (c) A Venturi tube. *(Courtesy of Central Scientific Company)*

Two examples of Bernoulli's principle. (left) The beach ball remains aloft even though it is not directly over the blower. The stream of air coming from the blower moves faster over the top of the ball, resulting in lower pressure that provides lift. *(© L. W. Black)* (right) Air blowing over and around the top of the Ping-Pong ball means the air pressure is less there. The air around the bottom of the Ping-Pong ball is nearly stationary so it provides greater pressure. This greater pressure on the bottom holds the ball up. When the air flows through the cone and around the ball stops, the ball falls. *(© L. W. Black)*

means the pressure exerted at the restriction is less, as indicated by the U-tube manometers in (a). If the fluid is a liquid, then straight-tube manometers as in (b) indicate the difference in pressure and thus the difference in speed.

Another example of Bernoulli's principle at constant height is the lift of an airplane (Fig. 11.10).[1] An airplane wing is shaped so that air flowing across the top travels faster than air flowing across the bottom. This means that the pressure on the top of the wing is lower than the pressure on the bottom. This difference in pressure means the force exerted by the air above the wing is less than the force exerted by the air below the wing. This net upward force is called **lift.**

[1] There are other ways of describing lift of an airplane wing. Airplane wings are complex structures, and applying Bernoulli's principle to understand their lift should be considered only a first approximation.

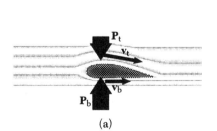

(a)

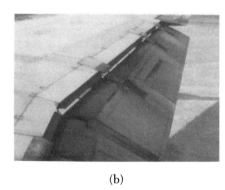

(b)

Figure 11.10

(a) An airplane wing is shaped so that air flowing across the top travels faster than air flowing across the bottom. Therefore the air pressure on the top is less than on the bottom, and this pressure difference causes a net upward force called *lift*.
(b) An aircraft wing with flaps down. Modern aircraft wings have several specialized hydraulically activated control surfaces that allow pilots to govern more precisely—and safely—the flight characteristics of the aircraft. *(© L. W. Black)*

Problem-Solving Tip

- Bernoulli's equation connects flow speed and pressure. The equation of continuity is often necessary to connect cross-sectional area and flow speed before Bernoulli's equation can be used.

Example 11.3

Figure 11.11 illustrates a Venturi flowmeter. If water flows through the wide part at 2.00 m/s with a gauge pressure of 12.0 mm Hg, what are the speed and pressure in the narrow part?

Reasoning

Since the height is constant, we use the form of Bernoulli's equation given by Equation 11.8. To find the flow speed in the narrow part, we use the Equation 11.3 form of the equation of continuity since density is constant.

Solution

Using subscript "W" for all variables pertaining to the wide part of the pipe and subscript "N" for the narrow part, we have

$$A_W v_W = A_N v_N$$

$$v_N = \frac{A_W v_W}{A_N} = \frac{\pi r_W^2 v_W}{\pi r_N^2}$$

$$= \frac{(2.62 \text{ cm})^2 (2.00 \text{ m/s})}{(1.75 \text{ cm})^2} = 4.48 \text{ m/s}$$

Knowing v_w and v_N, we can apply Bernoulli's equation once we convert pressures to newtons per square meter:

$$12.0 \text{ mm Hg} \left(\frac{133 \text{ N/m}^2}{1 \text{ mm Hg}} \right) = 1600 \text{ N/m}^2$$

$$\tfrac{1}{2}\rho v_W^2 + P_W = \tfrac{1}{2}\rho v_N^2 + P_N$$

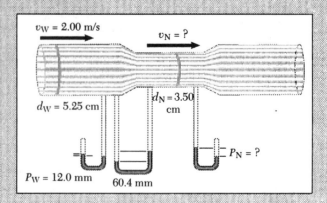

Figure 11.11
The Venturi flowmeter for the fluid flow in Example 11.3 shows a negative gauge pressure in the narrow area. If a hole were drilled there, air from the outside would be pulled into the tube.

$$P_N = \tfrac{1}{2}\rho v_W{}^2 + P_W - \tfrac{1}{2}\rho v_N{}^2$$

$$= \tfrac{1}{2}(1000 \text{ kg/m}^3)(2.00 \text{ m/s})^2 + 1600 \text{ N/m}^2 - \tfrac{1}{2}(1000 \text{ kg/m}^3)(4.48 \text{ m/s})^2$$

$$= 2000 \frac{\text{kg}}{\text{m} \cdot \text{s}^2} + 1600 \frac{\text{kg}}{\text{m} \cdot \text{s}^2} - 10035 \frac{\text{kg}}{\text{m} \cdot \text{s}^2}$$

$$= -6435 \frac{\text{kg}}{\text{m} \cdot \text{s}^2} \approx -6440 \frac{\text{N}}{\text{m}^2}\left(\frac{1 \text{ mm Hg}}{133 \text{ N/m}^2}\right) = -48.4 \text{ mm Hg}$$

Remember, this is the gauge pressure. The absolute pressure in the narrow part is 48.4 mm Hg less than atmospheric pressure. For a U-tube manometer connecting the two regions, the difference in the mercury height will be 60.4 mm. If a hole were drilled in the narrow part, air from the outside would be pulled into the tube. Perfume atomizers work on this principle.

Practice Exercise   What will the pressure be in the region of the narrow part or constriction which has a diameter of 3.0 cm?

Answer   −110 mm Hg

11.4 Applications of Bernoulli's Equation

CONCEPT OVERVIEW

■ Bernoulli's equation can be used in many situations to measure flow speeds.

An "atomizer" (Fig. 11.12) sprays perfume by causing air to flow across the top of a tube and thereby reducing the air pressure there. The air pressure in the bottle is greater because the air inside is at rest. The greater air pressure inside then pushes the liquid up the tube until it enters the air stream and is broken up into tiny droplets.

You usually see the word *atomizer* in quotation marks in physics books because the device does not really break the liquid up into its individual atoms. The drops produced are so tiny that the misnomer has stuck.

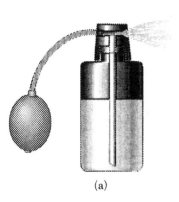
(a)

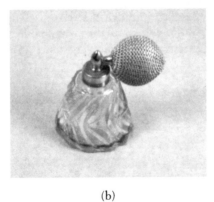
(b)

Figure 11.12
(a) An atomizer causes air to flow across the top of a tube, reducing the pressure there. The air pressure in the bottle then pushes liquid up the tube until it enters the air stream and is broken up into tiny droplets. (b) A perfume atomizer.   (© *L. W. Black*)

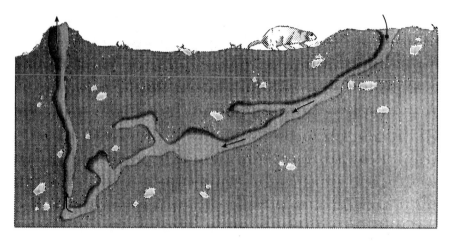

Figure 11.13
Differences in air speed across the entrances to a prairie dog's burrow are caused by differences in the size, shape, and elevation of the openings. In turn, these differences in air speed induce pressure differences, forcing air to flow through the burrow. Note that the burrow is roughly U-shaped.

Because of Bernoulli's principle, an unused fireplace can pull away room air that has been heated by a central heating system. In order not to waste fuel, therefore, you should be sure the damper is closed when the fireplace is not in use.

It is Bernoulli's fault the shower curtain keeps blowing against you when you are showering with the water on full force.

This same idea explains how fireplace chimneys draw. Although the air inside a house is still, the outside air will often be moving. Thus the air pressure at the top of a chimney, where the air is moving, is less than the air pressure inside. Even without a fire, air can be pulled up a chimney. Of course, the additional buoyancy of the heated air from a fire increases this effect.

This same effect causes air to circulate through the underground burrows of animals (Fig. 11.13). With two or more openings to a burrow, the air speed is usually different across the different openings because they are in different places. One opening may be exposed to the wind, while another may be hidden behind grass or bushes. One may face north, while another faces east. As the air flows past the openings with different speeds, the difference in pressure causes air to flow through the burrow.

In addition to the manometer arrangements of Figure 11.9, fluid speed can also be determined with the arrangement shown in Figure 11.14. Because

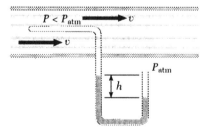

Figure 11.14
A Pitot tube. Because the fluid is flowing past the upper opening, the pressure is less there than at the lower opening, which is at atmospheric pressure. This difference in pressure shows up as a difference in the liquid heights in the two arms of the manometer.

the fluid is flowing past the upper opening, the pressure is less there than at the lower opening, which is at atmospheric pressure. This difference in pressure shows up as a difference in the height of the columns of fluid in the manometer. This device is known as a **Pitot tube.**

A Pitot tube is commonly used as an air speed indicator on airplanes. The arrangement of Figure 11.14 is modified as in Figure 11.15. Air rushes by the outer casing with speed v and causes a lower pressure P_2. Inside the tube, the air is still, with higher pressure P_1. This difference in pressure could be noted with a fluid manometer, as shown in the sketch. More realistically, though, it is usually measured electrically and the information appears on the instrument panel as air speed in kilometers per hour, miles per hour, or nautical miles per hour (knots).

Figure 11.16 shows water coming out a small hole at a depth y below the surface of water in a tank. We can apply Bernoulli's equation to determine the water's speed. We assume that the tank is so large—or the hole so small—that the rate at which the liquid level drops is negligibly small ($v_{top} \approx 0$). Then, we can use Equation 11.5,

$$\tfrac{1}{2}\rho_{top}v_{top}{}^2 + P_{top} + \rho_{top}gy_{top} = \tfrac{1}{2}\rho_{bottom}v_{bottom}{}^2 + P_{bottom} + \rho_{bottom}gy_{bottom}$$

where $v_{top} = 0$, $y_{top} = y$, v_{bottom} is the speed of the water leaving the hole, and $y_{bottom} = 0$. Because the water surface and the water flowing out the hole are at the same ambient air pressure, $P_{top} = P_{bottom}$, and of course $\rho_{top} = \rho_{bottom}$. Thus our expression reduces to

$$\tfrac{1}{2}v_{bottom}{}^2 = gy$$

$$v_{bottom} = \sqrt{2gy}$$

This result is known as **Torricelli's theorem.** Note that it is the same result we get using Equation 2.7 for the speed of an object dropped from a height y.

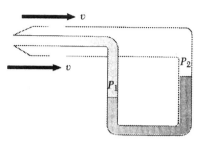

Figure 11.15
This arrangement of a Pitot tube is often used on airplanes as an air speed indicator. Air rushes by the outer casing with speed v_o and lower pressure P_2. Inside the tube the air is still ($v_i = 0$) and as the pressure P_1 is higher.

Example 11.4

Consider water flowing in a horizontal pipe that changes area from $25\ cm^2$ to $15\ cm^2$. In the $15\ cm^2$ portion, the flow velocity is 5.0 m/s. What is the pressure difference in the two sections?

Reasoning

We begin with the equation of continuity (Eq. 11.2) to determine the flow velocity v_1. Once both velocities are known, we may apply Bernoulli's equation (Eq. 11.5) to determine the pressure difference $P_1 - P_2$. Since this is a horizontal pipe, the heights will drop out of Bernoulli's equation, $y_2 - y_1 = 0$.

Solution

First, the equation of continuity,

$$A_1v_1 = A_2v_2$$
$$v_1 = [A_2/A_1]v_2$$
$$v_1 = [15\ cm^2/25\ cm^2]\ 5.0\ m/s$$
$$v_1 = 3.0\ m/s$$

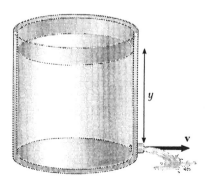

Figure 11.16
Torricelli's theorem. Water leaves the tank with a speed $v = \sqrt{2gy}$, where y is the distance between the water surface and the exit point. An object dropped from the height y will have the same speed v just before it hits the ground.

Now we can use Bernoulli's equation,

$$\tfrac{1}{2}\rho_1 v_1^2 + P_1 + \rho_1 g y_1 = \tfrac{1}{2}\rho_2 v_2^2 + P_2 + \rho_2 g y_2$$

$$P_1 - P_2 = \tfrac{1}{2}\rho_2 v_2^2 - \tfrac{1}{2}\rho_1 v_1^2 + \rho_2 g y_2 - \rho_1 g y_1$$

$$P_1 - P_2 = \tfrac{1}{2}(10^3 \text{ kg/m}^3)[(5.0 \text{ m/s})^2 - (3.0 \text{ m/s})^2] + \rho g(y_2 - y_1)$$

$$P_1 - P_2 = \tfrac{1}{2}(10^3 \text{ kg/m}^3)[(5.0 \text{ m/s})^2 - (3.0 \text{ m/s})^2]$$

$$P_1 - P_2 = 8000 \text{ Pa} = 8 \text{ kPa}$$

Practice Exercise   If the narrow section has an area of 10 cm² what will the pressure difference be?

Answer   16 000 Pa = 16 kPa

Example 11.5 ─────────────────────────

Water flows through an 8.0-cm-diameter pipe at a flow speed of 0.60 m/s under a pressure of 4.00×10^5 Pa (*i.e.*, four times atmospheric pressure). What is the flow speed of the water when it goes into a 5.0-cm-diameter pipe 9.0 m higher? What is the water pressure in the 5.0-cm pipe?

Reasoning

First, we apply the equation of continuity to find the new flow rate through the smaller pipe. The water must flow at a higher speed for the same amount of water to flow through the smaller diameter pipe. Then we apply Bernoulli's equation to determine the pressure. The pressure will be different because of the height of the pipe and because of the different flow rate.

Solution

Begin with the equation of continuity,

$$A_1 v_1 = A_2 v_2$$

$$v_2 = [A_1/A_2]v_1$$

$$v_2 = [\pi(4.0 \text{ cm})^2/\pi(2.5 \text{ cm})^2]\, 0.60 \text{ m/s}$$

$$v_2 = 1.5 \text{ m/s}$$

Now that we know the flow speed and the height of the fluid we know everything in Bernoulli's equation except for the pressure, so we may solve for the pressure P_2.

$$\tfrac{1}{2}\rho_1 v_1^2 + P_1 + \rho_1 g y_1 = \tfrac{1}{2}\rho_2 v_2^2 + P_2 + \rho_2 g y_2$$

$$\tfrac{1}{2}(10^3 \text{ kg/m}^3)(0.60 \text{ m/s})^2 + 4.00 \times 10^5 \text{ Pa} + (10^3 \text{ kg/m}^3)(9.8 \text{ m/s}^2)(0) =$$
$$\tfrac{1}{2}(10^3 \text{ kg/m}^3)(1.5 \text{ m/s})^2 + P_2 + (10^3 \text{ kg/m}^3)(9.8 \text{ m/s}^2)(9.0 \text{ m})$$

$$(180 + 4.00 \times 10^5 + 0) \text{ Pa} = 1125 \text{ Pa} + P_2 + 8.8 \times 10^4 \text{ Pa}$$

$$P_2 = (4.00 \times 10^5 - 0.88 \times 10^5 - 1125 + 180) \text{ Pa}$$

$$P_2 = 3.11 \times 10^5 \text{ Pa}$$

*11.5 Viscosity

CONCEPT OVERVIEW

■ Any real fluid exerts friction forces as it flows, and these forces are described by the fluid's viscosity.

Real fluids are not frictionless. Instead, there is friction between the walls of a pipe and the fluid that flows through it as well as friction between different parts of the fluid. Because of friction forces, not all parts of the fluid flow at the same speed.

To understand the source of this friction in a moving fluid, imagine the fluid as being made up of a number of thin layers stacked one upon the other. Friction results as these layers move past each other. In liquids these internal friction forces occur because of the forces between molecules. In gases they occur because of collisions between molecules.

These internal forces in a fluid are described by **viscosity,** which is different for different fluids and under different conditions. Cold molasses has greater viscosity than warm syrup, for example, and toothpaste has much greater viscosity than water. Liquids generally have greater viscosity than gases.

We can express the viscosity of a fluid by a **coefficient of viscosity** η (Greek lower-case *eta*). Figure 11.17 shows a volume of fluid between two plates, each of area A and a distance l apart. One plate is held fixed, and the other has a force **F** exerted on it parallel to its direction of motion; it moves with a velocity **v.** The very thin layer of fluid in contact with either plate is at rest with respect to that plate. In other words, the bottom layer is at rest and the top layer moves to the right with a speed v, the speed of the moving plate. The various layers of fluid between these two extremes move at different speeds, from zero on the bottom to v on the top. This change of speed as we move up the fluid is known as a **velocity gradient.** To be precise, the change in speed ($\Delta v = v - 0 = v$) divided by the distance over which this change occurs is the velocity gradient, v/l.

The bottom (stationary) layer of fluid exerts forces on the layer immediately above it. That layer exerts forces on the layer immediately above *it*. And so on, and upward, until the top layer exerts a force on the plate above it. That is why a force **F** is needed to keep the upper plate moving with a constant velocity **v.** (You can test this yourself by pulling a knife across a large drop of syrup at the breakfast table.)

Experiments show that **F** is proportional to the area A of the plates and to the speed v and is inversely proportional to the separation of the plates l, with the constant of proportionality being the coefficient of viscosity η:

$$F = \eta \frac{Av}{l}$$

Therefore

$$\eta = \frac{Fl}{Av} \qquad \qquad (11.9)$$

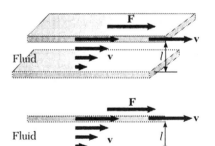

Figure 11.17
The coefficient of viscosity can be defined in terms of the force **F** necessary to keep a flat plate of area A moving with constant velocity **v** when there is a layer of fluid of thickness l between it and a stationary plate.

TABLE 11.1   Selected coefficients of viscosity

Substance	Temperature (°C)	Coefficient of Viscosity ($\times 10^{-3}$ Pa·s)
Air	30	1.9×10^{-2}
Water vapor	100	1.25×10^{-2}
Water	0	1.8
	20	1.0
	40	0.66
	60	0.47
	80	0.36
	100	0.28
Alcohol	20	1.2
Blood plasma	37	1.5
Whole blood	37	3
Oil, SAE 10	30 at 30°C	200
Glycerin	0	12 000
	15	2330
	30	600
Castor oil	0	5300
	40	230
	100	15

The viscosity of oil changes so much with temperature that people living in cold climates usually use a low-viscosity motor oil in winter. In more temperate climates, most people use a "multiweight" motor oil, where *multiweight* really means multiviscosity. Such oil contains an additive that keeps its viscosity from changing much with temperature.

Remember that *F/A* is the shear stress. The coefficient of viscosity, then, has SI units of N·m/(m/s)m² = N·s/m² = Pa·s. In an older and still common system of units, this coefficient is measured in units of *poise* (P), where 1 Pa·s = 10 P.

Table 11.1 lists coefficients of viscosity for several fluids. Notice that temperature is also listed, for that is quite important. To see why, try an experiment with molasses. First note how long it takes to pour cold molasses into a bowl. Place the bowl in a microwave oven and heat it for 20 s or so. Then note how long it takes to pour the warm molasses into another container. How does your initial pour compare with your second pour?

We have defined viscosity in terms of the plates of Figure 11.17 and Equation 11.9. A more useful situation is illustrated in Figure 11.18, where fluid flows through a circular pipe. The fluid flows fastest through the center of the pipe and is at rest wherever it touches the inner surface of the pipe, as shown by the arrows. This is as we expect, but the geometry is not as simple as with the flat plates.

Because of the friction force resulting from viscosity, a net force is necessary to keep fluid flowing through a pipe at constant speed. Because the deriva-

Figure 11.18
A viscous fluid flowing through a circular pipe. Fluid flows fastest in the middle of the pipe, and the layer of fluid next to the pipe wall is at rest.

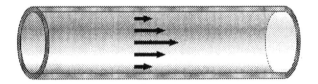

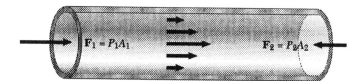

Figure 11.19
A net force of magnitude $F_{net} = F_1 - F_2$ is needed to cause a viscous fluid to flow through a pipe.

tion of this force requires calculus, we shall simply state the result. To cause fluid having a coefficient of viscosity η to flow through a pipe of length l with a speed v at the center, there must be a net external force on the fluid of magnitude

$$F_{net} = 4\pi\eta lv$$

This net external force is equal to the viscous force $\mathbf{F}_v$ opposing the motion.

Figure 11.19 shows the source of this net external force. The pressure at the left end of the pipe is P_1, and that at the right end is P_2. The pipe has constant cross-sectional area $A_1 = A_2 = A$. The forces at the two ends are

$$F_1 = P_1A_1 = P_1A \qquad \text{and} \qquad F_2 = P_2A_2 = P_2A$$

These forces act in opposite directions, so the net external force is

$$F_{net} = F_1 - F_2 = (P_1 - P_2)A$$

This must be the viscous force F_v:

$$F_v = (P_1 - P_2)A = 4\pi\eta lv$$

For a circular cross section, $A = \pi r^2$, so that

$$F_v = (P_1 - P_2)\pi r^2 = 4\pi\eta lv$$

$$P_1 - P_2 = \frac{4\eta l}{r^2}v \qquad\qquad \textbf{(11.10)}$$

It is often useful to look at the volume flow rate Q (Equation 11.4):

$$Q = \frac{\Delta V}{\Delta t} = \frac{A\Delta l}{\Delta t} = Av_{avg} \qquad\qquad \textbf{(11.11)}$$

For a pipe of circular cross section, the average speed of the fluid can be shown to be one half the speed of the fluid in the center of the pipe. Thus,

$$Q = Av_{avg} = \pi r^2\left(\frac{v}{2}\right) \qquad\qquad \textbf{(11.12)}$$

We can solve Equation 11.10 for v:

$$v = \frac{r^2}{4\eta l}(P_1 - P_2)$$

Substituting this expression for v Equation 11.12 gives

$$Q = \frac{\pi r^4}{8\eta l}(P_1 - P_2) \qquad\qquad \textbf{(11.13)}$$

This is known as **Poiseuille's law;** note the r-to-the-fourth dependence. The French physician Jean Louis Marie Poiseuille discovered this law *experimentally* in the 19th century!

Example 11.6

If the water pressure available at your house is twice atmospheric pressure (gauge pressure of 1.0 atm), what will the volume flow rate be through 15 m of garden hose having an inside diameter of 1.2 cm? Assume the water is at 40°C.

Reasoning
The flow through the hose is reduced because of viscosity, the frictional effect of rubbing against the wall of the hose. We can use Poiseuille's law directly.

Solution

$$Q = \frac{\pi r^4}{8\eta l}(P_1 - P_2)$$

$$= \frac{(3.14)(6.0 \times 10^{-3} \text{ m})^4(1.0 \text{ atm})}{(8)(0.66 \times 10^{-3} \text{ Pa} \cdot \text{s})(15 \text{ m})} \left(\frac{1.01 \times 10^5 \text{ Pa}}{1 \text{ atm}}\right)$$

$$= 5.2 \times 10^{-3} \text{ m}^3/\text{s}$$

Practice Exercise What will the volume flow rate be through 30 m of the same garden hose?
Answer $2.6 \times 10^{-3} \text{ m}^3/\text{s}$

Example 11.7

If you bought a cheaper garden hose, with a diameter of only 1.0 cm, what would the volume flow rate be under the same conditions as in Example 11.6? This new diameter is only 17 percent smaller.

Reasoning
We apply Poiseuille's law again.

Solution

$$Q = \frac{\pi r^4}{8\eta l}(P_1 - P_2)$$

Since the pressures, lengths, and viscosities are the same in the two cases, we can use a ratio rather than carry out the entire calculation again:

$$\frac{Q_2}{Q_1} = \frac{r_2^4}{r_1^4} = \frac{(1.0 \text{ cm})^4}{(1.2 \text{ cm})^4} = (1.2)^{-4} = 0.48$$

$$Q_2 = 0.48Q_1$$

Because of the r-to-the-fourth nature of viscous flow, this 17 percent reduction in radius results in a 52 percent reduction in flow rate!

Not only does the r-to-the-fourth nature of fluid flow affect garden hoses, but it also has the same effect—with much more serious consequences—on blood flow in human arteries.

Example 11.8

If cholesterol builds up in major arteries so that their diameter is reduced by 10 percent, what effect will this shrinkage have on blood flow?

Reasoning

As in Example 11.5, we can compare the flow when the radius is $r_f = 0.90r_i$ with the flow when the radius is at its initial value r_i.

Solution

$$\frac{Q_f}{Q_i} = \frac{r_f^4}{r_i^4} = \frac{(0.90r_i)^4}{r_i^4} = (0.90)^4 = 0.66$$

A 10 percent reduction in artery diameter reduces blood flow by 34 percent. This reduction of blood to the brain, muscles, and organs is serious enough in itself. In addition, though, the body often compensates by increasing the pressure difference in an attempt to maintain the blood flow rate, leading to all the serious problems associated with high blood pressure.

Practice Exercise   If cholesterol builds up enough to reduce the diameter by 20 percent, what effect does this have?

Answer $Q_2 = 0.41 \, Q_1$; the blood flow rate is reduced to 41 percent (or reduced by 59 percent).

SUMMARY

In streamline flow, the motion of all the particles of the fluid is identical at any particular point in the fluid. The path that such particles take is called a streamline. In turbulent flow, the particle motion is chaotic.

The mass flow rate of a fluid is the mass of fluid that flows past a point per unit time:

$$\text{Mass flow rate} = \frac{\Delta m}{\Delta t} = \rho A v \qquad \text{(11.1)}$$

The equation of continuity,

$$\rho_1 A_1 v_1 = \rho_2 A_2 v_2 \qquad \text{(11.2)}$$

describes the speed of the fluid as either the density or the cross section changes. For an incompressible fluid, this expression reduces to

$$A_1 v_1 = A_2 v_2 \qquad \text{(11.3)}$$

Bernoulli's equation,

$$\tfrac{1}{2}\rho_1 v_1{}^2 + P_1 + \rho_1 g y_1 = \tfrac{1}{2}\rho_2 v_2{}^2 + P_2 + \rho_2 g y_2 \qquad \text{(11.5)}$$

describes the interrelationships of speed, pressure, height, and density of a moving fluid. According to Bernoulli's equation, the pressure exerted by a fluid decreases as the speed of the fluid increases.

The volume flow rate Q through a pipe is

$$Q = \frac{\Delta V}{\Delta t} = A v_{\text{avg}} \qquad \text{(11.11)}$$

QUESTIONS

1. Why can we say that fluid never flows across the boundaries of a stream tube?
2. Explain the physics involved in the saying "still waters run deep."
3. As a stream of water comes out of a faucet, it gets narrower. Explain.
4. Explain how a baseball pitcher might throw a curve ball.
5. How can spin on a golf ball cause it to curve left or right after it is hit into the air?
6. In addition to escape from enemies, why do burrows that animals dig underground have more than one entrance?
7. If you blow across the top of a straw in a glass of soda, what happens to the level of the soda inside the straw?
8. How can a Pitot tube tell how fast an airplane is traveling when all it measures is pressure?
9. Why, during takeoff, must an airplane pilot be concerned with the altitude of the airport?
10. Why does viscosity change with temperature?
11. What happens to the human body when major arteries are reduced in diameter?
12. In a siphon (Fig. 11.20), a tube *initially filled with fluid* is placed in the higher container. Fluid then flows through the tube from the higher container to the lower one. Explain how this works.
13. During a tornado, the roof of a house may be lifted off. Use Bernoulli's principle to explain why.

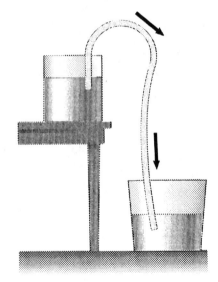

Figure 11.20
Question 12. A siphon transfers fluid from a higher container to a lower one.

14. Why do airplanes usually take off and land facing into the wind?
15. In a television commercial, identical ketchup is poured from two bottles onto identical forks. One ketchup is at room temperature, and the other has just come from the refrigerator. Which will drip off the fork first?

PROBLEMS

11.1 Fluid Flow

11.2 The Equation of Continuity

11.1 Water flows through a pipe at the rate of 0.35 m³/min. What is its speed where the pipe has a diameter of 1.5 cm? Where the diameter is 2.5 cm?

11.2 What is the volume flow rate of water flowing through a garden hose at 0.45 m/s if the hose has a diameter of 1.0 cm? What will the speed of the water be as it flows through another garden hose connected to this one if the second hose has a diameter of 1.2 cm?

11.3 Air flows into a room 5.2 m wide by 6.5 m long by 2.8 m high through an air vent 0.20 m by 0.40 m. Air exits the air vent at 1.25 m/s. How long does it take to change the air in the room completely?

11.4 Water flows at 1.5 m/s in a region of a river 3.0 m deep and 6.5 m across. What is the water speed through a rapids, where the river is 1.0 m deep and 8.0 m wide?

○11.5 In a wind tunnel, the cross-sectional area changes from 1.0 m² to 0.22 m². What must the flow speed in the wide region be to give a flow speed of 120 km/h in the narrow region?

11.3 Bernoulli's Equation

11.4 Applications of Bernoulli's Equation

11.6 What must the gauge pressure be at a fire hydrant if water from a fire hose is to reach a height of 12 m?

11.7 Air ($\rho = 1.3$ kg/m³) blows past a 1.0 m × 2.5 m window with a speed of 20 m/s. What is the difference in pressure on the two sides of the window? What is the net force on the window?

11.8 Air ($\rho = 1.3$ kg/m³) blows across a roof with a speed of 15 m/s. What is the difference in pressure on the two sides of the roof? What is the lift force on the roof if its dimensions are 8 m × 12 m?

11.9 What is the volume flow rate through a faucet with a 2.0-cm² opening if the pressure head is 12 m? (The "pressure head" of 12 m means the pressure is the same as that due to a column of water 12 m high.)

11.10 What lift force is experienced on an airplane wing of area 35.0 m² if air passes across the top at 300 m/s and across the bottom at 275 m/s?

○❖11.11 Below are data for the cross-sectional area of open pipes and the pressure head behind each. (The "pressure head" is the gauge pressure measured in meters of water.) Find the volume flow rate through each pipe and rank the seven in order of increasing flow rate.

Pipe	Cross-sectional Area (cm²)	Pressure Head (m)
a	1.0	2.5
b	1.2	2.5
c	1.5	2.5
d	1.8	3.0
e	2.0	4.0
f	3.0	5.0
g	5.0	6.0

○11.12 Air is blown across the top of a straw at a speed of 2.5 m/s. How high in the straw will water be drawn because of this?

○11.13 The top of a straw is 0.2 m above the surface of the water in a glass. With what speed must you blow air across the top of the straw in order to lift the water all the way to the top?

○11.14 In Figure 11.21, air flows at 2.0 m/s through a pipe having a cross-sectional area of 0.10 m². What is the air speed when the pipe narrows to 0.050 m²? What is the difference in height of the water in the two arms of the manometer?

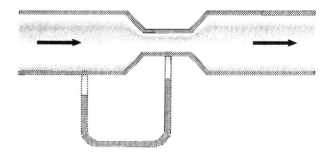

Figure 11.21 Problem 11.14.

○11.15 Suppose the wide and narrow areas in Figure 11.21 are 0.25 m² and 0.10 m², respectively. Find the flow speed in both regions if water flows through the pipe at a volume flow rate of 30 m³/min. What is the difference in height in the two arms of the manometer if it is filled with mercury?

○11.16 A hole with area 1.75 cm² is drilled in the side of a large tank 8.00 m below the surface of the water. Find the speed and the volume flow rate of the water coming out.

○11.17 Water is held in a sealed tank, with compressed air providing a gauge pressure of 2.5 atm at the water's surface. A hole is drilled in the side of the tank 5.0 m below the water's surface. With what speed does the water exit this hole?

○11.18 Water is held in a sealed tank, with compressed air providing pressure above the water's surface. A small hole is drilled in the side of the tank 1.0 m below the water's surface, and water exits with a speed of 3.2 m/s. What is the pressure at the water's surface?

○11.19 Seawater is held in a sealed tank, with compressed air providing a gauge pressure of 25 atm at the surface. A 1.5-cm² circular hole is drilled in the side of the tank 8.0 m below the water's surface. With what speed does the water exit this hole? What is the volume flow rate?

○11.20 The inside diameter of the larger part of the pipe in Figure 11.22 is 2.50 cm. Water flows through the pipe at a rate of 0.180 l/s (liters per second). What is the inside diameter of the constriction?

○ ❖ 11.21 Below are data for air flow through the tube shown in Figure 11.23. For each set of data, find the pressure at position 2 and the difference in height Δh of the two columns of water. Rank the seven in terms of increasing Δh.

Pipe	v_1 (m/s)	A_1 (cm²)	P_1 (atm)	A_2 (cm²)
a	1.0	3.5	1.0	1.0
b	1.5	4.0	1.5	1.0
c	2.0	5.0	1.5	2.0
d	2.5	7.5	2.0	2.0
e	3.0	10.0	3.0	3.0
f	3.0	12.5	3.0	3.0
g	3.0	15.0	3.0	3.0

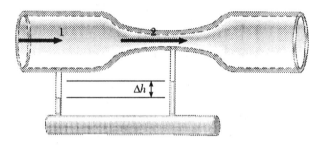

Figure 11.23 Problem 11.21.

○11.22 The dimensions given in Figure 11.24 are all inside diameters. When the faucet valve is wide open, it does not constrict the smaller pipe. If water enters the larger pipe under a pressure of 3800 torr, what volume emerges

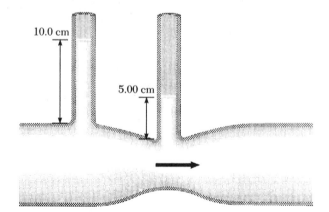

Figure 11.22 Problem 11.20.

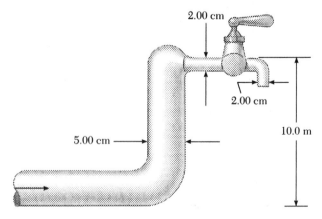

Figure 11.24 Problem 11.22.

from the faucet per second? Compare this exact result with the result obtained by disregarding the speed of the water in the larger pipe. Assume the pressure in the open faucet is 760 torr (1 atm).

○11.23 A jet of water squirts out from a leak in the side of the tank in Figure 11.25. If the hole has a diameter of 3.5 mm, what is the height h of the water level above the hole. Assume the tank is so large that the flow speed at the top of it is negligible.

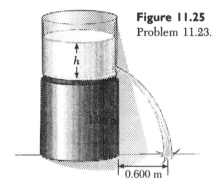

Figure 11.25
Problem 11.23.

0.600 m

○11.24 Blood is carried to the capillaries by small blood vessels called arterioles. Smoking causes arteriole diameter to decrease by about 5 percent. If the pressure remains the same, what change in volume flow rate does this reduction cause? If the flow rate remains the same, what pressure increase is necessary to counteract this reduction?

●11.25 The siphon in Figure 11.26 drains the aquarium. The inner radius of the hose is 6.0 mm.

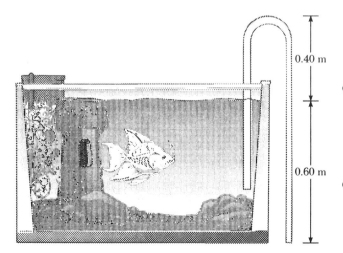

0.40 m

0.60 m

Figure 11.26 Problem 11.25.

(a) What must the gauge pressure be at the end of the hose not in the water in order to get the siphon started? (b) What is the initial volume flow rate from the tank?

*11.5 Viscosity

11.26 The block of ice (temperature 0°C) in Figure 11.27 is drawn over a level surface lubricated by a layer of water 0.1 mm thick. Determine the magnitude of the force **F** needed to pull the block with a constant speed of 0.5 m/s.

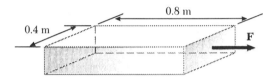

0.8 m

0.4 m

F

Figure 11.27 Problem 11.26.

11.27 The pulmonary artery, which connects the heart to the lungs, has an inner radius of 2.6 mm and is 8.4 cm long. If the pressure drop between heart and lungs is 3.0 torr, what is the average speed of blood in the artery? Assume a body temperature of 37°C.

○11.28 A pipe 1.5 m long with an inside diameter of 2.5 mm has water flowing through it at 0.050 m³/s. If the water temperature is 60°C, what must the pressure difference between the ends of the pipe be? What pipe diameter will allow twice the flow under the same conditions?

○11.29 What pressure difference is required between the ends of a section of transport pipeline that is 5.0 km long and 50 cm in diameter if it transports oil at a rate of 3.5 m³/h? Assume an oil density of 930 kg/m³ and a coefficient of viscosity of 0.25 Pa·s.

○11.30 Blood, during a transfusion, flows from a bottle 1.25 m above a needle that is 4.50 cm long and has an inside diameter of 0.500 mm. What is the volume flow rate? How long does it take to transfuse 0.5 l of blood?

○11.31 Water at 40°C flows through a pipe that is 1.5 m long and has an inside diameter of 1.5 cm. The water at the center of the pipe flows with a speed of 20 cm/s. Find the pressure difference between the ends of the pipe.

○11.32 An air duct 20 m long supplies air to a room that has dimensions 2.8 m × 3.5 m × 5.0 m. The duct and the vent at its end have a circular cross section 25 cm in diameter. What must the average air speed through this vent be in order to change the air in the room three times every hour? What must the pressure difference between the ends of the duct be to ensure this change rate? Assume the coefficient of viscosity of the air is 2.0×10^{-5} Pa·s.

○11.33 Blood, during a transfusion, flows from a bottle through a needle that is 3.5 cm long and has an inside diameter of 0.35 mm. How high above the needle must the bottle be held in order to transfuse 0.5 l of blood in 45 min?

○11.34 Glucose, during a transfusion, flows from a bag through a needle that is 3.0 cm long and has an inside diameter of 0.30 mm. If the transfusion rate is to be 0.5 l in 45 min, how high above the needle must the bag be held? Assume a density of 1040 kg/m³ and a coefficient of viscosity of 2.25×10^{-3} Pa·s for the glucose.

○11.35 An ice skate blade is 33.0 cm long and 4.00 mm wide. The skater, moving with a speed of 2.50 m/s, glides on one foot over smooth ice. A layer of water between blade and ice is 0.100 mm thick. If the skater weighs 485 N, determine the coefficient of friction between blade and ice due to the viscosity of the layer of water.

● 11.36 The radius of the shaft in Figure 11.28 is 3.00 cm. The inside radius of the bearing is 0.150 mm larger, and the length of the bearing is 7.50 cm. Lubricating oil with a coeffi-

cient of viscosity of 0.55 N·s/m² fills the gap. Determine the torque needed to keep the shaft rotating with a constant angular speed of 2.00 rad/s.

General Problems

○11.37 The heart of a person sitting quietly pumps blood at the rate of 4×10^3 cm³/min through an aorta of cross section 0.8 cm². (a) What is the average blood speed in the aorta? (b) The blood then spreads out into a network of about five billion tiny capillaries whose average radius is about 8×10^{-4} cm. What is the speed of the blood through these capillaries?

○11.38 A gardener tires of watering his garden with a hose that has an inside diameter of 0.50 in. How much less time will be required if he replaces it with a ¾-in. hose? Assume no change in hose length or water pressure.

● 11.39 The tank in Figure 11.29 is filled with water to a depth H. At a distance h_1 below the surface of the water, a small hole is drilled in the side of the tank. With what speed does the water exit this hole? Find the distance R from the side of the tank to the point where this water strikes the ground. Find the distance h_2 below the water surface that a second hole can be drilled such that the water from this hole hits the ground at the same distance R from the side of the tank. Assume the tank is so large (and the holes so small) that the water level does not significantly change as the water flows out.

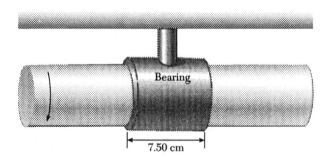

Figure 11.28 Problem 11.36.

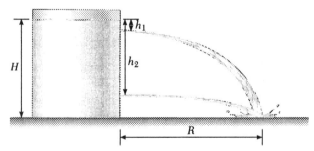

Figure 11.29 Problem 11.39.

●11.40 A horizontal pipe is 100 m long and has an inside radius of 3.5 cm. What must the power of a pump be in order for it to pump glycerine through the pipe at a rate of 80.0 l/min? The temperature of the glycerine is 15°C.

●11.41 Blood travels at 4.00 cm/s up a vein in the leg. The vein is 24.0 cm long and has an inside diameter of 2.60 mm. If the gauge pressure of the blood is 30.0 torr at the bottom of the vein, what is the gauge pressure at the top. (*Hint:* The pressure drop has two sources, the rise in level and the loss due to viscosity).

PHYSICS AT WORK

Many substances expand or contract as temperature rises or falls. Hot gases enlarge a balloon and heated liquids boil away, but expansion in solids is often less visible. Solids do expand or contract slightly, determined by the coefficient of thermal expansion.

During the day, objects on Earth absorb heat from the Sun and radiate that heat back into the atmosphere at night. For a large telescope, the heat coming off the dome and mirror can cause some turbulence in the surrounding air, which makes the images of faint stars and galaxies twinkle. Astronomers must wait for the observatory building and telescope to cool down to permit seeing clear, stable images.

The first large telescope mirrors were constructed from plate glass, which has an expansion coefficient almost as great as steel. Then Corning developed Pyrex with an expansion coefficient less than a third as great. The 5-m Hale Telescope was the first large telescope to have a Pyrex mirror. Stable images formed more quickly, giving astronomers more *seeing* time.

Astronomers continue to use Pyrex or compounds such as Cer-Vit or Zerodur to construct primary mirrors (*right*). The Keck 10-m telescope on Mauna Kea, Hawaii, is now the largest in the world. The Keck's main mirror is made of Pyrex but is assembled from 36 contiguous hexagonal segments.

(©1992 Roger Ressmeyer-Starlight, all rights reserved)

12

Temperature, Heat, and Internal Energy

Key Issues

- Heat is a form of energy due to the random motion of individual molecules in an object. Heat flows from a hotter body to a cooler one. Two objects in contact, initially at different temperatures, will come to a final temperature that is between their original temperatures.

- Measurable characteristics of an object—such as length or volume or density or pressure or electrical resistivity—change with temperature. These characteristics may be used to establish a temperature scale or a thermometer. The most common is the change in volume of a liquid; this is used in a common thermometer where the change in volume is made visible as the liquid expands through a small glass tube.

- The temperature of a material generally increases as heat is added to the material; the specific heat of a material describes how much heat (energy) must be added to the material to change its temperature.

- At a transition temperature, adding heat to a material will cause some of it to change state—from solid to liquid or from liquid to gas—while the temperature remains constant. The amount of heat required to change state is known as the heat of fusion for the transition from solid to liquid and the heat of vaporization for the transition from liquid to gas.

In our study of work, power, and energy in Chapter 6, we used the work-energy theorem to understand transformations in mechanical energy. We might also call this type of energy *macroscopic energy* because it is the energy due to large-scale motions that we can easily see. This and the following two chapters are concerned with energy transformations on a *microscopic* scale as the atoms or molecules of one object interact with the atoms or molecules of another object.

12.1 Temperature

CONCEPT OVERVIEW

■ Temperature is a description of how hot or cold an object is.

■ Temperature can be measured by observing a change in some thermometric property of an object.

When you hear the word *temperature,* you probably think of how hot or cold something is. Touching an object is one way to tell whether it is hot or cold. However, our sense of touch is not very accurate and can be fooled. If you reach into a freezer and pick up a metal ice-cube tray and a cardboard box of film you have stored there, the tray will feel colder than the box even though both are at the same temperature. The metal is a good heat conductor while the cardboard is a poor heat conductor. Soak your left hand in a bucket of warm water and your right hand in a bucket of cold water. Now reach for a can of soda. The can will feel cooler to your left hand and warmer to your right hand. For measuring temperature, therefore, we obviously need something more objective—and more quantitative—than our sense of touch.

Many measurable properties of materials change with temperature: liquid and gas volumes, gas pressure, and electrical resistance, to name just a few. Any of these properties may be used to define a temperature and are therefore known as **thermometric properties.**

The thermometer you most readily think of is probably a liquid-in-glass thermometer like the one shown in Figure 12.1. Mercury or colored alcohol is contained in a bulb at the base of a thin glass tube. Any change in the length of the tube as temperature changes is so small that we can ignore it and concentrate on the changes in the volume of the liquid. As this volume expands, the length of the liquid column in the tube increases.

This thermometer can be calibrated by placing it in contact with objects of known and reproducible temperature. As shown in Figure 12.1, we could place the thermometer in a beaker of ice water. The ice cubes and water are in thermal equilibrium at only one temperature; we call this equilibrium temperature zero degrees on the Celsius temperature scale (0°C). We place a mark on the glass tube indicating the length of the liquid column when the thermometer is in the ice water and label that mark 0°C. If the thermometer is now placed in a beaker of boiling water, the length of the liquid column increases. At 1 atm pressure, water boils at only one temperature, 100°C, so we now mark the tube to indicate the column's length when the thermometer is in the boil-

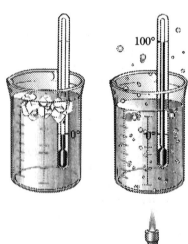

Figure 12.1
A liquid-in-glass thermometer is calibrated at two reproducible temperatures, such as the melting point of ice and the boiling point of water.

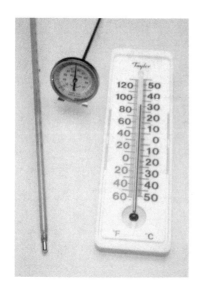

Liquid-in-glass thermometers. The liquid thermometers are on either side of a dial (bimetallic) thermometer. Commonly used liquids are mercury (*left*) and colored alcohol (*right*). (© *Charles D. Winters*)

ing water and label this mark 100°C. We then make equally spaced marks between these two calibrated marks to measure intermediate temperatures.

The temperature scale we have just defined—with 0° for the temperature of melting ice and 100° for the temperature of boiling water—is the **Celsius temperature scale.** It is used throughout most of the world.

In the United States, the **Fahrenheit temperature scale** is still commonly used. On this scale, the melting temperature of ice is 32°F and the boiling temperature of water is 212°F. That means the difference between these two temperatures is 180 F° (read as "one hundred eighty Fahrenheit degrees"). Knowing that the difference between these two temperatures is also 100 C°, we can write

$$100 \text{ C}° = 180 \text{ F}°$$

or

$$1 \text{C}° = \frac{9}{5} \text{ F}°$$

That the zero points of the two temperature scales do not coincide must be accounted for in converting from one scale to the other. Therefore the conversion equations are

$$T_C = \frac{5}{9}(T_F - 32)$$

and

$$T_F = \frac{9}{5}T_C + 32$$

The two temperature scales are compared in Figure 12.2.

In the past, the Celsius scale was known as the *centigrade* scale because it has 100 one-degree gradations between the two calibration temperatures. *Centum* is Latin for "hundred."

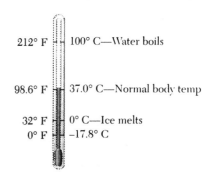

Figure 12.2
The Celsius temperature scale is the SI scale and is used throughout the world. The Fahrenheit scale is still commonly used in the United States.

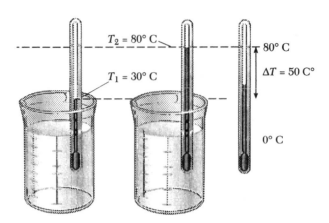

Figure 12.3
The difference between two temperatures is given as Celsius *degrees*, C° (or Fahrenheit degrees, F°).

Fahrenheit planned to have as his upper calibration point of 100°F "the limit of the heat in the blood of a healthy man," but, according to legend, he had a slight fever the day he used his own body temperature to calibrate his thermometer.

A bimetallic strip freezer thermometer showing the front dial (*top*) and the bimetallic coil visible from the rear (*bottom*). (© *L. W. Black*)

While the Celsius calibration temperatures, 0°C and 100°C, seem convenient and reasonable, the Fahrenheit calibration temperatures, 32°F and 212°F, are unusual enough that some historical background is called for. Those values were not the ones chosen initially. Gabriel Fahrenheit (1686–1736) intended the zero of his scale (0°F) to be the lowest temperature he could obtain by mixing water, ice, and salt; because salt lowers the freezing temperature of water (that's why we pour rock salt on icy sidewalks), Fahrenheit's zero point, being for an ice-water-salt mixture, was lower than the melting temperature of ice alone. He intended 100°F to be normal body temperature (thus even the Fahrenheit scale started as a *centi*grade scale!) but made a slight error in measuring, so that normal body temperature is not 100°F but 98.6°F.

A word on notation. A particular temperature is given as so many degrees Celsius; for example, $T_1 = 30°C$ or $T_2 = 80°C$. A temperature *difference* is given as so many Celsius degrees; for example, $\Delta T = T_2 - T_1 = 50$ C°, as sketched in Figure 12.3. This distinction will prove useful in solving problems.

Another common thermometer depends on the behavior of a **bimetallic strip.** Metals, like most other materials, expand when they are heated, but different metals expand at different rates, as we shall see in Section 12.3. If two types of metal are bonded together as in Figure 12.4, the resulting bimetallic strip will bend one way when heated and the other way when cooled. Such a strip can be used as a thermostat to turn a furnace or air conditioner on and off. This principle is also used in the automatic choke on an automobile. A bimetallic strip can be wound several times to get a larger motion of its end as it flexes; then a pointer and numeric scale can be added so that the device can be read as a thermometer.

Other thermometric properties can also be used to establish a temperature scale. The pressure in a constant-volume gas thermometer is a very accurate thermometric property. Such a device is sketched in Figure 12.5. The height of the liquid in the right side of the manometer is varied until the liquid in the left

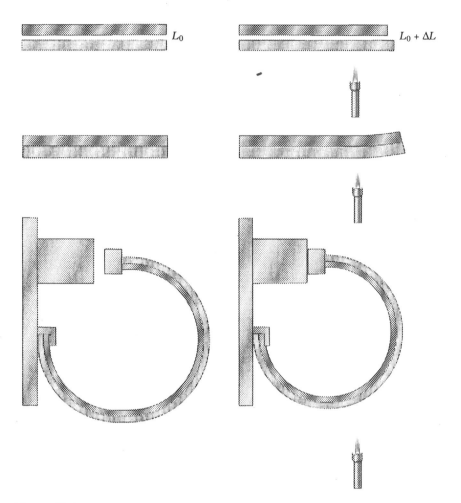

L_0

$L_0 + \Delta L$

Figure 12.4
Two strips made from different metals can be bonded together to form a bimetallic strip for use in a thermostat. The difference in the expansion rates of the two metals causes the strip to bend when its temperature changes. On the right, the two squares touch and complete an electric circuit that makes, say, an air conditioner run. On the left, the electrical contact is broken and the air conditioner is not running.

side is at a reference mark; this ensures constant volume in the gas reservoir. Then the pressure of the gas is determined by measuring h. This pressure is linearly related to temperature, as shown in Figure 12.6.

If a different density of gas is used—in other words, if more gas is added to the reservoir—the calibration curve will have a different slope but pressure and temperature will still be linearly related, as sketched in Figure 12.6(a). For any gas density, we can extrapolate to see where we expect the pressure to vanish. As long as the gas density and pressure are not too great, the pressure-versus-temperature line always intersects the horizontal axis (indicating zero pressure) at $-273.15°C$. Because the pressure in the gas-filled reservoir cannot be negative, the plots of Figure 12.6(b) and (c) imply that temperatures cannot be less than $-273.15°C$.

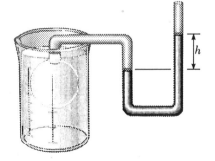

h

Figure 12.5
The pressure of a constant volume of gas can also be used as a thermometer.

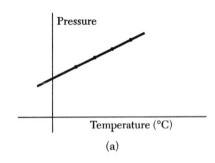

(a)

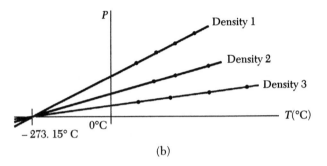

(b)

Figure 12.6
(a) Pressure in a constant-volume thermometer can be used to define temperature. (b) The slope of the pressure-versus-temperature curve for a constant-volume gas thermometer changes as the gas density in the reservoir changes, but regardless of slope, all the lines intersect the x axis at one temperature: $-273.15°C$. (c) This temperature of $-273.15°C$ is used to define the temperature scale used most often in scientific work, the Kelvin scale. This scale uses the x-intercept value of the pressure-versus-temperature curves as its zero point. Therefore, $0 K = -273.15°C$, ice melts at $0°C = 273.15 K$, and water boils at $100°C = 373.15 K$.

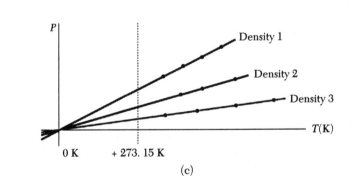

(c)

Example 12.1

In a constant-volume gas thermometer the pressure of the gas is 50 torr at 10°C and 65 torr at 95°C. What is the temperature when the pressure is 60 torr?

Reasoning

The pressure and temperature are linearly related as shown in Figure 12.7. Knowing the pressure and temperature for two sets of data allows us to find the equation of that line and then to calculate any other temperature when the pressure is known.

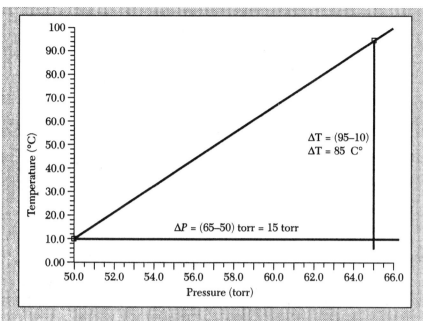

Figure 12.7
Temperature and pressure are linearly related in a constant-volume gas thermometer.

Solution

From the given data, we can construct the graph shown in Figure 12.7. The slope of this line is

$$\text{Slope} = \frac{\text{rise}}{\text{run}} = \frac{85 \text{ C}^\circ}{15 \text{ torr}} = 5.67 \frac{\text{C}^\circ}{\text{torr}}$$

Any straight line like this can be written as an equation with the form of

$$y = mx + b$$

where m is the slope and b is the y-intercept. In our case, with temperature T and pressure P, the equation is

$$T = mP + b$$

We have now found the slope $m = 5.67$

$$T = 5.67 \frac{\text{C}^\circ}{\text{torr}} P + b$$

so we can use that to find the y-intercept by evaluating b for either of the cases where T and P are known. Let us use $T = 10°C$ and $P = 50$ torr.

$$10°C = \left(5.67 \frac{\text{C}^\circ}{\text{torr}}\right)(50 \text{ torr}) + b$$

$$b = -273°C$$

Now we can write our temperature—pressure equation as

$$T = 5.67 \frac{C°}{torr} P - 273°C$$

and then use this to evaluate the temperature when the pressure is 60 torr.

$$T = \left(5.67 \frac{C°}{torr}\right)(60 \text{ torr}) - 273°C$$

$$T = 67°C$$

Because we know a lowest possible temperature exists, it is extremely useful to have a temperature scale with this absolutely lowest temperature defined as zero—**absolute zero.** On this scale, the **Kelvin temperature scale,** ice melts at +273.15 K (= 0°C) and water boils at +373.15 K (= 100°C). Notice that the degree symbol is not used with Kelvin temperatures. A temperature is given as so many kelvins, not so many degrees Kelvin.

12.2 Equilibrium Temperature and the Zeroth Law of Thermodynamics

CONCEPT OVERVIEW

■ Two bodies each separately in thermal equilibrium with a third body are in thermal equilibrium with each other.

In discussing an ice-water mixture in the previous section, the term *equilibrium temperature* was used because we have an intuitive understanding of it from everyday experience. If cold milk is poured into hot coffee, the temperature of the milk will go up, the temperature of the coffee will go down, and the milk-coffee mixture will have a final temperature somewhere between the initial low and high temperatures. This is the **equilibrium temperature** of the milk-coffee system.

Consider any two objects A and B, at different temperatures, placed in something like a styrofoam picnic chest, which *insulates* them from thermal interactions with their surroundings (Fig. 12.8). Yet let the two objects be in

Figure 12.8

Two objects initially at temperatures T_A and T_B, where $T_A \neq T_B$, are allowed to interact with each other thermally but remain thermally insulated from their surroundings. They will reach an equilibrium temperature T_f that is intermediate between the two initial temperatures.

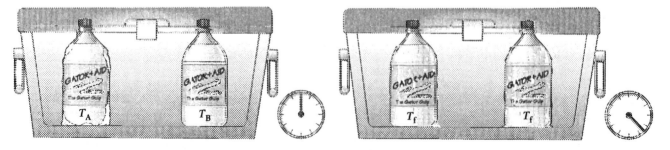

thermal contact with each other; that is, they do no work on each other but can affect each other because of the temperature difference between them. They will eventually come to an equilibrium temperature that is intermediate between their two initial temperatures.

Now consider three objects: A and B as before plus a third object C, which acts as a thermometer. Place A and C in an insulated container and let them come to their equilibrium temperature. Then place B and C in an insulated container and let them come to *their* equilibrium temperature. If the equilibrium temperature is the same in both cases, what will happen when A and B are placed in thermal contact? Nothing; A and B are already in thermal equilibrium. This result can be stated as

Two bodies each separately in thermal equilibrium with a third body are in thermal equilibrium with each other.

This means the temperature of C—the thermometer—can be taken as the temperature of any body in thermal equilibrium with it. This seemingly obvious statement, which provides the basis for using a thermometer, is known as the **zeroth law of thermodynamics.**

The zeroth law has its unusual name because its importance was not realized until after the first and second laws of thermodynamics had become well established.

This cooler made of Styrofoam is a poor thermal conductor or good insulator. Cold beverages and ice kept in such coolers will absorb heat very slowly from the outside. Knowing that Styrofoam is very light in weight, what do you conjecture is the reason for its exceptional insulating properties? *(© Charles D. Winters)*

12.3 Thermal Expansion

CONCEPT OVERVIEW

■ As the temperature of almost any substance increases, the increased motion of the individual atoms or molecules shows up as an increase in the length or volume of the substance.

Both the liquid-in-glass thermometer and the bimetallic-strip thermometer use the fact that most materials expand when heated. Expansion joints in bridges, highways, and sidewalks are necessary because of thermal expansion. The atoms that make up any object vibrate on a microscopic scale. As the object is heated, the atoms vibrate more vigorously and the average distance between them increases. This increase in atomic distances causes the whole object to expand.

Consider a long bar of metal such as that illustrated in Figure 12.9. For the moment, let us concentrate on its length, which is originally L_o. Experimentally, we find that if the bar is heated so that its temperature increases by a small amount ΔT, its length increases by an amount ΔL. That is, its new length is $L_{new} = L_o + \Delta L$, where

$$\Delta L = \alpha L_o \Delta T \qquad (12.1)$$

The change in length with temperature is directly proportional to the original length L_o and the temperature change ΔT. The constant of proportionality α is called the **coefficient of linear expansion** and varies from material to material. Note from Equation 12.1 that the units of α must be $1/C° = (C°)^{-1}$. Table 12.1 lists some representative values for this coefficient.

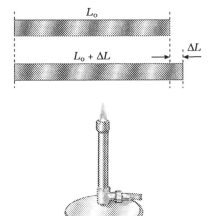

Figure 12.9
A bar expands lengthwise as it is heated. The amount of expansion ΔL is proportional to the bar's original length L_o and to the change in temperature ΔT.

TABLE 12.1　Coefficients of Expansion at Room Temperature (≈20°C)

Material	Coefficient of linear expansion α [$\times 10^{-6}$ (C°)$^{-1}$]	Coefficient of volume expansion β [$\times 10^{-6}$ (C°)$^{-1}$]
Solids		
Lead	29	87
Aluminum	24	72
Brass	19	56
Silver	19	56
Copper	17	51
Gold	14	42
Steel, iron	12	36
Concrete	≈12	≈36
Ordinary glass	10	30
Pyrex glass	3	9
Marble	≈2	≈6
Quartz	0.4	1
Liquids		
Benzene		1240
Carbon tetrachloride		1180
Ethyl alcohol		1100
Gasoline		950
Turpentine		900
Olive oil		680
Glycerin		500
Acetone		447
Water		210
Mercury		180
Gases (at 0°C and 1 atm)		
Air		367
Helium		366

Example 12.2

In a physics laboratory, a steel meter stick is heated from 23°C to 100°C with steam from boiling water. How much does the steel expand?

Reasoning

As the temperature of an object increases, its length also increases. This problem can be solved with Equation 12.1.

Solution

$$\Delta L = \alpha L_o \Delta T$$
$$= [12 \times 10^{-6} \, (\text{C}°)^{-1}](1.0 \text{ m})(77 \text{ C}°) = 9.2 \times 10^{-4} \text{ m} = 0.92 \text{ mm}$$

This is a small but measurable amount. Steel beams in a building or bridge, having an original length much greater than that of our meter stick, will have a much greater expansion ΔL. Expansion joints between beams take care of this length increase.

Practice Exercise How much will the steel meter stick expand if its temperature changes from 35° to 95°C?

Answer 0.72 mm

Example 12.3

Each section of a concrete highway is 15 m long. How great should the expansion gaps be to allow for changes in temperatures from -30°C in winter to 45°C in summer?

Reasoning

Again, as the temperature of each section increases, its length also increases. The gap must be large enough to accommodate the maximum length increase, and we can find this maximum increase from Equation 12.1.

Solution

$$\Delta L = \alpha L_o \Delta T = [12 \times 10^{-6} \, (\text{C}°)^{-1}](15 \text{ m})(75 \text{ C}°) = 0.014 \text{ m} = 1.4 \text{ cm}$$

Practice Exercise How great should the expansion gaps be if the expected temperature extremes are -20° and $+35$°C?

Answer 0.99 cm

Because the coefficient of thermal expansion of this cookware is quite small, it can withstand the extremes of heat and cold simultaneously. What would most likely happen to an ordinary drinking glass if you placed it in boiling water for several minutes and then plunged it into ice water? *(Courtesy of Corning Glass Works)*

Because heated metal expands more than heated glass, running a too-tight jar lid under hot water is a common trick for opening a stuck jar.

So far we have discussed expansion in a single dimension of a long, thin object like the bar in Figure 12.9, but every other dimension is also expanding. A rectangle will expand along its width as well as its length, as shown in Figure 12.10. This two-way expansion means that the dimensions of a hole cut in the rectangle will also expand according to Equation 12.1. Think of painting a circle on a metal plate and then heating the plate. The circle, along with the rest of the plate, will expand. Instead of painting a circle, we could cut one out to form a hole in the plate. If the plate is heated now, it will expand just as before, and consequently the hole will get larger. Instead of a plate, we could use a ring, as in Figure 12.11.

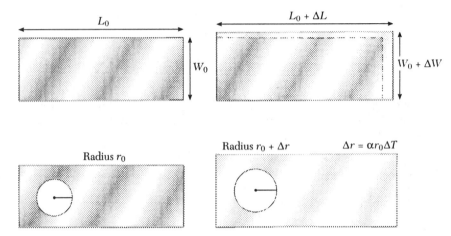

Figure 12.10
A plate expands in width as well as length when it is heated. Since every linear dimension on the plate expands, the radius of a hole cut in the plate also expands.

Expansion occurs in all three dimensions (Fig. 12.12). If a cube has an original edge length L_o, its original volume is $V_o = L_o{}^3$. Each dimension expands according to Equation 12.1 when the object is heated, and this expansion gives a new volume,

$$V_{new} = (L_o + \Delta L)^3 = L_o{}^3 + 3L_o{}^2\Delta L + 3L_o(\Delta L)^2 + (\Delta L)^3$$

Figure 12.11
(a) The ball passes easily through the ring at room temperature. (b) When heated, the ball expands—its radius increases along with its other linear dimensions—so that it can no longer pass through the ring. (© *Charles D. Winters*)

(a)

(b)

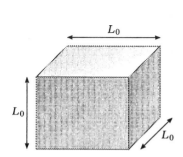 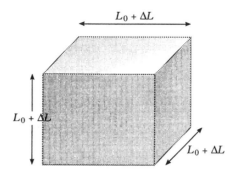

Figure 12.12
All three dimensions expand when a body is heated, causing an expansion of its volume.

Since $\Delta L \ll L_o$, we can ignore the terms in $(\Delta L)^2$ and $(\Delta L)^3$. Then

$$V_{new} = L_o^3 + 3L_o^2 \Delta L$$
$$= V_o + 3L_o^2(\alpha L_o \Delta T)$$
$$= V_o + 3\alpha L_o^3 \Delta T$$
$$= V_o + 3\alpha V_o \Delta T$$

Since $V_{new} = V_o + \Delta V$, it must be true that

$$\Delta V = 3\alpha V_o \Delta T = \beta V_o \Delta T \qquad \textbf{(12.2)}$$

The change in volume with temperature is directly proportional to the original volume V_o and to the temperature change ΔT. The constant of proportionality $\beta = 3\alpha$ is called the **coefficient of volume expansion.** The units of β are also $(C°)^{-1}$. Table 12.1 lists some typical values.

Because the coefficients α and β vary with temperature, Equations 12.1 and 12.2 are adequate as long as ΔT is not too great.

Example 12.4

The volume of mercury used in the bulb of a thermometer is 1.0 cm^3. How much does this volume increase when the temperature increases from room temperature (23°C) to normal body temperature (37°C)?

Reasoning

As the temperature increases, the volume of the mercury also increases; we can apply Equation 12.2 to solve for the increase in volume.

Solution

$$\Delta V = \beta V_o \Delta T$$
$$= [180 \times 10^{-6} \, (C°)^{-1}](1.0 \text{ cm}^3)(14 \text{ C}°) = 0.0025 \text{ cm}^3$$

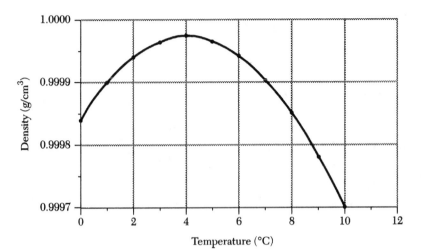

Figure 12.13
Unlike most other materials, water contracts in the temperature range from 0 to 4°C and becomes denser as its temperature increases!

Most materials expand when heated, and consequently their density decreases. Water is a very notable exception, however. Between 0 and 4°C, water *contracts* and becomes denser as its temperature increases! This is shown in Figure 12.13. This unusual behavior is important for its consequences. As the weather gets colder as winter comes on, a body of water—a lake, say—gets colder, too. As the air temperature and, consequently, the temperature of the surface water get down to 4°C, the surface water becomes denser than the warmer water below it. It therefore sinks to the bottom, and water warmer than 4° takes its place at the surface, drops to 4° when it comes in contact with the air, sinks, and new water takes *its* place. This water movement continues until all the water is at 4°. At that point, the surface water (assuming the air temperature is 0° or below, of course) drops to 0° and freezes. Because ice is less dense than liquid water at 4°, the ice floats on the surface and protects the 4° water below it from the frigid air. In other words, ice forms on the top of the lake, but fish are free to swim in the unfrozen water below.

If water behaved like most other materials, the coldest water would fall to the bottom, causing ice to form there first. Many of Earth's lakes and much of her oceans would turn completely to ice during winter! Life might not be sustainable beyond the tropics if it were not for this remarkable behavior of water between 0° and 4°C.

12.4 Thermal Stress

CONCEPT OVERVIEW

■ As the temperature of an object changes, its length changes if free to do so. If the ends are fixed, the change in temperature creates thermal stress in the object.

In Chapter 10 we found that a tensile stress (force per unit area) on a bar could elongate the bar and a compressive stress could compress it. From the definition of Young's modulus, Equation 10.4, we find

$$\Delta L = \frac{L_o(F/A)}{Y}$$

We also know, from Equation 12.1, that $\Delta L = \alpha L_o \Delta T$.

What happens when these changes conflict with each other? Figure 12.14 shows an aluminum bar held between two fixed mounts while it is heated. If it were free to expand lengthwise, we know its length would expand according to Equation 12.1. Therefore, the mounts must exert enough force, or compressive stress, on the bar to effectively shorten it by that amount. That is,

$$\frac{L_o(F/A)}{Y} = \Delta L = \alpha L_o \Delta T$$

$$\frac{F}{A} = \alpha Y \Delta T \qquad (12.3)$$

This is the compressive stress on the bar caused by its ends being fixed as its temperature changes. This stress is known as **thermal stress.**

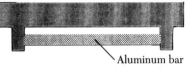
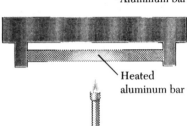

Figure 12.14
Internal stresses are created in a material when its temperature is increased while its ends are fixed. These are called thermal stresses.

Problem-Solving Tip

- In situations involving thermal stress, find the thermal expansion that would occur if the object were free to expand; then find the force and stress necessary to restore the object to its original length.

Example 12.5

A steel railroad rail 30 m long is firmly butted up against the adjoining rails. The rail is set in place on a cool day when the temperature is 10°C. What is the stress in it when the temperature reaches 40°C?

Reasoning

An increase in temperature would normally mean an increase in length, but there are no expansion joints. Therefore, if it is not to buckle, the expanded rail must be compressed to its original length. This necessary compressive stress is what we seek.

Solution

Young's modulus for steel is 20×10^{10} Pa (Table 10.1) and the coefficient of linear expansion is 12×10^{-6} (C°)$^{-1}$. We use these values in Equation 12.3 to determine the stress F/A:

$$F/A = \alpha Y \Delta T = [12 \times 10^{-6}(C°)^{-1}](20 \times 10^{10} \text{ Pa})(30 \text{ C°}) = 7.2 \times 10^{7} \text{ Pa}$$

Real-world effects of thermal expansion. The extreme heat of a summer day caused these railroad tracks in Asbury Park, New Jersey, to buckle. (*Wide World Photos*)

Figure 12.15
Heat is thermal energy that is transferred from one part of a system to another as a result of a temperature difference. The ice becomes warmer and melts while the water becomes cooler. The coffee cools down while the mug and spoon heat up.

The concept of heat has meaning *only* when there is a *transfer of internal energy*. A hot object doesn't "contain heat." We can talk about heat only when some of the internal energy that gives the object its high temperature is transferred to another object or to the surroundings. If there is no transfer of internal energy, there is no heat.

12.5 Heat

CONCEPT OVERVIEW

■ Heat is thermal energy that is exchanged between two objects as a result of a difference in their temperatures.

■ Heat always flows from a body at higher temperature to one at lower temperature.

The **internal energy,** or **thermal energy,** of an object is the total energy due to the motion of its molecules—kinetic energy due to rotation, vibration, and translation, and potential energy due to intermolecular interactions. A molecule has translational motion as it moves along a straight line with velocity **v.** Because it is made up of two or more atoms, it vibrates, and it can also rotate. All three types of motions occur simultaneously. When molecules collide with one another, energy is transferred from one to the other and also transformed from one form to another. As we shall see in a later chapter, the number of molecules in almost any sample is enormous. This means there are frequent collisions and frequent energy transfers. In any given sample, there are variations in energy—total energy and type of energy—from one molecule to another.

The internal energy of an object changes when work is done on the object. If you sand a piece of wood, it becomes warmer—its internal energy increases—because of the work you do on it. The internal energy also changes as thermal equilibrium is attained when the object is placed in thermal contact with another object at a different temperature. If ice is placed in a glass of room-temperature water the ice will become warmer and then melt while the temperature of the water decreases. The higher-temperature object loses internal energy (and its temperature decreases) and the lower-temperature object gains internal energy (and its temperature increases) until the two are in thermal equilibrium. This process, illustrated in Figure 12.15, is known as **heat transfer,** and heat is defined as follows:

Heat is internal energy that is transferred from one body to another as a result of a difference in temperature between the two bodies.

In Chapter 6, we used the work-energy theorem to relate work and macroscopic energy; this theorem is basically a statement of the conservation of energy and the equivalence of work and energy. External work W done on a system increases its total energy by an amount ΔE_{tot}:

$$W = \Delta E_{tot} = \Delta KE + \Delta PE$$

Now we can include the internal energy of the system E_{int} in our statement of conservation of energy. In addition to work from the outside, internal energy can come into the system through its contact with another system at a different temperature; this is called *heat* and is labeled Q. Our enhanced work-energy theorem then becomes

$$W + Q = \Delta E_{tot} = \Delta KE + \Delta PE + \Delta E_{int}$$

The work-energy theorem is general and can be expanded to include all other forms of energy. In all its forms, this theorem is also known as the **first law of**

thermodynamics; it will prove very useful in Chapter 14 when we study heat engines.

Heat, like work, has units of energy, that is, joules in the SI. However, many experiments were carried out on heat before scientists realized it was a form of energy. In this long process, other heat units were developed and are still used today. Probably the most common unit is the **calorie** (cal). One calorie is the amount of heat required to raise the temperature of 1 g of water from 14.5° to 15.5°C. As with all other units, a **kilocalorie** can be defined (1 kcal = 10^3 cal). One kilocalorie is the amount of heat required to raise the temperature of 1 kg of water from 14.5° to 15.5°C. The "Calorie" used to describe the energy available in foods is really this kilocalorie and is usually written with a capital C to make this distinction. A British thermal unit (Btu) is the amount of heat required to raise the temperature of 1 lb of water from 63° to 64°F. Of course, all these units are now defined in terms of joules:

$$1 \text{ cal} = 4.186 \text{ J}$$
$$1 \text{ Btu} = 1054.8 \text{ J}$$

These conversions, and the associated idea that heat and work are two forms of energy transfer, are known as the **mechanical equivalent of heat.** Work—as in rubbing two sticks together—can cause heat.

12.6 Specific Heat

CONCEPT OVERVIEW

■ The specific heat of a substance is the amount of heat needed to raise the temperature of 1 kg by 1 C°.

The amount of heat needed to increase the temperature of an object depends upon what substance the object is made of as well as on the mass of the object. The **specific heat** of a substance is defined as the amount of heat that must be added to 1 kg of the substance to raise its temperature 1 C°. That is, if an amount of heat Q is added to mass m of a substance and raises its temperature by ΔT, then the material has a specific heat c given by

$$c = \frac{Q}{m\Delta T} \qquad (12.4)$$

Equivalently,

$$Q = cm\Delta T \qquad (12.5)$$

That is, if the specific heat is known for a substance and we want to raise the temperature of a mass m of the substance by ΔT, the amount of heat Q given by Equation 12.5 must be added. If we want to lower the temperature of this mass m by ΔT, an amount of heat Q must be removed.

As you can see from Equations 12.4 and 12.5, specific heat has units of J/(kg · C°). Since the calorie is also commonly used as a heat unit, you may encounter specific heat given in units of cal/(g · C°). And, in British engineer-

TABLE 12.2 Specific Heats

Specific Heat c (J/kg · C°)

Solids		Liquids		Gases (at constant pressure)	
Sodium	123	Bromine	450	Carbon dioxide	830
Lead	130	Benzene	1740	Oxygen	915
Gold	130				
Silver	234				
Copper	385	Ethylene glycol	2380	Dry air	1000
Steel	450	Ethanol	2480	Nitrogen	1040
Sand	820	Methanol	2520	Steam	2020
Glass	840	Water	4186	Helium	5240
Brick	840			Hydrogen	14 200
Salt (NaCl)	880				
Concrete	880				
Aluminum	900				
Wood	1800				
Beryllium	1820				
Ice	2090				
Human body	3470				
Iron	470				

ing units, specific heat is measured in Btu/(slug · F°) or Btu/(lb · F°). Table 12.2 lists specific heats for various materials.

Example 12.6

A typical 40-gal home water heater contains about 155 kg of water. If the water comes in from the city main at 10°C, how much heat must be added to raise its temperature to 45°C?

Reasoning

The heat that must be added to something to raise its temperature a given amount depends upon the specific heat of the material the something is made of, its mass, and the change in temperature. We can use Equation 12.5.

Solution

$$Q = cm\Delta T = (4186 \text{ J/kg} \cdot \text{C°})(155 \text{ kg})(35 \text{ C°})$$
$$= 2.3 \times 10^7 \text{ J}$$

***

Practice Exercise   How much heat must be added to raise its temperature to 60°C?

Answer 3.2×10^7 J

12.7 Change of State

CONCEPT OVERVIEW

■ Most materials change state at a single, fixed temperature.

■ In any change of state, heat is either absorbed or given up. The ratio of this heat to the mass of the substance involved is called either the heat of fusion or the heat of vaporization of the substance.

Heat added to a solid causes the atoms or molecules of the solid to vibrate more vigorously; we see this microscopic change as a macroscopic temperature increase. Eventually the vibration becomes great enough to break some of the bonds that hold the solid together. At that point the solid melts and turns into a liquid. If heat is added to the resulting liquid, the atoms or molecules move even faster, increasing the temperature further. Eventually, the bonds holding the liquid together are broken, and the liquid vaporizes to a gas.

Thus we see that heat can cause a **change of state**—melting is a change from a solid to a liquid state, and vaporizing is a change from the liquid to a gas (or vapor).[1]

For most materials, at a particular pressure, a change of phase occurs at a single temperature. When we add heat to a solid, its temperature increases until the melting point is reached. Then, even though we continue to add heat, the temperature remains constant until all the solid has become liquid. All the heat now being added goes into performing this change of state. The heat required to melt 1 kg of the solid is called the **heat of fusion** of the material.

If we continue to add heat once all our solid has melted, the temperature of the liquid increases until the boiling point is reached. Then, just as before, even though we continue to add heat, the temperature remains constant until all the liquid has vaporized. The heat required to vaporize 1 kg of the liquid is called the **heat of vaporization** of the material. Both heats of fusion and heats of vaporization have units of joules per kilogram (J/kg). Some typical values are given in Table 12.3.

Because melting and solidifying are opposite processes, the same heat of fusion values apply to both. Thus, seeing that the heat of fusion for iron is 2.7×10^5 J/kg (Table 12.3), we know that 270 kJ of heat must be added to melt 1 kg of solid iron, and we also know that 270 kJ of heat must be removed from 1 kg of liquid iron in order to solidify it. (An outdated term for *to solidify* is *to fuse*, as in "molten iron fuses into the solid phase." This outdated term is where the name heat of fusion comes from.)

Because condensing and vaporizing are opposite processes, the same heat of vaporization values apply to both. Thus Table 12.3 tells us both that 854 kJ of heat must be added to vaporize 1 kg of liquid ethanol and 854 kJ of heat must be removed from 1 kg of ethanol gas in order to condense it to the liquid phase.

Freeze and *solidify* mean the same thing. We say "freeze" when a substance that is liquid at room temperature is chilled to become a solid, and "solidify" when a substance that is solid at room temperature but has been heated to melting returns to the solid phase.

---

[1] The word *gas* is sometimes taken to mean a substance that is in the gaseous phase under standard conditions—0°C and 1 atm—while *vapor* is reserved for a substance that is liquid or solid under those conditions. With those definitions, there may be water *vapor* in the air, but there is oxygen *gas* and nitrogen *gas* in the air. We do not need to worry about such distinctions here, however, and instead use the two terms interchangeably.

TABLE 12.3 **Heats of Fusion and Vaporization**

Material	Melting/Freezing Temperature (°C)	Heat of Fusion L_f ($\times 10^5$ J/kg)	Boiling/ Condensing Temperature (°C)	Heat of Vaporization L_v ($\times 10^6$ J/kg)
Helium	—	—	−269	5.97
Hydrogen	−259.31	0.586	−252.89	0.452
Oxygen	−218.83	0.138	−182.97	0.213
Nitrogen	−210.10	0.257	−195.81	0.201
Ethanol	−114	10.42	78	0.854
Water	0.00	3.33	100.00	2.256
Aluminum	658.7	3.984	2300	10.5
Copper	1083	1.34	2600	3.1
Iron	1530	2.721	2500	6.364
Gold	1063.00	0.644	2660	1.58

Heats of fusion and vaporization seem to be hidden inside a material, in the sense that we add heat but see no "proof" in the form of temperature increase. Therefore the early name was *latent* heat of fusion and *latent* heat of vaporization, *latent* being from the Latin word for "hidden." Thus the symbols L_f and L_v.

We do not always deal with exactly 1 kg of a material, of course, and therefore need a general expression for heats of fusion and vaporization. For any mass m of a material, the heat required for a change of phase from liquid to solid or from solid to liquid is

$$Q_f = m L_f \qquad (12.6)$$

and the heat required for a change of phase from vapor to liquid or from liquid to vapor is

$$Q_v = m L_v \qquad (12.7)$$

where L_f is the heat of fusion of the material and L_v is its heat of vaporization.

Figure 12.16
Heat added to a material causes its temperature to rise except at the boiling and melting points. At these two temperatures, all the added heat goes into changing the phase of the material. The heat is called heat of vaporization for the liquid-to-gas change and heat of fusion for the solid-to-liquid change.

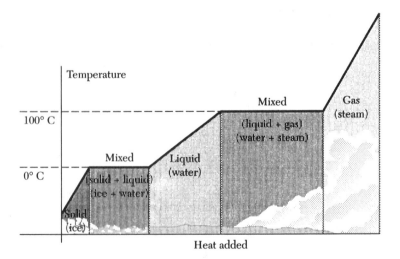

Figure 12.16 illustrates what happens when we start with a 1-kg block of ice below 0°C and add heat to it. As heat is added, the temperature of the ice increases until $T = 0°C$. The specific heat of ice is 2090 J/kg · C°, and therefore the slope of this part of the line is (1/2090) C°/J. At 0°C the ice begins to melt, and the temperature remains constant until all the ice has turned to water. From Table 12.3 we see that 3.33×10^5 J of heat must be added to melt the entire kilogram of ice before there is any further increase in temperature. Heat added after all the ice has melted causes the water's temperature to rise. The specific heat of water is 4186 J/kg · C°, and therefore the slope of this part of the line is (1/4186) C°/J. At 100°C the water begins to vaporize, and the temperature remains constant until all the water has turned to steam. From Table 12.3, we see that 2.256×10^6 J of heat must be added to vaporize the entire kilogram of water before there is any further increase in temperature. Heat added once all the water has become steam causes the temperature of the steam to rise (such steam is called "superheated steam").

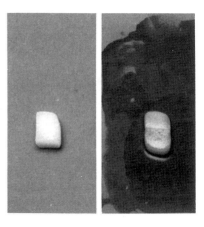

Change of state in the two types of ice. Dry ice (*left*), which is frozen CO_2, sublimes, leaving no puddle as the solid changes directly to a gas. Under normal conditions, water ice (*right*) passes through the liquid state, then evaporates. (© *David Rogers*)

Example 12.7

A 100-g ice cube is taken from the freezer at $-10°C$. How much heat must be added to it to end up with 100 g of water at 15°C?

Reasoning

We must think of this as a three-step process. First, how much heat is required to change the temperature of the ice cube from $-10°$ to $0°$? Then—step 2—how much heat is required to melt all the ice to water at $0°$? Finally,—step 3—how much heat is required to raise the 100 g of water from $0°$ to 15°C?

Solution

From Table 12.2, we find that the specific heat of ice is 2090 J/kg · C°. We use that in Equation 12.5:

$$Q_1 = cm\Delta T = (2090 \text{ J/kg} \cdot \text{C°})(0.100 \text{ kg})(10 \text{ C°})$$
$$= 2090 \text{ J}$$

At $0°$ the ice melts with the input of additional heat. For this we need Equation 12.6:

$$Q_2 = mL_f = (0.100 \text{ kg})(3.33 \times 10^5 \text{ J/kg}) = 33\,300 \text{ J}$$

$$Q_3 = cm\Delta T = (4186 \text{ J/kg} \cdot \text{C°})(0.100 \text{ kg})(15 \text{ C°}) = 6\,280 \text{ J}$$

The total heat required is the sum of these three:

$$Q_{tot} = Q_1 + Q_2 + Q_3 = 41\,670 \text{ J} \approx$$

Example 12.8

Consider a 0.80-kg cup of water initially at 15°C placed on a stove. How much heat has been absorbed by the time all the water has boiled away as steam?

Reasoning

As with Example 12.6, we must take this in steps. First, how much heat is required to change the temperature of the water from 15 to 100°? At 100°C, how much heat is needed to turn all the water to steam?

Solution

From Table 12.2, we find that the specific heat of water is 4186 J/kg · C°. We can use Equation 12.5:

$$Q_1 = cm\Delta T = (4186 \text{ J/kg} \cdot \text{C°})(0.80 \text{ kg})(85 \text{ C°}) = 0.28 \times 10^6 \text{ J}$$

At 100°C the water boils with the input of additional heat. To determine how much heat is needed, we use Equation 12.7:

$$Q_2 = mL_v = (0.80 \text{ kg})(2.256 \times 10^6 \text{ J/kg}) = 1.8 \times 10^6 \text{ J}$$

The total heat required

$$Q_{tot} = Q_1 + Q_2 = 2.1 \times 10^6 \text{ J}$$

12.8 Determining the Equilibrium Temperature

CONCEPT OVERVIEW

■ Heat flows from hotter objects to colder objects, changing both their temperatures until they reach a common equilibrium temperature.

As we learned in Section 12.2, if you pour cold milk into your morning cup of hot coffee, heat will flow from the coffee to the milk, and the milk-coffee system will come to an equilibrium temperature that is somewhere between the initial low temperature of the milk and the initial high temperature of the coffee. Let us now see how to calculate this equilibrium temperature.

As with any other system, energy is conserved in the milk-coffee system. Since we assume no motion and no change in position, there is no kinetic or potential energy to be concerned with; there is only internal energy to be conserved. Therefore if the milk gains heat Q_1 and the coffee loses heat Q_2, we can say

$$Q_1 + Q_2 = 0 \tag{12.8}$$

In working with internal heat changes, we will use the following sign convention. Heat that is absorbed by a body is positive; it causes either an increase in temperature or else melting or boiling. Heat that is given up by a body is negative; it causes either a decrease in temperature or else freezing or condensation. (Saying that the coffee loses heat Q_2 is equivalent to saying that $Q_2 < 0$.)

Heat given up by part of the system is absorbed by another part. This assumes, of course, that the system is isolated (or insulated) from its environment, an assumption we shall make in the situations we investigate.

Problem-Solving Tips

- Problems involving equilibrium temperature are energy-conservation problems, with the heat lost by one part of the system being gained by another part.
- The equilibrium temperature is always somewhere between the lowest and highest initial temperatures of the system.

Example 12.9

Pour 0.10 kg of 5.0°C milk into 0.80 kg of coffee at 95°C. What is the equilibrium temperature?

Reasoning

We know the equilibrium temperature will be between 5° and 95°C. We can take the specific heats of milk and coffee to be the same as that of water, 4186 J/kg · C°. The heat absorbed by the milk equals the heat given up by the coffee as they both reach an equilibrium temperature T_f.

Solution

If the equilibrium temperature is T_f, then the heat absorbed by the milk is

$$Q_{milk} = cm\Delta T$$
$$= (4186 \text{ J/kg} \cdot \text{C}°)(0.10 \text{ kg})(T_f - 5.0°C)$$

This heat value is positive because the temperature change $(T_f - 5.0°C)$ is positive.

The heat given up by the coffee is

$$Q_{coffee} = cm\Delta T$$
$$= (4186 \text{ J/kg} \cdot \text{C}°)(0.80 \text{ kg})(T_f - 95°C)$$

This heat value is negative because the temperature change $(T_f - 95°C)$ is negative. Remember that temperature differences, like all other differences, are always final value minus initial value: $\Delta T = T_f - T_i$.

If we add no heat from outside nor let any escape to the outside, then the sum of Q_{milk} and Q_{coffee} must be zero:

$$Q_{milk} + Q_{coffee} = 0$$
$$(4186 \text{ J/kg} \cdot \text{C}°)(0.10 \text{ kg})(T_f - 5.0°C) +$$
$$(4186 \text{ J/kg} \cdot \text{C}°)(0.80 \text{ kg})(T_f - 95°C) = 0$$

$$T_f = 85°C$$

Practice Exercise   Pour 0.10 kg of 15°C milk into 0.60 kg of coffee at 95°C. What is the equilibrium temperature?

Answer   84°C

This method can be used for any system composed of two or more parts having different initial temperatures (provided there is negligible expansion of the system so that $W = 0$). The final temperature always lies somewhere between the lowest and highest initial temperatures. Some care—and perhaps even trial and error—may be needed when a change of phase is involved. To understand how the procedure works, look at the following two examples.

Problem-Solving Tips

- Any possible change of phase must be accounted for by a term in Equation 12.8.
- Be sure to include a term for the heat required to bring part of the system to the temperature necessary for a change of phase to occur.

Example 12.10

Drop 0.200 kg of ice initially at -10.0°C into 0.500 kg of water initially at 40.0°C. What is the final temperature?

Reasoning
The final temperature will be between -10° and 40°C, but the calculations will differ depending on whether or not all the ice melts. Let us assume the ice melts completely so that the final temperature is between 0° and 40°C. (If we are wrong, we shall find out soon enough.) The ice will gain heat in coming up to 0°C, in melting, and in coming up to the final temperature.

Solution

$$Q_{ice} + Q_{water} = 0$$
$$Q_{ice} = Q_1 + Q_2 + Q_3$$

where Q_1 is the heat required to bring the ice to 0° (Equation 12.5), Q_2 is the heat needed to melt all the ice at 0° (Equation 12.6), and Q_3 is the heat required to bring the resulting water from 0° to T_f (Equation 12.5)

$$Q_1 = cm\Delta T$$
$$= (2090 \text{ J/kg} \cdot \text{C}°)(0.200 \text{ kg})(10.0 \text{ C}°) = 4180 \text{ J}$$

$$Q_2 = mL_f$$
$$= (0.200 \text{ kg})(3.33 \times 10^5 \text{ J/kg}) = 66\,600 \text{ J}$$

$$Q_3 = cm\Delta T$$
$$= (4186 \text{ J/kg} \cdot \text{C}°)(0.200 \text{ kg})(T_f - 0°C)$$
$$= 837 T_f \text{ J/C}°$$

$$Q_{ice} = 4180 \text{ J} + 66\,600 \text{ J} + 837 \, T_f \text{ J/C}° = 70\,800 \text{ J} + 837 \, T_f \text{ J/C}°$$

Now, to calculate the heat given up by the water:

$$Q_{water} = cm\Delta T$$
$$= (4186 \text{ J/kg} \cdot \text{C}°)(0.500 \text{ kg})(T_f - 40.0°C)$$
$$= 2090 \, T_f \text{ J/C}° - 83\,700 \text{ J}$$

Now we are ready to solve our original energy-conservation equation:

$$Q_{ice} + Q_{water} = 0$$
$$(70\,800 \text{ J} + 837 \, T_f \text{ J/C}°) + (2090 \, T_f \text{ J/C}° - 83\,700 \text{ J}) = 0$$
$$T_f = 4.4°C$$

Practice Exercise Drop 0.10 kg of ice, initially at $-20°C$, into 0.75 kg of water, initially at 50°C. What is the final temperature?

Answer 34°C

What if we start with more ice, say 0.400 kg, and do the same analysis? Small changes in initial conditions can lead to surprising changes in the final result. Notice that in the following example.

Problem-Solving Tip

- When several possibilities are available, you must make an assumption and then check to see if your results are consistent with your assumption. If they are not, make another assumption and check your results again.

Example 12.11

This time drop 0.400 kg of ice initially at $-10.0°C$ into 0.500 kg of water initially at 40.0°C. What is the final temperature?

Reasoning

Because this seems just like the previous example, we start our calculations in exactly the same manner.

Solution

$$Q_{ice} + Q_{water} = 0$$
$$Q_{ice} = Q_1 + Q_2 + Q_3$$

$$Q_1 = cm\Delta T$$
$$= (2090 \text{ J/kg} \cdot \text{C}°)(0.400 \text{ kg})(10.0 \text{ C}°) = 8360 \text{ J}$$

$$Q_2 = mL_f$$
$$= (0.400 \text{ kg})(3.33 \times 10^5 \text{ J/kg}) = 133\,000 \text{ J}$$

$$Q_3 = cm\Delta T$$
$$= (4186 \text{ J/kg} \cdot \text{C}°)(0.400 \text{ kg})(T_f - 0°\text{C}) = 1670\,T_f \text{ J/C}°$$

$$Q_{\text{ice}} = 141\,000 \text{ J} + 1670\,T_f \text{ J/C}°$$

Because the data for the water are exactly the same as in Example 12.10, we can take our value for Q_{water} from there:

$$Q_{\text{water}} = 2090\,T_f \text{ J} - 83\,700 \text{ J}$$

Now we are ready to apply energy conservation:

$$Q_{\text{ice}} + Q_{\text{water}} = 0$$
$$(141\,000 \text{ J} + 1670T_f \text{ J/C}°) + (2090T_f \text{ J/C}° - 83\,700 \text{ J}) = 0$$
$$57\,300 \text{ J} + 3760T_f \text{ J/C}° = 0$$
$$T_f = -15.2°\text{C}$$

Does this answer make sense? No, because it is lower than the lowest initial temperature, $-10.0°$C. A final temperature of $-15.2°$ means the whole system has turned to ice, a result that contradicts our initial assumption. When we used the Q_3 term in our ice equation, we were saying that the ice melts completely and the water it forms is heated to some temperature above $0°$. Arriving at a nonsensical answer tells us that this assumption was wrong. Therefore we need to try again, but this time let us assume that not all the ice melts—that only mass M_{melt} melts. Then the final temperature will be $0°$, and we can solve for M_{melt}.

$$Q_{\text{ice}} + Q_{\text{water}} = 0$$
$$Q_{\text{ice}} = Q_1 + Q_2$$

$$Q_1 = 8360 \text{ J} \quad \text{(as before)}$$

$$Q_2 = M_{\text{melt}} L_f = M_{\text{melt}}(3.33 \times 10^5 \text{ J})$$

$$Q_{\text{ice}} = 8360 \text{ J} + M_{\text{melt}}(3.33 \times 10^5 \text{ J})$$

where we have assumed that M_{melt} is measured in kilograms.

$$Q_{\text{water}} = cm\Delta T$$
$$= (4186 \text{ J/kg} \cdot \text{C}°)(0.500 \text{ kg})(-40.0 \text{ C}°) = -83\,700 \text{ J}$$
$$Q_{\text{ice}} + Q_{\text{water}} = 0$$
$$[8360 \text{ J} + M_{\text{melt}}(3.33 \times 10^5 \text{ J})] + (-83\,700 \text{ J}) = 0$$
$$M_{\text{melt}} = 0.226 \text{ kg} = 226 \text{ g}$$

In many equilibrium-temperature problems, it may not be obvious at first what assumptions about the final state of the system you should make. Make an assumption and then check to see if your results are consistent with that assumption.

SUMMARY

Temperature is a measure of how hot or cold an object is. The Celsius temperature scale has 0°C for the temperature of melting ice and 100°C for the temperature of boiling water.

The pressure in a constant-volume gas thermometer is linearly related to temperature. If we extrapolate the curve on a pressure-versus-temperature graph to find the temperature at which the pressure vanishes, the curve intersects the x axis, indicating zero pressure, at −273.15°C. This is the lowest possible temperature, and it is useful to define a new temperature scale, the Kelvin scale, with this temperature defined as zero—"absolute zero." On this scale, ice melts at 273.15 K and water boils at 373.15 K.

The zeroth law of thermodynamics states that any two bodies each separately in thermal equilibrium with a third body are in thermal equilibrium with each other.

A bar of original length L_o, when its temperature changes by a small amount ΔT, will change its length by ΔL, where

$$\Delta L = \alpha L_o \Delta T \qquad (12.1)$$

The coefficient of linear expansion α is different for different materials.

A material of original volume V_o, when its temperature changes by a small amount ΔT, will change its volume by ΔV, where

$$\Delta V = \beta V_o \Delta T \qquad (12.2)$$

The coefficient of volume expansion β is equal to three times the coefficient of linear expansion α.

The internal energy of an object (also called the thermal energy) is the energy due to the motion of its atoms or molecules. Heat is internal energy that is transferred as a result of a difference in temperature between two systems. Heat, like work, has the same units as energy and is measured in joules in the SI. The units calorie and Btu are still commonly used to describe heat.

The specific heat c of a material is the amount of heat that must be added to a unit mass to raise its temperature by one Celsius degree. For any mass m and temperature change ΔT and heat Q, we can define the specific heat c by:

$$c = \frac{Q}{m\Delta T} \qquad (12.4)$$

The amount of heat required to cause a unit mass of material to melt or to solidify (fuse) is known as the heat of fusion L_f. For any mass m, the heat Q_f required for this change of phase is

$$Q_f = mL_f \qquad (12.6)$$

The amount of heat required to cause a unit mass of material to vaporize or condense is known as the heat of vaporization L_v. For any mass m, the heat required for this change of phase is

$$Q_v = mL_v \qquad (12.7)$$

The heat released by one part of an insulated system is equal to the heat absorbed by another part of the system:

$$Q_1 + Q_2 = 0 \qquad (12.8)$$

QUESTIONS

1. How can two objects have the same temperature without having the same internal energy?
2. If an object is sitting still, with its center of mass having zero velocity, the average velocity of its molecules must be zero. Yet the average speed of its molecules is very high and gets even higher as its temperature increases. How can this be?
3. A steel measuring tape is accurately marked in the factory, where the temperature is 22°C, but is used on a summer day when the temperature

is 40°C. Explain any inaccuracies this difference in temperature might cause.

4. A steel measuring tape is accurately marked in the factory, where the temperature is 22°C, but is used on a winter day when the temperature is −5°C. Explain any inaccuracies this difference in temperature might cause.

5. How can running hot water over the metal lid on a glass jar make the lid easier to remove?

6. Oil pipelines and long steam pipes have expansion loops often shaped like a U or an Ω at regular intervals (Fig. 12.17). What purpose do these loops serve?

Figure 12.17
Question 12.6. An expansion loop (*right*) in the Trans-Alaska pipeline. (*© Cara Moore/The Image Bank*)

7. Why does an ordinary glass dish usually break when set on a hot stove? What characteristic of Pyrex glass prevents this breaking?

8. The balance wheel of a mechanical watch governs the frequency with which the watch ticks. A wheel cut from a single piece of metal would expand as its temperature increased, increasing its moment of inertia, and causing the watch to slow down. How might you design a balance wheel to keep more accurate time? (*Hint:* Use two kinds of metal with different coefficients of linear expansion arranged so that as the temperature changes and the metals expand, they continue to expand in a circle of constant radius so the moment of inertia remains constant.)

9. Explain how a bimetallic strip (Fig. 12.18) can be used in a thermostat to turn a furnace on and off.

Figure 12.18 Question 12.9.

10. A metal collar fits very tightly on a shaft. What could you do to collar and shaft to remove the collar?

11. Why might you damage the overheated engine of your car by quickly adding cold water to it?

12. Why do sidewalks and highways sometimes buckle? Does such buckling occur in winter or summer?

13. Will 1 kg of iron at 100°C and 1 kg of water at 0°C have an equilibrium temperature of 50°C? Explain your answer.

14. Explain how you might add heat to a system without increasing its temperature.

15. Why does your skin feel cool as moisture evaporates from it?

16. How does an ice cream freezer work? How can ice near 0°C and salt reduce the temperature of the ice cream mixture to well below 0°C?

17. Why can a burn caused by steam be so much worse than a burn caused by very hot water?

18. Blacksmiths often lower the temperature of a horseshoe rapidly by immersing it in water. Why does this work?

19. What is the temperature of your lemonade if it is well stirred and has ice cubes floating in it?

20. Do vegetables cook faster in vigorously boiling water than in water that is just barely boiling? Explain your answer.

PROBLEMS

12.1 Temperature

12.1 If room temperature is 70° on the Fahrenheit scale, what is it on the Celsius scale?

12.2 If you have a fever of 102° on a Fahrenheit thermometer, what is your temperature on a Celsius thermometer?

12.3 If a patient has a fever of 41° on a Celsius thermometer, what is her temperature on a Fahrenheit thermometer?

12.4 A mountain climber finds that water boils at 87°C. What temperature is that on the Fahrenheit scale?

12.5 What is normal body temperature, 37°C, on the Kelvin scale?

12.6 What is a typical room temperature, 72°F, on the Kelvin scale?

○12.7 In a constant-volume gas thermometer, the pressure is 45 torr at 0°C and 61.4 torr at 100°C. What is the temperature when the pressure is 75 torr?

○12.8 In a constant-volume gas thermometer, the pressure is 37 torr at 0°C and 53 torr at 100°C. What is the pressure when the temperature is 23°C?

○12.9 For what temperature will a Celsius thermometer and a Fahrenheit thermometer read the same numerical value?

12.2 Equilibrium Temperature and the Zeroth Law of Thermodynamics

12.3 Thermal Expansion

12.10 On a concrete road, how large should the expansion joints be between sections if each section is 15 m long? Consider a temperature range of −10° to 35°C.

12.11 How large should the expansion joints be between steel bridge girders (Fig. 12.19) that are each 20 m long? Consider a temperature range of −10° to 35°C.

12.12 Steel rails 18 m long are laid on a winter day when the temperature is −12°C. How much space must be left between the rails to allow for expansion on a hot summer day when the rails are at 40°C?

Figure 12.19 Problem 12.11.
Expansion joint in a bridge. *(Edward M. Wheeler)*

12.13 An aluminum power line is 100.0 km long when the temperature is 10°C. How long is it when the temperature is 40°C?

12.14 A copper telephone wire is 35.0 km long when the wire is strung on a 40°C day. How long is it when the temperature is 0°C?

12.15 The Trans-Alaskan pipeline is 1300 km long, reaching from Prudhoe Bay to the port of Valdez, and experiences temperatures from −73° to +35°C. How much does the steel pipeline expand due to this difference in temperature? How can this expansion be compensated for?

12.16 A radiator in the attic of a house is connected to the heating system in the basement by a straight, vertical iron pipe 11.5 m long. If the radiator rests snugly on the floor when the pipe is at 5°C, how high off the floor will it be lifted when the pipe contains steam at 100°C? Neglect any expansion of the house.

12.17 The Sears Tower in Chicago is the world's tallest office building at 443 m. It is constructed of concrete attached to an inner steel frame. How much does its height vary when the temperature changes from 38°C to −32°C?

12.18 Steel railroad rails 18 m long were formerly the standard length used. What size should the gap between sections be to allow for expansion over a temperature range from $-10°$ to $40°C$? Some railroads now use 800-m sections with special joints. What size must the expansion gaps be for these new rails?

12.19 What is the coefficient of volume expansion of a metal whose coefficient of linear expansion is 15×10^{-6} $(C°)^{-1}$?

12.20 A motorist purchases 40 l of gasoline when the temperature is $-20°C$. What is the volume of this gasoline after it has warmed up to $20°C$?

○12.21 An aluminum cup is brim-full, containing 200 cm^3 of glycerin at $20°C$. What volume of glycerin overflows when the cup is placed in a $125°C$ oven?

○12.22 To create a snug fit, steel rivets are sometimes cooled with dry ice (frozen carbon dioxide, CO_2) before being inserted into holes in steel plates. A steel rivet that has a diameter of 2.125 cm at $20.00°C$ is to be put into a hole that has a diameter of 2.122 cm at $20.00°C$. How cold must the rivet be before it will fit?

○12.23 In assembling a steel bridge on a day when the air temperature is $22.0°C$, a steel rivet 3.803 cm in diameter is to be placed in a hole 3.800 cm in diameter. Would it be possible to do the job by cooling the rivet in dry ice $(-78°C)$?

○12.24 A disk has moment of inertia I_o. If its temperature increases by ΔT, show that its moment of inertia increases by $\Delta I = 2\alpha I_o \Delta T$.

○12.25 The pendulum of a grandfather clock has a period of 2.00 s at $20.00°C$. The pendulum consists of a small, heavy bob on a thin brass rod. At $20.00°C$, the clock keeps accurate time. (a) What will the period be if the temperature rises to $30.00°C$? (b) At the end of one day at this higher temperature, how many seconds will the clock have gained or lost?

○12.26 If a steel gauge block is carefully machined to 10 cm $\pm$ 10^{-6} cm, how precisely must its temperature be controlled to keep this accuracy?

○12.27 A steel sphere has a diameter of 3.005 cm at $25.00°C$, and a brass ring has an inside diameter of 3.000 cm at $25.00°C$. To what temper-

ature must the ring be heated to allow the sphere to just pass through it? If both ring and sphere are heated, to what temperature must they be raised so the sphere can just pass through the ring?

○12.28 A 2000-cm^3 quartz beaker is filled to the brim with turpentine at $23°C$. If the beaker and contents are put into a refrigerator so that their temperature becomes $5.0°C$, what is the volume of the turpentine? Ignore the small contraction of the beaker.

●12.29 A steel surveying tape is carefully calibrated at $20.0°C$. However, it is used on a $38.0°C$ summer day. Are the distances measured too large or too small? What is the percentage of error introduced by this temperature change?

●12.30 A metal plate has length l_o and width w_o. It is heated through a temperature change ΔT. Show that its area changes by an amount approximately equal to $\Delta A = 2\alpha A_o \Delta T$.

●12.31 In the text we showed that the coefficient of volume expansion is three times the coefficient of linear expansion by looking at the expansion of a cube. Carry out this analysis by looking at a rectangular solid with length l_o, width w_o, and height h_o.

12.4 Thermal Stress

12.32 If concrete roadway sections are poured butted against each other when the temperature is $15°C$, what will the thermal stress be when the temperature reaches $40°C$? Young's modulus for concrete is about 20×10^9 Pa.

12.33 If steel girders in a bridge are set in place without expansion joints when the temperature is $10°C$, what will the thermal stress be when the temperature reaches $40°C$?

○12.34 A steel rail 20.0 m long at $-10.0°C$ is firmly clamped between two fixed supports 20.0 m apart. What is the compressive force in the rail when its temperature rises to $35°C$. The rail's cross-sectional area is 50.0 cm^2.

○12.35 A brass rod of length 78 cm and cross-sectional area 2.0 cm^2 is cooled from $100°$ to $30°C$. What force is needed to stretch it back to its original length?

●12.36 During an experiment measuring Young's modulus, weights were stretching a steel wire of length 1.05 m and cross-sectional area

2.00×10^{-7} m². During the experiment, room temperature dropped 10 C°. How did this temperature change affect the results of the experiment?

O 12.37 A band of iron is to be wrapped around a wagon wheel whose circumference at 15°C is 270.562 cm. At this temperature, the band has a cross-sectional area of 2.50 cm² and an inside circumference of 269.020 cm. (a) To what temperature must the band be heated to fit snugly on the wheel? (b) What tension will exist in the band after it cools to 15°C?

● 12.38 At 20°C, a steel rod of length 40.000 cm and a brass rod of length 30.000 cm, both of the same diameter, are placed end to end between two rigid supports 70.000 cm apart, with no initial stress in the rods. The temperature of the rods is then raised to 60°C. (a) What is the stress (the same in both rods) at the higher temperature? (b) How far from its old position is the new junction between the rods? (*Hint:* First find the total thermal expansion if the rods were free. Then find the stress by using the fact that the sum of the two compressions (due to stress) must equal the sum of the two expansions (due to temperature). Finally, find the contraction of each temperature-expanded rod due to the stress.)

12.5 Heat

12.39 How many calories are there in 1 Btu?

12.40 A home furnace is rated at 50,000 Btu/h. What is this power rating in kilowatts?

12.6 Specific Heat

12.41 A child's wading pool contains 1.2 m³ of water at 15°C. How much heat must be added to the pool to bring the temperature to 27°C?

12.42 A typical house contains about 500 m³ of air. How much heat is required to raise the temperature of that much air from 8.0° to 22°C?

12.43 A warmed brick used to be popular as a foot-warmer in horse-and-buggy days. How much heat does a 2.0-kg brick give up in going from 90° to 10°C?

12.44 A 0.250-kg steel camping cup is filled with boiling water. How much heat flows into the cup as its temperature goes from 20° to 95°C?

O 12.45 While on their honeymoon in the French Alps, James Prescott Joule and his bride carefully measured the temperature of water at the top of a waterfall and at the bottom. If the water fell a distance of 110 m and if all of its initial potential energy was converted to internal energy as a result of the friction forces involved in hitting the bottom, what temperature rise should the new Mr. and Mrs. Joule have found? (Actually, they found no rise because the water loses heat through evaporation.)

12.7 Change of State

12.46 Once its melting point is reached, how much heat is required to melt a 500-kg ingot of steel?

12.47 Once its melting point is reached, how much heat is required to melt a 0.500-kg bar of gold?

12.48 While heating 0.40 kg of coffee with a 300-W immersion heater (Figure 12.20), you get distracted by a morning news show. Once boiling starts, how long does it take to vaporize all the coffee? (If the immersion heater is left unattended that long, it will be damaged—probably melted.) Assume no heat is lost to the environment.

Figure 12.20 Problem 12.48.
A heating coil boils a mug of water. (© *Steve Nieddorf/The Image Bank*)

12.49 A 350-W electric immersion heater is placed in a cup containing 0.25 kg of water at 20°C. How long before the water starts to boil? How long until all the water has boiled away? Assume no heat is lost to the environment.

12.8 Determining the Equilibrium Temperature

12.50 If 0.60 kg of boiling hot coffee is poured into a 0.250-kg steel camping cup initially at 20°C, what is the final temperature of the cup-coffee system?

12.51 How much milk at 5.0°C should be added to 0.35 kg of coffee at 95°C to bring the mixture to 80°C?

12.52 A 0.200-kg metal block initially at 90.0°C is placed in a styrofoam cup containing 0.300 kg of water initially at 20.0°C. Eventually the system comes to an equilibrium temperature of 26.5°C. What is the specific heat of the metal? Assume no heat is lost to the environment.

○12.53 How much ice at −5.0°C should be added to 0.30 kg of tea (essentially water) at 95°C to bring the mixture to 15°C?

○12.54 If 0.20 kg of −10°C ice is put into 0.80 kg of 24°C lemonade (essentially water), what is the final temperature of the lemonade? Does any of the ice remain?

○12.55 If 0.25 kg of −15°C ice is put into 0.80 kg of 35°C tea (essentially water), what is the final temperature of the tea? Does any of the ice remain?

○12.56 After 0.25 kg of −15°C ice is put into 0.80 kg of 35°C tea (essentially water), the tea is stirred with a 0.10-kg steel spoon initially at 25°C. If the spoon is left in the glass, what is the final temperature of the tea? Does any of the ice remain?

●12.57 A 0.050-kg ice cube initially at −5.0°C is placed in 0.30 kg of water at 25°C. What is the final temperature of the water? Or, if

melting is not complete, how much ice remains unmelted at thermal equilibrium? Assume no heat is lost to the environment.

●12.58 A 0.15-kg ice cube initially at −5.0°C is placed in 0.30 kg of water at 25°C. What is the final temperature of the water? Or, if melting is not complete, how much ice remains unmelted at thermal equilibrium? Assume no heat is lost to the environment.

●12.59 A 0.15-kg ice cube initially at −15°C is placed in 0.30 kg of water at 25°C. What is the final temperature of the water? Or, if melting is not complete, how much ice remains unmelted at thermal equilibrium? Assume no heat is lost to the environment.

●12.60 An iron horseshoe is heated red hot to a temperature of 1500°C and then plunged into 5.0 kg of water at 28°C. The water and horseshoe eventually come to an equilibrium temperature of 35°C. What is the mass of the horseshoe? Assume no heat is lost to the environment.

General Problems

○12.61 The inside diameter of a steel lid and the outside diameter of a glass peanut butter jar are both exactly 11.50 cm at room temperature, 21.0°C. If the lid is stuck and you run 80.0°C hot water over it until the lid and the jar top both come to 80.0°C, what will the new diameters be?

○12.62 A brass cannon has a bore 10.000 cm in diameter at 20°C. A steel liner, in the form of a thin cylinder of outside diameter 10.020 cm at 20°C, is to be inserted. (a) To what temperature should the cannon be heated so that the liner, still at 20°, can be inserted? (b) To what temperature should cannon and liner be heated so that the liner becomes loose and can be removed?

○12.63 A mercury thermometer has a quartz bulb of volume 0.300 cm³ and a stem that has a bore of 0.0100 cm. How far does the mercury column move when the temperature changes from 30.0 to 40.0°C? Ignore any expansion of the bulb.

O 12.64 How long does it take a coffee-maker that uses a 600-W heater to heat 1.2 kg of water from 15° to 90°C? Assume that none of the heat escapes to the environment.

● 12.65 A 0.60-kg horseshoe is heated until it glows red at 1500°C and then is placed in 1.2 kg of water. The initial temperature of the water is 30°C. What is the final temperature of the

horseshoe-water system? Or, if boiling occurs, how much of the water will be left once thermal equilibrium is attained? Assume that none of the heat escapes from the water.

● 12.66 A 10°C ice cube is placed in 0.30 kg of 25°C water. The ice melts, and the equilibrium temperature of the system is 12°C. What was the mass of the ice cube? Assume that no heat is lost to the environment.

PHYSICS AT WORK

Sailplanes or gliders are unpowered aircraft with narrow fuselages and long wings that can stay aloft for hours riding thermal updrafts. Heat transfer or *thermal convection* in the atmosphere creates updrafts sought by sailplane pilots. These rising currents of warm air are often caused by features of the landscape. Freshly plowed fields, for example, tend to absorb more heat from the Sun than their surroundings do. The air over these fields then becomes warmer and rises, causing the updraft that allows a sailplane to gain altitude without any other power source. What other landforms can produce strong convective air currents favorable to soaring?

When rising air currents or columns of air reach an altitude at which condensation occurs, the water vapor in the air condenses as visible clouds (often cumulus clouds). Knowing this, sailplane pilots watch for the first wisps of forming cloud and pursue them as they mark the location of otherwise invisible updrafts.

Most sailplanes are towed into the air by a powered aircraft and then released. Control surfaces on the wings of the sailplane allow the pilot to bank, dive, climb, change direction, and ultimately land safely. Why do sailplanes use long, thin wings rather than shorter, wider wings to provide for lift (remember Bernoulli's principle)?

(© Toyohiro Yamada/FPG International Corporation)

13

Heat Transfer

- Heat may be transferred by conduction, convection, and radiation. The handle of a metal skillet placed on a stove gets hot by conduction as the random, thermal motion of individual molecules and free electrons is passed from one to another by individual collisions. A Franklin stove heats an entire room by convection because the warm air around it rises and cooler air moves in to replace it. The Sun warms our Earth through radiation as the energy in infrared (IR) radiation (long wavelength light waves) passes through the vacuum of space from the Sun to our Earth.

- Glass is transparent to visible light, allowing the energy in visible light to pass through and warm a room or the inside of a greenhouse. The objects inside a room or greenhouse radiate energy in the form of IR radiation. Glass is opaque to this IR radiation and does not allow it to pass through. This means the heat is trapped inside the room or greenhouse. This is known as the "greenhouse effect." Carbon dioxide (CO_2) and other gases in our atmosphere act in a similar manner controlling Earth's temperature.

Heat is internal energy that is transferred as a result of a difference in temperature, and *how* that energy is transferred is the topic of this chapter. Heat transfer has very important applications and consequences—from burning your hand on a skillet to heating your home in winter to keeping it cool on a summer day. So important are the latter two applications that ideas involving heat transfer are starting to be incorporated into building design.

13.1 Conduction

CONCEPT OVERVIEW

- In conduction, heat is transferred from one atom to the next through a material known as a thermal conductor.

- When heat is conducted from one region to another, the heat transfer rate is proportional to the difference in temperature between the two regions.

Place a metal skillet on the stove, as in Figure 13.1, and the handle soon becomes hot even though it is not directly in contact with the burner. Stir hot soup with a metal spoon and the spoon soon becomes hot all along its length even though only the bowl end is in contact with the soup. These are examples of heat transfer by **conduction,** and the metal of skillet and spoon is called a thermal **conductor.** Heat is conducted along the metal from atom to atom through interactions between the atoms. In conduction, there is no movement of the conducting material.

You can think of the atoms of the conducting material as being held together by imaginary springs that allow the atoms to vibrate. The faster-vibrating atoms in the hot part of the skillet and the spoon collide with their slower-vibrating neighbors, and during the collisions some kinetic energy is transferred from the faster atoms to the slower ones. These latter atoms, now moving faster (indicating a higher temperature), collide with their slower-moving neighbors and give some kinetic energy to them. And so it progresses from one end of the conductor to the other. In addition, electrons in the metal are free to move through conductors and carry internal energy.

Figure 13.1
A skillet handle becomes hot even though it is not directly heated by the stove. Likewise, the handle end of a metal spoon becomes hot when the bowl end is heated. In both cases, the heat-transfer process is conduction.

Usually we want to know how fast heat is transferred from one region to another (Fig. 13.2). The **heat-transfer rate** H is the amount of heat ΔQ transferred from region 2 at a higher temperature T_2 to region 1 at a lower temperature T_1 divided by the time Δt required for the transfer. The conduction heat-transfer rate H_{cd} is

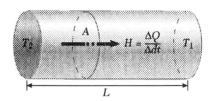

Figure 13.2
When $T_2 > T_1$, heat is transferred by conduction along a material of cross-sectional area A. The length L is the distance between the region where the temperature is T_1 and the region where the temperature is T_2. Because $T_2 > T_1$, the direction of heat flow is as indicated by the arrow.

$$H_{cd} = \frac{\Delta Q}{\Delta t} = \frac{KA(T_2 - T_1)}{L} \qquad (13.1)$$

where K is a proportionality constant called the **thermal conductivity** and depends upon the material being considered, A is the cross-sectional area of the material through which the heat is being conducted, T_1 and T_2 are the temperatures of the two regions, and L is the distance from region 1 to region 2. Some thermal conductivity values are given in Table 13.1. In Equation 13.1, the temperatures are in degrees Celsius, the area in square meters, and the distance in meters. Since H_{cd} is the rate at which energy is transferred, it has units of watts. Therefore, the thermal conductivity must have units of W/m·C°.

The temperature difference divided by the distance between the two regions,

$$\frac{T_2 - T_1}{L}$$

is called the **temperature gradient.** The greater the temperature gradient, the greater the heat flow rate. The gradient can be made large either by a large temperature difference or by a small distance between regions. Notice from Equation 13.1 that there is no heat flow if there is no temperature gradient. If the two ends of a metal rod are at the same temperature, there is no heat

TABLE 13.1	**Thermal Conductivities**		
Material	K(W/m·C°)	**Material**	K(W/m·C°)
Metals			
Aluminum	240	Glass	0.8
Brass	109	Ice	1.6
Copper	385	Styrofoam	0.016
Gold	314	Wood	0.08
Silver	420	Fat	0.02
Steel	80	Muscle, bone	0.04
Nonmetallic solids		Other	
Asbestos	0.08	Air	0.024
Dry soil	0.2	Goose down	0.023
Concrete, brick	0.8	Water	0.6

Potholders and oven mitts work because the cotton batting they are
filled with is a very poor heat conductor.

transfer along the rod. The atoms collide with each other, and energy is exchanged, but on the average there is an equal exchange of energy in both
directions.

Notice the different values of thermal conductivity in Table 13.1: very high
for metals because they are good conductors of heat but low for materials that
are good **thermal insulators.** What makes the difference between good thermal conductors and poor ones ("poor thermal conductor" is another way of
saying "good thermal insulator") is the free electrons characteristic of metals.
Materials with low thermal conductivity do not have such free electrons.

Problem-Solving Tip

■ Use the temperature gradient $(T_2 - T_1)/L$ in calculating the conduction
heat transfer rate.

Example 13.1

A steel rod 1.5 cm^2 in cross-sectional area and 25 cm long is held with
one end in a flame at 1700°C and the other end in 24°C water. At what
rate is heat conducted along the rod?

Reasoning

Heat moves through the rod as the thermal motion of atoms and electrons is transferred from the hot end to the cool end. This is conduction
and therefore we can apply Equation 13.1 directly.

Solution

$$H_{cd} = \frac{\Delta Q}{\Delta t} = \frac{KA(T_2 - T_1)}{L}$$

$$= \frac{(80 \text{ W/m·C}°)(1.5 \times 10^{-4} \text{ m}^2)(1700°C - 24°C)}{0.25 \text{ m}}$$

$$= 80.4 \text{ W}$$

Practice Exercise   What is the heat conduction rate for this rod if it has a
cross-sectional area of 3.5 cm^2?

Answer   190 W

Example 13.2

A camping coffee cup 8.0 cm in diameter is made of steel 2.0 mm thick.
When it is filled to a depth of 10 cm with piping hot coffee at 95°C, what
is the heat transfer rate through the steel to the air, which is at a temperature of 22°C?

Reasoning

The geometry is quite different from that of the rod of Example 13.1, but
the mechanism is exactly the same.

Solution

The distance L in Equation 13.1 is now the thickness of the steel. The cross-sectional area is the total cup area—sides and bottom—in contact with the coffee:

$$A = \pi r^2 + 2\pi rh = \pi(0.040\ \text{m})^2 + 2\pi(0.040\ \text{m})(0.10\ \text{m}) = 0.0302\ \text{m}^2$$

Thus,

$$H_{\text{cd}} = \frac{KA(T_2 - T_1)}{L}$$

$$= \frac{(80\ \text{W/m·C°})(0.0302\ \text{m}^2)(95°\text{C} - 22°\text{C})}{0.0020\ \text{m}} = 88\ \text{kW}$$

Practice Exercise　What is the heat transfer rate if the hot coffee is only 5 cm deep?

Answer　55 W

Example 13.3

A Styrofoam coffee cup 8.0 cm in diameter is 2.0 mm thick. When it is filled to a depth of 10 cm with piping hot coffee at 95°C, what is the heat transfer rate through the cup to the air, which is at 22°C?

Reasoning

This is the same as the previous example except that now the cup is made of Styrofoam instead of steel.

Solution

$$H_{\text{cd}} = \frac{(0.016\ \text{W/m·C°})(0.0302\ \text{m}^2)(95°\text{C} - 22°\text{C})}{0.0020\text{m}} = 18\ \text{W}$$

A Styrofoam cup is clearly more comfortable to hold in your hand than a steel one. In both these examples we have ignored heat loss to the air through the top surface of the coffee. We will discuss such losses in the next section, on convection.

Windows are a major source of heat loss in winter and heat gain in summer. The heat transfer rate is much greater through windows than through the surrounding walls for two reasons. One, glass is a better heat conductor than wood, wallboard, or plaster, and two, windows are far thinner than walls. However, in using Equation 13.1 to calculate heat transfer through a window, we must exercise some caution. Even though the inside room temperature is, say, 25°C and the outside temperature is −10°C, these may not be—and probably are not—the right values to use. Figure 13.3 shows a window under such circumstances. If both room air and outside air are calm, there will be a layer of still air on each side of the glass. The temperatures of these still-air layers are

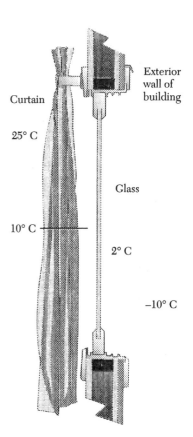

Figure 13.3
Layers of still air on either side of a window help reduce heat loss.

more moderate than the temperatures far away from the window. A curtain will moderate the inside still-air temperature even more. Of course, a strong wind may blow away the still, insulating layer of air on the outside, making the heat loss much greater on a windy day.

Example 13.4

Using the temperatures of Figure 13.3, determine the heat loss through a glass window 0.50 cm thick and 0.50 m wide by 1.25 m high.

Reasoning

This is yet another direct application of Equation 13.1. Heat is transferred by conduction because there is a temperature difference. This conduction is across an area $A = (0.50 \text{ m})(1.25 \text{ m}) = 0.625 \text{ m}^2$ and through a distance $L = 0.50$ cm.

Solution

$$H_{cd} = \frac{KA(T_2 - T_1)}{L}$$

$$= \frac{(0.8 \text{ W/m·C°})(0.625 \text{ m}^2)(10°\text{C} - 2°\text{C})}{0.0050 \text{ m}} = 800 \text{ W}$$

Problem-Solving Tip

- Heat conduction through multiple layers of different materials requires matching the boundary temperatures so that the heat flow rate is the same through all the layers.

Example 13.5

Determine the heat loss through a window 0.500 m wide by 1.25 m high that is covered by a storm window, as in Figure 13.4. The storm window traps 3.00 cm of air between the two panes of glass, and each pane is 0.500 cm thick.

Reasoning

Let us call the storm window pane S and the regular window pane R. There is a difference of 8.00 C° between the outside of pane S and the inside of pane R, but our heat-transfer equation requires the temperature at each end of each conducting material. We can solve for the intermediate temperatures because the heat-flow rates through pane R, the 3.00-cm air gap, and pane S must all be the same. (If these rates were not all equal, heat would accumulate somewhere and the temperature at that accumulation point would become higher than the inside temperature. Since that does not happen, we know all the transfer rates are equal.) A

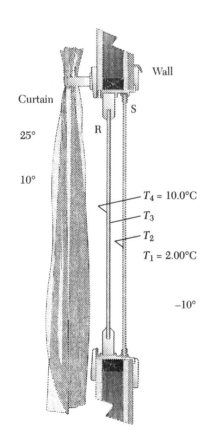

Figure 13.4
A storm window adds the insulation of a layer of still air between the two panes of glass.

situation like this, in which heat-flow rates and the temperatures at various locations remain constant over time, is called a *steady-state condition*.

Solution

Label the interface temperatures T_1, T_2, T_3, and T_4 as indicated in Figure 13.4. For all three layers, the area is 0.625 m^2. Then,

$$H_{\text{pane S}} = \frac{K_{\text{glass}}A(T_2 - T_1)}{L_{\text{glass}}}$$

$$= \frac{(0.8 \text{ W/m·C}°)(0.625 \text{ m}^2)(T_2 - 2.00°\text{C})}{0.00500 \text{ m}} = (100 \text{ W/C}°)T_2 - 200\text{W}$$

$$H_{\text{air}} = \frac{K_{\text{air}}A(T_3 - T_2)}{L_{\text{air}}}$$

$$= \frac{(0.024 \text{ W/m·C}°)(0.625 \text{ m}^2)(T_3 - T_2)}{0.0300 \text{ m}} = (0.500 \text{ W/C}°)(T_3 - T_2)$$

$$H_{\text{pane R}} = \frac{K_{\text{glass}}A(T_4 - T_3)}{L_{\text{glass}}}$$

$$= \frac{(0.8 \text{ W/m·C}°)(0.625 \text{ m}^2)(10.0°\text{C} - T_3)}{0.00500 \text{ m}}$$

$$= 1000 \text{ W} - (100 \text{ W/C}°)T_3$$

Setting these three values of the heat-transfer rate equal to each other, we can solve for T_2 and T_3; we know $T_1 = 2.00°\text{C}$ and $T_4 = 10.00°\text{C}$. Our two equations in two unknowns are

$$(100 \text{ W/C}°)T_2 - 200 \text{ W} = (0.50 \text{ W/C}°)(T_3 - T_2)$$

$$(0.50 \text{ W/C}°)(T_3 - T_2) = 1000 \text{ W} - (100 \text{ W/C}°)T_3$$

Multiply through by 2 to eliminate the factor 0.500 and drop the units for clarity:

$$200T_2 - 400 = T_3 - T_2 \tag{1}$$

$$T_3 - T_2 = 2000 - 200T_3 \tag{2}$$

Solve Equation 2 for T_3 in terms of T_2:

$$201T_3 = 2000 + T_2$$

$$T_3 = 9.95 + \frac{T_2}{201}$$

Substitute this value for T_3 into Equation 1:

$$200T_2 - 400 = 9.95 + \frac{T_2}{201} - T_2$$

$$200.995T_2 = 409.95$$

A cut-away view of a double-glazed window. The two glass panes separated by an air space reduce heat loss considerably. *(© L. W. Black/courtesy of The Window Factory, Inc., Haddon Heights, NJ)*

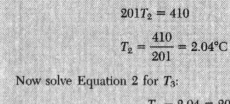

$$201T_2 = 410$$

$$T_2 = \frac{410}{201} = 2.04°C$$

Now solve Equation 2 for T_3:

$$T_3 - 2.04 = 2000 - 200T_3$$

$$201T_3 = 2002.04 \approx 2002$$

$$T_3 = \frac{2002}{201} = 9.96°C$$

$$H_{\text{pane S}} = (100 \text{ W/C°})(2.04°C) - 200 \text{ W} = 4.0 \text{ W}$$

$$H_{\text{air}} = (0.500 \text{ W/C°})(9.96°C - 2.04°C) = 3.96 \text{ W} \approx 4.0$$

$$H_{\text{pane R}} = 1000 \text{ W} - (100 \text{ W/C°})(9.96°C) = 3.96 \text{ W} \approx 4.0$$

The single-pane value we calculated in Example 13.4 was 800 W. Thus adding the storm window reduces the heat loss rate by a factor of 200!

Convection currents (see below) in the air gap change the temperatures in the gap and increase the heat transfer rate somewhat, but storm windows remain very important in reducing heat losses.

Notice that the temperature difference is the same across the two panes: $T_2 - T_1 = 0.04$ C° and $T_4 - T_3 = 0.04$ C°. The temperature differences ($T_2 - T_1$ and $T_4 - T_3$) are the same because the heat transfer rates are the same *and* because the thicknesses of the two panes of glass are the same and the two panes have the same K_{glass}.

The pattern of convection can be seen here as the paper bits on the right side of the beaker rise in heated water, cool at the top, and then fall on the left side of the beaker. *(© Richard Megna/Fundamental Photographs, NY)*

13.2 Convection

CONCEPT OVERVIEW

▬ In convection, heat is transferred from one place to another by a moving fluid.

Heat can be transferred by a moving fluid; this process is known as **convection**. A good example of convection is your automobile engine, which pumps hot cooling fluid from the hot engine block to the radiator, where the fluid is cooled and then returned to the block. Using a pump to move the fluid around is called **forced convection** (Fig. 13.5a). The circulatory system of the human body is another example of forced convection; heat from the interior of the body is carried by the blood to the skin. Another example of forced convection is a forced-air furnace, in which air is heated inside the furnace and then pumped by a fan throughout a building.

A building might also be heated by hot water or steam in a radiator. The heated air around the radiator is less dense than the cooler air farther away and therefore rises. Cooler air moves in to replace it. This is an example of **natural convection** (Fig. 13.5b).

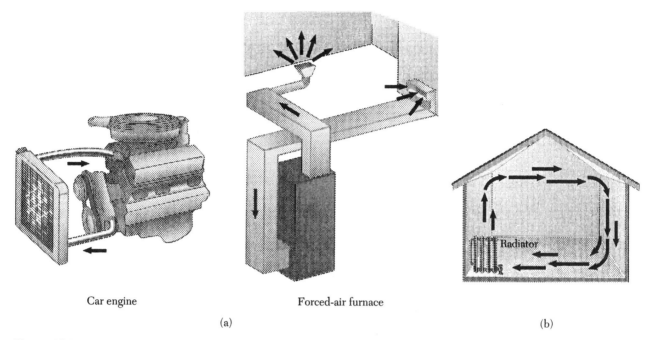

Car engine   Forced-air furnace

(a)   (b)

Figure 13.5
In convection, heat is transferred by a moving fluid from a higher temperature region to a lower temperature region. (a) In forced convection, the fluid is pumped,
as in an automobile engine or a forced-air furnace. (b) In natural convection, the
fluid moves because of buoyant forces caused by the difference in the density of
the hotter and cooler parts of the fluid.

In both types of convection, thermal energy is transferred by the movement of a fluid from a high-temperature region—engine block, your own
body's interior, furnace, room radiator—to a low-temperature region—car
radiator, skin, rest of house.

We can write an expression for the heat-transfer rate due to convection
similar to Equation 13.1:

$$H_{cv} = \frac{\Delta Q}{\Delta t} = hA(T_{surf} - T_{fluid}) \qquad (13.2)$$

where H_{cv} is the convection heat-transfer rate from a surface of area A, T_{surf} is
the temperature of the surface, and T_{fluid} is the temperature of the convecting
fluid. The proportionality constant h is the **convection coefficient** and is
found empirically for a given situation. The value of h depends upon the speed
of the fluid across the surface, the orientation of the surface, its temperature,
and even its location—a steam radiator is more effective on the floor than on
the ceiling, for example.

That the convection heat transfer rate is proportional to the temperature
difference was known to Sir Isaac Newton. Equation 13.2, in fact, is known as

Newton's law of cooling. This cooling rate means that, with the help of calculus, the temperature of any surface that is losing heat by convection can be written as a function[1] of time:

$$T_{surf} = T_{sur} + \Delta T e^{-t/\tau} \qquad (13.3)$$

where T_{surf} is the temperature of the surface at time t, T_{sur} is the temperature of the surroundings, and ΔT is the difference between T_{sur} and the original temperature of the surface at $t = 0$. The factor τ is known as the **time constant.** $e^{-t/r}$ is the exponential function.

Problem-Solving Tip

■ Evaluate an exponential function using a calculator just as you would any other function and then use its numerical value in your calculations.

Example 13.6

A cup of coffee initially at 95°C cools to 85°C in 5.0 min when sitting in a 20°C room. Assume the heat loss is entirely due to convection from the top surface of the coffee, and use Newton's law of cooling (Eq. 13.3) to determine how long it will take to cool to 50°C.

Reasoning
We can solve Equation 13.3 for the time constant and then use that value to find the time needed for the coffee to reach 50°C.

Solution
At $t = 0$, we have $T_{surf} - T_{sur} = \Delta T e^0$, which yields $\Delta T = 75$ C°. (Remember that $e^0 = 1$ because any number raised to the zeroth power is unity.) Because ΔT is by definition the difference at $t = 0$, we use this value of $\Delta T = 75$ C° for all values of t.
At $t = 5.0$ min, we have

$$T - T_{sur} = 85°C - 20°C = 65\ C° = (75\ C°)e^{-5.0\ min/\tau}$$

$$\frac{65\ C°}{75\ C°} = 0.8667 = e^{-5.0\ min/\tau}$$

$$\ln 0.8667 = \ln (e^{-5.0\ min/\tau})$$

$$-0.1431 = -5.0\ min/\tau$$

$$\tau = \frac{-5.0\ min}{-0.1431} = 35\ min$$

---

[1] If the exponential function is not familiar to you, read more about it in the appendix. You can use the exponential function of a calculator to provide answers as needed.

With a value for the time constant, we can solve Equation 13.3 for the time t required for the coffee to cool to 50°C:

$$T_{surf} = T_{sur} + \Delta T\, e^{-t/\tau}$$

$$50°C = 20°C + 75\ C° \, e^{-t/35\ min}$$

$$e^{-t/35\ min} = \frac{30\ C°}{75\ C°} = 0.40$$

$$\ln(e^{-t/35\ min}) = \ln 0.40$$

$$\frac{-t}{35\ min} = -0.916$$

$$t = 32\ min$$

Practice Exercise   How long will it take the coffee to cool to 40°C?

Answer   46 min

Example 13.7

In a room at 25°C, it takes 10 min for a cup of tea to cool from 95° to 60°C. Again, assume the heat loss is entirely due to convection from the top surface. How long will it take to cool to 42.5°C?

Reasoning

At first glance this appears to be nearly identical to Example 13.6, but there is one interesting difference. Look at the temperature *differences*. At $t = 0$, $\Delta T = T_{surf} - T_{sur} = 95°C - 25°C = 70\ C°$. At $t = 10$ min, $\Delta T = 60°C - 25°C = 35\ C°$. And $35\ C°$ is one-half the $t = 0$ value of $\Delta T = 70\ C°$. That means that the half-life for this system is 10 min: $T_{1/2} = 10$ min. Half-life is defined as the length of time needed for half of anything to disappear.

Solution

In another 10 min, the temperature *difference* will drop from 35 C° to half of *that*, or 17.5 C°. The temperature of the tea will now be 25°C + 17.5 C° = 42.5°C, the temperature given. Therefore, in another 10 min— 20 min from the time it was poured—the hot tea will have cooled to 42.5°C.

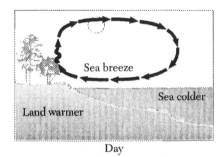

Sea breeze

Sea colder

Land warmer

Day

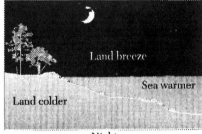

Land breeze

Sea warmer

Land colder

Night

Figure 13.6
Breezes, winds, and much of our weather are caused by convection. Near large bodies of water, natural convection cycles are particularly noticeable. During the day, the land warms up more quickly than the water, giving rise to a sea breeze. At night, the land cools more quickly than the water, and the cycle is reversed, giving rise to a land breeze.

Situations like that in Example 13.7, where information can be thought of in terms of half-lives, are usually easier to understand—and often more meaningful—than otherwise, as in Example 13.6.

Convection is very important in many situations. Figure 13.6 is a sketch of convection currents in air. On a sunny day the land becomes warmer faster than the water (because the specific heat of soil is much less than the specific

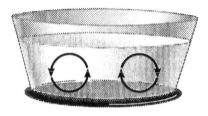

Figure 13.7
Convection currents carry water heated at the bottom up through the rest of the water.

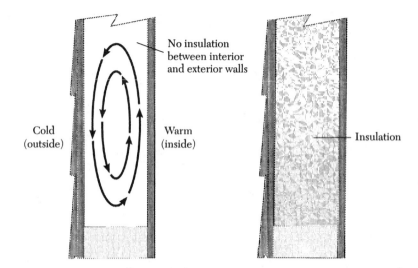

No insulation between interior and exterior walls

Cold (outside)

Warm (inside)

Insulation

Figure 13.8
Convection currents inside a hollow wall would carry heat from the inside of the building to the outside. Insulation traps the air so that it cannot move in convection currents.

A sea breeze blows from sea to land, a land breeze from land to sea. The name is always taken from where the breeze originates—for example, a north wind is one that comes from the north.

heat of water and because convection currents in the water help it retain more heat). Therefore air above the land becomes warm and rises. As it does so, it expands, cools, and falls toward the water surface. At night the land cools more quickly than the water and the direction of the cycle is reversed.

Convection distributes heat throughout the contents of a pot on a stove, as sketched in Figure 13.7. Sometimes we may need to reduce the convection, as in the case of air currents set up inside the hollow wall of a house, as sketched in Figure 13.8. Conduction heat losses through still air are very low because the thermal conductivity of air is quite small, but if the air is free to move inside a hollow wall, the convection heat losses can become substantial as warmed air moves from the inside to the outside of the space within the wall. Indeed, the purpose of insulation inside a wall is to trap the air and prevent its movement.

Figure 13.9
Moving air carries away heat by convection and therefore feels cooler than still air.
(© *Steve Niedorf/The Image Bank*)

TABLE 13.2 Wind Chill Temperatures

Still Air Temperature (°C)	Wind Speed (m/s)				
	2	5	10	15	20
	Wind Chill Temperature (°C)				
2	1.0	−6.6	−12	−16	−18
0	−1.3	−8.4	−15	−18	−20
−5	−7.0	−15	−22	−26	−29
−10	−12	−21	−29	−34	−36
−15	−23	−34	−44	−50	−52

From Alan VanHeuvelen, *Physics,* 2d ed. Boston: Little, Brown and Company, 1986, p. 211.

Air moving across your body carries away heat; that is the purpose of the fan (Fig. 13.9). You feel cooler in moving air than in still air. The "wind chill temperature" in the morning weather report is a description of how cold you are going to feel because of the moving air. It is the temperature of the still air that would cause the same heat loss you will experience with the wind blowing. Some wind chill temperatures for various wind speeds are given in Table 13.2.

Much of the insulating effect of winter clothing comes from trapped layers of air around your body. Several thinner layers of clothes will be warmer than one heavy coat because of this. Summer clothes are designed to be loose and allow air to circulate and carry away heat and perspiration.

13.3 Radiation

CONCEPT OVERVIEW

■ In radiation, heat is transferred from an object to its surroundings in the form of either infrared or visible light.

A third way of transferring heat is by **radiation.** If you sit in front of a fire as in Figure 13.10, you will be warmed. You are not touching the fire, and therefore the heat is not being transferred by conduction.[2] Convection currents carry heat up the chimney, not out into the room where you are sitting. Yet you are warmed; the mechanism for this heat transfer is radiation.

We shall study radiation in more detail when we take up electricity, magnetism, and optics. All we need to know now is that all objects emit heat energy in the form of radiation that travels at the speed of light. Nonluminous objects emit heat as infrared radiation, which the human eye cannot see, and luminous

Figure 13.10
A fire warms you by radiation. (© *Dann Coffey/The Image Bank)*

---

[2] Air is a very poor heat conductor, so we can ignore any slight conduction it may be responsible for here.

A lump of coal at room temperature gives off heat in the form of infrared radiation, which our hands are too insensitive to feel. A glowing ember in the fireplace gives off heat in the form of visible radiation, as well as a greatly increased amount of infrared radiation, which we both see and feel.

objects emit visible radiation, which the human eye can see. Even though you cannot see infrared radiation, you can feel it, as when you warm yourself near a steam radiator or feel the warm hood of an automobile.

The amount of radiation emitted by an object is proportional to the fourth power of its Kelvin temperature. This relationship is known as the **Stefan-Boltzmann law:**

$$R_{\text{emit}} = e\sigma A T_{\text{obj}}^{4} \tag{13.4}$$

where T_{obj} is the Kelvin temperature of the object emitting the radiation, A is the surface area of the object in square meters, σ is a constant known as the Stefan-Boltzmann constant ($\sigma = 5.67 \times 10^{-8}$ W/m^2·K^4), and e, the **emissivity,** is a number between 0 and 1 that is determined by the nature of the object's surface. The emissivity is very near unity for a dull, dark surface, with *blackbody radiation* being that given off by a flat-black surface whose emissivity is 1.0. The emissivity is near zero for a shiny, light-colored surface.

By convention, the symbol for the rate at which heat is transferred by radiation is R rather than the H we use for conduction and convection. If you prefer, the Stefan-Boltzmann law can as easily be written

$$H_{\text{rad}} = \frac{\Delta Q}{\Delta t} = e\sigma A T_{\text{obj}}^{4} \tag{13.4'}$$

In either form, the law tells us that any object that has a temperature above absolute zero radiates energy in the form of heat.

Example 13.8

The emissivity of the human body is approximately 0.95, and the surface area of an average man is 1.75 m^2. Normal body temperature is 37°C, but skin temperature is slightly lower, 35°C = 308 K. What is the rate at which an average man radiates heat?

Reasoning
This problem calls for a direct application of Equation 13.4.

Solution

$$
\begin{aligned}
R_{\text{emit}} &= e\sigma A T_{\text{obj}}^{4} \\
&= (0.95)(5.67 \times 10^{-8}\ \text{W/m}^2\cdot\text{K}^4)(1.75\ \text{m}^2)(308\ \text{K})^4 \\
&= 848\ \text{W} \approx 850\ \text{W}
\end{aligned}
$$

The amount of heat emitted by a body depends strongly on its temperature because of the T^4 dependence in Equation 13.4. Photographs like those in Figure 13.11 are taken with film sensitive to infrared radiation; they are called **thermographs** and can be used in medicine because diseased tissue has a higher temperature than healthy tissue. Having a higher temperature, a diseased region will radiate more infrared radiation and appear as a brighter area on the thermograph. Lack of circulation to fingers or toes can be determined

(a)

from a thermograph because these areas will be darker as a result of their lower temperature. Thermographs of a house show where heat is being radiated and therefore where additional insulation is needed.

Because every object radiates heat, we should expect the temperature of everything to decrease as time goes by. However, every object also *absorbs* heat from its surroundings. An object in a room with walls at temperature T_{sur} absorbs radiation at the rate

$$R_{absorb} = e\sigma A T_{sur}{}^4 \qquad (13.5)$$

where e, σ, and A are the emissivity, Stefan-Boltzmann constant, and object's surface area just as before. The rate at which energy is absorbed depends upon the fourth power of the Kelvin temperature of the surroundings $T_{sur}{}^4$. The emissivity is identical for emission and for absorption. A dull, dark object, which is a good emitter, is also a good absorber; sitting in the sun, a dark car gets hotter more quickly than a light one. A shiny, light-colored object, which is a poor emitter, is also a poor absorber; it reflects much of the energy that radiates to it from its surroundings. Light-colored clothing really is cooler in the summer.

The net rate of heat transfer by radiation is

$$R_{net} = R_{emit} - R_{absorb} = e\sigma A(T_{obj}{}^4 - T_{sur}{}^4) \qquad (13.6)$$

Example 13.9

In Example 13.8 we calculated that a typical man radiates heat at a rate of 850 W. If this same man sits in a room whose walls are at 25°C, at what rate does his body absorb heat and what is the net rate of heat transfer?

Reasoning

The heat radiated depends upon the temperature of the radiating body; we used that fact in solving Example 13.8. Now we need the rate at which the body *absorbs* heat, and that rate depends upon the temperature of the surroundings—in this case, the walls at 25°C.

Remember that it is the Kelvin temperature that goes into our heat-transfer equations.

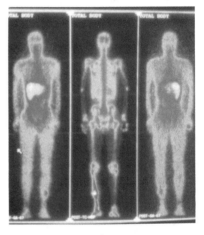

(b)

Figure 13.11
Thermographs are photographs taken with film sensitive to infrared radiation. Similar to thermographs are thermograms, which are images made with infrared scanners. Areas of higher temperature show up brighter. (a) The different colors in this thermogram, made during cold weather, indicate the amount of heat escaping from various parts of these houses. What part of each house is most responsible for heat loss? (VANSCAN® *Thermogram by Daedalus Enterprises, Inc.*) (b) A total body thermogram. Why does the spinal column (*middle view*) stand out as "hotter" than surrounding parts of the body? (© *Jay Freis/The Image Bank*)

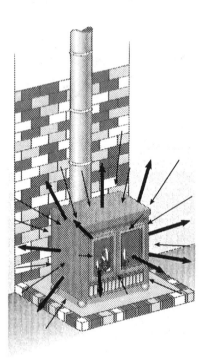

Figure 13.12

A stove emits more radiation than it absorbs because its temperature is higher than the temperature of its surroundings.

Solution

$$R_{absorb} = e\sigma A T_{sur}^4$$

$$= (0.95)(5.67 \times 10^{-8} \text{ W/m}^2\cdot\text{K}^4)(1.75 \text{ m}^2)(298 \text{ K})^4 = 740 \text{ W}$$

$$R_{net} = R_{emit} - R_{absorb} = 110 \text{ W}$$

This is the amount of power the man's metabolism must put out just to keep the skin temperature at 35°C.

Radiation absorbed by your body from the warmer surroundings of a room—walls, furniture, and floors—can make you feel warm even if the air is at a lower temperature. For this reason, some heating systems heat the floors or the ceilings; homes with such heating systems are especially comfortable in winter.

Example 13.10

The wood-burning stove in Figure 13.12 has 2.1 m² of surface area at 125°C. Calculate the rate at which it radiates heat into a room whose walls are at 18°C. Consider the emissivity of the stove to be 1.0.

Reasoning

This is much like Example 13.9, and we can apply Equation 13.6 directly.

Solution

$$R_{net} = e\sigma A(T_{obj}^4 - T_{sur}^4)$$

$$= (1.0)(5.67 \times 10^{-8} \text{ W/m}^2\cdot\text{K}^4)(2.1 \text{ m}^2)[(398 \text{ K})^4 - (291 \text{ K})^4]$$

$$= 2134 \text{ W} \approx 2100 \text{ W}$$

Example 13.11

A pan containing 0.25 kg of water is placed inside an oven whose walls are at 230°C (Fig. 13.13). The pan has a total outside surface area of 5.0×10^2 cm² and an emissivity of 0.90. Once boiling starts, how long does it take to boil off all the water?

Reasoning

Just as with the stove in Example 13.10, the pan both emits and absorbs radiation, so we must calculate the net radiation with Equation 13.6. As long as water remains in the pan, the temperature of water and pan is essentially 100°C; we take this as our value for T_{obj}.

Solution

As always, to use the Stefan-Boltzmann law (Eq. 13.4) or the net radiation heat transfer equation (Eq. 13.16) we must ensure that the temperatures are expressed in kelvins.

$$T_1 = T_{surr} = 230°C = 503 \text{ K}$$

$$T_2 = T_{obj} = 100°C = 373 \text{ K}$$

Because the area in the Stefan-Boltzmann law is in square meters, we must convert the area given in the problem statement. Then we can use Equation 13.6:

$$R_{net} = e\sigma A(T_{obj}{}^4 - T_{sur}{}^4)$$

$$= (0.90)(5.67 \times 10^{-8} \text{ W/m}^2 \cdot \text{K}^4)(0.050 \text{ m}^2)[(373 \text{ K})^4 - (503 \text{ K})^4]$$

$$R_{net} = -114 \text{ W} \approx -110 \text{ W}$$

The negative sign indicates that heat is flowing into the pan.

To boil off 0.25 kg of water requires an amount of energy given by the Equation 12.7:

$$Q = mL_v = (0.25 \text{ kg})(2.256 \times 10^6 \text{ J/kg}) = 5.6 \times 10^5 \text{ J}$$

We can solve $R_{net} = \Delta Q/\Delta t$ for Δt:

$$\Delta t = \frac{\Delta Q}{R_{net}}$$

$$= \frac{5.6 \times 10^5 \text{ J}}{110 \text{ W}} = 5100 \text{ s} = 85 \text{ min}$$

This calculation does not include the heat transferred by convection, which is considerable. Convection ovens speed cooking time by circulating heated air.

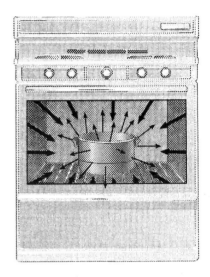

Figure 13.13
A pan of water inside a hot oven emits and absorbs heat. Since the oven walls are hotter than the pan, the pan absorbs more heat than it emits.

The Thermos Bottle

A device that effectively minimizes all three forms of heat transfer is the familiar **thermos bottle,** or **vacuum bottle,** sketched in Figure 13.14. The walls are made of two layers of glass separated by a vacuum; this arrangement reduces heat transfer by conduction. The glass is silvered to reduce its emissivity and, hence, the heat transfer by radiation. A lid reduces heat loss by convection.

The same type of container is also used to hold liquid nitrogen at $-169°C$ or liquid helium at $-270°C$. In scientific settings the container is usually called a Dewar flask. Sometimes a double Dewar flask—one flask inside another—is used for liquid helium.

The containers in Figure 13.15 illustrate the effects of removing the various design components of a thermos bottle. Hot water at the same temperature was placed in all four containers at the same time. The one on the left is a typical good vacuum bottle with silvered walls, a vacuum between them, and a lid (with a hole for the thermometer). The next one has no silver on its walls, increasing radiation heat transfer. The next has no silver on its walls and has air between them, increasing conduction heat transfer.[3] The one on the right has

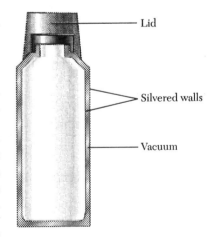

Figure 13.14
A thermos bottle or Dewar flask reduces all three forms of heat transfer.

[3] Even though air is a very poor heat conductor, as mentioned at the beginning of this section, it is still a better conductor than vacuum. Air between the walls also provides heat transfer by convection currents.

Vacuum bottles insulate liquids from the surrounding environment, keeping hot liquids hot or cold liquids cold for extended periods. *(© L. W. Black)*

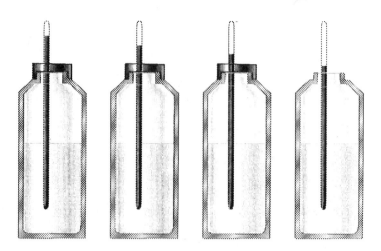

Figure 13.15
Hot water at the same temperature was placed in these four thermos bottles at the same time. The one on the left has silvered sides, a vacuum between its walls, and a lid. The next has no silver on its walls. The next has no silver and no vacuum. The one on the right has no silver, no vacuum, and no lid.

no silver on its walls, air between them, and no lid, increasing convection heat losses. The container on the left will keep the water hot for a long time. The one on the right will reach room temperature first.

The Greenhouse Effect

The **greenhouse effect** is an important consequence of radiative heat transfer and contains some very interesting physics.

Not only is the amount of radiation emitted by an object dependent upon temperature, but the nature of the radiation is temperature-dependent as well. For example, a horseshoe, just like any other object, radiates infrared radiation at room temperature. As a blacksmith heats the horseshoe, it emits more radiation and the type of radiation changes. When the temperature gets high enough, the shoe glows a dull red. Red light has a higher *frequency* than infrared radiation (we shall discuss frequency later when we study waves and light). At still higher temperatures, the shoe glows a bright orange-red that has a still higher frequency. The filament in a light bulb glows white hot at 1800°C and emits radiation at frequencies corresponding to all of visible light.

Sunlight, because it comes from an extremely hot source, contains infrared, visible, and ultraviolet radiation. The visible portion readily passes through the glass walls of a greenhouse, as shown in Figure 13.16a. This heat radiated by the Sun is absorbed by objects inside. These objects, at relatively low temperature, emit only infrared radiation, and glass is *not* transparent to it. This means that radiation from the Sun can come in and be absorbed but radiation

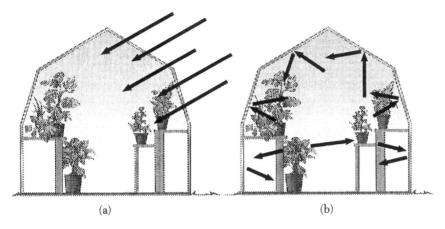

Figure 13.16
In a greenhouse, visible-light radiation from the Sun passes through the glass walls of a greenhouse, but the infrared radiation emitted by objects inside the greenhouse does not pass through the glass and is trapped inside. This trapped infrared radiation subsequently leads to a rise in the inside temperature.

emitted from within the greenhouse cannot escape (Fig. 13.16b). Hence, the temperature in the greenhouse rises.

On a global level, the water vapor and carbon dioxide in our atmosphere act like the glass of a greenhouse to retain solar energy. There is considerable concern that increasing the level of carbon dioxide in the atmosphere will increase Earth's greenhouse effect: objects on Earth's surface will retain more of the Sun's energy and the average temperature of Earth will increase. Any increase in Earth's average temperature will melt some of the polar ice caps and increase the level of the oceans with dire consequences to coastal areas.

The ultraviolet (UV) portion of sunlight is suspected of causing skin cancer. Upper-atmosphere ozone absorbs part of this radiation and keeps it away from us. There is considerable concern these days that ozone depletion is allowing more UV rays to reach Earth, with possibly deadly consequences.

13.4 Evaporation

CONCEPT OVERVIEW
- Because energy is required to convert liquid molecules to vapor molecules, evaporation causes cooling.

Water changes from liquid to vapor at 100°C when it boils, but it can also change from liquid to vapor at temperatures below the boiling point. Clothes hung out on a line will dry. A puddle left after a shower will soon be gone. Sweat on your skin will dry up. All of these are examples of **evaporation.**

The molecules of a liquid are always in motion, even at room temperature. As molecules at the liquid surface strike the liquid–air interface, most rebound and remain in the liquid. However, some of the more energetic surface molecules escape from the liquid and become molecules of vapor, as sketched in Figure 13.17. This escape process leaves the liquid with relatively fewer high-energy molecules and relatively more low-energy molecules. The result is that the temperature of the liquid drops. In other words, heat is taken from the liquid as it evaporates.

As with convection, the rate of evaporation depends upon many factors. Wet clothes dry faster on a warm day than on a cold day; they dry faster in the sunshine than in the shade. **Humidity** describes how much water vapor is in the air. If the humidity is high, some of the water molecules in the air strike the

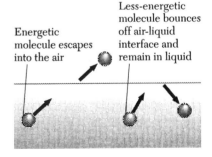

Figure 13.17
More-energetic molecules escape from the liquid surface to become vapor molecules. This escape leaves the liquid with a relatively larger number of less-energetic molecules, and the temperature of the liquid drops.

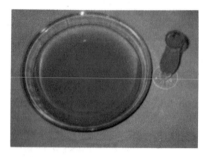

Figure 13.18
Temperature, humidity, air speed, and surface area all affect evaporation rate. Will the liquid evaporate faster from the bowl *(left)* or the vase *(right)*? Why? *(© L. W. Black)*

water-air interface and become part of the liquid, greatly slowing down the evaporation rate. If the air is very still, a region of high-humidity air forms near the interface and evaporation is slowed. However, if the air is moving, water molecules that escape from the liquid are carried away before they can return to the liquid and the evaporation rate is increased.

Surface area is important in evaporation. A glass of water evaporates very slowly, but that same amount of water evaporates much more quickly if poured out over a large area. A wet towel wadded into a ball remains wet, but one hung over a railing dries. These ideas are illustrated in Figure 13.18.

On a still, humid day, you remain sweaty with even mild exercise because the sweat does not evaporate readily. You are far more comfortable on a day with low humidity and a gentle breeze because then the sweat evaporates easily. Sweating is an important process for regulating body temperature. Sweat forms as a thin film of large surface area over the body. As it evaporates, it carries with it heat generated within the body and carried to the skin by the blood.

While details of the heat transfer rate due to evaporation are difficult to calculate, we can reasonably calculate the total amount of heat carried away by evaporation from the expression

$$\Delta Q_{ev} = mL'_v \tag{13.7}$$

where m is the mass of the liquid evaporated and L'_v is the heat of vaporization *at the temperature* at which the evaporation takes place. This value is usually *different* from the heat of vaporization at the boiling point L_v. For example, water's heat of vaporization at normal skin temperature, 35°C, is 2.40×10^6 J/kg.

Rubbing alcohol and ether are very volatile liquids and evaporate rapidly at room temperature. That means a layer of either liquid on the skin feels cool as body heat leaves with their evaporation. Likewise a wet T-shirt feels cool when you stand in front of a fan because of increased evaporation.

Example 13.12

How much heat is carried away when 1.0 g sweat evaporates from your body at 35°C?

Reasoning

Heat is required to turn the 1.0 g of sweat on your skin to 1.0 g of water vapor in the air. This is a direct application of Equation 13.7.

Solution

$$\Delta Q_{ev} = mL'_v = (0.0010 \text{ kg})(2.40 \times 10^6 \text{ J/kg}) = 2.4 \times 10^3 \text{ J}$$

*13.5 Solar Energy/Solar Homes

CONCEPT OVERVIEW

■ The Sun's radiation can be used to heat water and buildings.

Our Sun radiates enormous amounts of energy. Solar radiation reaches Earth's upper atmosphere at the rate of 1.353 kW/m². Some of that radiation is reflected by the atmosphere, and at sea level on a cloudless day, 0.95 kW/m² reaches Earth's surface. This is an enormous and abundant energy source, but before we see how to harness it, we must look at how Earth orbits our Sun.

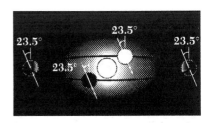

Figure 13.19
Earth orbits the Sun on an axis inclined 23.5° from a line perpendicular to the plane of the orbit.

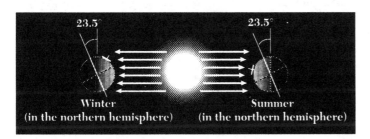

Winter
(in the northern hemisphere)

Summer
(in the northern hemisphere)

Figure 13.20
Because of the inclination of Earth's axis of rotation, sunlight falls more directly on the northern hemisphere in summer than in winter. This inclination is the cause of summer and winter.

Earth's axis of rotation is inclined 23.5° from the perpendicular to the plane of the orbit, as sketched in Figure 13.19. This inclination affects the amount of solar radiation reaching the ground and is the cause of our seasons. As sketched in Figure 13.20, as Earth orbits the Sun, solar radiation falls more obliquely on the northern hemisphere during one part of the orbit and more perpendicularly during another part of the orbit. When the radiation falls more perpendicularly, more solar energy is absorbed per unit area of Earth's surface and we have summer. When it falls more obliquely, less solar energy is absorbed and we have winter. The inclination of Earth's rotation axis also means the midday Sun is higher in the sky in summer than in winter, as sketched in Figure 13.21

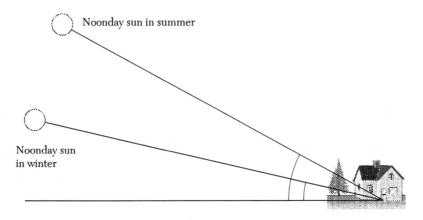

Noonday sun in summer

Noonday sun
in winter

Figure 13.21
Because of the inclination of Earth's axis of rotation, the noonday sun is higher in the sky in summer than in winter.

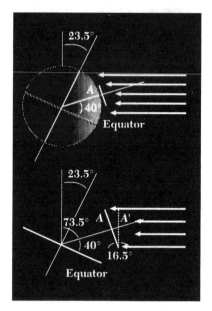

Figure 13.22
At 40° latitude, in summer, an area A is inclined 73.5° from the Sun's rays (16.5° from a perpendicular to the rays). All the sunlight that passes through area A' strikes area A. These areas are related by $A' = A \cos 16.5° = 0.96A$.

Example 13.13

The area of the 48 contiguous states of the United States is about $7\,840\,000$ km^2. These 48 states are located between 25° and 49° north latitude. As an average value, consider the total area as being located at 40° north latitude. What is the total solar power incident on the 48 states at noon on a clear summer day?

Reasoning

Solar radiation reaches Earth's surface at the rate $R = 0.95$ kW/m^2. This is the power P that passes through a *perpendicular* area A':

$$R = P/A' = 0.95 \text{ kW/m}^2$$

$$P = RA'$$

To calculate how much power reaches an area A that is flat on Earth's surface and therefore inclined to the Sun's rays rather than the perpendicular area A', we need the diagrams and angles of Figure 13.22. In mid-summer, a surface area A at 40° north latitude is tilted so that the Sun's rays make an angle of 16.5° with the perpendicular to this area ($16.5° = 40° - 23.5°$). All the sunlight that passes through area A' will strike area A. These areas are related by $A' = A \cos 16.5° = 0.96\,A$.

Solution

With the inclination, we must use A':

$$P = (0.95 \text{ kW/m}^2)(0.96 \times 7\,840\,000 \text{ km}^2)(10^6 \text{ m}^2/\text{km}^2)$$

$$= 7.2 \times 10^{12} \text{ kW}$$

Total power consumption in the United States is approximately 3×10^8 kW. Therefore, Example 13.13 illustrates that solar power is abundant; the sheer amount of it is enormous even though the value we calculated in the example applies only at noon and goes to zero at night. However, how to

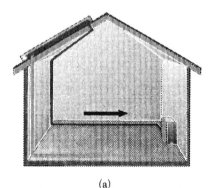

(a)

(b)

Figure 13.23
(a) A solar heating system may use water pumped through rooftop collectors.
(b) Solar collectors for heating water in a home. (© *L. W. Black*)

harness this power is the important question. Solar homes trap some of the Sun's radiation in winter and use it for heating. This radiation can also be trapped year-round for heating hot water.

Figure 13.23 illustrates an *active* solar heating system. The collectors may be glass enclosures with black pipes through which water is pumped. The pipes become warm as a result of the greenhouse effect. This heat is conducted from the pipes to the water and then carried throughout the house by forced convection.

Passive solar heating uses the greenhouse effect in the living area of a house. A south-facing wall may have large areas of glass through which sunlight can shine. Because heat is also lost through glass, the other outside walls—especially on the north side—have minimal areas of glass. Such a house is illustrated in Figure 13.24. Clerestory windows admit light and solar energy into the second story and provide additional ventilation in the summer. Skylights are sometimes used to admit light and solar energy through the roof.

As with a greenhouse, too many windows can make a house too warm for comfort. Of course, you could decrease the glass area, but a better solution may be to let the solar radiation heat rocks or concrete or water tanks as a way of storing this excess energy. These storage devices then re-radiate the heat after the Sun has set. Deciduous trees or shrubs can be planted in front of the south-facing windows. In the winter they offer little hindrance to the Sun's rays, but in summer, with their leaves on, they offer protection from the rays.

A carefully sized eave overhang also offers protection in summer. As sketched in Figure 13.25, sunlight is admitted in winter when the Sun is low in the sky and blocked in summer when the Sun is higher in the sky. Some little thought and effort in sizing eaves can greatly reduce heating and air-conditioning costs. The Anasazi, ancient peoples of the southwestern United States, used this technology 1000 years ago in their cliff homes (Fig. 13.26).

Figure 13.24
Passive solar designs make careful use of glass—ample on the south side and minimal elsewhere.
(© L. W. Black)

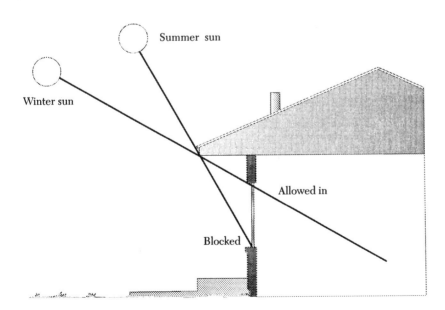

Summer sun

Winter sun

Allowed in

Blocked

Figure 13.25
A properly sized eave overhang allows sunlight in during the winter but not during the summer.

Figure 13.26
Anasazi cliff dwellings at Mesa Verde, Colorado, built around A.D. 1000. The builders knew about solar heating and cooling. Built under the overhang of a large cliff, the structures would receive the winter Sun's rays but would be protected from the intense summer Sun high in the sky. *(© Agnes Gruliow)*

Figure 13.27
A cut-away view of a double-glazed window. Such windows are now standard in new homes or installed as replacement windows in older homes. More energy-efficient homes will use triple-glazed windows. *(© L. W. Black/courtesy of The Window Factory, Inc., Haddon Heights, NJ)*

*13.6 Insulation

CONCEPT OVERVIEW
- Good insulation means poor heat transfer.
- *R* values describe the insulating capability of a material.

Solar homes are designed to *bring* solar heat into a house. At least as important are designs that *keep* heat inside a house once it is there—whatever its source. Some of these designs were mentioned in our study of convection.

Air is a good thermal insulator, but convection currents in the air inside a hollow wall let heat escape, as was shown in Figure 13.8. Most insulation materials contain great amounts of air, but air that is trapped in small spaces so that it cannot move in convection currents. Fiberglass insulation is mostly air; even styrofoam is mostly air.

As we discussed in Example 13.5, storm windows add an additional insulating layer of air. Again, convection currents can be created in this air space if it is too wide. Double-pane windows are windows with two panes of glass separated by a narrow layer of air—usually about 1 cm wide. Such windows are often called **thermopane** windows. Triple-glazing refers to storm windows in addition to double-pane windows (Fig. 13.27). Triple-glazing is now standard in any home that is intended to be energy-efficient.

Figure 13.28
An aerial view of an earth-berm structure, the Amity underground school in Boise, Idaho. *(© David R. Frazier Photolibrary/Contact Press Images)*

Soil is a very good insulator. Underground homes or **Earth-berm structure,** have a thick layer of soil around the home for insulation (Fig. 13.28). Furthermore, because the Earth remains much warmer than the air in winter, the temperature gradient across the walls of an Earth-berm house is reduced. In summer the Earth is much cooler than the air, so that the temperature gradient across the walls is again reduced and there is less need for air-conditioning.

"Thermal conductivity" is never mentioned when you talk to a home builder or remodeler. Rather, at least in the United States, you hear about the "R value" of various insulations or wall designs. What are R values, and what do they have to do with heat transfer?

With good insulation to reduce convection currents, most heat lost through a wall is lost by conduction. Equation 13.1 was easy to use for a single piece of material, but in Example 13.5, dealing with a compound material, the solution was no longer easy. Let us look at a multilayer wall again to see if there is some easier way to calculate heat loss.

Figure 13.29 shows a compound slab composed of several materials, each with thermal conductivity K_i and thickness L_i. It is fairly straightforward to show that the conduction heat-transfer rate through such a compound barrier is

$$H_{cd} = \frac{A(T_2 - T_1)}{\Sigma\,(L_i/K_i)}$$

(13.8)

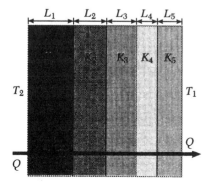

Figure 13.29
The R value for a compound slab is $R_{tot} = \sum_i R_i = \Sigma\,(L_i/K_i)$.

where A is the cross-sectional area and is the same for all the materials. Just as for a simple slab, T_2 and T_1 are the temperatures at the edges of the compound slab; we do not need to solve for the intermediate temperatures. The L_i and K_i are the thickness and thermal conductivity for the ith layer in the slab. In designing and building a house, this L/K term is referred to as the **R value.** R values are useful because they can be added together; adding another layer to a wall means adding its R value to the R value of the rest of the wall. In terms of R values, the conduction heat-transfer rate is

$$H_{cd} = \frac{A(T_2 - T_1)}{R} \tag{13.9}$$

so that a higher R value means lower heat conduction.

Table 13.3 gives R values for some common building materials. Notice the units. In the United States, R values are given in British engineering units of $ft^2 \cdot F° \cdot h/Btu$. What do such units mean? In 1 h, a layer of material with an R value of 1.0 will conduct 1 Btu across each square foot of area if there is a temperature difference of 1 F° across the layer. An R value of 2.0 will require a 2-F° difference to conduct heat at the same rate. Figure 13.30 shows two walls, one uninsulated and one insulated; the associated R values are given in Table 13.4.

TABLE 13.3 R Values for Common Building Materials

Material	R value ($ft^2 \cdot F° \cdot h/Btu$)
Air layer (outside, still air)	0.17
Sheetrock (0.5 in.)	0.45
Air layer (inside)	0.68
Wood shingles	0.81
Glass pane (0.125 in.)	0.89
Wood siding (1 in., hardwood)	0.91
Air space in walls (3.5 in.)	1.01
Wood siding (1 in., pine)	1.25
Exterior sheathing (0.5 in.)	1.32
Insulating glass (0.25-in. air gap)	1.54
Concrete block (filled core)	1.93
Cellulose fiber (1 in.)	3.70
Polystyrene (1 in., expanded)	3.85
Brick (4 in.)	4.00
Fiberglass board (1 in.)	4.35
Polyurethane (1 in., expanded)	5.88
Fiberglass (3.5 in.)	10.90
Fiberglass (5.5 in.)	18.80

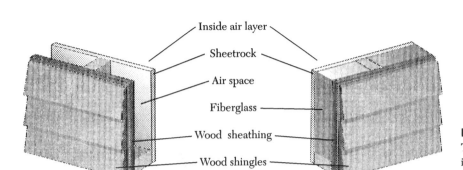

Uninsulated wall   Insulated wall

Inside air layer
Sheetrock
Air space
Fiberglass
Wood sheathing
Wood shingles
Outside air layer

Figure 13.30
The exterior walls of a house can be insulated or uninsulated. The differences in R value are given in Table 13.4.

Example 13.14

Compare heat-transfer rates through the insulated and uninsulated walls of a cabin 24 ft by 36 ft by 8.0 ft high when the inside temperature is 70°F and the outside temperature is 10°F. Use the R values given in Table 13.4.

Reasoning

This is another conduction heat-transfer problem, but now the characteristics of the conducting material (the walls and their insulation) are known in terms of R values rather than conductivities. Therefore, we use Equation 13.9.

TABLE 13.4 R Values for Figure 13.30

Material	R value (ft²·F°·h/Btu)	
	Uninsulated wall	Insulated wall
Inside air layer	0.68	0.68
Sheetrock	0.45	0.45
3.5-in. air space	1.01	—
3.5-in. fiberglass	—	10.90
Exterior wood sheathing	1.32	1.32
Wood shingles	0.81	0.81
Outside air layer	0.17	0.17
Total R value	4.44	14.33

Solution

With the R values given in British engineering units, it is easier to leave the dimensions of the house in feet and the temperatures in degrees Fahrenheit. We shall convert our H values to SI units, however, using the conversion factor we learned in Chapter 12: 1 Btu = 1054.8 J.

The combined surface area of the four walls is

$$A = 2 \times (24\ \text{ft} + 36\ \text{ft})(8.0\ \text{ft}) = 960\ \text{ft}^2$$

For the uninsulated wall,

$$H_{\text{cd}} = \frac{(960\ \text{ft}^2)(60\ \text{F}°)}{4.44\ \text{ft}^2 \cdot \text{F}° \cdot \text{h/Btu}} = 1.30 \times 10^4\ \text{Btu/h} = 1.37 \times 10^7\ \text{J/h}$$

$$= 3.80\ \text{kW} \approx 3.8\ \text{kW}$$

For the insulated wall,

$$H_{\text{cd}} = \frac{(960\ \text{ft}^2)(60\ \text{F}°)}{14.33\ \text{ft}^2 \cdot \text{F}° \cdot \text{h/Btu}}$$

$$= 4020\ \text{Btu/h} = 4.24 \times 10^6\ \text{J/h} = 1.18\ \text{kW} \approx 1.2\ \text{kW}$$

Practice Exercise   If 2×6 studs are used for walls so the thickness of the inside of the walls is increased from 3.5 in. to 5.5 in., what will the heat transfer rate be through the insulated walls?

Answer   2600 Btu/h = 0.76 kW

Higher R values mean lower heat losses (and lower energy bills). Two-by-four lumber, which is actually only 3.5 in. wide, is commonly used in building houses. Today, many builders are constructing exterior walls made of two-by-six lumber, which is 5.5 in. wide. More fiberglass insulation fits inside the thicker walls. In addition to insulation inside the wall, it is common to include a layer of insulation on the inside or outside of the house.

Air infiltration—basically convective heat transfer—is an enemy in insulating a home. Caulking around windows and doors reduces air infiltration. Tight-fitting windows and doors with good weather stripping to ensure a good fit are also important. An air-lock entry with two sets of doors, like that shown in Figure 13.31, reduces air exchange because the outer door can be closed before the inner door is opened.

In the northern hemisphere, winter winds come mostly from the north and northwest. Protection from these winds can reduce the air infiltration and the "wind chill effect" on a house. This can be done by planting a wind break of evergreen trees on the north and west, designing closets to lie along the north or west, or putting an attached garage on the north or west of a house. The house in Figure 13.24 uses such a garage to protect from these northwesterly winter winds.

Figure 13.3 showed the temperatures around a window. The layer of air between the curtain and the window offered additional insulation and reduced the heat losses through the window. To make this insulating effect of curtains

Melting snow patterns reveal differences in thermal conduction. Note that most of the snow has melted off this house's main roof. What does that observation tell you about the amount of insulation present in the house's attic area?   *(Courtesy of Albert Bartlett)*

Figure 13.31
An air-lock entry system reduces
heat transfer in winter and summer.
(© *L. W. Black*)

more effective, the curtains also need to reduce convection currents. This can
be accomplished by ensuring that the curtains touch the floor and ceiling,
stopping the convection cycle. Such curtains offer far more insulating effect
than similar curtains that miss the floor and ceiling by even a few inches. Little
things can have big effects in home insulation.

SUMMARY

In conduction, heat is moved from one atom to the next through a material; there is no bulk movement of the conducting material. The faster-moving atoms in the hot part of the material collide with their slower-moving neighbors and give some kinetic energy to them. The conduction heat-transfer rate is

$$H_{cd} = \frac{\Delta Q}{\Delta t} = \frac{KA(T_2 - T_1)}{L} \qquad (13.1)$$

Convection is heat transfer that involves the bulk movement of a fluid. We can *approximately* describe convection heat transfer from a surface by

$$H_{cv} = \frac{\Delta Q}{\Delta t} = hA(T_{surf} - T_{fluid}) \qquad (13.2)$$

Radiation is heat transfer from an object to its surroundings in the form of either infrared or visible light. Heat can radiate through empty space, as in the case of our Sun warming Earth. The radiation emitted by any object is

$$R_{emit} = e\sigma A T_{obj}^{4} \qquad (13.4)$$

The temperature T_{obj} must be measured on the Kelvin temperature scale. The factor σ is the Stefan-Boltzmann constant, $\sigma = 5.67 \times 10^{-8}$ W/m^2·K^4. The factor e, the emissivity, is a number between 0 and 1 determined by the surface.

Any object can also absorb radiation from its surroundings. The absorption rate depends upon the surface area of the object and its emissivity and upon the Kelvin temperature T_{sur} of the surroundings:

$$R_{absorb} = e\sigma A T_{sur}^{4} \qquad (13.5)$$

The net rate of heat transferred by radiation is

$$R_{net} = e\sigma A(T_{obj}^{4} - T_{sur}^{4}) \qquad (13.6)$$

Water can change from liquid to vapor at temperatures below the boiling point. This process is known as evaporation. Evaporation carries away heat and is an important—even vital—means used by the human body to control temperature.

QUESTIONS

1. Cooks who know their physics may insert a large nail into a baking potato before putting it in the oven. Why?

2. Why does a tile floor feel colder to your bare feet than a carpeted floor at the same temperature?

3. On a hot day, why will concrete around a pool feel hotter to your bare feet than a wooden step at the same temperature?

4. Why are the handles of some fireplace pokers made in the shape of a spring?

5. Why are many skillet handles made of plastic or wood?

6. If you fill a brown paper bag with water and carefully set it on an *outdoor* barbecue grill, you can boil the water without burning the bag. Explain how this can be. (And don't try it on an indoor stove, just in case you are not careful enough.)

7. Sailplane pilots, hawks, and eagles use updrafts for soaring (Fig. 13.32). Explain what causes an updraft.

Figure 13.32
Question 7. A sailplane soaring on an updraft near Lake Tahoe, California. *(© David Lissy/FPG International Corporation)*

8. Why is a Styrofoam ice chest far less effective when the lid is left off?

9. Highway signs near bridges warn "bridge may freeze before roadway." Why is that true?

Question 9. *(© Charles D. Winters)*

10. Cooling fluid carries heat from an automobile engine to the radiator. There the heat is transferred to the air moving through the radiator. Does a radiator really radiate? That is, by what means is most of the heat transferred from the cooling fluid to the air?

11. Insulating glass—two panes of glass with a layer of air in between—is more efficient (offers more insulation) as the air gap is increased to some thickness and then is less efficient as the air gap is increased beyond the optimal thickness. Explain.

12. Explain how you can tell which way the wind is blowing by wetting your finger and holding it up in the wind.

13. Blowing across the top of hot coffee cools it by two means. What are they?

14. Swimming pools warm up faster (and therefore can be used earlier and later in the season) if they are covered by a layer of plastic. Explain two ways the cover increases heating.

15. People cool themselves by sweating. Dogs do not have sweat glands. How can a dog's panting serve a similar purpose?

16. Bicycle riders and joggers often accumulate far more sweat when moving with the wind than against it. Yet moving with the wind requires less exertion than going against it. Explain the greater accumulation of moisture.

17. Some home air-cooling systems and most commercial ones blow air across large areas of water. How can evaporating this water cool the air? Will such systems function better in regions of low humidity or of high humidity?

Question 17. A large water-tower-style air conditioner atop a multistory apartment building.
(© L. W. Black)

18. Explain how the orientation of a house can effect heating and cooling bills.
19. Skylights are very beneficial to a solar home in winter. Why might they not be beneficial in summer?
20. How can ceiling fans make a room more comfortable both in summer and in winter?
21. Why would thicker walls be important in insulating a home?

PROBLEMS

13.1 Conduction

13.1 An aluminum rod 50 cm long and 2.5 cm^2 in cross-sectional area is held with one end in a flame at 1900°C and the other end in a block of ice at 0°C. What is the heat-transfer rate along the rod? Assume no heat loss to the surroundings.

13.2 A window pane is 1.0 m high by 0.35 m wide and 0.30 cm thick. Immediately next to it, the inside temperature is 18°C and the outside temperature is 3°C. What is the heat-transfer rate through this window?

13.3 A log house is made with solid wood walls 20 cm thick. What is the heat-transfer rate through them if the house is 7.5 m long by 6.0 m wide, the walls are 2.5 m high, the outside temperature is 5.0°C, and the inside temperature is 25°C?

13.4 A steel pan is 20 cm in diameter and 2.5 mm thick. It is filled to a depth of 15 cm with 90°C water. The pan sits on an insulating surface so that there is essentially no heat loss through the bottom. Calculate the heat-transfer rate through the sides when the surrounding temperature is 23°C.

13.5 A goose-down sleeping bag has a surface area of 2.25 m^2 and is filled with a layer of down 5.00 cm thick. What is the heat-transfer rate through it from a person with skin temperature of 35.0°C to the outside air at −5.00°C? How does this rate compare with the body's minimal metabolism rate, about 100 W?

13.6 A 50 cm × 50 cm × 75 cm Styrofoam ice chest must prevent a 5.0-kg block of 0°C ice from melting completely in 24 h when the surrounding air has a temperature of 27°C. How thick must the chest walls be?

○❖ 13.7 Data are given below for the temperatures T_1 and T_2 and the characteristics of the con-

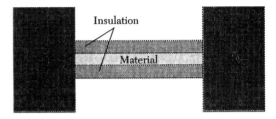

Figure 13.33   Problem 13.7.

ducting material in Figure 13.33. For each set of data, calculate the conduction heat-flow rate in the material and then rank your results in order of increasing flow rate.

	T_1 (°C)	T_2 (°C)	Length (m)	Cross-Sectional Area (cm^2)	Material
a	0	100	0.50	10	Aluminum
b	0	100	0.40	10	Brass
c	0	80	0.30	10	Steel
d	10	60	0.25	8.0	Aluminum
e	20	60	0.25	8.0	Aluminum
f	30	60	0.25	5.0	Steel
g	30	60	0.25	2.0	Brass

● 13.8 A thermopane window consists of two crown-glass panes 0.40 cm thick with a sealed layer of air 1.0 cm thick in between. If the inside temperature is 23°C and the outside temperature is 0°C, determine the conductive heat-flow rate through 1.0 m^2 of the window. Compare this rate with that through a single pane of glass 0.80 cm thick. Assume the temperature gradient in the air layer is linear.

13.2 Convection

13.9 For convection from a vertical surface, the

convection coefficient of Equation 13.2 is approximately $h = 1.75(\Delta T)^{1/4}$ W·m^2·(C°)$^{5/4}$. A radiator is made of several vertical panels through which air can easily pass. The total vertical area is 1.2 m^2. If these panels are kept at 80°C while the air in the rest of the room is at 22°C, calculate the heat-transfer rate.

13.10 The globe of a table lamp is cooled by convection in a 20°C room. When the lamp has a 60-W bulb, the globe has a surface temperature of 70°C. What is the temperature of the globe when a 100-W bulb is used? Assume 95 percent of the 60 W or 100 W is given out as heat.

13.11 In a 20°C room, a boiling cup of tea (100°C) cools to 60°C in 10 min. How long will it take to cool to 40°C?

O 13.12 On a 35°C day, a can of soda warms from 5.0° to 20°C in 15 min. How long does it require to reach 28°C?

13.3 Radiation

13.13 What is the net radiation heat-transfer rate of a nude person with skin area of 1.5 m^2 standing in a room at 25°C? Normal skin temperature is about 35°C. Take the emissivity to be 0.90.

13.14 What is the net radiation heat-transfer rate of a nude person with skin area of 1.3 m^2 standing in a sauna at 45°C? Normal skin temperature is about 35°C. Take the emissivity to be 0.90.

13.15 The temperature of the tungsten filament in a glowing light bulb is 2250 K. Its emissivity is 0.35. What is the surface area of the filament in a 100-W bulb? (*Hint:* The answer is more than you might expect.)

13.16 The tungsten filament in a photographic floodlight operates at 2800 K and has an area of 0.06 cm^2 and an emissivity of 0.35. What is its power?

O 13.17 A lamp bulb inside an ornamental spherical glass shell 20 cm in radius and 0.50 cm thick radiates 100 W of thermal power. What is the difference in temperature between the inner and outer surfaces of the glass?

13.4 Evaporation

13.18 How much energy is transferred as 0.10 kg of water evaporates from a towel at 35°C?

13.19 How much energy is taken away from a jogger as 0.5 kg of water evaporates from her skin at a skin temperature of 35°C?

13.20 Using the specific heat and heat of vaporization of water and the specific heat of steam, show that the value $L'_v = 2.40 \times 10^6$ J/kg is a reasonable estimate for the heat of vaporization of water at 35°C. In other words, calculate the energy required to raise 1.0 kg of water from 35° to 100°C, boil it, and then lower the temperature of the steam from 100° to 35°C (this would be called "supercooled" steam). Then calculate the energy required to evaporate 1.0 kg of water at 35°C. Compare these two values of energy.

*13.5 Solar Energy/Solar Homes

13.21 Victoria, Canada, is located at 48.5° north latitude. How high above the horizon is the Sun on the first day of summer and first day of winter?

13.22 Rio de Janeiro, Brazil, is located at 23° south latitude. How high above the horizon is the Sun on the first day of summer and first day of winter? When *are* the first day of summer and the first day of winter?

13.23 A solar home is to be built in Atlanta, 33.5° north latitude. What is the elevation of the Sun at midday on the first day of summer and on the first day of winter? An overhanging eave 75 cm wide is planned. How far below this eave will sunlight fall on the first day of summer and the first day of winter? Why is this information important in deciding on placement of the windows?

13.24 A solar home is to be built in Minneapolis, 45° north latitude. What is the elevation of the Sun at midday on the first day of summer and on the first day of winter? An overhanging eave 85 cm wide is planned. How far below this eave will sunlight fall on the first day of summer and the first day of winter?

*13.6 Insulation

13.25 Calculate the R value of a wall made of the following layers: 0.50 in. of sheetrock, 3.5 in. of fiberglass insulation, 0.50 in. of exterior sheathing, 4.0 in. of brick. Include the R values from the air layers on both sides. What is the conduction heat-transfer rate through this wall when the inside temperature is 70°F and the outside temperature is 10°F?

13.26 Calculate the R value of an insulated wall like that in Figure 13.30 if 2 × 6's are used for wall studs instead of 2 × 4's. The larger boards allow 5.5 in. of fiberglass insulation instead of 3.5 in. By what percentage does this increase in thickness change the R value? How does the increased thickness change the heat loss calculated in Example 13.14?

○13.27 An ice-cube container is in the form of a cube of outside edge 30 cm, made of plastic insulating material whose thermal conductivity is 0.040 W/m·C°. The base, walls, and lid are 2.0 cm thick. If 7.0 kg of ice at 0°C is placed in the container and taken on a picnic on a day when the temperature is 33°C, how long will it be before all the ice melts?

General Problems

○13.28 One wall of a brick house has an area of 3.0 m × 10 m and is 20-cm thick. How much heat per day flows through the wall if the inside temperature is 20°C and the outside temperature is −10°C? Assume no insulating layer of still air on either side.

○13.29 The crown glass in a storm window is 0.50-cm thick. The outside temperature is −10°C, and the temperature inside the storm window is 0°C. If 1.0 g of moisture condenses on 0.33 m² of the inside surface of the storm window, determine the time it takes for the water to freeze.

○13.30 A roasted turkey cools from 85° to 80°C in 10 min when sitting in a 25°C room. How long does it require to cool from 85° to 55°C?

○13.31 A solar collector has an area of 12.0 m². The collector is thermally insulated, so that conduction is negligible relative to radiation. On a cold but sunny winter day, the temperature outside is −20.0°C, and the sun irradiates the collector with a power per unit area of 300 W/m². Treating the collector as a black body, determine the steady-state temperature inside it.

●13.32 A passive solar home uses a wall of concrete to collect solar energy. The wall has an absorbing area of 20 m² and a mass of 1.0×10^4 kg. During the day, the wall is illuminated for 6.0 h with an average power per unit area of 400 W/m². The wall stores 30 percent of this energy. (*a*) What is the specific heat of the wall? (*b*) What is the increase in the wall's internal energy after 6.0 h of solar illumination? (*c*) If the initial temperature of the wall is 18°C, what is its temperature after 6.0 h of solar illumination?

●13.33 The skin temperature of an adult is about 35°C, maintained by a flow of heat from a central core at 37°C. Metabolism replaces heat lost by conduction through a layer of subcutaneous tissue 3.0-cm thick. The surface area of an adult is about 1.8 m², and the thermal conductivity of tissue is 0.0012 cal/cm·s·C°. (*a*) Estimate the basal metabolic rate in watts. (*b*) To treat severe burns, it is proposed to immerse a patient in a cold-water bath to bring the skin temperature to 25°C. To maintain the core at a life-sustaining 37°C, an additional heat source must be provided, perhaps by diverting some blood through a heat exchanger. Estimate the rate at which heat must be supplied over and above the basal rate calculated in (*a*).

The rather simple hot air balloon offers a good example of a real thermodynamic system. Energy stored in propane molecules is released as heat through burning. The burner's flame heats the air above it, and the hot gas and air mixture rises into the balloon's envelope, which stores the hot air and keeps it separate from the surrounding atmosphere.

As the burner's flame continues to heat the air in the envelope, the pressure of the gas inside increases according to well-known gas laws. This increased pressure pushes on the inside of the balloon inflating it. Once enough hot air is collected in the envelope, the balloon and gondola become buoyant relative to the surrounding atmosphere, allowing the balloon to rise against the force of gravity.

Put more hot air into the balloon and it rises. Turn off the burner, the gas inside cools, and the balloon descends. Unlike ideal thermodynamic systems, the hot air balloon loses some of the heat from burning to the outside environment. What are some of the ways in which hot air balloons lose this heat? How could you modify the balloon to prevent such heat loss? Why is it impossible to construct a hot air balloon that is a perfectly closed thermodynamic system?

(© Karl Hentz/The Image Bank)

14

Thermodynamics and Heat Engines

Key Issues

- An ideal gas can be described by PV/T = constant. P, V, and T are known as "state variables"; specifying two of them fully determines the "state" or condition of an ideal gas. The ideal gas law can be written as $PV = nRT$, where n specifies how much gas is being considered, and R is a universal constant.

- Gas does work as it expands, because it exerts a force (P times area) through a distance, and work is done on a gas as it is compressed. The change in internal energy, ΔU, of a system is equal to the heat Q flowing into it minus the work W done by the system on its surroundings: $\Delta U = Q - W$. Known as the first law of thermodynamics, this is a restatement of the work-energy theorem or the law of conservation of energy.

- Most processes are not reversible. A dropped light bulb shatters, but we do not see the pieces of glass reassembling themselves into a light bulb that bounces upward. The second law of thermodynamics describes this irreversibility of nature. A system's *entropy* is another way of describing the irreversibility of nature.

- Heat engines take the energy available in thermal energy at a higher temperature and turn that into mechanical work plus some thermal energy exhausted at a lower temperature. Ideal heat engines use processes that are conceptually and mathematically easier to handle than real heat engines.

The steam engine, seen here in the guise of a steam locomotive, is symbolic of the understanding of thermodynamic principles by nineteenth century physicists and the application of these principles to technology. (© *Paul Bodine*)

Thermodynamics is the study of the interrelationships among heat, work, and energy. Gas is a common substance that can be used in a heat engine—whether an idealized heat engine operating over idealized cycles or a real heat engine like the one in your car. We discuss the behavior of gases and, after developing the first and second laws of thermodynamics, describe the operation of heat engines, which turn internal energy into usable mechanical energy.

14.1 Gases: Boyle's Law and Charles's Law

CONCEPT OVERVIEW

■ For most gases and under a wide range of conditions, V/T = constant (Charles's law) and PV = constant (Boyle's law).

Perhaps the best description of what a gas really "looks like" is given by Alan Van Heuvelen as he describes the behavior of a gas in a very personal manner: "Suppose for a moment that you have been reduced to the size of a molecule—about 10^{-10} times your normal size. What would life be like if you, in your reduced state, became a molecule of the air? Other molecules would be constantly moving past you in all directions at speeds of about 1000 mi/h; some would be moving faster, others slower.

"On the average, you would be separated from the nearest molecule by a distance of about ten times your own size, a situation comparable to having twenty people spread over the area of a football field. At this separation, you feel little effect of the presence of other air molecules. Nevertheless, your isolation is interrupted by frequent collisions. The collisions are violent, since the molecules move at speeds of about 1000 mi/h. Each time a molecule crashes into you, you are knocked randomly in a new direction. After a collision you move only about 10^{-7} m before being hit by another molecule. During the course of a minute, you experience about 300 billion collisions, each causing your direction of motion to change. Because of the random nature of these collisions, you find after one minute that you have moved only about 4 cm from your starting position, even though your speed during that time has averaged about 1000 mi/h. During this time, you have had no meaningful relations with any particular air molecule, just a succession of abrupt collisions."[1]

In using a constant-volume thermometer to define the Kelvin scale in Chapter 12, we found that temperature and pressure are proportional for a constant volume of gas (at low pressure and well above the condensation temperature):

$$\frac{P}{T} = \text{constant} \qquad \text{(constant volume)} \qquad \textbf{(14.1)}$$

In practice, we often know the pressure and temperature for some original situation and want to find either the pressure or the temperature after the situation has changed. Then it is easier to think of this relationship as

$$\frac{P}{T} = \frac{P_\text{o}}{T_\text{o}} \qquad \text{(constant volume)} \qquad \textbf{(14.1')}$$

[1] Alan Van Heuvelen, *Physics*. Boston: Little, Brown and Co., 1986.

Both Equations 14.1 and 14.1′ mean the same thing, and both are true only for a constant amount of gas kept at a constant volume. Absolute pressure, not gauge pressure, is to be used in both equations.

In a similar way, if we hold the temperature of a gas constant and vary the pressure, we find that the volume changes. You already know that you can squeeze the air in an inflated balloon into a smaller volume by pressing on the balloon and thus increasing the pressure. Were you to measure carefully, you would find that the volume and absolute pressure are inversely proportional:

$$PV = \text{constant} \qquad \text{(constant temperature)} \qquad \textbf{(14.2)}$$

or, when we are comparing some original situation with a later one,

$$PV = P_o V_o \qquad \text{(constant temperature)} \qquad \textbf{(14.2′)}$$

This relationship is known as **Boyle's law** after Robert Boyle (1627–1691), an English physicist and chemist. Equations 14.2 and 14.2′ are for a gas at constant temperature and refer to absolute pressures, not gauge pressures.

If the pressure of a fixed amount of gas is held constant, the volume varies directly as the temperature:

$$\frac{V}{T} = \text{constant} \qquad \text{(constant pressure)} \qquad \textbf{(14.3)}$$

or

$$\frac{V}{T} = \frac{V_o}{T_o} \qquad \text{(constant pressure)} \qquad \textbf{(14.3′)}$$

This relationship is known as **Charles's law** after the French physicist and chemist Jacques Charles (1746–1823).

Charles was a hot-air balloonist and therefore very interested in the volume-temperature relationship.

Boyle's law and Charles's law may be combined as

$$\frac{PV}{T} = \text{constant} \qquad \textbf{(14.4)}$$

or

$$\frac{PV}{T} = \frac{P_o V_o}{T_o} \qquad \textbf{(14.4′)}$$

Temperatures must be in kelvins and pressures must be absolute. This restriction applies throughout our study of thermodynamics.

14.2 Ideal Gases

CONCEPT OVERVIEW

■ Pressure, volume, and temperature are known as state variables because specifying any two of them fully determines the state of an ideal gas.

■ The ideal gas law can be written $PV = nRT$, where n specifies how much gas is being considered and R is a universal constant.

The relationships expressed in Equations 14.1 through 14.4′ accurately describe most dilute gases over a wide range of conditions. As gas concentration

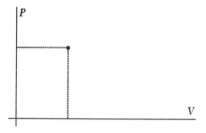

Figure 14.1
The state of a gas is specified by its "coordinates" on a PV diagram.

increases, however, the behavior of almost any gas begins to deviate from that predicted by the equations. Because Boyle's and Charles's laws are so useful, we define an **ideal gas** as one that obeys them exactly. Thus, we speak of ideal gases and *real* gases, the latter being all the gases we encounter in the real world, which do not obey Charles's and Boyle's laws exactly. Equations 14.4 and 14.4′ are called the **ideal gas law.**

Specifying pressure, volume, and temperature tells us what **state** a gas is in. We therefore call these three quantities **state variables,** and equations that describe how they are related to each other and how they change are known as **equations of state.** It is often useful to give information about state variables in terms of a diagram, such as the PV diagram shown in Figure 14.1. Each point on such a diagram (which could also be a PT diagram or a VT diagram) specifies the state of the gas.

The constant in Equation 14.4 is not arbitrary. We can rewrite this equation in the form

$$PV = Nk_BT \qquad (14.5)$$

where N is the **number of molecules** of the gas we are considering and k_B is a universal constant known as **Boltzmann's constant,** $k_B = 1.38 \times 10^{-23}$ J/K. Since the **number** of molecules in almost any sample is enormous, describing a gas in terms of N is cumbersome and measuring N accurately is nearly impossible. To avoid these difficulties, we usually report gas amounts in a unit called the mole. One **mole** (abbreviated mol) of any substance is the number of grams equivalent to the atomic or molecular mass of the substance. Atomic masses for all the elements are given in the Appendix. The atomic mass of carbon, for example, is 12 atomic mass units, so 1 mol of carbon has a mass of 12 g. As another example, adding atomic masses ($12 + 16 + 16$), we get for the molecular mass of CO_2—carbon dioxide—44 atomic mass units, so that 1 mol of CO_2 has a mass of 44 g. Likewise, 88 g of CO_2 represents 2 mol of CO_2, and 9 g of carbon represents 0.75 mol of carbon. In general, the number of moles n in a sample is

$$n = \frac{m}{M} \qquad (14.6)$$

Mole is a quantity word just the way *gross* and *dozen* are; a dozen is 12 of anything, a gross is 144 of anything, and a mole is 6.022×10^{23} of anything.

where m is the mass of the sample in grams and M, called the *molar mass,* is the number of grams in 1 mol of the substance making up the sample.

One **mole** of any material always contains 6.022×10^{23} molecules. This number is known as **Avogadro's number** and is given the symbol N_A; that is, $N_A = 6.022 \times 10^{23}$ molecules/mol. The total number of molecules N in any sample is related to the number of moles n in the sample and Avogadro's number by

$$N = nN_A \qquad (14.7)$$

We can use this relationship to rewrite Equation 14.5 as

$$PV = Nk_BT = n(N_Ak_B)T$$

$$PV = nRT \qquad (14.8)$$

where $R = N_A k_B$ is known as the **universal gas constant** and has the value

$$R = 8.314 \text{ J/mol·K}.$$

Equation 14.8 is the most common form of the **ideal gas law.**

Example 14.1

How many air molecules do you blow into a balloon when you fill it to a volume of 0.0042 m^3 (about 20 cm in diameter) at a pressure of 125 kPa and a temperature of 35°C? What is the mass of this number of air molecules?

Reasoning

Because we know the pressure, volume, and temperature, we can solve the ideal gas law for n, the number of moles of air in the balloon. Knowing n, we can use Equation 14.7 to determine N, the number of molecules. To determine the mass of this number of air molecules, we need to know the molar mass M of air so that we can use Equation 14.6. Air is about 80 percent nitrogen (as N_2 molecules) and 20 percent oxygen (O_2). Therefore

$$M = 0.80(28 \text{ g/mol}) + 0.20(32 \text{ g/mol}) = 29 \text{ g/mol}$$

Solution

$$n = \frac{PV}{RT} = \frac{(125 \text{ kPa})(0.0042 \text{ m}^3)}{(8.314 \text{ J/mol·K})(308 \text{ K})} \left(\frac{10^3 \text{ Pa}}{1 \text{ kPa}} \right) = 0.20 \text{ mol}$$

$$N = nN_A = (0.20 \text{ mol})(6.022 \times 10^{23} \text{ molecules/mol})$$

$$= 1.2 \times 10^{23} \text{ molecules}$$

$$m = (0.20 \text{ mol})(29 \text{ g/mol}) = 5.8 \text{ g}$$

You should check the units in the ideal gas law to ensure that

$$\frac{(\text{Pa})(\text{m}^3)}{(\text{J/mol·K})(\text{K})}$$

does indeed equal moles.

Practice Exercise How much air do you blow into a balloon when you fill it to a volume of 0.0042 m^3 (about 20 cm in diameter) at a pressure of 150 kPa and a temperature of 25°C?

Answer 0.25 mole, 7.3 g

Problem-Solving Tip

▬ The ideal gas law can be used to find a state variable (P, V, and T) at some later time if the state variables are known for an earlier time.

20° C

90° C

Figure 14.2
When pressure is held constant, an increase in temperature brings about an increase in volume.

Example 14.2 _____

Exactly 1.00 l of air at 20.0°C and 1.00 atm is heated to 90.0°C while the pressure is held constant, as sketched in Figure 14.2. To what volume does the air expand?

Reasoning

The ideal gas law in the form of Equation 14.4′ relates the volume, temperature, and pressure after the expansion to the original volume, temperature, and pressure.

Solution

$$\frac{PV}{T} = \frac{P_o V_o}{T_o}$$

$$V = \frac{P_o T}{P T_o} V_o = \frac{T}{T_o} V_o = \frac{363 \text{ K}}{293 \text{ K}} (1.00 \text{ l}) = 1.24 \text{ l}$$

Notice that the temperatures must be Kelvin temperatures.

_____

Practice Exercise   To what volume does the air expand if it is heated to 100°C?

Answer 1.27 l

14.3 The Kinetic Theory of Gases

CONCEPT OVERVIEW

- A gas has pressure because its many individual molecules are constantly colliding with the container holding the gas.

- The ideal gas law can be explained by the constant movement of gas molecules.

While an ideal gas is, by definition, one that obeys the ideal gas law, what is it in other terms? What are its other physical characteristics? An ideal gas is one in which the individual molecules do not interact with each other. They collide only with the walls of their container, and all such collisions are totally elastic. Real gases approach these conditions unless they are under great pressure or near their condensation temperature.

How does a gas—which is mostly vacant space with only sparsely scattered molecules in it—exert a pressure? Boyle's law and Charles's law—or their combination, the ideal gas law—describe gas behavior but offer no clue as to *why* gases behave this way. Over the years, many ideas were proposed to explain gas behavior, but none could provide a complete answer until the kinetic theory of gases was developed in the 19th century. Because this was the time of the Industrial Revolution, there was a great need for understanding the details of such things as steam engines.

Gas molecules are in constant random motion. They collide with the walls of their containers (Fig. 14.3), and it is these collisions we need to look at in

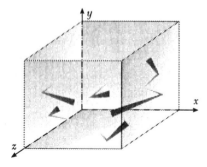

Figure 14.3
Pressure is caused by the collisions of gas molecules with the walls of the container holding the gas.

order to understand pressure. The gas molecules in a container move in all directions with a very wide range of speeds. At any moment, the average velocity is zero because there are as many moving with velocity components along the positive x axis as along the negative x axis and likewise for the y and z axes.

In developing the kinetic theory of gases, we must make some harsh assumptions in order to simplify our mathematical derivations. Assume a cubic container of edge length L with edges aligned along the x, y, and z axes, as in Figure 14.4. Assume that one third of the molecules move along the x axis, one third along the y axis, and one third along the z axis. Assume all molecules move with a *constant* speed v, and collisions with the walls are elastic.

With Figure 14.4 we can direct our attention to one molecule of mass m moving in the positive x direction with speed v. When this molecule collides with the rightmost wall at $x = L$, its momentum changes from $p_i = +mv$ to $p_f = -mv$ since the collision with the wall is elastic. Therefore $\Delta p = p_f - p_i = -2mv$. The wall must exert a force

$$F_{\text{molec,wall}} = \frac{\Delta p}{\Delta t} = \frac{-2mv}{\Delta t} \qquad (14.9)$$

during the time Δt that the molecule is in contact with the wall. The molecule then rebounds from the wall with speed v, moving in the negative x direction. It travels a distance L to the leftmost wall at $x = 0$. It undergoes a collision there and comes back along the positive x axis with speed v, traveling another distance L until it again strikes the wall at $x = L$, where the cycle begins again. The total time t_{tot} for this cycle is the time it takes the molecule to travel a total distance $2L$ at a speed v:

$$t_{\text{tot}} = \frac{2L}{v}$$

The *average* force exerted by the wall, then, is the force calculated in Equation 14.9 multiplied by the fraction of the time the molecule is in contact with the wall, $\Delta t/t_{\text{tot}}$:

$$F_{\text{avg}} = F_{\text{molec,wall}}\left(\frac{\Delta t}{t_{\text{tot}}}\right) = \left(\frac{-2mv}{\Delta t}\right)\left(\frac{\Delta t\, v}{2L}\right) = \frac{-mv^2}{L}$$

This is the average force exerted by the wall on *one* molecule moving in the positive and negative x directions. There are N molecules in the container, and one third of them, $N/3$, move just like this, while another $N/3$ move in the y direction and another $N/3$ in the z direction. Therefore, the total force exerted by the wall on any one of these groups of molecules is

$$F_{\text{molec,wall}} = \left(\frac{N}{3}\right)\left(\frac{-mv^2}{L}\right)$$

Of course we can employ Newton's third law to find the force exerted on the wall by these $N/3$ molecules:

$$F_{\text{wall,molec}} = \frac{N}{3}\frac{mv^2}{L}$$

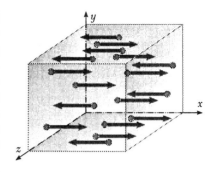

Figure 14.4
In our simplified model, we assume one third of the gas molecules move along the x axis at constant speed v.

Remember our subscript convention from Chapter 4: F_{AB} means the force exerted on body A by body B.

In dealing with gases, we usually need the pressure rather than the force. We can readily calculate pressure by dividing the force on the wall by its area L^2:

$$P = \frac{F}{A} = \frac{F_{\text{wall,molec}}}{L^2} = \frac{1}{3}\left(\frac{Nmv^2}{L^3}\right) = \frac{1}{3}\left(\frac{Nmv^2}{V}\right)$$

so that

$$PV = \tfrac{1}{3}Nmv^2$$

The kinetic energy of each gas molecule is $\tfrac{1}{2}mv^2$; so we can restate this relationship in terms of the molecular kinetic energy:

$$PV = \tfrac{2}{3}N(\tfrac{1}{2}mv^2)$$

Since we know from Equation 14.5 that

$$PV = Nk_{\text{B}}T$$

we can say

$$\tfrac{2}{3}N(\tfrac{1}{2}mv^2) = Nk_{\text{B}}T$$

$$\tfrac{1}{2}mv^2 = \tfrac{3}{2}k_{\text{B}}T$$

Our assumptions call for all molecules to have a single, constant speed v, but this relationship is also valid for a distribution of values if v^2 is replaced with $\langle v^2 \rangle$, the average of the square of the speed of the molecules:

$$\tfrac{1}{2}m\langle v^2 \rangle = \tfrac{3}{2}k_{\text{B}}T \tag{14.10}$$

This average kinetic energy is the average **internal energy** of a molecule in the gas. The total internal energy in the gas, containing N molecules is

$$E_{\text{int}} = \tfrac{3}{2}Nk_{\text{B}}T \tag{14.11}$$

In terms of number of moles rather than number of molecules, we have

$$E_{\text{int}} = \tfrac{3}{2}nN_{A}k_{\text{B}}T = \tfrac{3}{2}nRT \tag{14.12}$$

Equations 14.11 and 14.12 are true only for monatomic gases because we have looked only at the molecular energy due to *translational* motion. Diatomic molecules also have kinetic energy due to their rotational and vibrational motion, and this additional energy is not accounted for in these two equations.

This average of the square of the speed $\langle v^2 \rangle$ deserves some mention. The average *velocity* $\langle \mathbf{v} \rangle$ is zero (unless you throw the container across the room!). The average *speed* $\langle v \rangle$ is not zero but, except for very special cases, it is also *not* the square root of $\langle v^2 \rangle$. The square root of $\langle v^2 \rangle$ is called the **root mean square speed** v_{rms}:

$$v_{\text{rms}} = \sqrt{\langle v^2 \rangle} = \sqrt{\frac{3k_{\text{B}}T}{m}} \tag{14.13}$$

Example 14.3 _____

Consider a collection of ten particles with speeds of 1.0, 2.0, 3.0, 4.0, 5.0, 6.0, 7.0, 8.0, 9.0 and 10.0 m/s. Calculate the average speed $\langle v \rangle$ and the root mean square speed v_{rms}.

Reasoning

The average speed is just what we expect when we use the term *average* or *mean* in everyday conversation—we add up all the values and divide by the total number.

The root mean square speed is something different. First we square all the speeds and then take the average of those values. Doing so gives us the mean of the *squared* values of the speed, however, so we must take a square root of that value.

Solution

$$\langle v \rangle = \tfrac{1}{10} \sum_{i=1}^{10} v_i = \tfrac{1}{10}(1.0 + 2.0 + 3.0 + 4.0 + 5.0 + 6.0 + 7.0 + 8.0 + 9.0 + 10.0) \text{ m/s} = 5.5 \text{ m/s}$$

The root mean square is the square root of the mean (average) of the squares:

$$\langle v^2 \rangle = \tfrac{1}{10} \sum_{i=1}^{10} v_i{}^2 = \tfrac{1}{10}(1.0^2 + 2.0^2 + 3.0^2 + 4.0^2 + 5.0^2 + 6.0^2 + 7.0^2 + 8.0^2 + 9.0^2 + 10.0^2) \text{ m}^2/\text{s}^2 = 38.5 \text{ m}^2/\text{s}^2$$

$$v_{rms} = \sqrt{\langle v^2 \rangle} = 6.2 \text{ m/s}$$

Clearly $\langle v \rangle$ and v_{rms} are different things. If our data set included negative values, the difference between $\langle v \rangle$ and v_{rms} would be even more striking.

Example 14.4 _____

What is the root mean square speed of helium atoms at room temperature (25°C)?

Reasoning

Once we determine the mass of one helium atom, we can apply Equation 14.13 directly. Helium is monatomic, and we can find the mass of one atom by using the molar mass and Avogadro's number.

Solution

The mass of one helium atom is

$$m = \frac{M}{N_A} = \frac{0.0040 \text{ kg/mol}}{6.022 \times 10^{23} \text{ atoms/mol}} = 6.6 \times 10^{-27} \text{ kg/atom}$$

Unit check: $\left(\dfrac{J}{K}\right)\left(\dfrac{K}{kg}\right) = \dfrac{J}{kg} =$
$\dfrac{N \cdot m}{kg} = \dfrac{(kg \cdot m/s^2) \cdot m}{kg} = \dfrac{m^2}{s^2}$

Using this value in Equation 14.13, we get

$$v_{rms} = \sqrt{\frac{3(1.38 \times 10^{-23}\,\text{J/K})(298\,\text{K})}{6.6 \times 10^{-27}\,\text{kg}}}$$

$$= 1400\,\text{m/s} = 4900\,\text{km/h} = 3000\,\text{mi/h}$$

Speed in meters per second is preferable for solving physics problems, but most of us still *think* in kilometers per hour or—horrors!—even miles per hour.

Because of their very small mass, the speed of helium atoms is extremely high. So high, in fact, that helium atoms in the upper atmosphere often reach escape speeds and leave Earth's atmosphere.

Practice Exercise   What is the rms speed of neon atoms at 90°C?

Answer   670 m/s

Equation 14.11 was developed for molecules considered to be point particles and is valid for monatomic species like helium. Diatomic and polyatomic molecules, like O_2, N_2, CO_2, H_2O, and NH_3, can also have internal energy because of rotation. When rotational motion—referred to as an additional **degree of freedom**—is taken into account, Equation 14.11 becomes

$$E_{int} = \tfrac{5}{2}Nk_BT \tag{14.14}$$

In terms of n, the number of moles, we have

$$E_{int} = \tfrac{5}{2}nRT \tag{14.15}$$

Despite the additional energy due to rotation, the root mean square speed of polyatomic gases is the same as for monatomic gases:

$$\tfrac{1}{2}m\langle v^2 \rangle = \tfrac{3}{2}k_BT$$

$$v_{rms} = \sqrt{\langle v^2 \rangle} = \sqrt{\frac{3\,k_BT}{m}}$$

This Pratt & Whitney jet engine burns fuel (in the presence of air) to create a hot gas that is compressed by the large fan or turbine (at the front of the engine) and then exhausted at high speed out the rear. The expanding hot gas creates thrust that can power airplanes equipped with such engines. Though a complex device, this engine uses the basic principle of an expanding gas to do work.   *(Courtesy of Pratt & Whitney, Inc.)*

14.4 Work Done by a Gas

CONCEPT OVERVIEW

■ When a gas expands, it does work on its surroundings. That work is equal to the area under the curve on the PV diagram that describes the expansion.

Energy can be used to do work, and a gas has internal energy. Therefore a gas should be able to do work. How does it do work, and how can we calculate the amount of work done? As we answer these questions, we must be very careful to distinguish between work done *on* the system and work done *by* the system. Perhaps more than usual, we must carefully define what we mean by *system*. Everything else is "the outside" or "the surroundings" or "the environment."

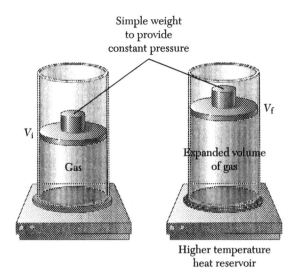

Figure 14.5
Work is done as a gas expands under constant pressure.

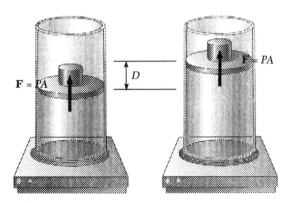

Figure 14.6
The work done by a gas as it expands at constant pressure is $W = P \, \Delta V$.

Figure 14.5 shows a cylindrical container of gas at some pressure. A piston free to move up and down keeps the gas in the container. In addition to the atmospheric pressure that may be present, a mass sits on the piston so that there is some additional force exerted. The gas exerts a force on the piston to balance this force. We can describe the gas by specifying its initial state—in other words, by giving its pressure P_i and volume V_i. We shall assume throughout that we are dealing with an ideal gas. Therefore, if we know P_i and V_i (and n), we also know T_i from the ideal gas law.

Isobaric Processes

As illustrated in Figure 14.5, let a gas expand to a new volume V_f while the pressure remains constant; that is, $P_i = P_f = P$. How much work has been done by the gas, which is our system? If the cross-sectional area of the container is A, then the force on the piston is $F = PA$. If the piston moves a distance D, as shown in Figure 14.6, the work done *on* the piston *by* the gas is

$$W = FD$$

since F and D are in the same direction. As usual when dealing with fluids, it is more useful to talk of pressure rather than forces:

$$W = PAD$$

Since AD is the *change* in volume ΔV, we see that, at constant pressure P, the work done by a gas as it expands and its volume changes by ΔV is

$$W_{\text{isobaric}} = P \, \Delta V \qquad \textbf{(14.16)}$$

Isobaric has two Greek roots: *iso-* meaning "the same" and *bar-* meaning "weight (or pressure)."

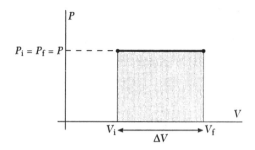

Figure 14.7
An expansion at constant pressure is represented on a PV diagram by an isobar, a straight, horizontal line indicating constant pressure. The work done by the gas is the area under the isobar: $W = P(V_f - V_i) = P\,\Delta V$.

A constant-pressure process like this is known as an **isobaric** process.

Figure 14.7 displays this information on a PV diagram, where the work done is the area under the curve. The curve connects the initial, final, and all intermediate states through which the system passes. In our present example, that of expansion at constant pressure, the curve is a straight line at constant pressure. Because it describes an isobaric process, this curve is called an **isobar.**

Plotting this curve on a PV diagram implies that the process is reversible and that P and V are well defined at each point along the path of the curve. Many real expansions—such as an explosion—are not reversible and therefore cannot be plotted on a PV diagram.

For more complicated processes, such as the one shown in Figure 14.8, the work done by the gas is still the area under the PV curve. Each point on the curve represents an equilibrium state, and each curve represents an equilibrium path (which means we move reversibly along the curve). You can approximate the area under a PV curve for *any* process by a series of little rectangles,

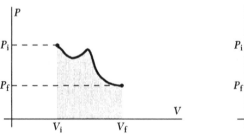
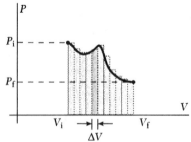

Figure 14.8
The work done by a gas in going from an initial state to a final state is the **area** under the curve on a **PV** diagram. The curve passes through all intermediate states of the system.

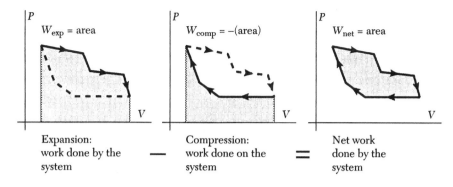

Figure 14.9
During one cycle, the net work done by the system is equal to the area bounded by the PV curve.

each having as one side a constant pressure P. The work done by the gas is $P\,\Delta V$ for each rectangle, and the total work is the sum of these, $\Sigma\,P\,\Delta V$, which is the area under the curve. The work done by the gas is positive if $V_f - V_i > 0$ and negative if $V_f - V_i < 0$ (negative work *by* the gas is positive work *on* the gas).

Work is done *by* the system during an expansion and *on* the system during a compression. If the system undergoes a complete cycle, as shown in Figure 14.9, the net work done by the system is equal to the area bounded by the PV curve describing the cycle. During the expansion part of the cycle, when $\Delta V > 0$, work is done *by* the system and $W_{exp} > 0$. During the compression part of the cycle, when $\Delta V < 0$, work is done *on* the system and $W_{comp} < 0$. The net work for one cycle is the sum of these, and this sum is the area bounded by the PV curve.

Example 14.5

A cylindrical container with a movable piston, like the ones in Figures 14.5 and 14.6, has a cross-sectional area 0.15 m². A constant pressure of 3.5×10^5 N/m² is exerted by the piston on a gas in the container. As the gas is heated, the piston rises 15 cm. How much work has been done by the gas?

Reasoning

The work done in expanding a gas at constant pressure is given by Equation 14.16. We are given the pressure and must find the volume change.

Solution

$$\Delta V = AD = (0.15\ m^2)(0.15\ m) = 0.0225\ m^3$$

$$W = P\,\Delta V = (3.5 \times 10^5\ N/m^2)(0.0225\ m^3) = 7875\ J \approx 7900\ J$$

Practice Exercise How much work is done if the constant pressure is 4.5×10^5 N/m² and the piston expands 25 cm?

Answer $16\,875\ J \approx 17\,000\ J$

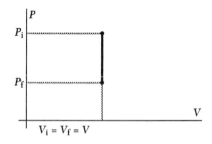

Figure 14.10
No work is done by a gas in an isometric process.

The Greek word is *isometros*, "of equal measure."

Isometric Processes

In the **isometric** (constant volume, also sometimes called **isochoric**) process sketched in Figure 14.10, the area under the curve is zero. There is *no work* done by a gas held at constant volume. (The vertical line is called an **isomet**.)

Isothermal Processes

Figure 14.11 sketches an expansion at constant temperature, called an **isothermal** expansion. The container is placed in a **heat reservoir**, which is a large, massive body that can provide heat to (or absorb heat from) the gas in the container without any significant change in its own temperature. Throughout the expansion, the heat reservoir provides or absorbs just enough heat to keep the gas at a constant temperature. You might think of a huge lake as a heat reservoir. With such a large mass of water, we can transfer heat to it without making a measurable change in its temperature. This idea of a heat reservoir will be important in our study of heat engines later in this chapter.

Figure 14.12 shows an isothermal process on a PV diagram. This curve, called an **isotherm**, is a hyperbola since

$$PV = nRT = \text{constant}$$

As always, the work done by the gas as it goes from initial volume V_i to final volume V_f along the isotherm is the area under the isotherm between the initial and final points. Calculating this amount of work requires integral calculus, so we simply state the result:

$$W_{\text{isotherm}} = nRT \ln\left(\frac{V_f}{V_i}\right) \tag{14.17}$$

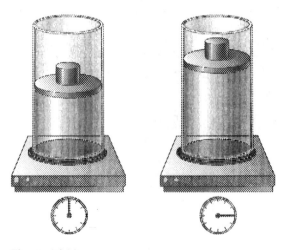

Figure 14.11
An isothermal expansion is an expansion at constant temperature. The temperature remains constant because the gas container is sitting in a heat reservoir.

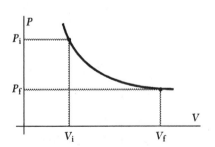

Figure 14.12
The curve on a PV diagram for an isothermal process is a hyperbola.

The function $\ln x$ is positive for $x > 1$ but negative for $x < 1$. Therefore $\ln (V_f/V_i)$ is positive for an expansion ($V_f > V_i$) and negative for a compression ($V_f < V_i$). That is, the work done by the system is positive in an isothermal expansion and negative in an isothermal compression. (The negative work in the compression is another way of saying positive work is done on the system by the outside.)

Example 14.6

If 1.00 l of helium at 23.0°C and 1.00 atm expands isothermally to 1.50 l, how much work does the helium do on its surroundings?

Reasoning

The work done in an isothermal expansion is given by Equation 14.17. Since we are given the temperature and the initial and final volumes, all we must find is n, the number of moles of helium in the system, and then we can use Equation 14.17. To find n, we use the ideal gas law.

Solution

$$n = \frac{PV}{RT} = \frac{(1.00 \text{ atm})(1.00 \text{ l})}{(8.31 \text{ J/mol} \cdot \text{K})(296 \text{ K})} \left(\frac{1.013 \times 10^5 \text{ Pa}}{1 \text{ atm}} \right) \left(\frac{0.001 \text{ m}^3}{1 \text{ l}} \right)$$

$$= 0.0412 \text{ mol}$$

$$W_{\text{isotherm}} = nRT \ln \left(\frac{V_f}{V_i} \right)$$

Take the natural logarithm of 1.50 by using the "ln" key on your calculator:

$$W = (0.0412 \text{ mol})(8.31 \text{ J/mol} \cdot \text{K})(296 \text{ K}) \ln(1.5) = 41.1 \text{ J}$$

Note that in using Equation 14.17, we did not need to convert volumes to cubic meters because the two volumes are used in a ratio and therefore the units cancel. When working with the ideal gas law in the first part of this solution, of course, we did have to make the conversion because our R value is for pascals, cubic meters, and kelvins.

Practice Exercise If 2.00 l of helium at 27.00°C and 1.00 atm expand isothermally to 4.00 l, how much work does the helium do on its surroundings?

Answer 140 J

Adiabatic Processes

Figure 14.13 sketches another expansion of interest to us: an **adiabatic** expansion, defined as one in which the system is thermally insulated from its surroundings. Therefore no heat can flow into or out of the system ($\Delta Q = 0$). The

Yet another Greek origin: *abiabatos* is Greek for "impassable." An adiabatic system is one through which no heat can pass.

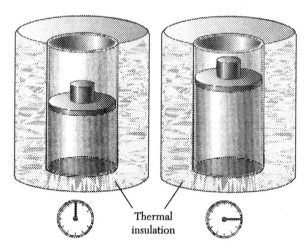

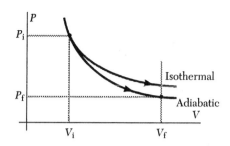

Figure 14.14
When an isothermal expansion and an adiabatic one both begin at P_i and V_i, the isothermal expansion produces more work than the adiabatic expansion. The area under the isotherm is greater than the area under the adiabat, or, to say the same thing another way, the adiabat is steeper than the isotherm.

Figure 14.13
In an adiabatic expansion, the system is thermally insulated from its surroundings. Therefore no heat can flow into or out of the system.

Adiabatic and *thermal* are not the same thing. There can be heat flow in an isothermal process and temperature changes in an adiabatic process, but not vice versa. This harks back to Chapter 12, where we learned that temperature and heat are not the same thing.

curve in Figure 14.14 that corresponds to this adiabatic process is called an **adiabat.** Since the gas is thermally insulated ($\Delta Q = 0$) when it expands and thereby does work ($\Delta W > 0$) on its surroundings, it must use its own internal energy ($\Delta E_{int} < 0$) and therefore its temperature decreases ($\Delta T < 0$). By Charles's law, a decrease in temperature means a decrease in volume. Therefore, on a *PV* diagram, an adibat will always be steeper than an isotherm that passes through the same state; this can be seen in Figure 14.14.

Problem-Solving Tip

- Finding the work done by a gas almost always involves finding the area under a *PV* curve.

Example 14.7

How much work is done by the gas undergoing the process shown in Figure 14.15 as it expands from 1.00 to 3.00 l?

Reasoning
The work done is equal to the area under the curve. Because this particular curve is made of straight lines, the calculation is straightforward.

Solution
By inspection, we can see that the area under the curve is the sum of the areas of two rectangles and a triangle. The area of each rectangle is

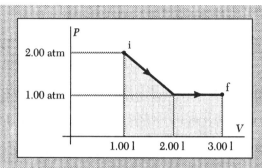

Figure 14.15
How much work is done by the gas as it expands from $V_i = 1.00\,l$ to $V_f = 3.00\,l$?

$1.00\,l \cdot atm$, and the area of the triangle is $0.500\,l \cdot atm$, so that the total area is $2.50\,l \cdot atm$. We need to change these units to joules:

$$W = 2.50\,l \cdot atm\left(\frac{0.001\ m^3}{1\,l}\right)\left(\frac{1.013 \times 10^5\ Pa}{1\ atm}\right)\left(\frac{N/m^2}{1\ Pa}\right) = 253\ J$$

14.5 The First Law of Thermodynamics

CONCEPT OVERVIEW

- According to the first law of thermodynamics, the change in internal energy of a system is equal to the heat that flows into the system minus the work done by the system on its surroundings: $\Delta E_{int} = Q - W$.

- The first law of thermodynamics is a restatement of the work-energy theorem and the conservation of energy concept.

A system has internal energy as a result of the random motion of its molecules. We defined this internal energy in Equations 14.11, 14.12, 14.14, and 14.15. Heat Q can flow into a system and increase its internal energy or out of a system and decrease its internal energy. A system can do work W on its environment, or work can be done on a system by its environment. Heat and work are examples of energy transfer. The concept of conservation of energy leads us to believe that, if heat Q flows into a system and the system does work W on its environment, the internal energy of the system changes by an amount ΔE_{int} given by

$$\Delta E_{int} = Q - W \qquad (14.18)$$

This is known as the **First Law of Thermodynamics,** as first mentioned in Section 12.5. In defining the first law this way, we are using the convention that Q is positive if heat flows into the system and W is positive if work is done by the system.

The first law of thermodynamics truly is a grand law of physics, on the level of Newton's laws of motion and his law of universal gravitation. No exceptions to the first law have ever been found. It is a restatement of the work-energy theorem and of the conservation of energy concept.

Problem-Solving Tips

- The change in internal energy of a system, the heat flowing into or out of the system, and the work done by or on the system are related by the first law of thermodynamics. Any one of these variables can be determined when the other two are known.
- The work can often be found as the area under a *PV* curve.

Example 14.8 _____

A gas is confined to a cylindrical container with a movable piston. While the pressure is maintained constant at 2.00 atm, the volume increases by 0.0150 m³ and 589 J of heat flows into the system, as sketched in Figure 14.16. What is the change in the internal energy of the system?

Reasoning

The change in internal energy is the difference between the heat that flows into the system and the work done by the system (Equation 14.18). We are given the value of the heat and can calculate the amount of work done.

0.0150 m^3

$Q = 589 \text{ J}$

Figure 14.16
What is the change in the internal energy of this system?

Solution

$$W = P\,\Delta V = (2.00\text{ atm})(0.0150\text{ m}^3)\left(\frac{1.013 \times 10^5\text{ Pa}}{1\text{ atm}}\right) = 3039\text{ J}$$

$$\Delta E_{int} = Q - W = 589\text{ J} - 3039\text{ J} = -2450\text{ J}$$

The internal energy *decreases* by 2450 J. That means the gas is cooled by this process.

Practice Exercise If the pressure is maintained constant at 3.00 atm while the volume increases by 0.0250 m³ and 1650 J of heat flows into the system, what is the change in the internal energy of the system?

Answer −5950 J

Example 14.9

One liter of air at standard temperature (0°C or 273 K) and standard pressure (1.00 atm) is enclosed in a rigid, insulated container (Fig. 14.17). The container is in thermal contact with a heat reservoir through its bottom surface but is otherwise thermally insulated from its surroundings. If 10.0 J of heat is added to this system, what is the final temperature?

Reasoning

This is another application of the first law of thermodynamics. Because the container is rigid, the volume of the gas does not change. Therefore no work is done on or by the system, so $W = 0$ and all the heat goes into changing the internal energy.

Solution

$$\Delta E_{int} = Q - W = 10.0\text{ J} - 0 = E_{int,f} - E_{int,i}$$

$$E_{int,f} = E_{int,i} + 10.0\text{ J}$$

Air is mostly a mixture of nitrogen and oxygen, both diatomic gases. Therefore, the internal energy is given by Equation 14.15. From the ideal gas law, we find n, the number of moles of air, and then we can use Equation 14.15:

$$n = \frac{PV}{RT} = \frac{(1.00\text{ atm})(1.00\text{ l})}{(8.31\text{ J/mol·K})(273\text{ K})}\left(\frac{1.013 \times 10^5\text{ Pa}}{1\text{ atm}}\right)\left(\frac{0.001\text{ m}^3}{1\text{ l}}\right)$$

$$= 0.0446\text{ mol}$$

The initial internal energy is

$$E_{int,i} = \frac{5}{2}nRT_i = \frac{5}{2}(0.0446\text{ mol})(8.31\text{ J/mol·K})(273\text{ K}) = 253\text{ J}$$

Remember from Figure 14.10 that there is no "area under the curve" in an isometric process.

Figure 14.17
What is the final temperature of this system?

So the final internal energy is

$$E_{int,f} = E_{int,i} + 10.0 \text{ J} = 263 \text{ J}$$

Now that we know $E_{int,f}$, we can use Equation 14.15 again to obtain the final temperature:

$$E_{int,f} = \frac{5}{2} nRT_f$$

$$T_f = 284 \text{ K} = 11°C$$

Practice Exercise   If 20.0 J of heat is added, what is the final temperature?

Answer   295 K = 22°C

Now we can return to the adiabatic process of Figure 14.14 and see how the adiabat compares with the isotherm. In the isothermal expansion, the temperature remains constant, so the internal energy is also constant, $\Delta E_{int} = 0$. Therefore, $Q = W_{iso}$; the heat that flows into the system equals the work done by the system. In the adiabatic expansion, the system is thermally insulated so that no heat flows into it, $Q = 0$. Therefore, $-\Delta E_{int} = W_{ad}$; the internal energy decreases by an amount equal to the work done by the system. Since the internal energy decreases, the temperature decreases during the adiabatic expansion. If both expansions involve the same initial and final volumes, as shown in Figure 14.14, the lower final temperature requires a lower final pressure for the adiabatic expansion, as shown.

Figure 14.18 shows an isothermal compression and an adiabatic one. At the same final volume V_f, the final pressure is higher for the adiabatic compression, meaning that, in accordance with the ideal gas law, $PV/T = $ constant, the temperature rises in an adiabatic compression.

A prime example of this is the compression stroke in a Diesel engine. The volume is compressed by a factor of about fifteen or more. The compression occurs so rapidly that there is essentially no heat exchanged with the environment; hence we can treat this as an adiabatic process. This compression greatly increases the pressure and the temperature; the temperature increases so much that the fuel will ignite without the need of a spark plug.

Figure 14.18
The temperature increases during an adiabatic compression. For a given volume, an increase in pressure requires an increase in temperature, according to the ideal gas law, $PV/T = $ constant.

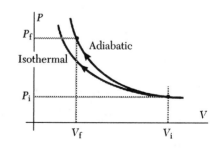

Air usually expands as it rises, and this expansion can be nearly adiabatic so that the air becomes cooler. This is why it gets cooler as you drive up into the mountains. This air movement can also act in reverse. If weather conditions are such that air rushes down a mountainside, it will be compressed as a result of the greater air pressure at lower elevation. If this is a nearly adiabatic compression, the air is warm when it reaches the foot of the mountains. The Chinook winds in the Rocky Mountains and the Santa Anna winds in southern California are examples of this warming.

14.6 Irreversibility and the Second Law of Thermodynamics

CONCEPT OVERVIEW

■ Most natural processes are not reversible, and the second law of thermodynamics describes this irreversibility.

Many of the processes we describe in physics we treat as if they were completely reversible. For instance, if we videotape a swinging frictionless pendulum and show the tape backwards, it will be indistinguishable from the tape running forward. However, most processes in nature are *not* reversible. A real pendulum encounters the friction of air resistance, and consequently its amplitude gets smaller and smaller. A videotape of a real pendulum run in reverse would show an increasing amplitude, and therefore, it would be easy to tell whether the tape were running forward or in reverse. Drop a glass and, when it strikes the floor, it breaks. An ice cube placed in a cup of hot tea melts. Likewise, although the first law allows a pendulum's amplitude to increase with time, as air molecules give up their energy to the pendulum, we always see a pendulum winding down to smaller and smaller swings. And the first law would not be violated if the floor molecules gave some of their internal energy to the pieces of your broken glass so that the shards came back together and the now-whole glass jumped up from the floor and back into your hand.

Irreversible processes like these abound. We *never* see a cup of lukewarm tea re-ordering the internal energy of its molecules so that some of them give up energy and turn into an ice cube while others move more rapidly and become hotter, even though such an occurrence is in no way forbidden by the first law of thermodynamics and energy would be conserved.

There is another law, the second law of thermodynamics (what else?), that explains this one-way-only behavior. Throughout the remainder of this chapter, it will be stated in several different but equivalent ways. The simplest statement, given by Rudolph Clausius (1822–1888), is that

> *Heat does not spontaneously flow from a low-temperature region to a high-temperature region.*

This is shown schematically in Figure 14.19.

In the discussion of heat engines that follows, we are going to assume *reversible* processes even though we know they almost never exist in the real world. Making this assumption will greatly simplify our calculations. To get an

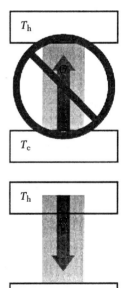

Figure 14.19
Heat never flows spontaneously from a cold body to a hot body. This statement is one form of the second law of thermodynamics.

Figure 14.20
As an attempt to create a truly reversible process, we pile sand onto a piston one grain at a time. The pressure increases so slowly that the gas in the system is always in equilibrium. When the sand is removed one grain at a time, the process is reversed.

idea of what a real-world reversible system might be, consider Figure 14.20, where a *frictionless* piston contains gas in a cylinder. Sand is added to the piston one grain at a time, very slowly increasing the pressure on the gas. The temperature of the gas changes very slowly also, so at every moment the gas is in equilibrium. In the reverse process, the sand can be removed one grain at a time. The gas will remain in equilibrium at every moment, with its pressure and temperature uniform throughout.

14.7 Heat Engines

CONCEPT OVERVIEW

- A heat engine converts internal energy to mechanical energy.

- The operation of a reversible heat engine can be described by a *PV* diagram.

- The efficiency of a reversible heat engine depends upon the temperatures between which it operates.

It is easy to turn mechanical work into internal energy. Rub a piece of sandpaper over a piece of wood and they both get warm. Two sticks rubbed together vigorously get hot enough to burn kindling. Rubbing your hands together warms them, and on and on. Turning internal energy into mechanical

In a nuclear power plant enriched uranium is brought together in a "critical mass" that starts and sustains a chain reaction. During this chain reaction the uranium atoms undergo fission, releasing large quantities of heat and radiation in the form of high-speed neutrons. This heat is used to boil water to produce steam that turns generators that produce electricity. The nuclear reactor is housed in a containment building (*above*). Water used to cool the reactor usually comes from a nearby river or lake, which acts as a heat sink. (Why do real thermodynamic processes always produce some amount of rejected or waste heat?) The large cooling tower (*center*) cools down this recycled water before it is returned to the environment, which helps to protect the temperature-sensitive ecology of rivers and lakes. (*Courtesy of U.S. Department of Energy*)

work is much more difficult, however. For that you need a **heat engine**—a device that converts internal energy to usable mechanical work. Examples include converting the internal energy of burning coal or fissioning uranium to turn the shaft of a generator to create electricity and converting the internal energy of burning gasoline to turn the tires on your car.

A schematic of a heat engine is shown in Figure 14.21, with the engine represented by an oval. The heat engine absorbs heat Q_h from a hot reservoir at temperature T_h, produces work W, and expels heat Q_c to a cold reservoir at temperature T_c. We consider only cyclic heat engines—that is, those that run a **working substance** through a repeatable cycle. Because the working substance returns over and over to its initial state, the change in its internal energy must be zero. Therefore, from the first law of thermodynamics, the work done by the engine must equal the net heat that flows into the engine:

$$W = Q_h - Q_c$$

Efficiency, in physics as in ordinary conversation, describes how well work is done or compares how much is accomplished to how much effort is put into something. The efficiency of a heat engine is defined as the ratio of the work produced to the heat absorbed:

$$Eff = \frac{W}{Q_h} = \frac{Q_h - Q_c}{Q_h} = 1 - \frac{Q_c}{Q_h} \tag{14.19}$$

To have an efficiency of 1.00, or 100 percent, there must be no heat exhausted ($Q_c = 0$), but, as we shall see later, having $Q_c = 0$ is impossible. Since exhausted heat includes frictional losses, our definition of efficiency means that reducing friction increases efficiency.

The French engineer Sadi Carnot (1796–1832) proposed that maximum efficiency is obtainable from a heat engine whose processes are reversible. A reversible process is illustrated in Figure 14.22. A cylinder with a frictionless piston is placed in thermal contact with the surroundings. Gas inside the cylinder absorbs heat Q from the surroundings and does work W on the surroundings as it expands. If the gas then returns to its original state, work W is done by the surroundings on the gas (or the gas does work $-W$) and the same amount of heat that initially flowed into the cylinder now flows out. No work is lost to friction; the system is back in its original state, and the surroundings are back in their original state. (Remember, friction always makes a process irreversible.)

Carnot's proposal is now known as **Carnot's principle:**

An irreversible heat engine operating between two heat reservoirs at constant temperatures cannot have an efficiency greater than that of a reversible heat engine operating between the same two temperatures.

Carnot's principle carries a corollary:

All reversible heat engines operating between the same temperatures have the same efficiency.

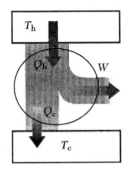

Figure 14.21
A heat engine absorbs heat Q_h from a hot reservoir, produces work W, and expels heat Q_c to a cold reservoir.

Sadi Carnot (1796–1832), who died suddenly of cholera, is considered to be the founder of thermodynamics. He was the first to show the quantitative relationship between work and heat.

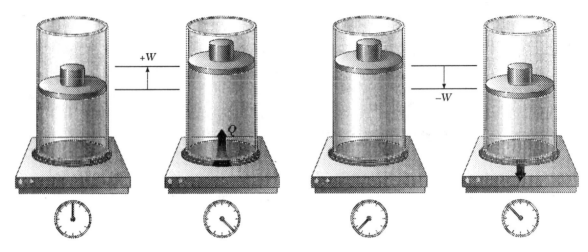

Figure 14.22
A reversible process returns both the system and its surroundings to their initial states.

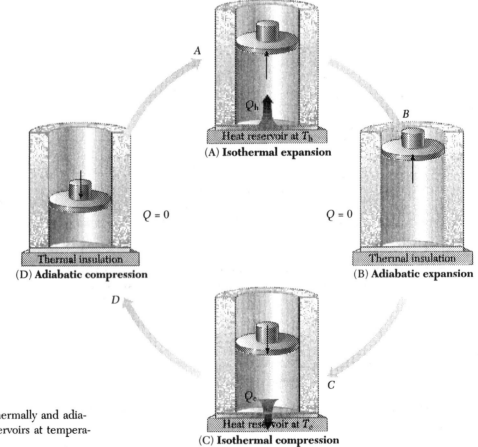

Figure 14.23
A Carnot engine operates isothermally and adiabatically between two heat reservoirs at temperatures T_h and T_c.

Carnot's principle turns out to be another statement of the second law. If this principle were violated—if an irreversible heat engine had an efficiency greater than that of a reversible heat engine—heat engines could be connected such that the net result would be spontaneous heat flow from a cold body to a hot body.

A **Carnot engine** (Figs. 14.23 and 14.24) is one that is reversible and whose heat transfers are isothermal. From state A, the working substance (for example, a gas) in a Carnot engine undergoes an isothermal expansion at a hot temperature T_h until it reaches state B. The engine is then thermally insulated from the environment so that the gas can undergo an adiabatic expansion to state C. As the gas expands adiabatically, the engine cools to a temperature T_C. The gas then undergoes an isothermal compression while in contact with this cold reservoir at T_c; this compression brings the gas to state D. From there it is thermally insulated and compressed adiabatically so that it returns to its original state A.

By definition, no heat flows into or out of the Carnot engine during its two adiabatic processes. It can be shown that the amounts of heat Q_h and Q_c that flow into and out of the engine during the nonadiabatic parts of the cycle are proportional to the absolute temperatures of the two reservoirs. Thus the efficiency of a Carnot engine is

$$Eff = \frac{W}{Q_h} = \frac{Q_h - Q_c}{Q_h} = \frac{T_h - T_c}{T_h}$$

$$Eff = 1 - \frac{T_c}{T_h}$$

(14.20)

Efficiency depends upon the absolute temperatures of the two heat reservoirs. The only way the engine could have an efficiency of 1.00 would be if the cold temperature were absolute zero!

The Carnot cycle is important because it uses only isothermal and adiabatic processes, which make calculations relatively easy. However, it can be shown that Equation 14.20 gives the efficiency for *any* reversible heat engine operating between two temperatures T_h and T_c. Because efficiency of *any* heat engine can never be greater than this efficiency for a reversible heat engine, Equation 14.20 gives a *maximum efficiency*. The efficiency becomes greater as T_h becomes greater.

Another form of the second law of thermodynamics is

It is not possible to make a heat engine whose only effect is to absorb heat from a high-temperature region and turn all that heat into work.

Such an impossible heat engine is illustrated in Figure 14.25.

Is this statement really equivalent to the statement given by Clausius and illustrated in Figure 14.19? Let us take the work output of the forbidden heat engine of Figure 14.25 and use that as the *input* to an allowable heat engine, as shown in Figure 14.26. The net effect of doing so is that heat Q_c flows from a cold region to a hot region. Since this direction of heat flow is forbidden by the second law, our two statements are equivalent.

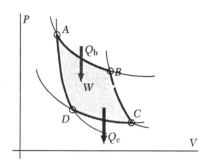

Figure 14.24
The work done by a Carnot engine during one cycle is the area bounded by the isotherms and adiabats that describe the cycle.

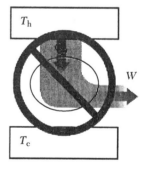

Figure 14.25
It is not possible to make a heat engine whose only effect is to absorb heat from a high-temperature reservoir and turn all of that energy into work. This statement is another form of the second law of thermodynamics.

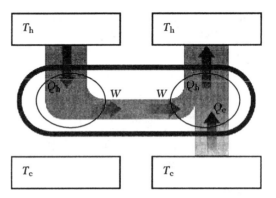

Figure 14.26
Using the work output of a heat engine that violates the second law produces a situation in which the net effect is that heat Q_c flows spontaneously from a reservoir at a cold temperature to a reservoir at a hot temperature, a violation of the second law.

Example 14.10

A Carnot engine operates with a cold reservoir of 350 K with an efficiency of 0.35. What is the temperature of the hot reservoir?

Reasoning

In a Carnot engine the efficiency is determined by Equation 14.20 which depends only upon the temperatures of the hot and cold reservoirs.

Solution

$$Eff = 1 - \frac{T_c}{T_h}$$

$$0.35 = 1 - \frac{350 \text{ K}}{T_h}$$

$$\frac{350 \text{ K}}{T_h} = 1 - 0.35 = 0.65$$

$$T_h = \frac{350 \text{ K}}{0.65}$$

$$T_h = 540 \text{ K}$$

Remember that the temperatures in Equation 14.20 must be the absolute temperatures—temperatures measured in kelvins. If the temperature of the cold reservoir had been given as 77°C, we would have had to convert it to 350 K before using it in Equation 14.20.

*14.8 Real Heat Engines

CONCEPT OVERVIEW

■ The operation of real heat engines can be approximated by *PV* diagrams.

How do the Carnot engine and the *PV* diagram of Figure 14.24 compare with the real engines used in automobiles? Figure 14.27 illustrates a four-stroke internal combustion engine—the gasoline engine most of us have in our automobiles. In (a) the intake valve opens and the piston drops down, increasing the volume in the cylinder and bringing in an air-fuel mixture. In (b) the valves are closed and the piston moves up, compressing the air-fuel mixture. The *compression ratio* is the ratio of the maximum volume at the beginning of this stroke to the minimum volume at the end of this stroke; typical compression ratios for gasoline engines range from about 8:1 to 10:1. This compression also increases the temperature. In (c), the spark plug ignites the air-fuel mixture, and in (d) the burning gases expand, doing work on the piston as it moves down. In (e) the exhaust valve opens and the piston moves up and exhausts the burned gases from the cylinder.

The gasoline-powered internal combustion engine is an example of a modern heat engine. (© *David Rogers*)

The processes taking place in an internal combustion gasoline engine are difficult to show accurately on a *PV* diagram, and it is more difficult still to calculate the heat flow and work done. Figure 14.28 shows a *PV* diagram for an **Otto cycle.** This is not "auto" but "Otto," after Nicholas Otto (1832–1891), a German engineer who invented the four-stroke engine in 1876. The Otto cycle is similar to the cycle in a gasoline engine. The compression stroke corresponds to the process from point *B* to point *C*. From *C* to *D*, the volume does not change but the pressure increases greatly; this corresponds to ignition as the temperature and pressure of the burning gas increase. The expansion from *D*

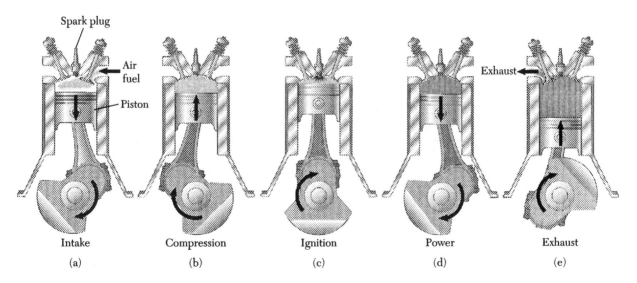

| Intake | Compression | Ignition | Power | Exhaust |
| (a) | (b) | (c) | (d) | (e) |

Figure 14.27
A four-stroke internal combustion engine.

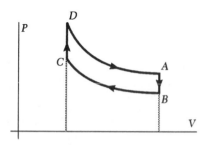

Figure 14.28
The *PV* diagram for an Otto cycle heat engine, which resembles an ordinary gasoline engine.

Diesel engines are most often used to power large trucks that carry or pull heavy loads. The high compression and high combustion temperature present in a diesel engine make it more efficient than a gasoline engine of similar size. Note that this modern truck also has a wind deflector atop the cab to reduce air drag and increase fuel efficiency. *(Courtesy of Roadway Express, Inc.)*

to *A* corresponds to the power stroke. The process from *A* to *B* roughly corresponds to the exhaust *and* intake strokes.

The Otto cycle describes a heat engine with a working substance that is used over and over again. A real gasoline engine draws in new air and fuel and exhausts the burned gases. The Otto cycle is easier to analyze and can be used for making calculations, but it is only an approximation to the gasoline engine.

The diesel engine, named for its inventor Rudolf Diesel (1858–1913), is a four-stroke design similar in operation to the gasoline engine. The big differences are in the compression ratio and in the changes that difference causes. In a diesel engine, only air is brought in during the intake stroke. Compression ratios for diesel engines range from about 15:1 to 20:1. This extremely high compression heats the air so much that, if fuel were present, it would ignite. At maximum compression, fuel is injected into the cylinder at high pressure and ignites immediately, even though no spark plug is present. The expanding gases push the piston down on the power stroke. Figure 14.29 is a *PV* diagram for a diesel engine.

A refrigerator is basically a heat engine run in reverse, as sketched in Figure 14.30. Work *W* is done on the engine by its surroundings, (specifically, by the electric motor). Heat Q_c is absorbed from the cool region and heat Q_h is exhausted to the hot region. Instead of efficiency, the rating we used for heat engines, we describe a refrigerator by its **coefficient of performance** (*CoP*), the ratio of the useful heat transferred to the input work. For an ordinary refrigerator, the useful heat transferred is Q_c, the heat taken from the cold region. Therefore we have

$$CoP = \frac{Q_c}{W} \qquad \text{refrigerator} \qquad \text{(14.21)}$$

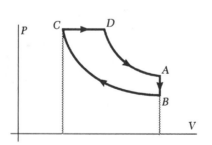

Figure 14.29
The *PV* diagram for a Diesel engine.

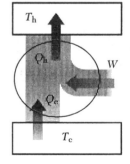

Figure 14.30
A refrigerator is a heat engine run in reverse. Work is done on the engine by the surroundings (an electric motor); heat Q_c is absorbed from the cold region and heat Q_h is given up to the hot region.

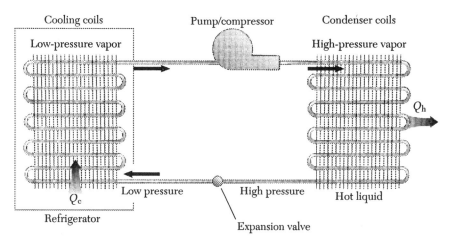

Figure 14.31
Details on the operation of a refrigerator.

Figure 14.31 shows how a refrigerator works. A working substance, the *refrigerant*, circulating inside a set of coils, absorbs heat inside the refrigerator and discharges this heat to the outside. From the second law of thermodynamics, we know that work must be done on the system to cause this heat flow from cold region to hot region. A pump, usually called a *compressor*, compresses low-pressure refrigerant vapor coming from the cooling coils of the refrigerator. This compression heats the refrigerant, which is still a vapor. This hot vapor passes through a set of coils on the back of the refrigerator. As air flows over these coils, heat is taken away from the hot vapor by convection and the vapor condenses into a liquid. Hence, these are called *condenser* coils. The hot, high-pressure liquid now passes through an expansion valve. The change in pressure across this expansion valve causes the refrigerant to partially vaporize as it passes through, and this evaporation cools the remaining liquid so that the refrigerant that comes through the valve is now a mixture of vapor and cold liquid. This cold liquid absorbs heat from the interior of the refrigerator until the liquid heats up to its boiling temperature and vaporizes. Once all the refrigerant in the cooling coils is in the vapor state, it passes into the compressor and the cycle begins again.

An air conditioner is just a whole-house refrigerator (Fig. 14.32). The cooling coils may be part of a central air-handling system. The compressor and condensing coils sit outside and give off heat to the outside air. You can *feel* warm air coming from an air conditioner's outside condensing unit. Such an air conditioning system is illustrated in Figure 14.32.

Notice the great similarity between a refrigerator and an air conditioner. Whether the object to be cooled is the relatively small volume inside a refrigerator or the relatively large volume of a room, heat is transferred from inside the volume to outside the volume.

The back of a refrigerator showing the array of condensation coils. A compressor powered by an electric motor pumps liquid Freon through the cooling compartment of the refrigerator. As the Freon absorbs heat from the compartment it becomes a gas. The gaseous Freon is then directed toward the coils where it condenses, releasing heat to the environment. The cycle then begins again. Why are modern refrigerators able to keep foods cold without having the compressor running continuously? (© *Paul Silverman/Fundamental Photographs, NY*)

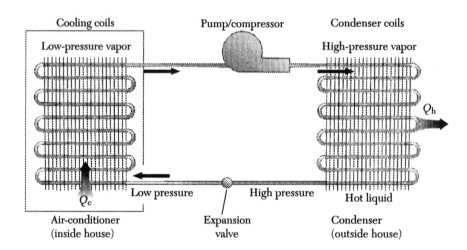

Figure 14.32
An air conditioner is just a whole-house refrigerator.

As long as the air temperature is above 0 K, the air has some internal energy that the heat pump can extract.

In a refrigerator or an air conditioner, we do not care about the heat Q_h exhausted to the surroundings since the goal is to cool either food or rooms. A **heat pump** is used to heat homes in winter and cool them in summer (Fig. 14.33), and here we do care about Q_h. Configured with the cooling coils outside the house and the condenser coils inside, a heat pump run in winter absorbs heat Q_c from the outside air on a cold day and gives up heat Q_h to the warm air inside the house. In the summer, it absorbs heat Q_c from inside the house and exhausts heat Q_h to the outside. As shown in Figure 14.33, a four-way valve simply reverses the direction of the flow of refrigerant through the

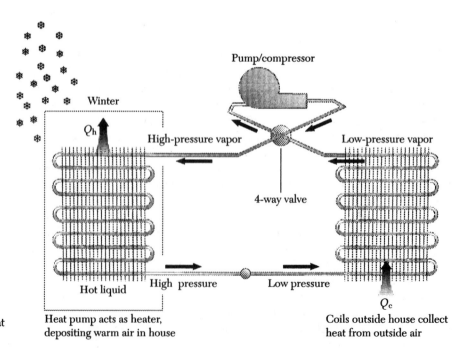

Figure 14.33
A heat pump can either cool or heat a house.

Heat pump acts as heater, depositing warm air in house

Coils outside house collect heat from outside air

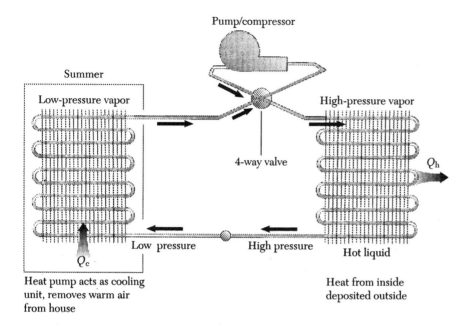

Heat pump acts as cooling
unit, removes warm air
from house

Heat from inside
deposited outside

system. The outdoor coils now become colder than the outside air and so can
absorb heat from the air. (Because frost on the coils is a problem, heat pumps
also have a defrost cycle that is necessary for operation but not for our present
understanding.) Some heat pumps use outdoor coils placed in a pond or a well
so that heat is exchanged with water instead of with the outside air. The indoor
coils are now the condenser coils and therefore become hot—or at least warm.
Air blown across these warm coils is distributed throughout the house.

Because the useful heat in the operation of a heat pump is Q_h (rather than
the Q_c useful heat in refrigerator operation), the coefficient of performance for
a heat pump is

$$CoP = \frac{Q_h}{W} \quad \text{heat pump} \tag{14.22}$$

Example 14.11

The efficiency of a particular gasoline engine is 0.42. Estimate the temperature of the combustion gases if the exhaust gases leave the engine at
140°C.

Reasoning

We can only do the calculations for idealized, reversible heat engines
such as the Carnot engine. The efficiency of a Carnot engine is given by
Equation 14.20. Therefore, we will make our estimate based upon that.
The combustion gases act as the hot reservoir. Work is done by the

engine and heat is exhausted to the surroundings. The temperature of this exhaust system will be the temperature of our cold reservoir.

Solution

$$Eff = 1 - \frac{T_c}{T_h}$$

Remember that the temperatures in Equation 14.20 must be the absolute temperatures—temperatures measured in kelvins.

$$T_c = 140°C = (140 + 273)K = 413 \text{ K}$$

$$Eff = 0.42 = 1 - \frac{413 \text{ K}}{T_h}$$

$$\frac{413 \text{ K}}{T_h} = 1 - 0.42 = 0.58$$

$$T_h = \frac{413 \text{ K}}{0.58}$$

$$T_h = 712 \text{ K}$$

$$T_h = (712 - 273)°C = 439°C \approx 440°C$$

Example 14.12

A refrigerator is a heat engine with its hot reservoir at room temperature and its cold reservoir at the temperature of the freezer. If room temperature is 27°C and the freezer is at −10°C, what is the coefficient of performance of the refrigerator?

Reasoning

The coefficient of performance of a refrigerator is given by Equation 14.21 which is stated in terms of Q_c, the heat absorbed from the cold reservoir and W, the work supplied.

$$CoP(ref) = \frac{Q_c}{W}$$

We will have to find these since we know data about this refrigerator only in terms of the temperatures.

Solution

For an ideal heat engine (which we will assume), we know that the ratio of the heats transferred at the cold and hot temperatures is equal to the ratio of the absolute temperatures; that is,

$$\frac{Q_h}{Q_c} = \frac{T_h}{T_c}$$

We also know that the work supplied is equal to the net heat exchanged,

$$W = Q_{net} = Q_h - Q_c$$

$$\frac{W}{Q_c} = \frac{Q_h}{Q_c} - 1$$

$$\frac{W}{Q_c} = \frac{T_h}{T_c} - 1$$

$$\frac{Q_c}{W} = \frac{1}{(T_h/T_c) - 1}$$

Now we can use this in Equation 14.21,

$$CoP(ref) = \frac{Q_c}{W}$$

$$CoP(ref) = \frac{1}{(T_h/T_c) - 1}$$

Remember, T_h and T_c must be measured in kelvins,

$$T_h = 27°C = (27 + 273)K = 300 \ K$$

$$T_c = -10°C = (-10 + 273)K = 263 \ K$$

$$CoP(ref) = \frac{1}{\dfrac{300 \ K}{263 \ K} - 1}$$

$$CoP(ref) = \frac{1}{1.14 - 1}$$

$$CoP(ref) = \frac{1}{0.14}$$

$$CoP(ref) = 7.1$$

*14.9 Entropy

CONCEPT OVERVIEW

■ For all natural processes, the net entropy of a system and its surroundings increases, and this increasing entropy is an indication of the irreversibility of natural processes.

In 1865 Rudolf Clausius introduced a new thermodynamic quantity called **entropy.** This new quantity gives us a quantitative method of describing the second law of thermodynamics and the irreversibility of all natural processes.

For a reversible process, the **change in entropy** ΔS of a system is defined as the heat ΔQ that flows *into* the system at absolute temperature T divided by

Irreversibility in nature is often described by the phrase "the arrow of time points in one direction only." Nothing (and nobody!) ever gets any younger.

Rudolf Clausius (1822–1888) coined the term "entropy," and did much original research on the Second Law. *(AIP Niels Bohr Library, Lande Collection)*

that absolute temperature:

$$\Delta S = \frac{\Delta Q}{T}$$

(14.23)

Like pressure, volume, internal energy, and temperature, entropy is a state variable we can use to describe the state of a system. This definition, with its single temperature value, assumes an isothermal process. If that is not the case, we can sum over numerous intermediate processes for which the temperature change is miniscule. For our present purposes, we can use an average temperature.

Now we can state the second law of thermodynamics in yet another form:

For all natural processes, the total entropy of a system and its surroundings either increases or does not change.

Example 14.13

When 0.500 kg of water at 98°C is added to 0.500 kg of water at 22°C, we know from Chapter 12 that the final temperature will be 60°C [(22°C + 98°C)/2]. Find the entropy change.

Reasoning

This is clearly not an isothermal process. For our purposes, however, it will suffice to use an average temperature. The average temperature of the initially hot water is (98°C + 60°C)/2 = 79°C = 352 K. That of the initially cool water is (60°C + 22°C)/2 = 41°C = 314 K.

Solution

The heat absorbed by the initially cool water is, from Equation 12.5,

$$\Delta Q = mc\Delta T$$
$$= (0.500 \text{ kg})(4186 \text{ J/kg·C}^0)(38 \text{ C}^0) = 79\,534 \text{ J} \approx 80\,000 \text{ J}$$

Of course, this is also the heat given up by the initially hot water. In calculating the entropy change of the initially hot water, however, we use the value $\Delta Q = -80\,000$ J because positive heat gained is negative heat given up.

Now we can calculate the entropy change of each portion of water and the net change. Call ΔS_1 the change in entropy of the initially hot water at $T_1 = 352$ K:

$$\Delta S_1 = \frac{\Delta Q_1}{T_1}$$

$$\Delta S_1 = \frac{-80\,000 \text{ J}}{352 \text{ K}}$$

$$\Delta S_1 = -227 \frac{\text{J}}{\text{K}}$$

Because Equation 14.23 defines the entropy change in terms of heat flowing *into* a system, we need the minus sign here because 80 000 J flowed *out of* the system.

Call ΔS_2 the change in entropy of the initially cold water at $T_2 = 314$ K:

$$\Delta S_2 = \frac{\Delta Q_2}{T_2}$$

$$\Delta S_2 = \frac{80\ 000\ \text{J}}{314\ \text{K}}$$

$$\Delta S_2 = 255\ \frac{\text{J}}{\text{K}}$$

The total entropy change is the sum of these,

$$\Delta S_{\text{Tot}} = -227\ \frac{\text{J}}{\text{K}} + 255\ \frac{\text{J}}{\text{K}}$$

$$\Delta S_{\text{Tot}} = -27.5\ \frac{\text{J}}{\text{K}} > 0$$

Suppose we have that 1.000 kg of water at 60°C from Example 14.10. Can we somehow separate out the slower molecules into one beaker and the faster ones into another beaker? Can we do this in such a way that we end up with half the water at 23°C and half at 98°C? There would be no violation of energy conservation in doing that, and the process would satisfy the first law of thermodynamics. However, the process would *decrease* the total entropy! According to the second law of thermodynamics, therefore, this process will not happen. According to our own experience, of course, we would not expect it to happen.

Entropy is associated with disorder. Entropy increases as a system becomes more disordered. When an ice cube melts, its molecules lose the spatial order they had as a solid and their entropy increases. For water to spontaneously crystallize into an ice cube would decrease entropy and violate the second law. Therefore, the process of freezing water, because it decreases entropy, requires a larger increase in entropy somewhere else. In a refrigerator, for example, heat is absorbed from the water and given off to the environment. The entropy increase of the environment is greater than the entropy decrease of the water. The net entropy change will always be an increase (or zero for a reversible process).

With this correspondence between entropy and disorder, the second law can be given in yet another form:

All natural systems, if left to themselves, tend to go from order to disorder.

We have now seen several different statements of the second law. They sound very different but are all equivalent.

SUMMARY

Boyle's law,

$$PV = P_o V_o \quad \text{(constant temperature)} \quad (14.2')$$

and Charles's law,

$$\frac{V}{T} = \frac{V_o}{T_o} \quad \text{(constant pressure)} \quad (14.3')$$

can be combined as

$$\frac{PV}{T} = \frac{P_o V_o}{T_o} \quad (14.4')$$

to describe the behavior of gases under many conditions. Equation 14.4′ is one form of the ideal gas law.

Pressure, volume, and temperature are called state variables because they *define the state a gas is in*, which means they specify all information about the gas. The state of a gas is specified by its "coordinates" on a PV diagram.

A gas that follows the behavior predicted by Equation 14.4′ is known as an ideal gas, and another form of the ideal gas law is

$$PV = nRT \quad (14.8)$$

where $R = 8.314$ J/mol·K is the universal gas constant.

The internal energy—the sum of all the kinetic energy due to the random motion of the molecules—of a monatomic gas is

$$E_{int} = \tfrac{3}{2} N k_B T = \tfrac{3}{2} n R T \quad (14.11, 14.12)$$

In addition to translational motion, a diatomic gas can also vibrate and rotate. Each of these types of motion—or degrees of freedom—can have kinetic energy. The internal energy of a diatomic gas is therefore

$$E_{int} = \tfrac{5}{2} N k_B T = \tfrac{5}{2} n R T \quad (14.14, 14.15)$$

At constant pressure, the work done by a gas as its volume changes is

$$W_{isobaric} = P \, \Delta V \quad (14.16)$$

The work done by a gas is equal to the area under the curve on a PV diagram that describes the process the gas undergoes.

From conservation of energy we know that, if heat Q flows into a system and the system does work W on its environment, the internal energy of the system changes by an amount

$$\Delta E_{int} = Q - W \quad (14.18)$$

This is the first law of thermodynamics.

The second law of thermodynamics is stated in diverse but equivalent ways. One statement is that heat does not spontaneously flow from a cool region to a hot region. Another form is that it is not possible to make a cyclic heat engine that produces no other effect than absorbing heat from a hot reservoir and turning all of that energy into work.

A heat engine absorbs heat Q_h from a hot reservoir, produces work W, and expels heat Q_c to a cold reservoir. The efficiency for a heat engine is

$$Eff = \frac{W}{Q_h} = \frac{Q_h - Q_c}{Q_h} = 1 - \frac{Q_c}{Q_h} \quad (14.19)$$

No heat engine can be more efficient than a Carnot engine, the efficiency of which is

$$Eff = 1 - \frac{T_c}{T_h} \quad (14.20)$$

The change in entropy ΔS of a system is

$$\Delta S = \frac{\Delta Q}{T} \quad (14.23)$$

Entropy is associated with disorder. Another statement of the second law is that all natural systems tend to go from order to disorder.

QUESTIONS

1. How can the average velocity of the air molecules in a room be zero when the average speed is so enormously high?
2. What happens to the volume of a balloon when you carry it with you as you go scuba diving?
3. Why does the pressure in automobile tires increase after the car has been driven several kilometers?
4. Why do air bubbles released under water expand as they come to the surface?
5. Why does compressing a gas increase its internal energy? Use conservation of energy in your explanation.
6. Why does expanding a gas decrease its internal energy? Use conservation of energy in your explanation.
7. Why is air cooled in escaping from an aerosol can or a tire?
8. In starting from an initial pressure P_i and expanding from an initial volume V_i to a final volume V_f, will a gas do more work if the expansion is isothermal or adiabatic?
9. Can heat flow into a system even though the temperature of the system decreases?
10. Can the temperature of a system rise even though heat flows out of the system?
11. Can a refrigerator be used to cool a room? Explain.
12. For an isothermal process, compare the work done and the heat absorbed.
13. Can mechanical work be completely converted to heat?
14. Can heat be completely converted to mechanical work?
15. The first law of thermodynamics is sometimes humorously paraphrased as "you can't get something for nothing" and the second law as "you can't even break even." Explain what lies beyond the humor in these statements.
16. A steam turbine is used to turn a generator and generate electricity. Why will the efficiency of the steam turbine be increased if superheated steam is used?
17. Would you buy stock in a company that advertises it has devised a way to burn coal and generate electricity with absolutely no exhaust heat?
18. Compare the change in entropy for a gas expanded isothermally with that for a gas expanded adiabatically between the same two volumes.
19. What is meant by a sign that says "Conserve energy by turning off lights as you leave"? Is this conserving effort related to the energy conservation we learn about in physics?

PROBLEMS

14.1 Gases: Boyle's Law and Charles's Law

14.2 Ideal Gases

14.1 An ideal gas is sealed in a rigid container at 25°C and 1.0 atm. What will its temperature be when the pressure is increased to 2.0 atm? What will its pressure be when the temperature is increased to 50°C?

14.2 A cylinder with a movable piston contains 1.0 l of an ideal gas at 20°C and 1.0 atm. What is the pressure in the cylinder when the piston has been moved so that the volume is 0.80 l? Assume no temperature change.

14.3 A 1.00-cm³ air bubble is released from a submarine at a depth of 220 m. What is the bubble's volume when it reaches the surface?

14.4 How many moles of nitrogen are in a 22.4-l container at 0°C and 1 atm?

14.5 If 2.0 mol of oxygen is compressed to 22.4 l at 0°C, what is the pressure?

14.6 What is the mass of 1 mol of (a) nitrogen, (b) hydrogen, (c) water, and (d) table salt (NaCl)?

○14.7 Neon is a monatomic gas. What is the internal energy of 20 g of neon gas at 23°C and 1.0 atm?

○14.8 Argon, like all the other noble gases, is monatomic. What is the internal energy of 5.0 g of argon gas at 55°C and 2.5 atm?

●14.9 Using the ideal gas law, develop an equation for the density of a gas as a function of pressure and temperature.

14.3 The Kinetic Theory of Gases

14.10 How many air molecules are there in 1.0 m³ of air at 23°C and 1.0 atm? An extremely good vacuum might be 10^{-12} Pa. At this pressure and 23°C, how many air molecules occupy 1.0 m³?

14.11 At 1-s intervals, 10-g rubber bullets are fired at a 3-m² target. The bullets collide with the target elastically and bounce off. If the bullet speed is 100 m/s, what are the average force and average pressure on the target?

14.12 At 23°C, and 1.0 atm what is the root mean square speed of   (a) a helium atom,   (b) a hydrogen molecule,   (c) nitrogen molecule, (d) an oxygen molecule, and   (e) a water vapor molecule?

14.13 Repeat Problem 14.12 using 100°C as the temperature.

○14.14 A pressure of 1.0×10^{-9} torr is a high vacuum. If a 1.0-l vessel contains helium at this pressure and 300 K, how many atoms are in the vessel? What is their root mean square speed?

●14.15 At what temperature is the root mean square speed of helium equal to the escape speed of Earth (40 300 km/h, Equation 6.23)? At what temperature is it equal to the escape speed of the Moon (2.38×10^3 m/s)?

●14.16 At what temperature is the root mean square speed of oxygen equal to the escape speed of Earth? At what temperature is it equal to the escape speed of the Moon? (See Problem 14.15.)

14.4 Work Done by a Gas

14.17 Gas at 1.5 atm expands from 2.5 l to 3.5 l. How much work is done by the gas?

14.18 A gas is compressed from 150 l to 120 l at constant pressure, the process requiring 1.2×10^4 J of work. What is the pressure of the gas?

14.19 Steam enters a steam engine's cylinder at a pressure of 3.5×10^5 Pa, which remains constant. The radius of the cylinder is 12 cm. If the steam pushes a piston through a 28-cm stroke, how much work is done?

○14.20 Gas in a container expands from 5.0 l to 6.0 l while the pressure is held constant at 1.8 atm. Then the volume is held constant at 6.0 l while the pressure increases to 2.0 atm. Then the pressure is held constant at 2.0 atm while the volume decreases to 5.5 l. What is the net work done by the gas?

○❖14.21 To compress a gas from V_i to V_f at constant pressure requires work W. Find this constant pressure for each of the following situations and rank them in order of increasing pressure:

	V_i(l)	V_f(l)	$W(\times 10^4$ J)
a)	150	120	1.20
b)	150	100	1.50
c)	120	85.0	1.30
d)	100	85.0	1.10
e)	120	85.0	1.60
f)	120	100	1.50

○14.22 When 2.0 mol of an ideal gas expands isothermally at 200°C, from 1.0 atm to 2.0 atm how much work is done by the gas?

○14.23 A gas can go from initial state I to final state F via three different processes, as indicated in Figure 14.34. Calculate the work done by the gas in each process.

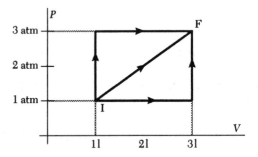

Figure 14.34 Problem 14.23.
Different amounts of work are done in going from state I to state F along different paths, representing different processes.

○14.24 Sketch the following processes on a PV diagram: (a) a gas expands from 3 l to 5 l at 2 atm,   (b) the gas is heated until its pressure increases to 3 atm while the volume remains at 5 l,   (c) the gas is then compressed to 4 l, still at 3 atm,   (d) the pressure is reduced to 2 atm while the volume remains at 4 l.   How much work is done by the gas in each process? What is the net work done during all four steps?

14.25 A monatomic gas is compressed to one half its original volume at a constant pressure of 1.5 atm. If 150 J of work is done on the gas during this compression, what was its original volume?

14.5 The First Law of Thermodynamics

14.26 Under a constant pressure of 1.5 atm, a gas expands from 2.0 l to 3.0 l while 500 J of energy flows into it.   (a) How much work is done by the gas?   (b) What is the change in its internal energy?

14.27 Under a constant pressure of 2.25 atm, a gas is compressed from 2.00 l to 1.50 l while 375 J of energy flows out of the gas.
(a) How much work is done on the gas?
(b) What is the change in its internal energy?

14.28 A person takes in a breath of 0°C air and holds it until it warms to 37.0°C. The air has an initial volume of 0.600 l and a mass of 7.70×10^{-4} kg. Determine   (a) the work done on the lungs by the air if the pressure remains constant at 1.00 atm,   (b) the heat added to the air,   (c) the change in internal energy of the air.

❖ 14.29 A gas expands under a constant pressure P from an initial volume V_i to a final volume V_f while an amount of heat Q flows out of the gas. For each data set,   (a) how much work is done by the gas, and   (b) what is the change in its internal energy?   Rank the results in order of increasing amount of work done by the gas.

	P(atm)	V_i(l)	V_f(l)	Q(J)
a)	1.00	1.50	2.00	800
b)	1.00	1.50	2.60	800
c)	1.50	2.00	3.00	800
d)	2.00	2.00	3.00	1000
e)	2.00	2.25	3.50	1000
f)	2.50	2.25	3.75	1200

14.30 While expanding isothermally at 50.0°C, 1.00 mol of a gas does 500 J of work. If its initial volume is 2.25 l, find its final volume.

14.31 When boiled at 1.000 atm, 1.000 cm^3 of water becomes 1671 cm^3 of steam.   (a) How much work is done by the water in expanding into steam?   (b) How much heat must be added to vaporize the water?   (c) What is the change in internal energy during this change of phase?

14.32 If 1.0 mol of an ideal gas expands isothermally at 30°C to twice its original volume, how much work is done by the gas? How much heat flows into it during this expansion?

14.33 If 1.0 mol of a monatomic ideal gas initially at 30°C and 1 atm expands adiabatically to twice its original volume, how much work is done by the gas? What is its final temperature?

14.34 If 1.00 mol of a diatomic ideal gas initially at 25.0°C and 1.25 atm expands isobarically to twice its original volume, how much work is done by the gas? How much heat flows into it during this expansion? What is the final temperature?

14.35 For 3.0 mol of gas expanding from 5.0 l to 15 l at 30°C, calculate   (a) the work done by or on the gas,   (b) the change in its internal energy,   (c) the heat that flows into or out of it.

14.36 For 2.5 mol of gas being compressed from 12 l to 5.0 l at 40°C, calculate   (a) the work done by or on the gas,   (b) the change in its internal energy,   (c) the heat that flows into or out of it.

14.37 After 1.0 mol of an ideal gas initially at 0°C and 1.0 atm expands isothermally to twice its initial volume, it is compressed isobarically back to its original volume. What is the net work done by or on the gas? What is its final temperature?

14.38 While the pressure is held constant at 5.00 atm, 40.0 J of heat is added to hydrogen gas in a cylinder, causing the gas volume to increase from 10.0 cm^3 to 30.0 cm^3. What is the change in internal energy of the gas?

14.39 Plot a PV curve for an ideal gas that expands isothermally from a state of $P_1 = 2.0$ atm, $V_1 = 1.0 \times 10^{-3}$ m^3 to a state $P_2 = 1.0$ atm, $V_2 = 2.0 \times 10^{-3}$ m^3.   (a) Divide the area under the curve into ten equal segments and compute the work done in each segment. Determine the total work done by adding your ten values and compare your answer with the value you find from Equation 14.17   (b) Determine the change in internal energy and the heat added during the process.

● 14.40   A gas initially occupies 1 l at 1 atm and 0°C. First the pressure is increased to 3 atm at constant volume. Then the gas expands isothermally to 3 l as the pressure drops to 1 atm. Finally the gas is compressed at constant pressure back to 1 l.   (a) Sketch these processes on a PV graph.   (b) How much gas is present (in moles)?   (c) What is the temperature during the isothermal expansion?   (d) How much work is done by or on the gas during each part of the cycle?   (e) What is the net work by or on the gas during the complete cycle?

● 14.41   At 1 atm and −29°C, 1 mol of an ideal gas has a volume of 20 l. Its pressure increases to 2 atm at constant volume. It then expands to 40 l at constant pressure. Then the pressure is reduced to 1 atm at constant volume. Finally the gas is returned to its original volume by compression at constant pressure.   (a) Draw this cycle on a PV diagram.   (b) Find the temperature for each end point of each process.   (c) Calculate the work done for each process and the net work done over the cycle.   (d) Calculate the heat entering or leaving the system during each process.   (e) What is the efficiency of this cycle?   (f) What is the efficiency of a Carnot engine operating between these temperature extremes?

14.6 Irreversibility and the Second Law of Thermodynamics

14.42   Should you invest in a company that claims to have a heat engine that does 40 000 J of work for every 50 000 J of heat absorbed at 500°C while exhausting 10 000 J of heat at 100°C?

14.43   Should you invest in a company that claims to have a heat engine that does 40 000 J of work for every 50 000 J of heat absorbed at

500 K while exhausting 10 000 J of heat at 300 K?

14.7 Heat Engines

14.44   A heat engine absorbs 6 J from a source of heat and does 2 J of mechanical work. What is its efficiency?

14.45   A heat engine performs 2000 J of work and rejects 8000 J of heat to a cold reservoir. What is its efficiency?

14.46   A reversible heat engine whose efficiency is 15 percent does 180 J of work. How much heat does it absorb from the hot reservoir?

14.47   A Carnot engine exhausts heat at 100°C. What must its high temperature be if its efficiency is 0.33?

14.48   A Carnot engine produces 1 kW of mechanical work while absorbing heat at the rate of 1.8 kW from a hot reservoir at 650°C. What is the temperature of the cold reservoir?

14.49   An engine whose efficiency is half the Carnot efficiency operates between 100°C and 500°C and rejects heat to the cold reservoir at the rate of 25 kW. What is the rate of heat input to the engine?

14.50   A reversible heat engine operates between a 600 K reservoir and a 1000 K reservoir. What is its efficiency? How much work is done when it absorbs 1000 J of heat at 1000 K? How much heat is given up at 600 K?

14.51   A heat engine operates between a 300 K reservoir and a 500 K one. What is its maximum efficiency? What is the maximum amount of work that can be done when it absorbs 1000 J of heat at 500 K?

○ 14.52   A Carnot engine operates with a hot reservoir of 650°C at 0.30 efficiency. To have 0.35 efficiency, what must the temperature of this reservoir be?

○ 14.53   A heat engine has an efficiency of 0.37 when exhausting heat to a cold reservoir at 37°C. What is the minimum temperature of the hot reservoir?

○ 14.54   A reversible heat engine has an efficiency of 0.27 when exhausting heat to a cold reservoir at 33°C. What is the temperature of the hot reservoir? What must the temperature of the hot reservoir be to increase the efficiency to 0.30? Leaving the high temperature at its initial value, what must the low temperature be to increase the efficiency to 0.30?

○14.55 A coal-fired power plant operates with super-heated steam at 850°C for its hot reservoir and has a lake at 30°C for its cold reservoir. What is its maximum possible efficiency?

●14.56 A power plant with 0.35 efficiency produces 500 MW of electrical energy. At what rate is heat exhausted? This heat is exhausted to the air through cooling towers. If the air temperature is allowed to change 10 C°, what must the air flow rate through the towers be? How much air flows through these towers in one day? If this heated air formed a layer 100 m high, how much area would it cover?

*14.8 Real Heat Engines

14.57 A gasoline engine has an efficiency of 0.52 and exhausts 130°C gas to the atmosphere. Treat it as an ideal heat engine and find the temperature of the hot "reservoir" during the combustion stroke.

14.58 The efficiency of a gasoline engine is 0.55. Estimate the temperature of the combustion gases if the exhaust gases leave the engine at 125°C.

14.59 A residential heat pump extracts heat from outside air at 0°C and exhausts it into the central air-handling system at 40°C. What is the maximum coefficient of performance for this heat pump?

○14.60 A refrigerator absorbs heat from its interior at 5°C and exhausts it to the room at 24°C. What is its maximum coefficient of performance? What is the minimum amount of work necessary to cool 1.0 kg of water from 24°C to 5°C using this refrigerator?

○14.61 A freezer absorbs heat from the interior of the freezer at −5°C and exhausts it to the room at 24°C. What is its maximum coefficient of performance? What is the minimum amount of work necessary to cool 0.25 kg of water (about what it takes to fill one ice cube tray) from 24°C to 0°C, freeze it, and lower its temperature to −5°C using this freezer?

○14.62 A heat engine absorbs 350 J of heat and does 250 J of work during each cycle. What is its

efficiency? How much heat is exhausted during each cycle? If it is a reversible heat engine with a high temperature of 700 K, what is the low temperature?

○14.63 A heat engine absorbs 700 J of heat and exhausts 400 J of heat during each cycle. What is its efficiency? How much work is done during each cycle? If it is a reversible heat engine with a high temperature of 700 K, what is its low temperature?

○14.64 A refrigerator is a heat pump with the hot reservoir at room temperature (25°C) and the cold reservoir at the temperature of the freezer (−20°C).   (a) Calculate the maximum coefficient of performance.   (b) Calculate the actual coefficient if the refrigerator is one fourth as efficient as theoretically possible.   (c) If the cost of electrical energy is $0.16/kW·h, what is the cost of freezing 1.0 kg of ice cubes by changing water at 25°C to ice at −20°C?

*14.9 Entropy

14.65 What is the change in entropy when 1.0 kg of ice melts?

14.66 What is the change in entropy when 1.0 kg of water boils?

14.67 What is the approximate entropy change when 1.0 kg of water is heated from 0 to 100°C?

○14.68 If 0.10 kg of ice at 0°C is placed in 0.50 kg of water at 35°C, find the temperature of the water after the ice melts. What is the entropy change in the ice as it melts? What is the approximate entropy change in the water from the ice as its temperature rises from 0° to the equilibrium temperature? What is the approximate entropy change in the initially warm water as it cools to the equilibrium temperature? What is the net entropy change in the system?

○14.69 If 0.20 kg of copper at 125°C is placed in 0.80 kg of water at 25°C, find the equilibrium temperature of the copper-water system. Calculate the approximate entropy change for the copper, for the water, and for the system.

General Problems

● 14.70 For the gas represented by Figure 14.35, $P_i = 1.50$ atm, $V_i = 0.250$ m³, and $T_i = 238$ K. Its final volume is $V_f = 0.500$ m³, and the $T = T_f$ notation means that the curved part of path 2 takes place at the final temperature of the system. For the three paths shown, find (a) the heat added, (b) the work done, and (c) the change in internal energy.

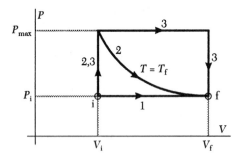

Figure 14.35 Problem 14.70.

○ 14.71 A monatomic gas is expanded to twice its original volume at a constant pressure of 1.5 atm. If 250 J of work is done by the gas during this expansion, what was its original volume?

● 14.72 A gas initially has pressure P_o, volume V_o, and temperature T_o. Its pressure increases to $2P_o$ at constant volume. It then expands to $2V_o$ at constant pressure. Then the pressure is reduced to P_o at constant volume. Finally the gas is returned to its original volume by compression at constant pressure. (a) Draw this cycle on a PV diagram. (b) Find the temperature for each end point of each process. (c) Calculate the work done for each process and the net work done over the cycle. (d) Calculate the heat entering or leaving the system during each process.

(e) What is the efficiency of this cycle?
(f) What is the efficiency of a Carnot engine operating between these temperature extremes?

○ 14.73 In the Caribbean, the ocean water near the surface has a year-round temperature near 25°C, while the water at a depth of 750 m is constant at 10°C. Using this temperature difference to power heat engines for generating electric energy has been suggested. What is the maximum efficiency for a heat engine operating between these two temperatures? How great would the heat flow need to be to provide 1000 kW of mechanical work?

● 14.74 A power plant with 0.30 efficiency produces 750 MW of electrical energy. At what rate is heat exhausted? This heat is exhausted to water in a nearby river. If the water temperature is allowed to change 5 C°, what must the flow rate in the river be? How much water flows through this river in one day? If this warmed water formed a lake 10 m deep, how much area would it cover?

○ 14.75 A heat pump is designed to deliver air at 37°C into the room (warmer than room temperature to avoid feeling chilly). For efficient exchange of heat to the air, the hot reservoir should be 8 C° warmer than the air (that is, 45°C). The temperature outdoors is 2°C. (a) Calculate the maximum coefficient of performance of the heat pump operating between 2°C and 45°C. (b) Calculate the actual coefficient if the heat pump is only one third as efficient as theoretically possible. (c) What power must be supplied by the motor if heat is supplied to the house at the rate of 24 kW?

○ 14.76 If 0.25 kg of iron at 250°C is placed in 2.5 kg of water at 25°C, find the equilibrium temperature of the iron-water system. Calculate the approximate entropy change for the iron, for the water, and for the system.

PHYSICS AT WORK

Simple harmonic motion is a particular kind of motion that we see all around us all the time. You have seen and experienced simple harmonic motion since you were a young child. As just one example, the child on a swing shown here is undergoing simple harmonic motion.

The motion of a pendulum is also simple harmonic motion. This motion has some interesting and important characteristics. Perhaps the most interesting—and certainly the most useful—is that the time a pendulum takes to go through one complete cycle is the same whether it moves through a small distance or a large distance. This means a pendulum can be used as a timing device; that is, it can be used in a clock to keep accurate time.

Many other systems have motion that is approximately simple harmonic motion. The wiggling of water molecules in food when excited in a microwave oven is simple harmonic motion. The bouncing of your car as it goes over a pothole (especially if your shock absorbers are bad) is nearly simple harmonic motion. The vibration of a violin string or a piano string or a clarinet reed—all these are examples of simple harmonic motion.

What do a playground swing, a car's suspension system, and the tuning circuit of your radio have in common? Once we understand the characteristics of simple harmonic motion, we can easily answer this question.

(© Michael Hart/FPG International Corp.)

15

Periodic Motion

- The prototype for simple harmonic motion (SHM) is an object of mass m attached to a spring of spring constant k. When moved from equilibrium, the net force on the object is proportional to its distance from equilibrium ($F = -kx$). The oscillating motion that results is periodic with a period that depends only upon the mass and the spring constant—the period is independent of the amplitude, the maximum displacement from equilibrium. SHM is important because many systems—from mechanical systems like the suspension of your car to electrical systems like the tuning circuit of your radio—behave with SHM.

- As a simple harmonic oscillator (SHO) continues its motion, energy is transferred from kinetic energy generated by the moving mass to potential energy stored in the stretched or compressed spring. But the total energy of the system remains constant.

- A simple pendulum—mass swinging from a string—behaves very nearly as a SHO. This observable motion means its period is independent of its amplitude, which makes a simple pendulum very useful as a time-keeping device (as the pendulum in a clock).

The St. Louis Gateway Arch, designed by Eero Saarinen, has a span of 192 m (630 ft) and a height of 192 m (630 ft). Wind will set the Gateway Arch oscillating in simple harmonic motion, and the top of the structure can sway back and forth a distance of several feet in a stiff breeze. The Gateway Arch is designed to allow for this oscillation without breaking apart and collapsing.

In mathematics, *harmonic* means not something musical but rather "expressible as a sine or cosine function"—as we shall see, an apt description of the vibratory motion we are studying.

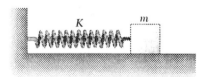

Figure 15.1
Our prototype for understanding simple harmonic motion is a mass attached to a spring.

Vibrations occur throughout nature. In this chapter, we focus on the vibrations of a mass attached to a spring, but the ideas developed for this simple system help us understand other vibrating systems.

Nearly everything vibrates when disturbed. The head of a drum, reeds in a bagpipe, strings on a violin all vibrate to create an air disturbance that we call sound (or even music). In addition, many nonmusical objects vibrate: playground swings, suspension bridges, tall buildings, cars on a bumpy road, atoms in a crystal. Electric circuits vibrate in a manner that can be described with the ideas we develop in this chapter, and light is the vibration of electric and magnetic fields. Sometimes we want to minimize vibration, as in the suspension springs of an automobile or the movement of a building. At other times we want to maximize it, as in a piano or a playground swing.

15.1 Simple Harmonic Motion

CONCEPT OVERVIEW

- A mass attached to a horizontal spring provides the prototype for a simple harmonic oscillator.

- Hooke's law states that the force exerted by a spring is proportional to the distance the spring has been stretched or compressed from equilibrium.

- The acceleration of a simple harmonic oscillator is proportional to its displacement.

Any motion that repeats itself at regular intervals is called either **periodic motion** or **harmonic motion** (the two terms are synonyms). In studying this type of motion, we concentrate first on the motion of a mass that sits on a horizontal, frictionless surface and is attached to a spring, as sketched in Figure 15.1.

The unstretched length of any spring is called its *equilibrium length*. We can align an x axis alongside a spring with the $x = 0$ mark at the end of the unstretched, uncompressed spring as in Figure 15.2. If we push to the left with an external force $\mathbf{F}_{ext}$, the spring is compressed and the end that was at $x = 0$ at equilibrium is now at some negative value of x. The spring responds to this leftward force by pushing to the right. If we pull to the right with an external force $\mathbf{F}_{ext}$, the spring is stretched and the end that was at $x = 0$ at equilibrium is now at some positive value of x. The spring responds to this rightward force by pulling to the left. If the external force is increased, the compression or stretch of the spring will be greater and the reaction force the spring exerts will also increase. Because the spring force acts in the direction opposite the stretch or compression, we call it a **restoring force**.

As we learned in Section 6.1, the magnitude of the force exerted by a spring is directly proportional to the stretch or compression of the spring, a relationship known as **Hooke's law**:

$$F_{\text{spring}} = -kx \qquad (15.1)$$

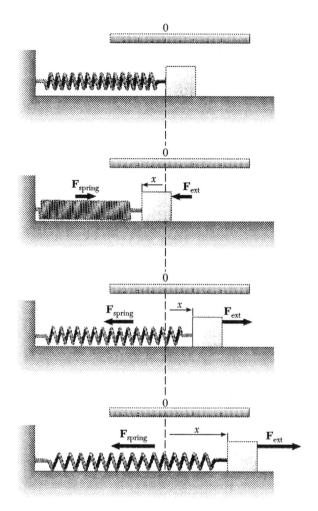

Figure 15.2
The restoring force exerted by a spring is proportional to the external force stretching or compressing the spring and is in the opposite direction.

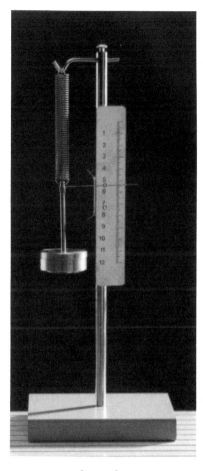

An apparatus for performing experiments confirming Hooke's law and simple harmonic motion. *(Courtesy of Central Scientific Company)*

Figure 15.3 is a graph of the absolute value of the spring force $|\mathbf{F}|$ as a function of the displacement from equilibrium. This graph is a straight line the slope of which is the spring constant k.

Example 15.1

A force of 1.2 N is required to stretch a spring 3 cm. What is the spring constant?

Reasoning

Equation 15.1 describes the connection between force, displacement, and spring constant. Because we are interested in the spring constant, which has no direction associated with it, we can ignore the minus sign in Equation 15.1 and use the absolute value of the force.

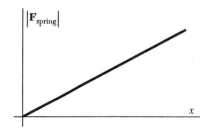

Figure 15.3
The force-displacement graph for a spring that obeys Hooke's law.

Solution

$$|F| = kx$$

$$k = \frac{|F|}{x} = \frac{1.2 \text{ N}}{0.03 \text{ m}} = 40 \text{ N/m}$$

Practice Exercise   A force of 4.2 N is required to stretch a spring 6 cm. What is the spring constant?

Answer   70 N/m

"To vibrate" and "to oscillate" mean the same thing, and physicists use the two terms interchangeably.

A **simple harmonic oscillator** is a vibrating (oscillating) system that obeys Hooke's law. Instead of talking of the force of such a system, we can use Newton's second law and talk in terms of the acceleration,

$$a = -\left(\frac{k}{m}\right)x \qquad (15.2)$$

which tells us that

> *the acceleration of a simple harmonic oscillator is directly proportional to its displacement and in the opposite direction.*

For our prototypical mass-spring system, the proportionality constant is the spring constant divided by the mass.

What does the motion of a mass-spring system—or any other simple harmonic oscillator—look like? To find out, let us pull the mass to the right as sketched in Figure 15.4, and let go. This maximum displacement from equilibrium is the **amplitude** A of the oscillation. At the instant we let go, the mass starts from rest with the spring force pulling it to the left, causing it to accelerate. It continues to pick up speed and has maximum speed as it passes through the equilibrium position $x = 0$. There the spring force, which has been decreasing as x decreases, vanishes. The moving mass, due to its inertia, overshoots the equilibrium position. Now, as the spring compresses, it exerts a restoring force to the right, causing an acceleration to the right that slows the mass. We shall see shortly that the leftward-moving mass comes to rest when it has moved a distance A from equilibrium. Because the displacement is now in the opposite direction, however, $x = -A$. The spring force, which is now directed to the right and has been increasing since the mass passed through equilibrium, is at its maximum when the mass is at $x = -A$. After being momentarily at rest at $x = -A$, the mass begins to accelerate to the right and again reaches maximum speed as it passes through equilibrium and overshoots again. And so the oscillation continues.

We've met *period* before: in the opening comments in Chapter 5, where we talked about a satellite orbiting Earth with a period of 24 h.

The time required for one cycle of this motion is known as the **period** T of the oscillation. (Note that the time needed to move from $x = +A$ to $x = -A$ is only *half* a period.) Sometimes it is easier to describe periodic motion by how many oscillations occur in a unit time; this is the **frequency** f of the oscillations. Frequency and period are related by

$$f = \frac{1}{T} \qquad (15.3)$$

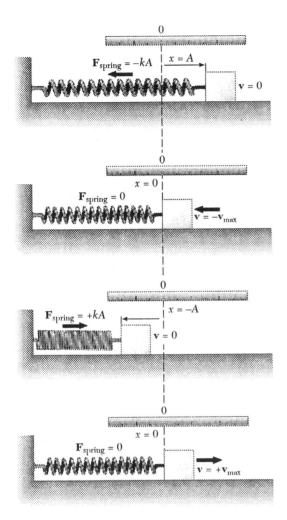

Figure 15.4
A mass-spring system as a simple harmonic oscillator. The maximum displacement is known as the amplitude A. The mass (the oscillator) is at rest when its displacement is A and has maximum velocity as it goes through its equilibrium position ($x = 0$).

The period is measured in units of time, usually seconds, and frequency is measured in reciprocal seconds or cycles per seconds. This unit is called the **hertz** (Hz):

$$1 \text{ Hz} = 1 \text{ cycle/s} = 1 \text{ s}^{-1}$$

Frequency and period convey the same information, but one unit is often more convenient than the other. The period of the tides is about 12 h, and that of the Moon is 28.3 days. The frequency of household electricity is 60 Hz, and an average heart rate for humans is about 74 beats/min, equal to 1.2 Hz.

The hertz is named for Heinrich Hertz (1857–1894), a German physicist who was the earliest experimenter with radio waves.

Problem-Solving Tip

▬ If you have trouble converting from frequency to period (or the other way round) for very large or very small numbers, think first of an equivalent problem involving small whole numbers. For example, if a periodic motion has a frequency of 4 cycles per second ($f = 4$ Hz), then its period is $\frac{1}{4}$ s. Thinking this way makes it easier to recall that $T = 1/f$.

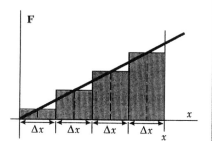

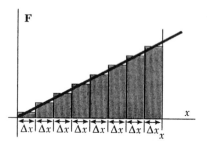

Figure 15.5
The work done by a varying force can be calculated from $\Delta W = \Sigma F \Delta x$. This work is the area under the curve on a force-distance graph.

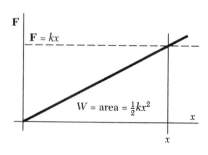

Figure 15.6
The work done in stretching or compressing a spring a distance x is the area under the curve on a force-distance graph: $W = \text{area} = \frac{1}{2}kx^2$.

Example 15.2

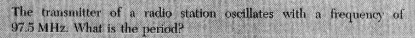

The transmitter of a radio station oscillates with a frequency of 97.5 MHz. What is the period?

Reasoning

We can apply Equation 15.3 directly.

Solution

$$f = \frac{1}{T} = 97.5 \text{ MHz} = 9.75 \times 10^7 \text{ Hz}$$

$$T = \frac{1}{f} = \frac{1}{9.75 \times 10^7 \text{ s}^{-1}} = 1.02 \times 10^{-8} \text{ s} = 10.2 \text{ ns}$$

Practice Exercise What is the period of oscillation of a radio station broadcasting at 1270 kHz?

Answer $7.87 \times 10^{-7} \text{ s}$

15.2 The Energy of a Simple Harmonic Oscillator

CONCEPT OVERVIEW

■ The energy of a simple harmonic oscillator continuously changes form, but the total energy of the system remains constant.

■ A mass on a spring undergoing simple harmonic motion momentarily stops when its displacement equals its amplitude; at that point all its energy is potential energy.

■ The potential energy is momentarily zero as the mass passes through its equilibrium position; at that point all its energy is kinetic energy.

You already know that the work ΔW done by a constant force $\mathbf{F}$ as the force moves an object through a distance Δx (in the same direction as the force) is $\Delta W = F\Delta x$, but what if the force is *not* constant? In particular, how much work is done on a spring as it is stretched or compressed a distance x from equilibrium? We looked at this varying force in Section 6.1 in deriving Equation 6.4. Figures 15.5 and 15.6 show the external force $F_{\text{ext}} = kx$ that must be applied to a spring to change its length by a distance x. As in Section 6.1 the work done is equal to the area under the curve on a force-distance graph. Since the force-distance curve for a Hooke's law spring is a straight line, that area is the area of a triangle:

$$W = \text{area} = \frac{1}{2}\text{height} \times \text{base}$$

$$= \frac{1}{2}(F_{\text{max}})(x) = \frac{1}{2}(kx)(x) = \frac{1}{2}kx^2$$

This result is the same for compressing as for stretching a spring since in either case the external force doing the work on the spring is in the same direction as the motion. Once work is done on the spring, the spring can do work on something else. That is, this work done on the spring is **spring potential energy** (also called **elastic potential energy**), which we saw as Equation 6.14:

$$PE = \frac{1}{2}kx^2 \qquad \textbf{(15.4)}$$

An oscillator also has kinetic energy

$$KE = \frac{1}{2}mv^2$$

This equation assumes that the mass of the spring is negligible compared with the mass that it is attached to; this is often the case. The total energy is the sum of potential energy stored in the stretched or compressed spring and the kinetic energy stored in the moving mass,

$$E = PE + KE = \frac{1}{2}kx^2 + \frac{1}{2}mv^2 \qquad \textbf{(15.5)}$$

In Figure 15.7, the spring has been stretched to $x = +A$ and the mass is at rest. Therefore, the total energy of the system at this moment is given by Equation 15.5 with $x = A$ and $v = 0$:

$$E = \frac{1}{2}kA^2 + \frac{1}{2}m(0)^2 = \frac{1}{2}kA^2$$

Because energy is conserved, the energy at any position x other than $x = +A$ is still $\frac{1}{2}kA^2$. Once the mass is released from rest, we can use that fact to solve for its speed as a function of position:

$$E = \frac{1}{2}kx^2 + \frac{1}{2}mv^2 = \frac{1}{2}kA^2$$

Multiplying through by 2 gives

$$kx^2 + mv^2 = kA^2$$

Subtracting kx^2 yields

$$mv^2 = k(A^2 - x^2)$$

$$v = \pm\sqrt{\frac{k}{m}\left(A^2 - x^2\right)} \qquad \textbf{(15.6)}$$

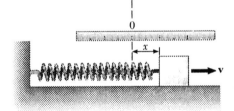

Figure 15.7
The total energy of a simple harmonic oscillator is $E = \frac{1}{2}kx^2 + \frac{1}{2}mv^2$.

Notice that, for any value x, the speed is the same as the mass moves to the right or the left. You have to decide on the sign of $\mathbf{v}$ from other details.

The speed of the mass is zero at $x = \pm A$. That is, the mass oscillates exactly as far to one side of equilibrium as to the other side; the oscillation is symmetric about the equilibrium position. From Equation 15.6, we can solve for v_{max}, the speed of the mass as it passes through equilibrium:

$$v_{max} = A\sqrt{\frac{k}{m}}$$

At $x = 0$, all the energy of the system is in the form of kinetic energy of the moving mass, and the spring has zero potential energy.

Problem-Solving Tip

- Energy conservation is useful in solving simple harmonic motion problems. The total energy remains constant and is equal (a) to the maximum kinetic energy as the oscillator passes through equilibrium and (b) to the maximum potential energy when the displacement equals the amplitude.

Example 15.3 _____

A 0.250-kg mass is attached to a spring whose spring constant is 60.0 N/m. To start oscillation, the mass is pulled to one side 8.00 cm and released. Find (a) the speed of the mass at $x = 4.00$ cm, (b) the speed of the mass at $x = 0$, and (c) the position(s) at which the speed is half the maximum speed.

Reasoning

The initial stretch to $x = 8.00$ cm requires that work be done on the spring, and that work becomes the total energy of the system, equal to the maximum potential energy when $x = A = 8.00$ cm. From conservation of energy, we know that the sum of the spring's potential energy and the mass's kinetic energy always equals this total energy.

Solution

(a) The total energy is

$$E = \frac{1}{2}kA^2 = \frac{1}{2}(60.0\ \text{N/m})(0.0800\ \text{m})^2 = 0.192\ \text{J}$$

When the mass is at $x = 4.00$ cm, the spring has potential energy of

$$PE = \frac{1}{2}kx^2 = \frac{1}{2}(60.0\ \text{N/m})(0.0400\ \text{m})^2 = 0.0480\ \text{J}$$

The rest of the energy must be kinetic energy:

$$KE = E - PE = 0.192\ \text{J} - 0.0480\ \text{J} = 0.144\ \text{J}$$

$$KE = \frac{1}{2}mv^2 = \frac{1}{2}(0.250\ \text{kg})v^2 = 0.144\ \text{J}$$

$$v^2 = 1.15\ \text{J/kg} = 1.15\ \text{m}^2/\text{s}^2$$

$$v = 1.07\ \text{m/s}$$

(b) When the mass is $x = 0$, all its energy is kinetic energy:

$$KE = \frac{1}{2} m v_{max}^2 = E = 0.192 \text{ J}$$

$$v_{max}^2 = 1.54 \text{ m}^2/\text{s}^2$$

$$v_{max} = 1.24 \text{ m/s}$$

(c) Now we want to find the positions x such that $v = v_{max}/2$. Because energy is still conserved, we can say

$$E = \frac{1}{2} k x^2 + \frac{1}{2} m v^2$$

$$x = \pm \sqrt{\frac{2E - mv^2}{k}}$$

Now we substitute $v = v_{max}/2 = 0.620$ m/s:

$$x = \pm \sqrt{\frac{2(0.192 \text{ J}) - (0.250 \text{ kg})(0.620 \text{ m/s})^2}{60.0 \text{ N/m}}}$$

$$= \pm 0.0693 \text{ m} = \pm 6.93 \text{ cm}$$

Practice Exercise   Answer the questions in the example if the mass is 0.500 kg.

Answer   0.76 m/s, 0.88 m/s, 6.93 cm

15.3 Simple Harmonic Motion and Circular Motion

CONCEPT OVERVIEW

■ The frequency of a mass-spring system in simple harmonic oscillation is $f = \frac{1}{2\pi} \sqrt{\frac{k}{m}}$.

The frequency of a mass-spring system undergoing simple harmonic motion is

$$f = \frac{1}{2\pi} \sqrt{\frac{k}{m}} \qquad (15.7)$$

Since $T = 1/f$, the period of this system is

$$T = 2\pi \sqrt{\frac{m}{k}} \qquad (15.8)$$

We shall derive these results shortly. These equations tell us that if you increase the mass of an oscillator, it becomes more difficult to move: the frequency decreases or, to say the same thing another way, the period becomes longer. If you increase the spring constant by using a stronger spring, the mass moves more quickly; the frequency increases and the period decreases.

Notice that the amplitude does not appear in these equations. If you give a gentle push so that the amplitude is small, the mass oscillates with a frequency

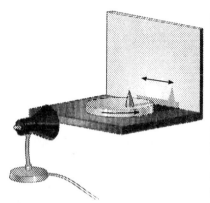

Figure 15.8
The projection of uniform circular
motion is simple harmonic motion.

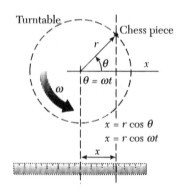

Figure 15.9
The relationships between uniform
circular motion and simple harmonic
motion.

given by Equation 15.7 and a period given by Equation 15.8. If you give a very large push so that the amplitude is large, the frequency and period remain unchanged! The mass moves through a greater distance at higher speeds, but the time required for each oscillation remains the same. It is sometimes reported that Galileo noticed this while watching chandeliers sway in the cathedral at Pisa. This amplitude-independence of the frequency is the basis for the use of pendula and balance wheels in clocks and watches. It explains why a piano or guitar can play the same note softly or loudly. It is a most useful property.

Rather than simply pulling Equations 15.7 and 15.8 out of thin air by saying, "It can be shown by calculus . . . ," we can derive them in an interesting but indirect manner. Figure 15.8 shows an oblique view of an object undergoing uniform circular motion. Light from far away shines on this object from the side, and we see the motion of the object projected on a screen. Rather than dealing with only a thought experiment or a mathematical operation, think of setting a chess piece on a rotating turntable, shining a light on the turntable from the side, and watching the shadow of the chess piece moving on the screen. As the piece is carried in a circular path by the turntable, the shadow of the piece moves back and forth on the screen in a straight, horizontal line.

Figure 15.9, which is an overhead view of our rotating chess piece, shows the connection between uniform circular motion and simple harmonic motion. The piece moves in a circle of radius r with angular speed ω. (In simple harmonic motion, ω is also called the **angular frequency.**) The x position of the piece is given by $x = r \cos \theta$, where θ is measured in radians. In our sketch, the piece has rotated past the x axis to an angle $\theta = \omega t$, where ωt is also measured in radians. This value of θ assumes the piece was on the x axis and therefore its shadow on the screen had maximum displacement when the clock started at $t = 0$. Thus,

$$x = r \cos \omega t \qquad (15.9)$$

The chess piece has a tangential speed $v = r\omega$. The x component of its velocity is $v_x = -v \sin \theta$, or

$$v_x = -v \sin \omega t = -r\omega \sin \omega t \qquad (15.10)$$

This is the speed of the shadow of the chess piece as it moves back and forth across the screen; details are shown in Figure 15.10. Figure 15.11 shows the *acceleration* of this motion. Circular motion always has a centripetal acceleration directed toward the center of $a_c = r\omega^2$ (Equation 9.10). (Or we can say the net force on the object is a centripetal force directed toward the center $F_c = mr\omega^2$.) The x component of this acceleration is $a_x = -a_c \cos \theta$ or

$$a_x = -r\omega^2 \cos \omega t \qquad (15.11)$$

When we compare Equations 15.9 and 15.11, we see that

$$a_x = -\omega^2 x$$

This is exactly the form of Equation 15.2, which shows the acceleration equal to a negative constant multiplied by the position. This identity between the two equations means the motion of the shadow on the screen—a projection of

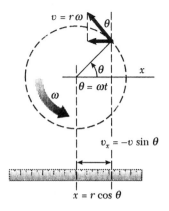

Figure 15.10
The x component of the velocity of an object undergoing uniform circular motion is the same as the velocity of an object undergoing simple harmonic motion.

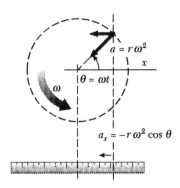

Figure 15.11
The x component of the acceleration of an object undergoing uniform circular motion is the same as the acceleration of an object undergoing simple harmonic motion.

circular motion—is indeed simple harmonic motion. Comparing this result for a_x with Equation 15.2 means

$$\omega^2 = \frac{k}{m}$$

or

$$\omega = \sqrt{\frac{k}{m}} \qquad \textbf{(15.12)}$$

How is this rotational angular frequency ω, in radians per second, related to the oscillation frequency f of the shadow on the screen, in cycles per second? In one cycle or one revolution, the turntable rotates through 2π rad; that is,

$$2\pi \text{ rad} = 1 \text{ cycle}$$

To convert radians per second of ω to cycles per second of f, we must divide ω by 2π:

$$f = \frac{\omega}{2\pi}$$

Therefore,

$$f = \frac{1}{2\pi} \sqrt{\frac{k}{m}} \qquad \textbf{(15.7)}$$

Of course, we can immediately get the period of the oscillation since $T = 1/f$:

$$T = 2\pi \sqrt{\frac{m}{k}} \qquad \textbf{(15.8)}$$

Example 15.4

Find the frequency and period of a simple harmonic oscillator composed of a 0.20-kg mass attached to a spring having a spring constant of 60 N/m.

Reasoning

We can apply Equation 15.7 directly.

Solution

$$f = \frac{1}{2\pi} \sqrt{\frac{k}{m}} = \frac{1}{2\pi} \sqrt{\frac{60 \text{ N/m}}{0.20 \text{ kg}}} = 2.8 \text{ s}^{-1} = 2.8 \text{ Hz}$$

$$T = \frac{1}{f} = 0.36 \text{ s}$$

Practice Exercise   What are the period and frequency of a 0.50-kg mass attached to a 50 N/m spring?

Answer $f = 1.6 \text{ s}^{-1} = 1.6 \text{ Hz}; \; T = 0.63 \text{ s}$

Example 15.5

A 20.0-g mass attached to a spring oscillates with a frequency of 5.00 Hz. What will the frequency be if the 20.0-g mass is removed and a 30.0-g mass is attached instead?

Reasoning

If we knew the spring constant, we could use Equation 15.7 to solve for the frequency. Therefore, we first use Equation 15.7 to find the spring constant. Then we can use that same equation again, this time with $m = 0.0300$ kg, to find the frequency.

Solution

$$f = \frac{1}{2\pi} \sqrt{\frac{k}{m}}$$

$$k = 4\pi^2 f^2 m = 4\pi^2 (5.00 \text{ s}^{-1})^2 (0.0200 \text{ kg}) = 19.7 \text{ N/m}$$

$$f = \frac{1}{2\pi} \sqrt{\frac{19.7 \text{ N/m}}{0.0300 \text{ kg}}} = 4.08 \text{ Hz}$$

Using this connection between simple harmonic motion and circular motion, let us do even more than just find the frequency and period. In Equation 15.9, r is the radius of the circle, but it is also the amplitude of the simple harmonic motion. Therefore, for our simple harmonic motion discussion, we rewrite Equation 15.9 as

$$x = A \cos \omega t \tag{15.13}$$

And this equation gives the position of the oscillator as a function of time. Likewise, we can rewrite Equations 15.10 and 15.11 for the velocity and acceleration of the simple harmonic oscillator as

$$v = -A\omega \sin \omega t$$

$$a = -A\omega^2 \cos \omega t$$

(15.14)

Problem-Solving Tip

- Whenever you need to calculate a trigonometric function of the angle ωt, remember that the angle is in radians and so you must switch your calculator from the "deg" mode to the "rad" mode.

Example 15.6

A simple harmonic oscillator is made of a 0.100-kg mass attached to an 18.0-N/m spring. If the mass is pulled to the right 15.0 cm from equilibrium and released, find its position and velocity 0.200 s later.

Reasoning

The amplitude is 15.0 cm. Knowing the mass and spring constant, we can find the angular frequency using Equation 15.12. Then Equations 15.13 and 15.14 give the position and velocity as functions of time.

Solution

$$\omega = \sqrt{\frac{k}{m}} = \sqrt{\frac{18.0 \text{ N/m}}{0.100 \text{ kg}}} = 13.4 \text{ rad/s} = 13.4 \text{ s}^{-1}$$

$$x = A \cos \omega t$$
$$x_{0.200 \text{ s}} = (15.0 \text{ cm}) \cos [(13.4 \text{ s}^{-1})(0.200 \text{ s})]$$
$$= (15.0 \text{ cm}) \cos (2.68)$$

Set your calculator to "rad" to evaluate cos (2.68 rad):

$$x_{0.200 \text{ s}} = (15.0 \text{ cm})(-0.895) = -13.4 \text{ cm}$$

This position is approaching maximum displacement to the left of equilibrium. That is to be expected since the time $t = 0.200$ s is close to half the period, which is $T = 2\pi/\omega = 0.469$ s.

For the velocity, we have

$$v = -A\omega \sin \omega t$$
$$v_{0.200 \text{ s}} = -(15.0 \text{ cm})(13.4 \text{ s}^{-1}) \sin (2.68)$$
$$= -(201 \text{ cm/s})(0.445) = -89.5 \text{ cm/s}$$

The negative sign indicates the mass is still moving to the left, that it has not yet stopped at the maximum displacement to the left of the equilib-

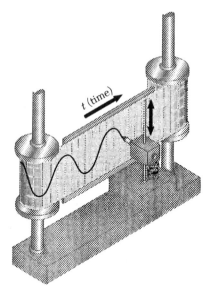

Figure 15.12
Simple harmonic motion is also known as sinusoidal motion because a pen attached to the oscillating mass traces out a sine or cosine curve.

rium position. This, too, is consistent with what we expect since the time $t = 0.200$ s is less than half a period, $T/2 = 0.234$ s.

Practice Exercise Where is it 0.10 s later, when $t = 0.30$ s?
Answer -9.58 cm

The motion described by Equation 15.13,

$$x = A \cos \omega t$$

is also known as **sinusoidal motion.** (Both the cosine and sine functions are referred to as sinusoidal functions.) Figure 15.12 illustrates some apparatus we might create to draw a graph of this motion as a mass oscillates back and forth and the graph paper moves along at a constant rate. The curve traced out by the pen is just $x = A \cos \omega t$ or $x = A \sin \omega t$, depending on when we choose to start measuring by setting $t = 0$.

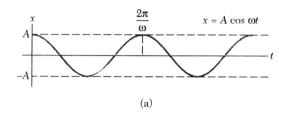

(a)

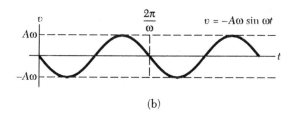

(b)

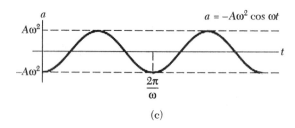

(c)

Figure 15.13
For simple harmonic motion, graphs of (a) displacement, (b) velocity, and (c) acceleration as functions of time are cosine functions (initial conditions: $x_i = A$ and $v_i = 0$ at $t = 0$).

As we see from Equation 15.13, a graph of position versus time is just the cosine of ωt multiplied by the amplitude A. This is sketched in Figure 15.13, along with the velocity-versus-time and acceleration-versus-time graphs.

*15.4 A Vertical Spring and Hanging Mass

CONCEPT OVERVIEW

- The frequency or period of a mass-spring system is the same for vertical motion as for horizontal motion.

So far we have considered only the motion of a horizontal spring attached to a mass resting on a horizontal, frictionless surface. Even then we looked only at the force exerted by the spring, $F = -kx$, ignoring any vertical forces. It is all right to ignore vertical forces because the only ones acting on the mass are its weight mg and the normal force $\mathbf{F}_N$ exerted by the surface, and these two exactly cancel, as sketched in Figure 15.14.

But what of a spring and mass suspended vertically, as sketched in Figure 15.15? Is the vertical motion still simple harmonic motion? When we rotate our horizontal spring to the vertical position, the attached mass makes the spring stretch and a new equilibrium position is attained. At this new equilibrium position, the weight of the mass must be supported by the force of the spring. The spring is stretched a distance Δx in this new equilibrium position, so that

$$\Sigma F = -k\Delta x + mg = 0$$
$$k\Delta x = mg$$

$$\Delta x = \frac{mg}{k}$$

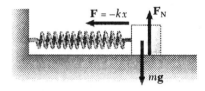

Figure 15.14
The weight mg and the normal force $\mathbf{F}_N$ cancel each other. The net force is the spring force of $F = -kx$.

When the mass is moved a distance x' from this new equilibrium position, the spring exerts an upward force

$$F_{spring} = -kx = -k(x' + \Delta x)$$

Because gravity pulls down with a force equal to the weight of the mass, the net force is

$$F_{net} = F_{spring} + mg$$
$$F_{net} = -kx' - k\Delta x + mg$$

Since $k\Delta x = mg$, we have

$$F_{net} = -kx'$$

And that is exactly the form of Equation 15.1, which defines simple harmonic motion. Therefore, we see that a vertical mass-spring system executes simple harmonic motion about an equilibrium position. The period and frequency remain the same as in the horizontal case. The motion is exactly like that for the horizontal case except, of course, the equilibrium length of the spring is longer.

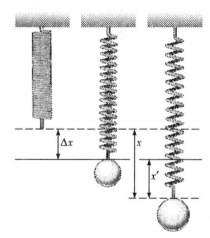

Figure 15.15
A mass m causes a vertical spring to stretch a distance Δx; $k\Delta x = mg$ or $\Delta x = mg/k$. The spring can be stretched an additional length x' beyond this new equilibrium position; $x = \Delta x + x'$.

Example 15.7

When a 50.0-kg sack of cement is placed in the trunk of your car, the car sits 2.00 cm lower. What is the spring constant of the suspension system?

Reasoning

A car's suspension system is more complicated than just a single spring, but treating the whole system as if it were a single spring is still a reasonable approximation. The additional weight compresses the springs of the system, and the compressed springs exert an additional force ΔF that balances the weight of the cement.

Solution

$$k\Delta x = mg$$
$$= k(0.0200 \text{ m}) = (50.0 \text{ kg})(9.80 \text{ m/s}^2)$$
$$k = 24\,500 \text{ N/m}$$

15.5 The Simple Pendulum

CONCEPT OVERVIEW

- A simple pendulum is a simple harmonic oscillator when the amplitude of its swing is small.

A **simple pendulum** is defined as one in which the swinging bob is a point mass. Becoming familiar with this idealized system will help us understand the behavior of a real pendulum, more commonly called a physical pendulum, later in the chapter. Figure 15.16 shows a playground swing and a small bob attached to a string, two systems that are good approximations of an ideal (point mass) simple pendulum. Let us for now concentrate on the small bob of mass m attached to a string of length l. If this bob is pulled to one side and released,

A stroboscopic photograph of a simple pendulum swinging through its period. Note that the pendulum moves faster at the bottom of its arc. (© *Paul Silverman/Fundamental Photographs, NY*)

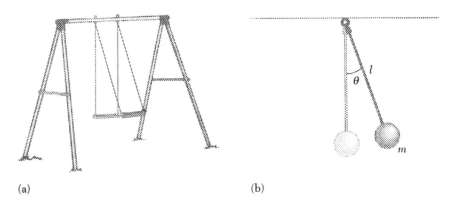

(a)   (b)

Figure 15.16
(a) A child's swing acts like a simple pendulum. (b) A simple pendulum.

we know it will swing back and forth in some type of periodic motion. Is this motion simple harmonic motion?

Figure 15.17 illustrates the forces on a simple pendulum. When the bob is displaced an angle θ from equilibrium, the restoring force pulling it back to the equilibrium position is $mg \sin \theta$. Because this is a restoring force, it pulls to the left if the bob is displaced to the right and to the right if the bob is displaced to the left. We indicate the restoring nature of this force with a negative sign, so that

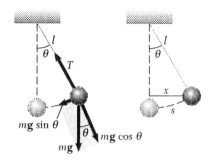

$$F = -mg \sin \theta \qquad (15.15)$$

Figure 15.17
The restoring force pulling a simple pendulum back to equilibrium is $mg \sin \theta$.

Since $a = F/m$, this expression tells us that

$$a = -g \sin \theta$$

Evidently this simple pendulum does *not* undergo simple harmonic motion for this acceleration equation is not Equation 15.2, which tells us that in simple harmonic motion the acceleration is a negative constant multiplied by a displacement.

From watching a simple pendulum, we know it looks so similar to a simple harmonic oscillator that we shall continue our analysis and see what we can find out. From Figure 15.17

$$\sin \theta = \frac{x}{l}$$

So the net tangential force, the restoring force, can be written as

$$F = -mg\left(\frac{x}{l}\right) = -\left(\frac{mg}{l}\right)x$$

Now our $a = F/m$ relationship gives us

$$a = -\left(\frac{g}{l}\right)x$$

which looks very much like Equation 15.2.

There is one problem with having this expression for acceleration in terms of the bob displacement x: the restoring force does not act along the distance x. Rather, it acts tangentially along the arc length s. Since for small angles, the arc length s is approximately equal to the distance x, we can write

$$\sin \theta \approx \frac{s}{l}$$

Then, to the extent that this approximation is valid, we can write the restoring force on the bob as

$$F = -\left(\frac{mg}{l}\right)s \qquad \text{(small angles)} \qquad (15.16)$$

and our expression for the acceleration becomes

$$a = -\left(\frac{g}{l}\right)s \qquad \text{(small angles)}$$

The constant g/l plays the same role as the constant k/m in Equation 15.2. Therefore, we can immediately find the frequency, period, and angular frequency of the pendulum. From Equation 15.7,

$$f = \frac{1}{2\pi} \sqrt{\frac{k}{m}}$$

we get

$$f = \frac{1}{2\pi} \sqrt{\frac{g}{l}} \qquad \textbf{(15.17)}$$

and Equation 15.8 gives us

$$T = 2\pi \sqrt{\frac{l}{g}} \qquad \textbf{(15.18)}$$

Replacing the k/m in Equation 15.12, we get for the angular frequency of a simple pendulum

$$\omega = \sqrt{\frac{g}{l}}$$

Notice that the mass does not appear in any of these equations (which are different forms of the same information). Only the length of the pendulum and the acceleration due to gravity determine the frequency, period, and angular frequency. If you watch a small child and an adult on swings of the same length, you will see that they swing with the same period. And, of course, the frequency and period are independent of the amplitude.

Example 15.8

What is the length of a pendulum whose period is 1.00 s?

Reasoning

We can solve Equation 15.18 for the length.

Solution

$$T = 2\pi \sqrt{\frac{l}{g}}$$

$$l = T^2 \frac{g}{4\pi^2}$$

$$= \frac{(1.00 \text{ s}^2)(9.80 \text{ m/s}^2)}{4\pi^2} = 0.248 \text{ m}$$

Careful measurement of the length and period of a pendulum can be used to calculate the acceleration due to gravity. The local value of g is greater over a large ore deposit and smaller over an underground pool of oil or water. Therefore geologists often use pendulum measurements to help determine what is underground. It is easier to carefully construct and make measurements on a pendulum than to do deep Earth drilling.

*15.6 Torsional Oscillations

CONCEPT OVERVIEW

■ A rotating shaft behaves like a spring in that the restoring torque it produces is proportional to the angle through which it has been twisted, $\tau = \kappa\theta$.

■ A disk attached to a rotating shaft undergoes rotations that are simple harmonic.

Figure 15.18 shows a horizontal disk supported by a flexible shaft. This shaft may be something exotic like a quartz filament or something common like a simple rope. When the disk is rotated one way, the shaft exerts a restoring torque directed the other way. This torque exhibits the same type of behavior as the force in a spring and can be written as

$$\tau = -\kappa\theta$$

where τ is the torque exerted by the shaft and κ, lower-case Greek kappa, is called the **torsion constant.** The negative sign indicates the torque tries to rotate the system back to equilibrium. The torsion constant must have units of torque per angular measure. It can be given in newton-meters per degree or, more commonly, newton-meters per radian. Just as with a spring, a higher value of κ means a stiffer or stronger shaft and a lower value of κ means a weaker shaft.

You should recall from Equation 9.13 that the rotational equivalent of Newton's second law is

$$\tau = I\alpha$$

where I is the moment of inertia of the rotating object and α is the angular acceleration. (This expression ignores the moment of inertia of the shaft.) Therefore,

$$I\alpha = -\kappa\theta$$

$$\alpha = -\left(\frac{\kappa}{I}\right)\theta$$

And this has exactly the same form as Equation 15.2; the angular acceleration is equal to a negative constant multiplied by the angular displacement. Immediately, we know that the rotational motion of the disk in Figure 15.18 is simple harmonic motion. The frequency and period are

$$f = \frac{1}{2\pi}\sqrt{\frac{\kappa}{I}} \qquad (15.19)$$

and

$$T = 2\pi\sqrt{\frac{I}{\kappa}} \qquad (15.20)$$

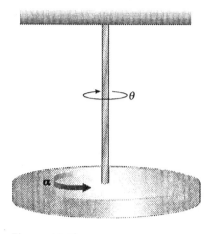

Figure 15.18
A torsional pendulum undergoes simple harmonic motion as it rotates back and forth.

A balance wheel and spring (*right*) inside of an early 1900's Elgin pocket watch containing a 17 jewel mechanical movement. Note that the balance wheel has small weights attached to its outside rim. What function do these weights perform?
(© *Dale E. Boyer/Photo Researchers, Inc.*)

Example 15.9

A rope holds a horizontal, solid disk that has a radius of 10.0 cm and a mass of 0.500 kg. A force of 1.25 N must be applied tangentially to the edge of the disk to rotate it 0.250 rad. If the disk is released from this position, what are the period and frequency of its oscillation?

Reasoning

The rope provides a restoring torque when the disk is rotated. This restoring torque—like the restoring force of a spring—causes the disk to oscillate. We must find the torsion constant κ and the moment of inertia of the disk and then use those values to calculate the frequency and period of the oscillations.

Solution

A force of 1.25 N applied tangentially to the edge of the disk produces a torque of (Equation 9.24)

$$\tau = rF = (0.100 \text{ m})(1.25 \text{ N}) = 0.125 \text{ N·m}$$

This torque is just balanced by the rope:

$$\tau = \kappa\theta = \kappa(0.250 \text{ rad}) = 0.125 \text{ N·m}$$

$$\kappa = 0.500 \text{ N·m/rad} = 0.500 \text{ N·m}$$

The moment of inertia for a solid disk is (Table 9.2)

$$I = \frac{mr^2}{2} = \frac{1}{2}(0.500 \text{ kg})(0.100 \text{ m})^2 = 0.00250 \text{ kg·m}^2$$

Now we can use these values in calculating the frequency:

$$f = \frac{1}{2\pi}\sqrt{\frac{\kappa}{I}} = 2.25 \text{ cycles/s} = 2.25 \text{ Hz}$$

$$T = 0.444 \text{ s}$$

A model of the Foucault pendulum showing Earth's rotation below its motion in a fixed plane.   (© D. Davis)

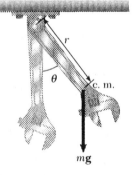

Figure 15.19
A physical pendulum is any extended body that oscillates.

*15.7 The Physical Pendulum

CONCEPT OVERVIEW

■ At small amplitudes, a physical pendulum—an extended body used as a pendulum—undergoes simple harmonic oscillations.

Figure 15.19 illustrates a **physical pendulum,** any extended object that can oscillate back and forth about a pivot point. The center of mass of the object is located a distance r from the pivot. The restoring torque on the object is

$$\tau = -mgr \sin \theta$$

This has exactly the same form as Equation 15.15 for the force on a simple pendulum. For small values of θ, we can make the small-angle approximation that $\sin \theta \approx \theta$ and write the torque as

$$\tau = -mgr\,\theta$$

which is analogous to Equation 15.16. (This small-angle approximation of $\sin \theta \approx \theta$ is equivalent to using $x = s$ in Figure 15.17, as we did in that previous section.)

Following the same procedure we used in the previous section for the torsional pendulum, we find

$$\tau = I\alpha = -mgr\,\theta$$

$$\alpha = -\left(\frac{mgr}{I}\right)\theta$$

which means the frequency and period are

$$f = \frac{1}{2\pi}\sqrt{\frac{mgr}{I}} \qquad \textbf{(15.20)}$$

and

$$T = 2\pi\sqrt{\frac{I}{mgr}} \qquad \textbf{(15.21)}$$

It is interesting to note that these two equations reduce to Equations 15.17 and 15.18 for the simple pendulum. For a point mass m on the end of a string of length r, the moment of inertia is $I = mr^2$, so Equation 15.20 becomes

$$f = \frac{1}{2\pi}\sqrt{\frac{mgr}{mr^2}} = \frac{1}{2\pi}\sqrt{\frac{g}{r}}$$

which is just Equation 15.17 with the pendulum length l replaced by r.

A full size Foucault pendulum at the Franklin Institute in Philadelphia, Pennsylvania. This type of pendulum was first used by the French physicist Jean Foucault to verify the Earth's rotation experimentally. As the pendulum swings, its plane of oscillation appears to rotate. Actually, the pendulum's plane of motion is fixed; it is the Earth that rotates beneath the swinging pendulum. (© D. Davis)

*15.8 Damped and Driven Harmonic Motion

CONCEPT OVERVIEW

- Friction decreases the amplitude of an oscillator.

- Driving an oscillator at its resonance frequency gives it the largest amplitude.

As a real mass-spring system oscillates, the amplitude does not remain constant as our equations would indicate. There is always some friction present—perhaps only a little and perhaps a great deal. That means energy is dissipated as work is done by the friction force. This frictional effect is often called **damping**—the motion is "damped out." Figure 15.20 shows the motion of a harmonic oscillator with a little damping. The oscillations continue with a constant frequency, but the amplitude decreases. This motion is described by the equation

$$x = Ae^{-t/\tau}\cos\omega t \qquad \textbf{(15.23)}$$

where the constant amplitude A of the simple harmonic oscillator of Equation

Figure 15.20
The amplitude of a damped harmonic oscillator decreases exponentially.

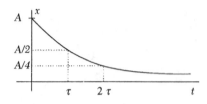

Figure 15.21
Overdamped harmonic motion has a great deal of friction.

Figure 15.22
Critically damped harmonic motion returns the system to equilibrium as quickly as possible, without overshooting equilibrium.

A motorcycle spring and shock absorber assembly. How do shock absorbers damp excess oscillatory motions such as those created by a bumpy road surface? *(© Richard Megna/Fundamental Photographs, NY)*

15.13 has been replaced by the expression $Ae^{-t/\tau}$. Once again we see the exponential decay we first encountered in Equation 13.3 and which we will see again later in this book. As before, τ is the **time constant** and is a measure of how rapidly something approaches a limiting value (zero, in this case).

In particular, this oscillatory motion with decaying amplitude is known as **underdamped harmonic motion.** For an example, you might think of running a mass-spring oscillator under water. The frictional drag of the water will cause the motion to dampen out. If we wait for one half-life, the amplitude will be half its original value. If we wait for another half-life, the amplitude will be half of that value (one fourth the original value). That is always the behavior of a decaying exponential, $e^{-t/\tau}$.

If the friction forces increase enough, there will be no oscillation. Think of putting our mass-spring system in a tank of molasses! If the mass is pulled to one side and released, it will very s-l-o-w-l-y return to the equilibrium position with no overshooting; it will not oscillate at all. This motion is illustrated in Figure 15.21 and is known as **overdamped motion.**

There is an interesting intermediate type of motion known as **critically damped motion** (Fig. 15.22). This motion brings the system back to equilibrium as quickly as possible without overshooting the equilibrium position. A prime example is the motion of a door-closing mechanism. The simplest device for closing a door is an ordinary spring. This closes the door quickly, but like any other mass attached to a spring, the door is moving fastest as it goes through equilibrium—wham! This slamming damages the door and the door jamb. A "shock absorber" can be added to the door. If too much damping is added, the door will stay open longer than necessary, allowing heated or cooled air out and insects in. Adjusting the shock absorber so that the door closes as quickly as possible without slamming into the jamb means we have achieved critical damping.

Critical damping is exactly the role of shock absorbers on automobile suspension systems. If the absorbers are worn and too weak, there will be multiple oscillations and the car will bounce up and down several times after hitting a bump. If the shocks are too strong or too heavy, it will be difficult for the wheels to move relative to the car and bumps will be transferred directly to the passengers. Shocks are "tuned" to the rest of the car to achieve critical damping.

You probably know from pushing a friend on a swing that you get a large amplitude if you apply an external force at just the right time (or with just the right frequency). The **natural frequency** of an oscillating system is the frequency it has when disturbed and left to oscillate on its own without any external force applied to it. Think of a playground swing with a natural frequency of, say, one-half cycle per second. If you stand behind the swing and push on it every 1.5 s or every 2.8 s, not much happens. However, if you synchronize your pushes so you push every 2.0 s as the swing is going forward—you will find your friend going higher and higher. This is an example of **resonance**—a very large output for a small input at just the right frequency. This **resonance frequency** is the particular frequency of an external force which gives the greatest amplitude.

We talk of an external force applied to an oscillator as **driving** the oscillator, or we say that this is a **driven** oscillator. If the driving force is sinusoidal, it varies in time the way a sine or cosine function does and can be written in the form

$$F_{ext} = F_o \cos \omega_{ext} t$$

or

$$F_{ext} = F_o \sin \omega_{ext} t$$

where F_o is the maximum value of the external force and ω_{ext} is the angular frequency of the external force. If the external force has a frequency that is near the natural frequency of the oscillator it will cause the system to oscillate with a large amplitude; this is another example of resonance. This may be good or bad; it may be important to enhance resonance on one project but to avoid resonance on another, as the following examples show.

The sound board of a violin or guitar or piano is designed to resonate along with the strings. An isolated vibrating string would be very difficult to hear, but if the vibrations cause the sound board to vibrate with great enough amplitude, the sound will be easily heard. An automotive engineer does not want the expansion joints in a highway to cause the suspension system of your car to resonate. Sometime, however, you may have experienced that. You may have been in a certain car on a certain road where all was quiet at 30 mi/h but at 40 mi/h vibration started and increased until it became terrible at 50 mi/h. Increasing your speed to 60 mi/h reduced the problem and perhaps the problem even disappeared at 70 mi/h.

The classic example of unwanted resonance is the Tacoma Narrows Bridge. In the late 1930s, a suspension bridge was built across the Tacoma Narrows in the state of Washington. From the time it was opened, the bridge seemed to swing and sway more than it should, but when the wind was just right—not fast and hard but at just the right speed and from just the right direction—the bridge twisted violently. The bridge was made from steel girders about 2 m tall strung on both sides of the roadway with the connecting horizontal steel beams covered by a thick layer of concrete. Steel and concrete can flex only so much before brittle fracture occurs. On November 7, 1940, this resonance response was especially large, and Figure 15.23 shows the bridge just before collapse.

The sound board of a piano helps to amplify the sound made as the strings are struck. (© *Mark Gottlieb/ FPG International Corp.*)

Figure 15.23
The Tacoma Narrows Bridge collapsed on November 7, 1940, as a result of wind-induced resonance oscillations. (*UPI/Bettmann Newsphotos*)

Resonance phenomena, like the rest of harmonic oscillation ideas, are not limited to mechanical systems. Every time you tune a radio, you are making use of resonance. Electric circuits can behave like oscillators, and an external driving force—a radio station's broadcast frequency in this case—can cause a large response if the circuit has a natural frequency very close to that of the external driving force.

SUMMARY

When either stretched or compressed from its equilibrium position, a spring exerts a restoring force that is proportional to the displacement. This is known as Hooke's law,

$$F = -kx \tag{15.1}$$

where k is the spring constant and x is the displacement from equilibrium. An object of mass m attached to this spring experiences an acceleration

$$a = -\left(\frac{k}{m}\right)x \tag{15.2}$$

The work done on the spring in an oscillating mass-spring system is stored as spring potential energy:

$$PE = \tfrac{1}{2}kx^2 \tag{15.4}$$

A moving oscillator also has kinetic energy $KE = \tfrac{1}{2}mv^2$. Its total energy is

$$E = \tfrac{1}{2}kx^2 + \tfrac{1}{2}mv^2 \tag{15.5}$$

and remains constant. Energy may change form—from potential to kinetic and back again—but the total remains constant.

The amplitude of a simple harmonic oscillator is its maximum displacement. A simple harmonic oscillator has maximum speed v_{max} as it passes through equilibrium at $x = 0$. The time required for one cycle is known as the period of the oscillation T. The frequency f, the number of oscillations per unit time, of a mass-spring system is

$$f = \frac{1}{2\pi}\sqrt{\frac{k}{m}} \tag{15.7}$$

and its period is

$$T = \frac{1}{f} = 2\pi\sqrt{\frac{m}{k}} \tag{15.8}$$

For a given m and k, the frequency and period are the same for horizontal or vertical motion.

The position of a simple harmonic oscillator as a function of time can be written as

$$x = r\cos\omega t \tag{15.9}$$

This motion is also known as sinusoidal motion.

The restoring force on a simple pendulum is

$$F = -mg\sin\theta \tag{15.15}$$

but if we limit ourselves to small angles, this restoring force can be written as

$$F = -\left(\frac{mg}{l}\right)s \quad \text{(small angles)} \tag{15.16}$$

which is in the form of the force exerted by a simple harmonic oscillator—a negative constant multiplied by the displacement. The frequency and period of a simple harmonic oscillator are

$$f = \frac{1}{2\pi}\sqrt{\frac{g}{l}} \tag{15.17}$$

$$T = 2\pi \sqrt{\frac{l}{g}} \qquad (15.18)$$

Resonance describes a situation in which the amplitude of an oscillator becomes very large as the frequency of the external driving force approaches a particular value, called the resonant frequency.

QUESTIONS

1. If a spring is cut in half, as in Figure 15.24, what happens to its spring constant?

Figure 15.24 Question 1.

2. If two identical springs are attached end to end, as in Figure 15.25, what is the spring constant of the resulting spring?

Figure 15.25 Question 2.

3. If two identical springs are attached side by side, as in Figure 15.26, what is the spring constant of the combination?

Figure 15.26 Question 3.

4. Where is the velocity zero for a simple harmonic oscillator?

5. Where is the acceleration zero for a simple harmonic oscillator?

6. In studying the mass-spring system, we ignored the mass of the spring. How does the mass of a real spring affect the period of a harmonic oscillator?

7. In terms of amplitude, what total distance is traveled by a simple harmonic oscillator during one period?

8. What happens to the period of a mass-spring oscillator if the mass is doubled?

9. Will doubling the amplitude of a simple harmonic oscillator also double the maximum velocity? Explain.

10. How does doubling the amplitude of a simple harmonic oscillator change the maximum velocity, the maximum acceleration, the frequency, and the period?

11. Will a pendulum clock that is accurate at sea level gain or lose time when it is taken to higher altitudes?

12. What happens to the period of a simple pendulum if the mass of the bob is doubled?

13. What happens to the period of a simple pendulum if the length is doubled?

14. What must be done to the length and bob to double the period of a simple pendulum?

15. Why will a 440-Hz tuning fork cause another 440-Hz tuning fork to vibrate, even though separated by some distance, but have no effect on a 512-Hz tuning fork?

16. Why is it important that the speaker cone in an audio system be damped?

PROBLEMS

15.1 Simple Harmonic Motion

15.1 An oscillator has a period of 1.2 s. What is its frequency?

15.2 An object oscillates with a frequency of 5.2 Hz. What is its period?

15.3 A radio station gives its broadcast frequency as 91.7 MHz. What is the corresponding period?

○15.4 A mass-spring simple harmonic oscillator has a mass of 2.0 kg and a spring constant of 72 N/m. Find its period and frequency. Can you determine its amplitude from this information?

○15.5 A mass-spring simple harmonic oscillator has a mass of 0.75 kg and a period of 1.1 s. What is the spring constant?

○15.6 A mass-spring simple harmonic oscillator has a mass of 0.75 kg and a frequency of 3.5 Hz. What is the spring constant?

○15.7 A mass-spring simple harmonic oscillator has a spring constant of 120 N/m and a period of 1.2 s. How much mass is attached to the spring?

○15.8 A mass-spring simple harmonic oscillator has a spring constant of 96 N/m and a frequency of 5.2 Hz. How much mass is attached to the spring?

○❖15.9 For the following sets of data, calculate the frequency of the mass-spring simple harmonic oscillator and rank the systems in order of increasing frequency:

Oscillator	Mass (kg)	Spring Constant (N/m)
a	1.0	2.5
b	2.5	1.0
c	3.0	7.2
d	50	100
e	80	200
f	100	250
g	120	250
h	200	500

●15.10 If a hole were drilled through Earth along a diameter, the force on an object of mass m dropped into this hole would be $-mgr/R$, where g is the acceleration due to gravity at Earth's surface, r is the distance of the object away from Earth's center, and $R = 6.4 \times 10^6$ m is the radius of Earth. Explain how this setup would produce simple harmonic motion. Calculate the period of this motion.

●15.11 Find the frequency for the three arrangements of simple harmonic oscillators illustrated in Figure 15.27.

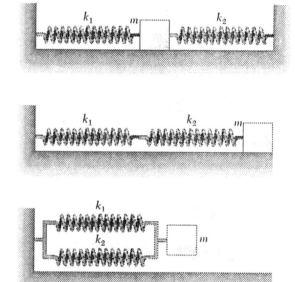

Figure 15.27
Problem 15.11. Different arrangements of springs produce different "effective spring constants" and thus different frequencies.

15.2 The Energy of a Simple Harmonic Oscillator

15.12 A spring has a spring constant of 80 N/m. How much work is done in compressing the spring 15 cm from equilibrium?

15.13 A spring having a spring constant of 25 N/m is stretched 5.0 cm from equilibrium. How much work is done in this process?

15.14 A spring that has a spring constant of 36 N/m is stretched 9.0 cm from equilibrium. How much work is required to stretch it an additional 3.0 cm?

○**15.15** A spring that has a spring constant of 40 N/m is attached to a 200-g mass. The mass is moved 10 cm from its equilibrium position and released from rest. What speed will it have as it moves through the equilibrium position?

○**15.16** A spring that has a spring constant of 25 N/m is attached to a 150-g object. The object is moved 8.0 cm from its equilibrium position and released from rest. What speed will it have as it moves through the equilibrium position?

○**15.17** A spring for which the spring constant is 120 N/m is attached to a 750-g object. The object is moved 10 cm from its equilibrium position and released from rest. What speed will it have when it is 5.0 cm from the equilibrium position?

○**15.18** A spring for which the spring constant is 100 N/m is attached to a 250-g object. The object is moved 10 cm from its equilibrium position and released from rest. What speed will it have when it is 4.0 cm from the equilibrium position?

○**15.19** The tines of a tuning fork vibrate with simple harmonic motion at a frequency of 512 Hz. The tip of each tine moves through 2.00 mm, 1.00 mm on either side of the center. What are the maximum velocity and maximum acceleration of the tip?

○**15.20** An object undergoes simple harmonic motion with an amplitude of 12 cm. At a point 8.0 cm from equilibrium, its speed is 20 cm/s. What is the period?

○**15.21** An object undergoes simple harmonic motion with an amplitude of 15 cm. At a point 8.0 cm from equilibrium, its speed is 10 cm/s. What is the period? How fast will it be going when it is 4.0 cm from equilibrium?

○**15.22** An object undergoes simple harmonic motion with an amplitude of 15 cm. At a point 8.0 cm from equilibrium, its speed is 12 cm/s. What is the period? If the object has a mass of 150 g, what is the spring constant of the spring to which it is attached?

○**15.23** A spring with a spring constant of 50 N/m is attached to a 750-g object. The object is moved 12 cm from its equilibrium position and given an initial speed of 8.0 cm/s. What is the amplitude of the resulting motion? What is the maximum speed of the object?

○**15.24** A 0.50-kg object is attached to a 40-N/m spring and moved 15 cm from equilibrium. There it is given an initial speed of 20 cm/s. What is the amplitude of the resulting motion? What is the maximum speed of the object?

● ❖ **15.25** Data for several mass-spring simple harmonic oscillators are given below. For each one, determine the speed at the position given. Then rank the oscillators in order of increasing speed.

Oscillator	Mass (kg)	Spring Constant (N/m)	Amplitude (m)	Position (m)
a	0.180	3.50	0.200	0.100
b	0.180	3.50	0.200	0.150
c	0.280	3.50	0.200	0.180
d	0.350	5.00	0.250	0.0500
e	0.350	5.50	0.250	0.100
f	0.500	5.50	0.250	0.100
g	0.500	5.50	0.250	0.150
h	0.500	5.50	0.300	0.150

*15.3 Simple Harmonic Motion and Circular Motion

15.26 A chess piece sits 10 cm from the center on a phonograph turntable that turns at 33 rpm. What is the angular velocity of the piece in radians per second? The x position of the piece is 10 cm when the angle of the turntable is 0°. What are the x component of its position and the x component of its velocity for angles of 45°, 90°, 135°, 150°, 210°, 270°, 300°, and 330°?

● 15.27 The pistons of an automobile engine undergo approximately simple harmonic motion. If the stroke (which is *twice* the amplitude) is 12 cm and the tachometer indicates 4200 rpm, what is the maximum acceleration of a piston? If the piston has a mass of 1.8 kg, what net force is exerted on it at the end of each stroke? (Be careful to change rpm into radius per second.)

*15.4 A Vertical Spring and Hanging Mass

15.28 When 200 g is suspended from a vertical spring, the spring stretches 5 cm. What is the spring constant? How far will the spring stretch if an additional 150 g is added?

15.29 When 300 g is suspended from a vertical spring, the spring stretches 7.5 cm. How far will it stretch if it supports 425 g?

15.30 When 400 g is suspended from a vertical spring, the spring stretches 8 cm. If this spring is now placed in the horizontal position, what force is necessary to stretch it 6 cm?

15.31 A 50-g mass hanging from a spring oscillates with a frequency of 2 Hz. What is the spring constant?

15.32 A 500-g mass hanging from a spring oscillates with a period of 2.5 s. What is the spring constant?

15.33 An object hangs from a 35-N/m spring and oscillates with a period of 2.8 s. What is the mass of the object?

15.5 The Simple Pendulum

15.34 Find the length of a "seconds pendulum," which is a pendulum whose period is exactly 2.00 s. Such a pendulum requires 1.00 s for each half of its swing.

15.35 What is the period of a pendulum 1.0 m long?

15.36 A child rides a playground swing that is 2.5 m long. What is the period of the swing?

If she swings with an amplitude of 10°, what is her maximum linear speed?

15.37 A Foucault pendulum is one used to demonstrate that Earth rotates on its axis. The first such pendulum, used by Jean Foucault in Paris in 1851, was 61 m long. What was its period?

○ 15.38 Two playground swings start out together. After ten complete oscillations by the slower swing, the two are out of step (or out of phase) by half a cycle. What is the percent difference in their length?

○ 15.39 If a petroleum geologist uses a seconds pendulum (see Problem 15.34) that normally has a period of 2.000 s and finds that its period is now 2.001, by how much will she conclude that the local value of **g** has changed?

○ 15.40 The acceleration due to gravity on the Moon's surface is about one sixth that on Earth's surface. How will this difference affect the frequency of a pendulum? In particular, what will the frequency be on the Moon for a pendulum that has a frequency of 2.5 Hz on Earth?

*15.6 Torsional Oscillations

15.41 A disk that has a moment of inertia of 2.3 kg·m^2 is suspended by a cord that has a torsion constant of 0.50 N·m/rad such that the flat surfaces of the disk are parallel to the floor. If the disk is rotated slightly and allowed to oscillate, what will its period of oscillation be?

15.42 A disk that has a moment of inertia of 3.9 kg·m^2 is suspended by a cord that has a torsion constant of 2.8 N·m/rad such that the flat surfaces of the disk are parallel to the floor. If the disk is rotated slightly and allowed to oscillate, what will its frequency of oscillation be?

15.43 A uniform 4.0-kg disk that has a diameter of 0.30 m is suspended so that its two surfaces are parallel to the floor. The suspending cord has a torsion constant of 3.5 N·m/rad. What is the moment of inertia of the disk? If the disk is rotated slightly and allowed to oscillate, what will its period of oscillation be?

○15.11 A disk that has a moment of inertia of 0.85 kg·m² is suspended by a shaft such that the two surfaces of the disk are parallel to the floor and the disk-shaft system acts as a torsional pendulum. Its oscillations have a period of 0.33 s. What is the torsion constant of the shaft?

○15.45 Find the period of the oscillations of a 1.2-kg sphere 10 cm in diameter when it is suspended by a wire that has a torsion constant of 0.20 N·m/rad.

○15.46 A mechanical watch keeps time by means of an oscillating wheel called a balance wheel. A certain balance wheel can be approximated by a hoop ($I = mr^2$) with a diameter of 1.20 cm and a mass of 0.850 g. What should the torsional constant of the balance spring attached to this wheel be if the wheel has a period of 0.500 s?

○15.47 If the mechanical watch described in the previous problem loses 2.0 min every day, how should the moment of inertia be changed to correct the problem? Balance wheels have screws along the rim so that mass can be moved closer to or farther from the axis of rotation to change the moment of inertia. Should the screws on this watch be moved toward or away from the center?

●15.48 A Cavendish balance (Figure 5.16) is used to measure the gravitational force between two objects of known masses. The dumbbell of a particular Cavendish balance is made of two 30-g masses attached to the ends of a lightweight rod 0.30 m long. Find the moment of inertia of this dumbbell. When allowed to oscillate, the dumbbell has a period of 4.2 s. Find the torsional constant of the supporting fiber.

*15.7 The Physical Pendulum

15.49 Calculate the period of a meter stick pivoted at one end.

15.50 Calculate the period of a meter stick pivoted at 25 cm.

○15.51 A hoop that has a diameter of 0.50 m is supported on its top edge so that it can swing as a physical pendulum (Fig. 15.28). What is its period of oscillation?

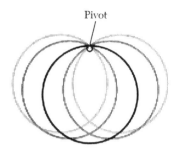

Figure 15.28
Problem 15.51. The moment of inertia of a hoop, for rotation about a point on its edge, is $I = 2mr^2$.

○15.52 A disk that has a diameter of 0.50 m has holes drilled along one radius so that it can oscillate about an axis placed at the center or at r/4, r/2, or 3r/4 from the center (Fig. 15.29). Find the period for oscillations about each axis.

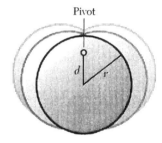

Figure 15.29
Problem 15.52. The moment of inertia for a disk of mass m and radius r pivoted about a point a distance d from the center is $I = mr^2/2 + md^2$.

Figure 15.30 Problems 15.53 and 15.54.

● ❖ 15.53 A movable-pivot physical pendulum is made of a 200-g meter stick and a 500-g mass, as shown in Figure 15.30. Find its period when the mass and pivot are at the following positions, and rank the pendula in order of increasing period:

Pendulum	Pivot Position (cm)	Mass Position (cm)
a	10	80
b	10	70
c	20	90
d	20	75
e	30	70
f	40	60
g	50	90
h	50	80

● ❖ 15.54 A movable-pivot physical pendulum is made of a meter stick and mass, as shown in Figure 15.30. Find its period for the following situations, and rank the situations in order of increasing period:

Pendulum	Mass of Meter Stick (g)	Size of Added Mass (g)	Position of Added Mass (cm)	Position of Pivot (cm)
a	200	500	80	10
b	200	750	70	10
c	200	1000	90	20
d	300	800	75	20
e	150	500	70	30
f	250	400	60	40
g	250	400	90	50
h	250	600	80	50

● 15.55 In walking, the leg swings back and forth somewhat like a physical pendulum pivoted at the hip. Assume such a pendulum is made of two rods rigidly attached (at the knee). The upper rod (thigh) has a mass of 7.0 kg and a length of 45 cm, and the lower rod (lower leg and foot) has a mass of 4.0 kg and a length of 50 cm. Calculate the period of this leg/pendulum. How does this calculated period compare with the period you measure for your own leg or that of other students?

The legs of a person walking move back and forth with a motion similar to that of a physical pendulum. (© *Stephen Simpson/FPG International Corp.*)

*15.8 Damped and Driven Harmonic Motion

15.56 A pendulum is set in motion with an initial amplitude of 15°. After 15 min the friction of air resistance has reduced the amplitude to 5°. What is the value of the time constant τ in Equation 15.23 for this pendulum?

● 15.57 The front spring of a car has a spring constant of 1.39×10^6 N/m and supports a weight of 2000 N. The wheel underneath it is 40.0 cm in radius. The wheel is slightly unbalanced and starts to vibrate strongly at a certain speed. What is that speed?

General Problems

● 15.58 A 0.40-kg mass is attached to a 25-N/m spring and moved 12 cm from equilibrium. There it is given an initial speed of 25 cm/s. What is the amplitude of the resulting motion? What is the speed of the mass when it is 15 cm from equilibrium?

○ 15.59 An object vibrates horizontally along the x axis with a frequency of 30 Hz. The total extent of its travel, from one extreme to the other, is 2.80 m. At time $t = 0$, the object is released from the point that is farthest to the right of center. (a) Write an equation for its motion. (b) What is its maximum velocity, and where does this velocity occur? (c) What is its maximum acceleration, and where does this acceleration occur? What are the (d) displacement, (e) velocity, and (f) acceleration of the object 1.0 and 1.5 s after being released?

○15.60 When an 85-kg person sits in a 1200-kg car, the car becomes lower by 0.8 cm. What is the spring constant of the car's suspension system? What is its natural frequency of vibration when it hits a pothole?

●15.61 The springs of a 1000-kg car compress vertically 7.0 mm when a 100-kg man gets in. With the man in the car, how many vibrations per second does the car make after going over a bump? (Ignore the effects of the shock absorbers.)

●15.62 If a grandfather's clock that keeps good time in the Great Plains is carefully moved to the Rocky Mountains, where **g** is 0.25% less, how long will it take for the time to be incorrect by 1 min? Will it be 1 min early or late?

●15.63 A rope suspends horizontally a solid disk that has a radius of 15.0 cm and a mass of 1.20 kg. A force of 2.25 N must be applied tangentially to the edge of the disk to rotate it 0.350 rad. If it is released from this position, what will the period and frequency of its oscillation be?

PHYSICS AT WORK

What happens to the air as music is played? How can we describe the music we hear? These are the ideas we discuss in this chapter.

A sound wave is simple harmonic motion that moves through the air from its source (an instrument in this orchestra, say) to a detector (our ears). Musicians describe music with words like loudness, pitch, and quality. Physicists describe the same characteristics with such words as amplitude, fundamental frequency, and harmonics.

An orchestra is a collection of instruments creating music of many different frequencies and characteristics. When this music is recorded and played back through a sound system, the speakers reproduce the different frequencies and characteristics so that our ears (and brains) hear a re-creation of the original music.

What do we hear when middle C is played by a flute and by a trumpet? Something must be the same for we can tell that both instruments are playing the same note. And something must surely be different for we can distinguish the flute from the trumpet. The common characteristic is the fundamental frequency (or pitch). The distinguishing characteristics are the harmonics (or overtones or quality or timbre).

(© Dick Luria/FPG International Corp.)

16

Waves and Sound

- A pulse or disturbance can travel along a medium such as a rope without the movement of mass from the source to the receiver. A continuous train of pulses moving through a medium is called a *wave*. A wave is described in terms of speed, frequency, wavelength, and amplitude.

- Waves can pass through each other without impeding their movement, which is known as *superposition*. Waves can interfere with each other, creating regions of constructive interference with large amplitude or regions of destructive interference with small amplitude.

- A wave moving down a string is reflected from the far end, and then the two waves interfere. For particular frequencies, an interference pattern is established for which there are "nodes" where the string does not move and "antinodes" where the amplitude is large.

- In transverse waves individual disturbances are perpendicular to the wave's general motion, like the up and down movement of a rope as a wave moves to the right or left. In longitudinal waves individual disturbances are in the same direction as the wave's general motion, like the back and forth movement of an air molecule as a sound wave passes by.

We encounter waves constantly. We see by light, which is a wave. We hear by sound, which is a wave. Part of the beauty of an ocean is the waves on its surface. We make frequent use of radio and television waves. We even sing of the "amber waves of grain" of a wheat field on the prairie.

What is a wave? With the amber waves of grain, for instance, we can watch *something* move across a wheat field, but what? The pattern or disturbance that is the wave moves from horizon to horizon, and yet each individual head of wheat moves only a very limited distance. This two-part movement—one part widespread, the other local—is characteristic of all waves.

There are two main categories of waves: mechanical and electromagnetic. **Mechanical waves** are those that require a medium to pass through. Sound waves, ocean waves, and wind waves across a wheat field are examples of mechanical waves. Without the medium—air, water, wheat—there can be no wave. **Electromagnetic waves** are those that can propagate through a vacuum. Light, radio waves, televison waves, microwaves, and x-rays are examples of electromagnetic waves.

In this chapter we focus on mechanical waves only. Electromagnetic waves we shall take up in Chapter 22 after we have studied the basics of electricity and magnetism.

16.1 Wave Motion

CONCEPT OVERVIEW

- A wave travels along a medium without any mass moving from the source of the disturbance to the receiver.

- A wave is described in terms of its speed, frequency, wavelength, and amplitude.

Figure 16.1 shows a rope that you and a friend might hold at each end. If you strike the rope near one end—simply hit it once with your free hand—you will see something—a disturbance, or pattern, called a **pulse**—move along the rope. This movement may happen so rapidly that all you notice is that the rope swayed or moved. If you and your friend pull harder on the rope—that is, increase the tension in it—the pulse moves even faster. Get a heavier rope and the pulse slows down. If the rope is heavy enough, you should be able to see the moving pulse clearly. To get an even denser medium, a spring may be used, as in Figure 16.2, where a strobe light has photographed the pulse at successive times as it moves along the spring. The pulse speed is related to the density of the medium by the expression

$$v = \sqrt{\frac{T}{m/L}} \qquad (16.1)$$

where T is the tension in the medium and m/L is its linear density.

Instead of giving the rope a single blow, you could attach one end to a simple harmonic oscillator so that the end undergoes simple harmonic motion. Now the pattern that travels along the rope is a succession of pulses, as shown in Figure 16.3. This succession of pulses is what we call a **wave,** defined as the transfer of energy from one point to another without the transfer of any mass.

A passing wave disturbs its medium momentarily but never causes any molecules of the medium to travel.

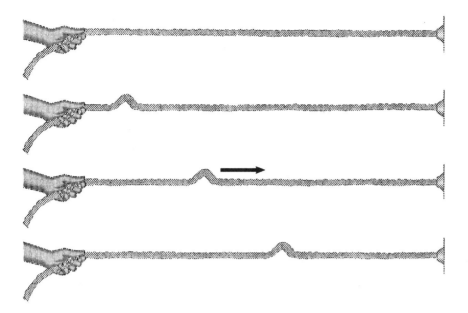

Figure 16.1
A pulse moves along a rope. If the rope is heavy enough, the pulse travels slowly enough to be seen.

Figure 16.2
A strobe light shows a pulse moving along a spring. *(Courtesy of Educational Development Center, Newton, MA)*

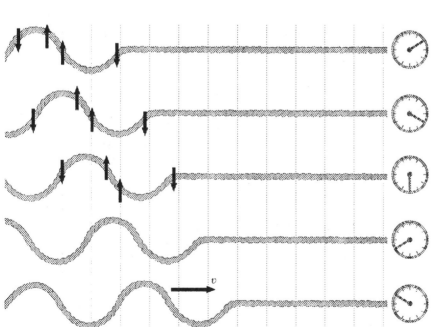

Figure 16.3
If the end of a rope is moved by a simple harmonic oscillator, a sinusoidal wave is created and travels along the rope.

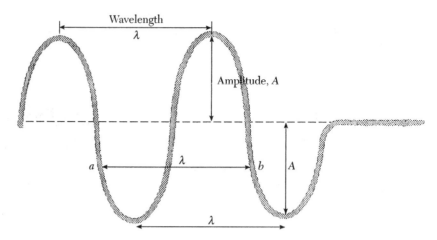

Figure 16.4
The amplitude of a wave is the maximum displacement of the medium from equilibrium. The wavelength is the distance from any point on the wave to the nearest point at which the wave pattern repeats itself.

The pulse of Figures 16.1 and 16.2, for example, moves along the entire length of the rope or spring, but each particle of the rope or spring moves only a small distance, as the pulse passes by, and then returns to its equilibrium position.

When the source of a wave is a simple harmonic oscillator, the wave is sinusoidal, as in Figure 16.3. The sinusoidal wave in Figure 16.4 shows the terms we must define in order to study wave motion. The **amplitude** A of the wave is the maximum displacement of the medium from its equilibrium position. As the wave moves along, its shape continues to repeat itself. The **wavelength** λ (Greek lower-case *lambda*) is the distance between repetitions. Wavelength is often measured from **crest** (highest part) to crest along a wave or from **valley** (lowest part) to valley. You should remember, however, that it is the distance from *any* point on a wave to the nearest repeat point, as the wavelength labeled *ab* in Figure 16.4 indicates.

If we station ourselves at some point alongside a rope through which a wave is passing—such as position x in Figure 16.5—we observe the part of the rope at that point undergoing simple harmonic motion (in the plane of the page) with some period T and frequency f. The **period** is the amount of time required for one complete oscillation. For example, if we say $t = 0$ when a valley is passing our observation point, as one is in Figure 16.5, the period of the wave is the length of time until the next valley passes our observation point. **Frequency** is the *number* of oscillations per unit time, usually per second. Just as in simple harmonic motion, wave period and wave frequency are related by Equation 15.3:

$$f = \frac{1}{T}$$

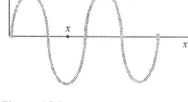

Figure 16.5
As a wave moves by it, a point on the rope undergoes simple harmonic motion in the plane of the page.

The wave moves along with **wave speed** v, which is related to frequency and wavelength. If the wave has a frequency of, say, 5 Hz, then five valleys pass by an observation point in 1 s. If the wavelength is, say, 4 m, then after 1 s, the valley that was at an observation point at $t = 0$ has moved 20 m to the right. In general, any wave travels with a speed

$$v = f\lambda \qquad (16.2)$$

Example 16.1

The wave speed of sound in air is about 345 m/s. The musical note middle C has a frequency of 262 Hz. What is the wavelength of the sound wave created when the middle C key on a piano is struck?

Reasoning

We can use Equation 16.2.

Solution

$$v = f\lambda$$

$$\lambda = \frac{v}{f} = \frac{345 \text{ m/s}}{262 \text{ Hz}} = \frac{345 \text{ m/s}}{262 \text{ s}^{-1}} = 1.32 \text{ m}$$

Practice Exercise The note "A" on the staff has a frequency of 440 Hz. What is the wavelength of a sound wave when "A" is heard?

Answer 0.78 m

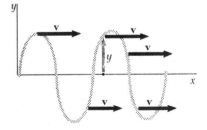

A transverse wave motion demonstrator. *(Courtesy of PASCO Scientific)*

16.2 Types of Waves

CONCEPT OVERVIEW

- In a transverse wave, the movement of individual particles of the medium is perpendicular to the wave velocity.

- In a longitudinal wave, the movement of individual particles of the medium is parallel to the wave velocity.

The waves and pulses we have sketched so far are examples of **transverse** waves or pulses. In this type of wave, the motion of each particle of the medium is perpendicular to the wave velocity, as sketched in Figure 16.6. A wave on a rope is a good example of a transverse wave.

In **longitudinal** waves, the displacement of the particles of the medium is either parallel or antiparallel to the wave velocity. In the example sketched in Figure 16.7, a longitudinal pulse moves from left to right (time progresses as you move down the figure). The vertical lines, which represent the particles of the medium, move first to the right and then back to the left. Sound waves in

Figure 16.6

In a transverse wave, the displacement y of the medium is perpendicular to the wave velocity. A wave on a rope is a good example of a transverse wave.

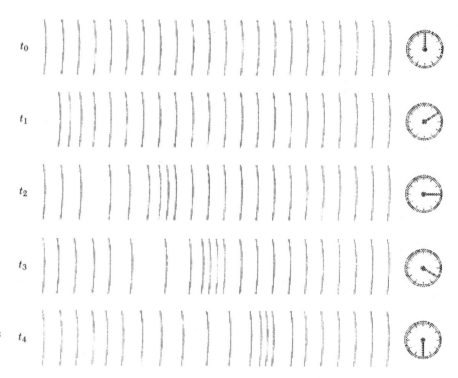

Figure 16.7
In a longitudinal wave, the displacement of the medium is either in the same direction as the wave velocity or in the opposite direction. This sketch shows the motion of a longitudinal pulse to the right. The vertical lines represent medium particles, and time progresses as you move down the sketch. As the pulse moves to the right, each line moves first to the right and then to the left.

air are an example of a longitudinal wave. As you speak to a friend down the hall, the air molecules oscillate back and forth along a line originating at your mouth and going down the hall to your friend's ear. The sound wave travels along this same direction. A Slinky toy can support a longitudinal wave, as sketched in Figure 16.8. The individual coils move first to the right and then to the left as the wave moves from left to right.

The wavelength of a longitudinal wave can be measured from a point of greatest medium density (called a **condensation**) to the next point of greatest

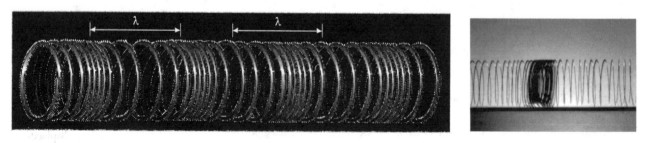

Figure 16.8
A longitudinal wave travels down a Slinky toy as the individual coils move first in the direction of the wave motion and then in the opposite direction. *(Photo © Richard Megna/Fundamental Photographs, NY)*

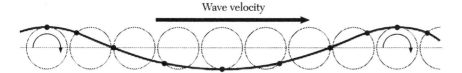

Figure 16.9
The motion of a water molecule as a wave passes by is more complex than the motion of a transverse or longitudinal wave.

A model boat in a wave tank clearly shows the crests and troughs present in water waves. In the open ocean what factors can affect the symmetry or complexity of water waves? (© L. W. Black)

density or from a point of lowest medium density (called a **rarefaction**) to the next point of lowest density.

An ocean wave looks like a transverse wave because we see it only on the water's surface, which seems to move up and down as the wave passes. Closer inspection, however, shows that a water wave is more complex than what appears on the surface. We can watch the surface wave more closely by sprinkling dry leaves on the surface. Then we find that the leaves move back and forth as well as up and down. They move in ellipses or in circles, as sketched in Figure 16.9. Water molecules below the surface undergo elliptical or circular motion that varies with the depth. Waves break as these underwater movements interact with the ocean floor. Water waves are beautiful and enjoyable to watch, but they are more complicated than the transverse and longitudinal waves we have discussed. We shall limit our discussions to simple transverse and longitudinal waves.

16.3 Superposition

CONCEPT OVERVIEW

■ Two waves pass through each other without the motion of either one being impeded.

■ When two waves overlap, the resulting disturbance is the algebraic sum of the individual disturbances.

You know what happens when two billiard balls or two cars run into each other. Velocities are altered. There is a collision. You have probably heard the saying, "No two pieces of matter can occupy the same space at the same time." A collision is just a verification of that fact.

Waves behave differently; waves pass through each other without alteration. For example, you cannot push on the light from one flashlight by shining light from a second flashlight into the first beam. The light waves pass through each other without effect, as sketched in Figure 16.10. Water waves created by a water skier pass right through the waves created by a passing boat. You can listen to the sound waves from a car radio and hear the waves emitted from an ambulance siren at the same time.

That intersecting waves pass through each other is illustrated in Figure 16.11, and the phenomenon is known as **superposition.** Superposition means that waves pass through each other, as illustrated by the water waves in Figure 16.12.

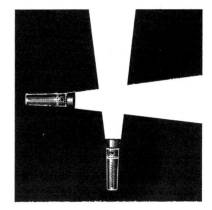

Figure 16.10
Light waves from a flashlight pass undeviated through the light waves from a second flashlight. Light is never bent by other light "pushing" on it from the side.

Figure 16.11
You can simultaneously hear sound from two sources because one sound wave never blocks out another. This ability of waves to pass through one another is known as superposition.

Figure 16.12
Water waves on a lake surface pass through each other unaffected. This is another example of superposition. (© Robert Mathena/Fundamental Photographs, NY)

Figures 16.13 and 16.14 show two pulses on a rope that pass through each other. It is interesting to look closely at the region where the two pulses overlap. In Figure 16.13 the two pulses are both up; that is, the medium is displaced in the positive y direction in both cases. The displacements add, and the result is a very large displacement. In Figure 16.14 one pulse is up and the other is down; that is, one displaces the medium in the positive y direction and the other displaces it in the negative y direction. Again, the displacements add as the two pulses pass through a common region. Since they add *algebraically*, however, the resulting displacement is small. If the two pulses are identical except that one is positive and the other is negative, the displacement momentarily vanishes as they pass through each other. After this interaction, the two pulses go their separate ways undisturbed!

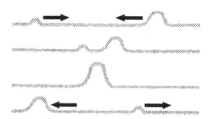

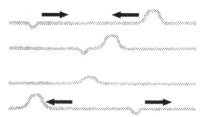

Figure 16.13
Two pulses pass through each other without hindrance. If the two displacements are positive—as with the two pulses shown here—there is a very large displacement when the two pulses pass through each other because their displacements add.

Figure 16.14
If the displacements of two overlapping pulses are opposite each other, there is a small displacement as the two pulses pass through each other because the displacements add algebraically.

Superposition is characteristic of all waves. We shall use it in the following section to understand standing waves and later in the chapter to understand beats. As discussed in Section 16.7, the interaction between two waves or between two pulses is called *interference*. Figure 16.13 is a case of *constructive* interference, where two waves or pulses add together to momentarily form a larger amplitude. Figure 16.14 illustrates *destructive* interference, where two waves or pulses cancel each other's effect and momentarily form a smaller amplitude.

16.4 Transverse Standing Waves

CONCEPT OVERVIEW

▬ At particular frequencies, standing waves can be formed as a result of constraints on a system. The *x* coordinate of the position of each node and antinode of a standing wave remains constant.

If you call out near a solid wall—a rock canyon is prettier but a brick wall will work as well—you will hear an echo. This echo is the sound wave your yell created being reflected from the wall. Waves are reflected at any boundary where there is some change in the medium carrying the wave. This reflection occurs for all waves and is illustrated in Figures 16.15 and 16.16 for a familiar pulse on a rope.[1] Figure 16.15 shows a pulse traveling along a rope toward a *fixed end*, which is an end held tightly in place so that it cannot move. The pulse reflected from this end comes back reversed: if the initial pulse was up, the reflected pulse is down; if the initial pulse was down, the reflected pulse is up.

[1] We use a single pulse in these illustrations for clarity. Everything that is true for a pulse is true for a wave as well.

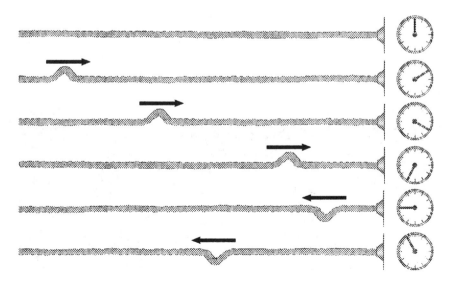

Figure 16.15
A pulse reflected from a fixed end of a rope is reversed. That is, an initially upward pulse is reflected from the fixed end as a downward pulse.

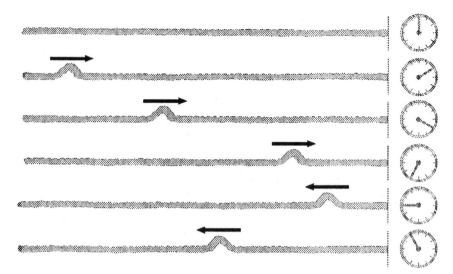

Figure 16.16

A pulse reflected from a free end of a rope is not reversed. That is, an initially upward is reflected as an upward pulse.

Figure 16.16 shows a pulse traveling along a rope toward an end that is *free to move.* (While you cannot easily have a horizontal rope with a free end, you can suspend a rope vertically with its free end hanging down.) The pulse reflected from this free end comes back looking just like the original: if the initial pulse was up, the reflected pulse is up; if the initial pulse was down, so is the reflected pulse.

Instead of single pulses on the rope, let us send out a wave. As the wave moves along the rope, the reflected part interacts with the oncoming part. In general, the result is a jumbled mess! No observable pattern is repeated in time or in space; we call such a jumbled mess **noise.** However, if we wiggle the end of the rope at certain frequencies, a very interesting pattern appears, as shown in Figure 16.17. This pattern is called a **standing wave** because it does not move along the rope—it stands in one place. A standing wave results when a wave moving in one direction interferes with a reflected wave moving in the opposite direction. The points on the rope that do not move at all are called **nodes;** at those points, the two traveling waves interfere *de*structively. The

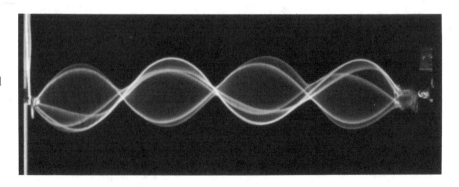

Figure 16.17

When one end of a rope is oscillated at certain frequencies, a standing wave appears on the rope. It is called a standing wave because the pattern does not move along the rope.   (© *Richard Megna/Fundamental Photographs/NY*)

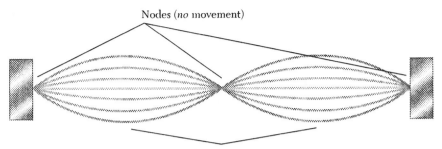

Nodes (*no* movement)

Antinodes (*maximum* movement)

Figure 16.18
The points at which there is no medium movement at all are called nodes. The points of maximum medium displacement are called antinodes. The x coordinate of position remains constant for both nodes and antinodes. The y coordinate is also constant (at $y = 0$) for nodes but oscillates between $y = A$ and $y = -A$ for antinodes.

regions of maximum displacement—the center of the loops that make up the standing-wave pattern—are called **antinodes;** at those places the two traveling waves interfere *con*structively. Nodes and antinodes are illustrated in Figure 16.18.

Figure 16.19 shows several standing waves that can be present on a string fixed at both ends. That is, these standing waves have *nodes at both ends*. The distance from node to node is half a wavelength and an integral number n of these distances must fit between the ends of the rope of length L. That is, a standing wave occurs when

$$L = \frac{\lambda}{2}, \ 2\frac{\lambda}{2}, \ 3\frac{\lambda}{2}, \ 4\frac{\lambda}{2}, \ \ldots \ n\frac{\lambda}{2} \qquad \textbf{(16.3)}$$

This condition restricts standing waves to the wavelengths

$$\lambda = 2L, \ \frac{2L}{2}, \ \frac{2L}{3}, \ \frac{2L}{4}, \ \ldots$$

or, in general,

$$\lambda = \frac{2L}{n} \qquad n = 1, 2, 3, \ldots \qquad \textbf{(16.4)}$$

where L is the distance between the two end nodes and n is an integer.

Of course, the wavelength restrictions of Equation 16.4 can easily be written as frequency restrictions. From Equation 16.2, we know the frequency in terms of wave speed and wavelength, $f = v/\lambda$. Therefore, Equation 16.4 leads to frequency restrictions given by

$$f_n = n\frac{v}{2L} \qquad n = 1, 2, 3, \ldots \qquad \textbf{(16.5)}$$

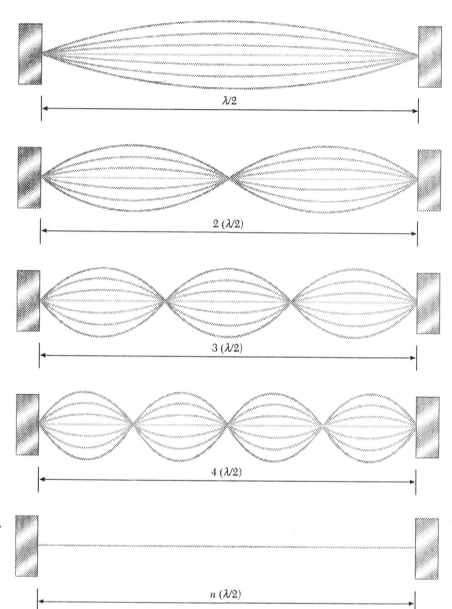

Figure 16.19
In order for a standing wave to form, the rope must have a node at each end and the total rope length must equal an integer multiplied by half the wavelength. This condition restricts standing waves to particular wavelengths—namely, $\lambda = 2L/n$.

where n is an integer $(1, 2, 3, \ldots)$, L is the length of the string, and v is the speed of the velocity of the wave along the string (v is *not* the velocity of the resulting sound in air).

The special frequencies that create standing waves are called either **harmonic frequencies** or simply **harmonics.** The lowest frequency that corresponds to a standing wave is called the **fundamental frequency** or the **first harmonic** and is labeled f_1. In terms of the wave speed and rope length, this fundamental frequency is

$$f_1 = \frac{v}{2L}$$ **(16.6)**

The next frequency that corresponds to a standing wave is called the **second harmonic** and is labeled f_2. As you can see from Equation 16.5, the nth harmonic is n times the fundamental frequency:

$$f_n = nf_1 \qquad (16.7)$$

Harmonics above the fundamental are also called **overtones.** The first harmonic is the fundamental frequency and therefore has no overtone designation. The second harmonic is called the first overtone, the third harmonic is the second overtone, and so on.

Any particular note produced by a musical instrument, such as the violin in Figure 16.20, is a combination of a fundamental frequency plus many of its overtones. Changing the length of a string changes its fundamental frequency. To play different notes on a violin (or guitar or any other stringed instrument), the musician presses a finger down on the string in different positions to lengthen or shorten the part of the string that is free to vibrate.

Figure 16.20
The frequency of an instrument string can be varied by changing its length. A violinist plays different notes by placing her fingers on the strings at different positions to shorten or lengthen the part of each string that is free to vibrate. (© Comstock, Inc.)

Problem-Solving Tip

■ Equation 16.2, $v = f\lambda$, can be used to switch back and forth between frequency and wavelength. It is usually more convenient to refer to sounds by frequency. However, it is usually easier to understand the restraints on a vibrating system in terms of wavelength.

Example 16.2

The part of an A string that vibrates on a particular guitar is 63.5 cm long. Being an A string, it has a fundamental frequency of 110 Hz when plucked. What is the wavelength of the fundamental standing wave in this string? What is the wave speed in it? What is the tension in the string? The mass per length of the string (m/L) is 0.00337 kg/m.

Reasoning

The fundamental frequency for a string fixed at both ends is the frequency that gives a standing wave having a wavelength that is twice the length of the string. Once we determine this wavelength, we can apply Equation 16.2 to find the wave speed in the string. For the tension, we use Equation 16.1.

Solution

The wavelength of the fundamental frequency is

$$L = \lambda/2$$
$$\lambda = 2L = 2(63.5 \text{ cm}) = 127 \text{ cm} = 1.27 \text{ m}$$

The wave speed is

$$v = f\lambda = (110 \text{ s}^{-1})(1.27 \text{ m}) = 140 \text{ m/s}$$

This is the speed of a pulse or wave moving along the guitar string. It has nothing to do with the speed of sound in air.

Now we solve Equation 16.1 for the tension:

$$v = \sqrt{\frac{T}{m/L}}$$

$$T = v^2(m/L)$$

$$T = (140 \text{ m/s})^2(0.00337 \text{ kg/m}) = 66.0 \text{ N}$$

Example 16.3

The high C (C on the treble staff) string in a piano has a fundamental frequency of 524.0 Hz. What are the frequencies of its second, third, and fourth harmonics?

Reasoning

According to Equation 16.7, the higher harmonics are related to the fundamental frequency, or first harmonic, by

$$f_n = nf_1$$

Solution

$$f_2 = 2(524.0 \text{ Hz}) = 1048 \text{ Hz}$$
$$f_3 = 3(524.0 \text{ Hz}) = 1572 \text{ Hz}$$
$$f_4 = 4(524.0 \text{ Hz}) = 2096 \text{ Hz}$$

Since *third overtone* is another name for the fourth harmonic, the value of this overtone is 2096 Hz.

Practice Exercise   A on the staff has a frequency of 440 Hz. What are the frequencies of the second and fourth harmonics of the A string on a piano?

Answer   880 Hz, 1760 Hz

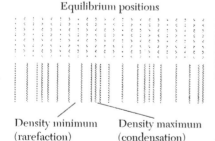

Equilibrium positions

Density minimum   Density maximum
(rarefaction)   (condensation)

Figure 16.21
As air molecules move first in the direction of wave motion and then in the opposite direction as a longitudinal wave passes through, they create regions of greater density called condensations and regions of lesser density called rarefactions.

16.5 Longitudinal Standing Waves

CONCEPT OVERVIEW

■ Standing sound waves can be formed in a pipe closed at one end and open at the other; the closed end is a node, and the open end is an antinode.

As we learned in Section 16.2, sound is a longitudinal wave. Figure 16.21 illustrates what a sound wave in air "looks" like. As the wave passes along from left to right, the air molecules move first in the direction of wave motion and then in the opposite direction. As they move to the right, they bunch together—

this is called a condensation. The formation of each condensation causes the formation of a rarefaction—a region where the molecules are spread apart—right behind the condensation.

Since transverse waves are easier both to visualize and to sketch, Figure 16.22 shows a transverse wave drawn to represent a longitudinal wave. As before, the y direction represents displacement of the medium and the x direction represents the direction of wave motion.

In Figure 16.23 a vibrating tuning fork is held above a pipe whose length can be changed. The fork sends a sound wave into the air just as an oscillating rope end sends a wave along the rope. When the wave inside the pipe strikes the piston face, which acts as a fixed end, it is reflected just as a wave on a rope is. The reflected parts of the wave superpose on the oncoming parts coming from the tuning fork. As with a wave on a rope, there are certain values of pipe length and wave frequency for which standing waves are produced. As before, this means large amplitudes—you can *hear* the sound increase! This is yet another example of a **resonance**—at just the right frequency, called the resonance frequency, the amplitude becomes quite large.

Figure 16.24 illustrates how we might visualize a standing wave for this system. The air at the open end—near the tuning fork—is free to move and vibrates with a large amplitude; this point is therefore an antinode. The air immediately next to the closed end—at the piston face—cannot move; this point is therefore a node. Sound is a longitudinal wave, but it is easier to sketch a transverse wave. Figure 16.24 shows a standing wave with a node at the closed end plus three additional nodes.

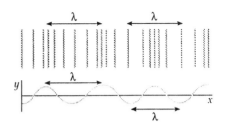

Figure 16.22
For convenience, a transverse wave is sometimes sketched to represent a longitudinal wave. The distance y is the displacement from equilibrium for a particle of the medium, and the x direction is the direction of wave motion.

Pipe organs may be constructed of both open and closed tubes. Standing sound waves in tubes of differing lengths and diameters produce the range of tones necessary to create music. Why do most organ pipes have a horizontal slot cut into them near their bases? Will the pitch of an organ pipe go higher or lower as the length of the pipe increases? (*© Vanessa Vick/Photo Researchers, Inc.*)

Figure 16.23
A tuning fork vibrating above a pipe of adjustable length can be used to set up standing longitudinal waves.

Figure 16.24
A standing wave is established if the wavelength is correctly related to the length of the pipe. The closed end is a node, and the open end is an antinode.

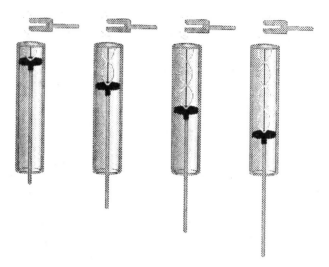

Figure 16.25
The shortest length of pipe that produces a standing wave is $L = \lambda/4$. Additional standing waves are produced each time L is increased by an amount equal to $\lambda/2$.

Figure 16.25 illustrates several standing waves that form as the length of the pipe is gradually increased. The sound becomes noticeably louder as these waves are produced. The shortest length of pipe that produces a standing wave is $\lambda/4$. Another standing wave is produced each time the length is increased by an amount equal to $\lambda/2$. That is, a standing wave is found whenever

$$L = n\frac{\lambda}{2} + \frac{\lambda}{4} \qquad n = 0, 1, 2, 3, \ldots \qquad \textbf{(16.8)}$$

Problem-Solving Tip

■ The resonant frequencies of a pipe closed at one end and open at the other are those frequencies that correspond to wavelengths that give a node at the closed end and an antinode at the open end.

Example 16.4

What is the fundamental frequency of the sound produced when you blow across the top of a soda bottle 24.3 cm high? What are the frequencies of the first two overtones? Take the speed of sound in air to be 340 m/s.

Reasoning

A soda bottle can be treated like a pipe closed at one end and open at the other. The fundamental frequency is the lowest frequency for which a standing wave can fit in the bottle with the bottom being a node and the top being an antinode. This is just the situation described by Equation 16.8 with $n = 0$ and Figure 16.25.

The first and second overtones are obtained from Equation 16.8 with $n = 1$ and $n = 2$.

Solution

The wavelength of the standing wave for $n = 0$ is

$$L = \frac{\lambda}{4}$$
$$\lambda = 4L = 4\,(24.3\text{ cm}) = 97.2\text{ cm}$$

The fundamental frequency therefore is

$$f = \frac{v}{\lambda} = \frac{340\text{ m/s}}{0.972\text{ m}} = 350\text{ Hz}$$

The first overtone corresponds to $n = 1$:

$$L = \frac{1\,\lambda}{2} + \frac{\lambda}{4} = \frac{3\lambda}{4}$$

$$\lambda = \frac{4L}{3} = \frac{4\,(24.3\text{ cm})}{3} = 32.4\text{ cm}$$

$$f = \frac{v}{\lambda} = \frac{340\text{ m/s}}{0.324\text{ m}} = 1\,050\text{ Hz}$$

The second overtone corresponds to $n = 2$:

$$L = \frac{2\lambda}{2} + \frac{\lambda}{4} = \frac{5\lambda}{4}$$

$$\lambda = \frac{4L}{5} = \frac{4\,(24.3\text{ cm})}{5} = 19.4\text{ cm}$$

$$f = \frac{340\text{ m/s}}{0.194\text{ m}} = 1\,750\text{ Hz}$$

Practice Exercise What is the fundamental frequency that will be excited if you blow across the top of a soda bottle that is 32.5 cm high?
Answer 262 Hz

Figure 16.26
Opening or closing the holes in this wind instrument determines the frequency produced by determining the location of nodes and antinodes in the vibrating air column. *(© Horst Schafer/Peter Arnold, Inc.)*

The common 12-tone scale is called an *octave* because the name refers to the eight white keys on a piano from any note of frequency *f* to the note of frequency *2f. Oct-* is the Latin root for "eight."

Figure 16.26 shows a wind instrument. A mouthpiece at the top of the column of air causes air in the column to vibrate, just the way the tuning fork of Figure 16.24 does. As the musician fingers different holes, the positions of nodes and antinodes in the vibrating air column change and sound of a different frequency is produced. In other words, different notes are played.

16.6 Characteristics of Sound

CONCEPT OVERVIEW

▬ Pitch is related to the dominant frequency of a sound.

▬ Quality describes the collection of frequencies in a sound and the behavior of the sound over time.

▬ Loudness is related to the power per unit area carried by a sound and is measured in decibels.

Human ears are sensitive to sound frequencies from about 50 to 20 000 Hz. The highest frequency a person can detect decreases as the person gets older. Dogs can hear frequencies well above 20 000 Hz, which go unheard by human ears. Sounds with frequencies above the range of human hearing are called **ultrasonic** (not to be confused with **supersonic,** which means *faster* than the speed of sound).

The speed of sound in air at 1.00 atm and 0°C is 331 m/s. The density of air decreases as temperature increases, causing the speed of sound to increase by about 0.60 m/s per Celsius degree. That is, the speed of sound at some temperature T is[2]

$$v = [331 + (0.60/C°)\,T]\ \text{m/s} \qquad \textbf{(16.9)}$$

Sound can be transmitted through solids and liquids. As in the old Western movies, put your ear to the ground and you can hear the sounds of an approaching buffalo stampede (or a marching band or an oncoming garbage truck). One scuba diver can hear the sound produced when another diver strikes two metal rods together under water. The speed of sound in water is about 1440 m/s.

An instrument playing middle C produces a fundamental frequency of 262 Hz; playing A above middle C produces a fundamental frequency of 440 Hz. Table 16.1 lists the frequencies for one octave of musical notes from a piano; this scale is called the *chromatic,* or *even-tempered,* scale. A note an octave higher than any given note has a frequency exactly twice as great. This scale uses 12 evenly spaced frequencies in an octave, corresponding to twelve adjacent white and black keys on a piano. Other scales have been used in history and are in use today. A pentatonic scale—with five equally spaced notes in an octave—has been common in many cultures.

---

[2]The "/C°" appears in this equation merely to cancel the units on T and thereby give a result in meters per second. To calculate the speed of sound in air at 25°C, for instance, write

$$v = \left[331 + \left(\frac{0.60}{C°}\right)(25°C)\right]\text{m/s} = 346\ \text{m/s}$$

Pitch, quality, and loudness are important characteristics in describing musical sound. Because these three characteristics involve physiological and psychological sensations that occur in the ear and brain, they are subjective and thus difficult to define. However, each is closely associated with a physical quantity that we can easily define and measure.

Pitch is related to frequency, as we learned earlier. A plucked or bowed or struck string vibrates in a very complex manner, and many levels of harmonics are produced. The pitch that is heard is determined primarily by the fundamental frequency, which can be altered by changing the tension in the string. Therefore a stringed instrument is tuned by adjusting the tension in each string. The fundamental frequency can also be altered by changing the length of the string. Therefore different notes can be played on the same guitar or violin string by placing the finger so that only a certain part of the string can vibrate. The fundamental frequency also depends upon the linear density of the string. Different strings on a guitar or violin or piano are made differently, some thick and some thin, with varying linear densities.

Quality, which describes the characteristic sound of different instruments, is related to frequencies other than the fundamental. If you hear A above middle C played by a piano, clarinet, flute, and trumpet, you recognize that all four instruments are playing the same note because for each sound the dominant frequency is the fundamental frequency: 440 Hz. Figure 16.27 shows the sound waves as you would see them on an oscilloscope of four different instruments[3] playing A above middle C. There you can see that the basic pattern of each wave is repeated with the same period or the same frequency. This is because the fundamental frequency is the same for all four—440 Hz. But the basic pattern of each wave is quite different—the four waves look quite distinctive just as the four instruments sound quite distinctive. This is because of the different set of harmonics that are present in each wave. Figure 16.28 shows the different frequencies that are present.

Two other factors that affect sound quality are *attack* and *decay*. Both terms refer to how a sound changes over time. Attack describes how a note begins on an instrument; decay describes how a note ends. A piano has a short or quick attack, meaning the note begins quickly. An organ's decay is much longer than a piano's since the organ can sustain a note indefinitely.

In order to understand loudness, recall from Equations 6.14 and 15.4 that the energy stored in a stretched or compressed spring is proportional to the square of its displacement from equilibrium:

$$PE = \tfrac{1}{2} kx^2$$

In a similar manner, the energy associated with a wave is proportional to the square of the amplitude:

$$E = \tfrac{1}{2}kA^2$$

This equation is useful for a one-dimensional wave such as a wave on a string. For a three-dimensional wave, however, such as a sound wave, we are inter-

TABLE 16.1 Notes and Frequencies of the Chromatic Scale	
Note	**Frequency (Hz)**
C (middle C)	262
C♯/D♭	277
D	294
D♯/E♭	311
E	330
F	349
F♯/G♭	370
G	396
G♯/A♭	415
A	440
A♯/B♭	466
B	494
C′ (on the staff)	524

[3] Actually, these waves are of synthesized sounds produced by an electronic keyboard.

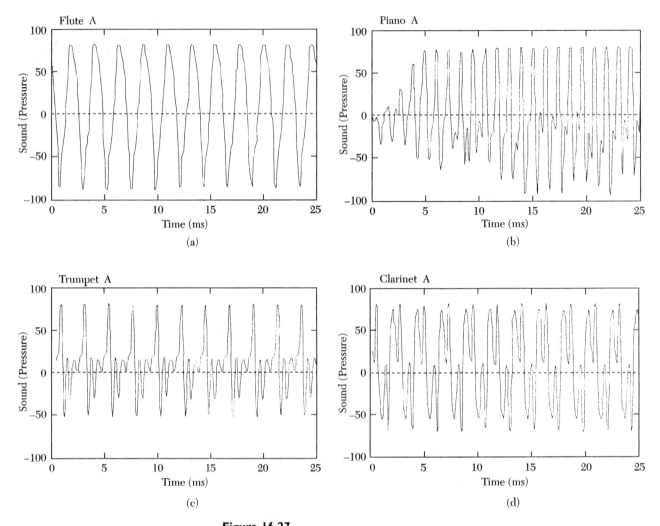

Figure 16.27
Sound waves for A above middle C (440 Hz) played on (a) flute, (b) piano,
(c) trumpet, and (d) clarinet. Notice the similarity: all four have the same fundamental frequency, which determines the pitch. Note also the difference in detail:
the overtones are different in each case. These differences in overtone determine
the quality of the sound and provide the distinguishing characteristics of each instrument.

ested not so much in the wave energy as the **wave intensity**, the energy
carried across a unit area in a unit time[4]:

$$I = \frac{E}{at} = \frac{kA^2}{2at} \qquad (16.10)$$

Since this definition also describes the power (E/t) carried across a unit area,
intensity has units of power divided by units of area, usually watts per square

---

[4]In this discussion, A is amplitude and a is area (*not* acceleration).

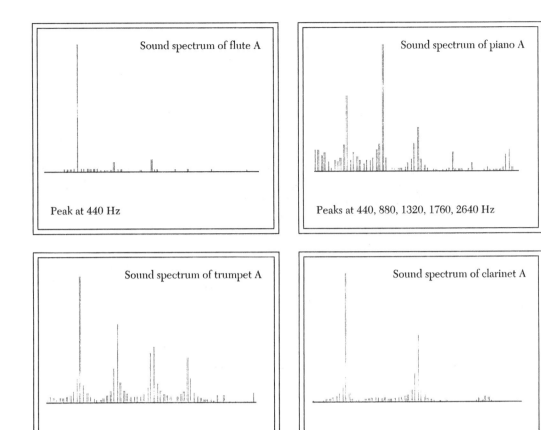

Figure 16.28
The sound waves in Figure 16.27 are the result of superposing—playing together—several waves of different frequency. These diagrams show the relative amplitudes of the different waves. In all cases, the sound produced is a combination of the fundamental frequency plus some overtones.

meter. We see from Equation 16.10 that the intensity of a wave is proportional to the square of the amplitude.

If you listen to two similar sounds, you can probably tell if one is louder or softer than the other, but when would you decide that one sound is twice as loud as another? Or ten times as loud? Or 7.38 times as loud? It is very difficult (or impossible) to make such decisions with the human ear. We can readily develop an instrument like a microphone that will measure either the amplitude or the intensity of a sound wave, both of which are certainly related to the loudness of a sound. A loud sound has a larger amplitude and a large intensity while a soft, quiet sound has a small amplitude and a small intensity. Yet loudness is perceived in the ear and brain and is surprisingly difficult to determine quantitatively.

The sound-level intensity in decibels is measured with a sound-level meter. *(Courtesy of Simpson Electric Company)*

Perceived loudness, which depends upon the frequencies involved and the duration of the sound, varies with people and with age. An average ear is sensitive to an enormous range of intensities, from 10^{-12} W/m^2 all the way up to 1 W/m^2. Sound intensities greater than this can also be heard but are painful. This is an enormous range—12 powers of ten.

When people are asked to decide when one sound is twice as loud as another, the results, averaged over many people, show that, to a first approximation, *when the intensity of one sound is ten times that of another sound, the former is perceived as being twice as loud as the latter.* This means that the human ear and brain perceive loudness approximately as the logarithm of the intensity. Therefore, it is useful to create a parameter called the **intensity level β**, measured in units of **bels**[5] (B), in terms of intensity I:

$$\beta \text{ (in B)} = \log \frac{I}{I_0} \tag{16.11}$$

where I is the intensity of the sound being measured and I_0 is the intensity of the quietest sound that a human can hear. This latter intensity is called the **threshold of hearing** and is commonly taken to be $I_0 = 10^{-12}$ W/m^2. Units of **decibels** (dB) are more common, where

$$1 \text{ B} = 10 \text{ dB}$$

This relationship means that Equation 16.11 can be rewritten as

$$\beta \text{ (in dB)} = 10 \log \frac{I}{I_0} \tag{16.12}$$

Consider two sounds, one having ten times the intensity (power per unit area) of the other; that is $I_2 = 10 \, I_1$. Sound 2 will be perceived as being *twice as loud* as sound 1, but its intensity level is 10 dB higher than the intensity level of sound 1. We can see this relationship by applying Equation 16.12:

$$\beta_1 \text{(in dB)} = 10 \log \frac{I_1}{I_0}$$

$$\beta_2 \text{(in dB)} = 10 \log \frac{I_2}{I_0} = 10 \log \frac{10 \, I_1}{I_0} = 10 \log \left(10 \, \frac{I_1}{I_0} \right)$$

Since $\log (ab) = \log a + \log b$, we have

$$\beta_2 = 10 \left(\log 10 + \log \frac{I_1}{I_0} \right)$$

Since $\log 10 = 1$, we have

$$\beta_2 = 10 \left(1 + \log \frac{I_1}{I_0} \right) = 10 + 10 \log \frac{I_1}{I_0} = 10 + \beta_1$$

Table 16.2 lists intensities and intensity levels for some common sounds.

Pitch, quality, and loudness are *subjective* characteristics of sound. Their *objective* counterparts are fundamental frequency, overtones, and intensity level.

[5]The unit of bels is named in honor of Alexander Graham Bell.

TABLE 16.2 Intensities and Intensity Levels for Common Sounds

Sound	Intensity (W/m^2)	Intensity Level (dB)
Rupture of eardrum	400	160
Jet airplane 30 m away	100	140
Jackhammer, machine gun	10	130
Threshold of pain	1	120
Loud indoor rock concert	1	120
Police siren 30 m away	1×10^{-3}	100
Busy traffic	1×10^{-5}	80
Typical auto, inside, 90 km/h	3×10^{-5}	75
Vacuum cleaner	1×10^{-6}	70
Ordinary conversation 0.5 m away	3×10^{-6}	65
Quiet music	1×10^{-8}	40
Whisper	1×10^{-10}	20
Threshold of hearing	1×10^{-12}	0

Example 16.5

A particular dot-matrix printer produces sound that has an intensity level of 50 dB at a distance of 3 m. What is the intensity of this sound?

Reasoning

Equation 16.12 relates the intensity level β in decibels to the intensity I and the reference intensity I_0.

Solution

$$\beta \text{ (in dB)} = 10 \log \frac{I}{I_0}$$

$$50 = 10 \log \frac{I}{10^{-12} \text{ W/m}^2}$$

Dividing both sides by 10, we have

$$5 = \log \frac{I}{10^{-12} \text{ W/m}^2}$$

Raising 10 to the power of each side (in other words, taking the antilogarithm of each side), we have

$$10^5 = \frac{I}{10^{-12} \text{ W/m}^2}$$

$$I = (10^5)(10^{-12} \text{ W/m}^2) = 10^{-7} \text{ W/m}^2$$

Practice Exercise What is the sound intensity of an 80-dB sound?

Answer $I = 1.0 \times 10^{-4}$ W/m^2

Example 16.6

If a second dot-matrix printer identical to the one in Example 16.5 is added to the office, what will the resulting intensity level be when both are printing?

Reasoning

The two printers running together means that twice the sound power is produced. That is, the intensity I is doubled. We use Equation 16.12 to calculate the intensity level in decibels.

Solution

$$I = 2 \times 10^{-7} \text{ W/m}^2$$

$$\beta(\text{in dB}) = 10 \log \frac{I}{I_0} = 10 \log \frac{2 \times 10^{-7} \text{ W/m}^2}{10^{-12} \text{ W/m}^2} = 10 \log (2 \times$$
$$10^5) \text{ dB} = 10 (\log 2 + \log 10^5) \text{ dB} = 10 (0.30 + 5) \text{ dB} = 53 \text{ dB}$$

Practice Exercise What is the intensity level of a 50-dB sound and a 60-dB sound when both are heard together?

Answer 60.4 dB

As you can see from this example, doubling the power of a sound wave means increasing its intensity level by 3 dB. In other words, doubling the power of a sound wave makes it seem louder but hardly twice as loud. The power of a sound wave must be increased by a factor of 10 before most people perceive the sound as being twice as loud. Increasing the power by a factor of 10 increases the intensity level by 10 dB (*not* by a factor of 10). In other words, most people will describe a 60-dB sound as being twice as loud as a 50-dB sound.

*16.7 Interference

CONCEPT OVERVIEW

- As two waves overlap, they may interfere constructively so that the result is a larger disturbance.

- As two waves overlap, they may interfere destructively so that the result is a smaller disturbance.

Earlier in this chapter we discussed superposition. Two waves can pass through each other and create a wave that is the algebraic sum of the two initial waves. This happens for all waves, not just the simple one-dimensional waves on a rope we discussed then.

Figure 16.29 shows two speakers driven by a single signal generator. The curved lines coming from the speakers represent the crests or condensations of the sound waves; the centers of the open spaces between lines represent the valleys or rarefactions. That is, the lines show where the air molecules have maximum displacement in one direction, and the centers of the open spaces show where the air molecules have maximum displacement in the opposite direction. Regions where two crests or two valleys of the separate waves coincide are regions of **constructive interference.** Here both waves are moving the air molecules in the same direction; the two movements reinforce each other so that a large amplitude results. Here the sound is loud. Regions where a crest from one wave and a valley from the other coincide are regions of **destructive interference.** Here the two waves are attempting to move the air molecules in opposite directions, and the movements cancel each other so that a small (or zero) amplitude results. Here the sound is faint.

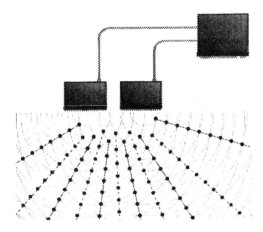

Figure 16.29
Sound coming from two speakers creates an interference pattern consisting of regions of constructive interference, where the sound is loud, and regions of destructive interference, where the sound is very soft.

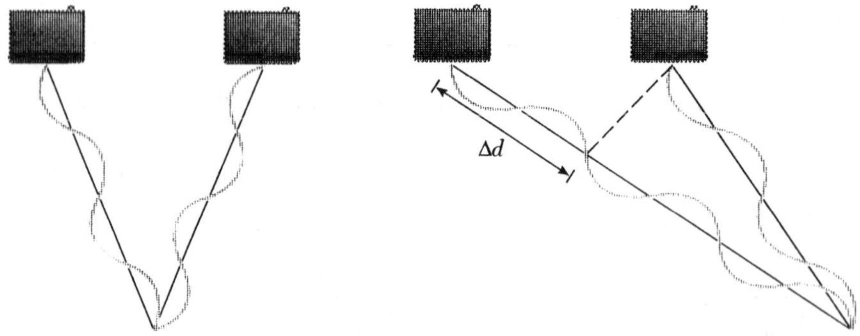

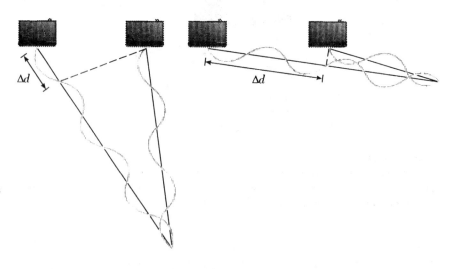

Figure 16.30
Constructive interference occurs when the distances traveled by the sound waves from two sources differ by an integral number of wavelengths (including $n = 0$).

Figure 16.30 illustrates this same idea in a different way. Constructive interference occurs wherever the two waves are **in phase**—both having displacements in the same direction. If the sound waves leave the two speakers in phase, constructive interference occurs at points where the two distances from the two speakers either are the same or else differ by an integral number of wavelengths. That is, we find constructive interference for

$$\Delta d = d_2 - d_1 = n\lambda \qquad n = 0, 1, 2, 3, \ldots \qquad \textbf{(16.13)}$$

where Δd is the difference in the length the sound travels for the two paths from the two sources.

Figure 16.31 illustrates destructive interference, which occurs wherever the two waves are **out of phase**—the displacement of one is opposite the displacement of the other. If the sound waves leave the two speakers in phase, destructive interference occurs at points where the two distances from the two speakers differ by one-half wavelength, three-halves wavelengths, or, in general, by an odd integral number of half wavelengths. That is, we find destructive interference for

$$\Delta d = d_2 - d_1 = n\frac{\lambda}{2} \qquad n = 1, 3, 5, \ldots \qquad \textbf{(16.14)}$$

Figure 16.31
Destructive interference occurs when the distances traveled by waves from two sources differ by an odd integral number of half wavelengths ($\lambda/2$, $3\lambda/2$, $5\lambda/2$, . . .).

Example 16.7

A listener sits 3.0 m from one speaker and 4.0 m from a second speaker. The two speakers are powered by a single signal generator that produces a sound of 345 Hz. Is the listener in a region of constructive interference (loudness increased) or in a region of destructive interference (loudness decreased)? Take the speed of sound in air to be 345 m/s.

Reasoning

Because the same signal is coming from both speakers, the sound waves interfere with each other. Given the distances involved, will our listener find destructive interference (and quietness), constructive interference (and loudness), or something in between?

Solution

From Equation 16.2 we can find the wavelength:

$$\lambda = \frac{v}{f} = \frac{345 \text{ m/s}}{345 \text{ s}^{-1}} = 1 \text{ m}$$

The difference in path length for the two waves

$$\Delta d = 4.0 \text{ m} - 3.0 \text{ m} = 1.0 \text{ m}$$

is one full wavelength. That means the listener is sitting at a point of constructive interference; the sound heard is louder than that produced by either speaker alone.

Example 16.8

The listener in Example 16.7 moves and is now 3.5 m from one speaker and 4.0 m from the second speaker. The two speakers are still powered by a single signal generator that produces a sound of 345 Hz. If the speed of sound is 345 m/s, what will the listener hear?

Reasoning

Because the same audio signal is coming from both speakers, the sound waves generated interfere with each other. With a frequency of 345 Hz, we have seen that the waves have a wavelength of 1 m, but what has happened to the difference in path length?

Solution

The difference in path length is now

$$\Delta d = 4.0 \text{ m} - 3.5 \text{ m} = 0.5 \text{ m}$$

which is half a wavelength. That means the listener is now at a point of destructive interference; the sound heard is quieter than that produced by either speaker alone.

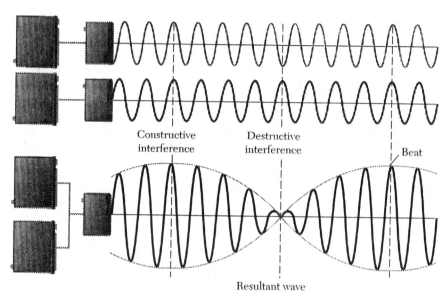

Figure 16.32
The term *beats* refers to the increasing and decreasing loudness heard as two sound waves come into phase and go out of phase.

*16.8 Beats

CONCEPT OVERVIEW

■ Sound of two different but close frequencies will alternately be in phase and out of phase, causing the loudness to alternately rise and fall; this phenomenon is known as beats.

Examples 16.7 and 16.8 deal with spatial interference—interference at various points in space. It is also interesting to look at *temporal* interference—interference at various moments in time.

Figure 16.32 shows two signal generators, each producing a sine wave. The two waves have nearly the same frequency. When these two signals are used to drive a single speaker, they will be sometimes in phase and sometimes out of phase. In phase, they interfere constructively and the loudness increases. Out of phase, they interfere destructively and the loudness decreases and even vanishes. That is, standing at any one point you will hear the loudness alternately increase and decrease as time goes by. This increasing and decreasing loudness is known as **beats.** The frequency with which the loudness rises and falls is known as the **beat frequency** and is equal to the difference between the two frequencies that cause it:

$$f_{\text{beat}} = f_2 - f_1 \qquad\qquad (16.15)$$

Listening for and eliminating beats is an effective way of tuning musical instruments. Piano and organ tuners use this technique, as do players of such instruments as guitars, violins, and bagpipes, as illustrated in Figure 16.33.

Figure 16.33
Listening for and eliminating beats is an effective method of tuning multipart musical instruments or of adjusting two or more instruments to be in tune with each other. *(© David R. Frazier Photolibrary)*

Example 16.9

The two tenor drone pipes on a set of Great Highland bagpipes both play an A note one octave below the staff, or 220 Hz. Suppose they are out of tune so that one is correctly sounding 220 Hz but the other is sounding 217 Hz. What will you hear?

Reasoning

With the two frequencies close but not identical, you will hear beats—the loudness will increase and decrease—with a frequency determined by Equation 16.15.

Solution

$$f_{beat} = 220 \text{ Hz} - 217 \text{ Hz} = 3 \text{ Hz}$$

The loudness will rise and fall three times per second. As the two pipes are tuned closer to each other, the beat frequency will get smaller and smaller until it finally vanishes when the two pipes are sounding the same note.

We talk of beats only when the beat frequency is fairly small. If two sounds have quite different frequencies—perhaps 220 Hz and 295 Hz—your ear cannot distinguish such rapid rising and falling of the loudness—75 times per second in this case. You can distinguish that the two sounds are quite different; you may hear a separate tone at the beat frequency if it is within the range of your hearing. A beat frequency of 75 Hz can be measured by a microphone and oscilloscope but is heard by the ear only as a low-frequency hum.

*16.9 The Doppler Effect

CONCEPT OVERVIEW

- The frequency heard by an observer changes if the source, the observer, or both are moving.

You may have listened to a car horn as the car approached you and then departed into the distance. The horn has a higher pitch as the car approaches and a lower pitch as it departs. This phenomenon is called the **Doppler effect.** We will now see how and why it occurs.

Sound from a stationary source—like the horn or siren on the stationary car in Figure 16.34—moves out as a spherical wave with all crests and valleys centered on the source. (Because Figure 16.34 is a two-dimensional rendering of a three-dimensional situation, of course, the concentric spheres are drawn as concentric circles. Just as in Figure 16.29, the circles represent crests of the spherical sound wave and the midpoints between adjacent circles represent valleys.) The wave moves with a speed v, the speed of sound in air. The distance between successive circles in the figure is one wavelength λ, and the amount of time between successive circles is one period T.

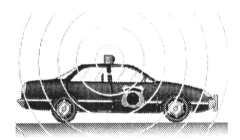

Figure 16.34
Sound from a stationary source moves out as a spherical wave centered on the source. The concentric circles represent crests of the sound wave. The midpoints between circles represent valleys.

Christian Johann Doppler (1803–1853) was an Austrian physicist and mathematician.

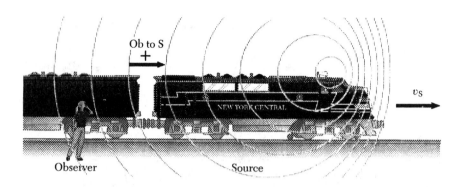

Figure 16.35
Sound from a moving source moves out as a spherical wave, each crest centered on the position of the source at the moment the crest was created.

Sound from a *moving* source also moves out as a spherical wave, but now each sphere is centered on the point at which the source was located when that particular crest was created. This is sketched in Figure 16.35, where the source is moving to the right with speed v_S. Because the source is moving, the wave crests are closer together on one side and farther apart on the other side. We shall take the direction *from* observer *to* source to be *positive*. In Figure 16.35, therefore, $v_S > 0$. Because the source moves, the **observed wavelength** λ' heard by the observer is different from λ, the wavelength when source and observer are both stationary. During one period T as measured by the source, the source moves a distance $v_S T$, or v_S/f. This means that, for a source moving away from a stationary observer, λ' is longer than λ by an amount v_S/f:

$$\lambda' = \lambda + \frac{v_S}{f} \qquad \text{(16.16)}$$

For a source moving toward a stationary observer, λ' is shorter by the same amount. Since $v_S < 0$ for that case, Equation 16.16 is applicable for both possibilities.

If the wavelength observed is changed, the frequency observed is also changed. Starting with Equation 16.2, we can write

$$f'\lambda' = v$$

where f' is the observed frequency just as λ' is the observed wavelength and v is the speed of sound in air. Then

$$f' = \frac{v}{\lambda'}$$

Using the Equation 16.16 value for λ', we get

$$f' = \frac{v}{\lambda + v_S/f}$$

Multiplying numerator and denominator by f gives

$$f' = f\left(\frac{v}{f\lambda + v_S}\right)$$

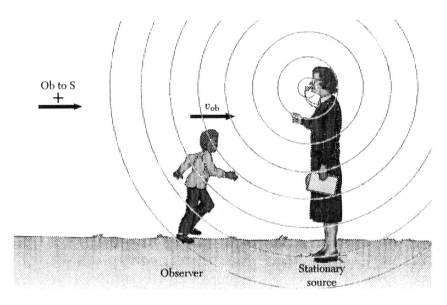

Figure 16.36
An observer moving relative to a source hears a wavelength or frequency different from that emitted by the source because the observer is running either away from or into the crests of the sound wave.

$$f' = f\left(\frac{v}{v + v_S}\right) \qquad (16.17)$$

This expression tells us that a departing source ($v_S > 0$) is observed to have a lower frequency, and an approaching source ($v_S < 0$) is observed to have a higher frequency.

In a similar manner, a moving observer observes a wavelength and frequency different from those emitted because the observer is running into or away from the wave crests produced by the source. This is sketched in Figure 16.36, where the observer is moving toward the source with a speed $v_{ob} > 0$. In time t, the observer moves a distance $v_{ob}t$ and runs into an additional number of crests equal to this distance divided by the wavelength:

$$\text{Additional crests} = \frac{v_{ob}t}{\lambda}$$

In terms of frequency, this additional number of crests encountered means an increase in frequency. Since the increase in crests observed *per second* is v_{ob}/λ, the observed frequency f' is

$$f' = f + \frac{v_{ob}}{\lambda}$$

Since

$$\frac{v_{ob}}{\lambda} = v_{ob}\left(\frac{f}{v}\right)$$

the observed frequency can be written

$$f' = f\left(\frac{v + v_{ob}}{v}\right)$$

(16.18)

An approaching observer ($v_{ob} > 0$) hears a higher frequency; an observer moving away from the source ($v_{ob} < 0$) hears a lower frequency.

If both source and observer are moving, Equations 16.17 and 16.18 can be combined as

$$f' = f\left(\frac{v + v_{ob}}{v + v_s}\right)$$

(16.19)

This equation takes care of all situations if we use the sign convention that the direction *from* observer *to* source is positive.

Problem-Solving Tip

▬ If observer and source are approaching each other $f' > f$; if they are getting farther apart, $f' < f$.

Example 16.10 ─────────────────

While standing still, an observer measures the frequency of a tuning fork to be 524 Hz (C above middle C). What frequency does he measure as he approaches the fork at 5.00 m/s? Take the speed of sound in air to be 340 m/s.

Reasoning
Approaching the tuning fork, the observer runs into more of the crests and therefore hears a higher frequency. We can apply Equation 16.18 directly.

$$f' = f\left(\frac{v + v_{ob}}{v}\right)$$

$$= (524 \text{ Hz})\left(\frac{340 \text{ m/s} + 5.00 \text{ m/s}}{340 \text{ m/s}}\right) = 532 \text{ Hz}$$

─────────────

Practice Exercise   What frequency will the observer measure if the observer moves away from the source at 15.0 m/s?

Answer   501 Hz

Example 16.11

When the car is standing still, the siren on a police car has a dominant frequency of 650 Hz. If you are driving down the highway at 75.0 km/h and are overtaken by this police car doing 110 km/h, what frequency do you hear as it approaches and then as it goes on past? Take the speed of sound in air to be 340 m/s. This situation is sketched in Figure 16.37.

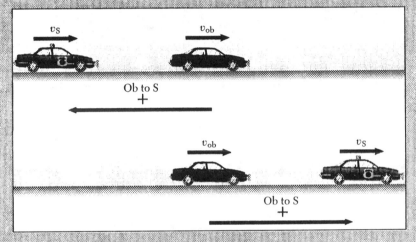

Figure 16.37
How does the frequency of a siren change as the source passes you?

Reasoning

As the siren approaches, the crests are being compressed and consequently you hear a higher frequency. Once the car passes you so that the source is moving away from you, the distance between crests is being stretched and you consequently hear a lower frequency.

Solution

We need the speeds in meters per second in order to use them in Equation 16.19:

$$v_{ob} = 75.0 \frac{km}{h} \left(\frac{1000 \ m}{1 \ km} \right) \left(\frac{1 \ h}{3600 \ s} \right) = 20.8 \ m/s$$

$$v_S = 110 \frac{km}{h} \left(\frac{1000 \ m}{1 \ km} \right) \left(\frac{1 \ h}{3600 \ s} \right) = 30.6 \ m/s$$

Remember that the direction from observer to source is positive. Initially both you and the source are moving to the right while the positive direction is to the left, so $v_{ob} < 0$ and $v_S < 0$. Therefore

$$f' = (650 \ Hz) \left(\frac{340 \ m/s - 20.8 \ m/s}{340 \ m/s - 30.6 \ m/s} \right) = (650 \ Hz)(1.03) = 670 \ Hz$$

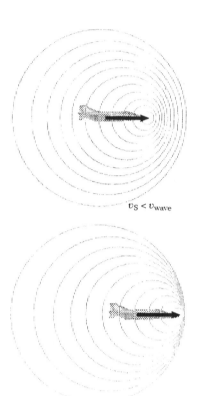

$v_S < v_{wave}$

$v_S = v_{wave}$ (Mach 1)

Figure 16.39

The crests of the sound wave from a moving source pile up in the direction in which the source is moving.

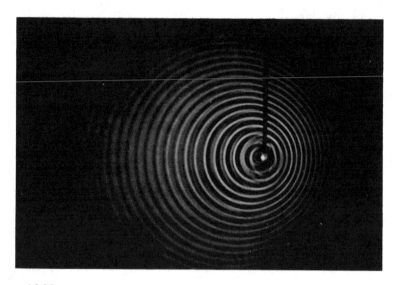

Figure 16.38

Water waves created in a ripple tank by a moving source illustrate the Doppler effect. *(Courtesy of Education Development Center, Newton, MA)*

After the police car passes you, both you and it are moving to the right, and now the positive direction is to the right. Therefore $v_{ob} > 0$ and $v_S > 0$:

$$f' = (650 \text{ Hz})\left(\frac{340 \text{ m/s} + 20.8 \text{ m/s}}{340 \text{ m/s} + 30.6 \text{ m/s}}\right) = (650 \text{ Hz})(0.974) = 633 \text{ Hz}$$

Radar speed-detection units use the Doppler effect with electromagnetic waves. Equation 16.19 applies only to sound waves, so a different equation is necessary for radar (or other electromagnetic) waves, but the general idea of the Doppler effect is the same for all waves. Figure 16.38 illustrates the Doppler effect with water waves in a ripple tank.

*16.10 Sonic Boom

CONCEPT OVERVIEW

- Sound waves generated by an object moving faster than the speed of sound pile up onto each other and create a region of very high pressure.

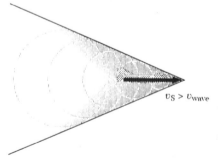

$v_S > v_{wave}$

Figure 16.40

When a source is traveling faster than the speed of the wave it creates, the crests pile up to create a shock wave. For a source traveling faster than the speed of sound, this shock wave creates a sonic boom.

Figure 16.39 shows crests of a wave emitted by a moving source. As the source speed gets closer and closer to the wave speed, the crests pile up on each other more and more. Figure 16.40 illustrates what happens when the source moves faster than the wave speed. New crests are created before the previous ones can move away from the source. This severe piling up creates a region where the wave has a very large amplitude; this is called a **shock wave**.

Figure 16.41
A boat traveling faster than the speed of the water waves it creates forms a bow wave. The wave crests pile up because they are being created faster than they can move away from the source. *(© Comstock, Inc.)*

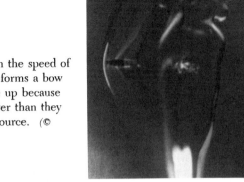

Figure 16.42
An object in air traveling faster than the speed of sound creates a shock wave. The crests of the waves pile up because they are being created faster than they can move away from the source. *(Harold E. Edgerton, Courtesy of Palm Press, Inc.)*

For a boat moving across the water, as in Figure 16.41, such a wave is called a *bow wave.* You can see the characteristic V shape of this wave in this figure.

An airplane flying faster than the speed of sound—that is, at *supersonic* speed—forms a shock wave, as shown in Figure 16.42. An airplane is said to be traveling at supersonic speeds when it travels faster than the speed of sound. When the shock wave reaches the ground, it produces a **sonic boom** heard as loud noise and felt as a pressure wave that may rattle or break windows, as sketched in Figure 16.43. Once established, this shock wave travels at the same speed as the source.

Mach number, named after the German physicist Ernst Mach, is pronounced with the *a* sounded as in yacht, not as in back.

An airplane's speed is often given in terms of the ratio of its speed to the speed of sound in air; this ratio is known as the **Mach number.** Thus, an airplane traveling at half the speed of sound is at Mach 0.5, while one traveling at the speed of sound is at Mach 1.0. The supersonic *Concorde,* shown in Figure 16.44, travels at Mach 2.3.

Figure 16.43
When it reaches the ground, a shock wave from a supersonic airplane produces a sonic boom.

Figure 16.44
A supersonic airplane, like this *Concorde,* can fly faster than the speed of sound—in this case, at a maximum speed of Mach 2.3 (761 m/s or 1,700 mi/h). *(© Comstock, Inc.)*

SUMMARY

A wave is the transfer of energy from one point to another without the transfer of any mass. The wave amplitude is the maximum displacement of the wave medium. The wavelength is the distance between repetitions or cycles. The period is the time required for one complete repetition of the wave. The frequency is the number of oscillations per unit time, usually per second.

In a transverse wave, the motion of the medium is perpendicular to the wave velocity. In a longitudinal wave, the motion of the medium is parallel (and antiparallel) to the wave velocity. As a longitudinal wave passes, the particles of the medium are pushed together into regions called condensations. Formation of every condensation causes formation of a low-particle-density region called a rarefaction.

Waves pass through each other without alteration; this is known as superposition.

A standing wave causes a wave pattern that stands in one place. Regions of no medium displacement are called *nodes;* regions of maximum medium displacement are called antinodes. Standing waves occur only for particular frequencies. The lowest standing-wave frequency is called the fundamental. Higher standing-wave frequencies are called either harmonics or overtones. For a string fixed at both ends, standing waves have wavelengths of

$$\lambda = \frac{2L}{n} \qquad n = 1, 2, 3, \dots \qquad \text{(16.4)}$$

where L is the length of the string and n is an integer. In terms of the speed of the wave along the string, the fundamental frequency is

$$f_1 = \frac{v}{2L} \qquad \text{(16.6)}$$

and the nth harmonic frequency is

$$f_n = \frac{nv}{2L} \qquad n = 1, 2, 3, \dots \qquad \text{(16.5)}$$

The speed of sound in air at temperature T is

$$v = [331 + (0.60/\text{C}°)T] \text{ m/s} \qquad \text{(16.9)}$$

Pitch corresponds to the frequency of a sound. Quality refers to the total collection of different frequencies that compose a sound. Loudness is connected with the amplitude of the sound wave. The intensity of a three-dimensional wave is the energy it carries across a unit area in a unit time:

$$I = \frac{E}{at} = \frac{kA^2}{2at} \qquad \text{(16.10)}$$

The intensity level in decibels is defined by

$$\beta \text{ (in dB)} = 10 \log \frac{I}{I_0} \qquad \text{(16.12)}$$

where I is the sound intensity being measured and I_0 is the faintest audible sound intensity, 10^{-12} W/m^2.

Waves of the same frequency from two sources may form an interference pattern, showing regions of constructive and destructive interference. Constructive interference occurs when the path difference from the two sources is an integral number of wavelengths,

$$\Delta d = n\lambda \qquad n = 0, 1, 2, 3, \dots \qquad \text{(16.13)}$$

Destructive interference occurs for

$$\Delta d = \frac{n\lambda}{2} \qquad n = 1, 3, 5, \dots \qquad \text{(16.14)}$$

When two sounds of different but nearly equal frequencies are heard together, they alternately interfere constructively and destructively. The increasing and decreasing of the loudness is known as beats. The beat frequency is

$$f_{\text{beat}} = f_2 - f_1 \qquad \text{(16.15)}$$

If an observer, a sound source, or both are moving relative to each other, the frequency heard by the observer is altered. The observed frequency is

$$f' = f\left(\frac{v + v_{ob}}{v + v_s}\right)$$ **(16.19)**

This is known as the Doppler effect. The equation assumes that the positive x direction is the direction from observer to source.

QUESTIONS

1. If you wiggle a rope to cause a wave, how does the frequency of the wave compare with the frequency of your hand?
2. How does the speed of a wave on a rope compare with the speed of individual particles of the rope, which move as the wave comes by?
3. What is the difference between the speed of a wave on a piano string and the speed of sound in air?
4. What would an astronaut hear if another astronaut played a piano on the moon?
5. If one end of a heavy rope is attached to one end of a light rope, the speed of a wave will change as the wave goes from the heavy rope to the light one. Will it increase or decrease? What happens to the frequency? To the wavelength?
6. If all else remains the same, what happens to the intensity of a wave when the amplitude is doubled?
7. Three identical tuning forks sit near the front of the laboratory. A piece of clay is struck to one so that it rings at a slightly lower frequency when struck. One of the other forks is struck, allowed to ring for a moment, and then silenced. The ringing is still heard. Further investigation shows that the identical tuning fork is ringing and the one with the clay is not. Explain what happened and why.
8. You can estimate how far you are from a lightning bolt by counting the seconds between your seeing the lightning and hearing the thunder. Explain. If lightning strikes 1 km away, how long is it before you hear the thunder?
9. How does air temperature affect the tuning of a wind instrument?

10. Why does a guitar have frets—ridges that tell a player where to place the fingers (Fig. 16.45)?

Figure 16.45
Questions 10 and 12. Frets on a guitar help a player make the right notes. (© *Martha Cooper/Peter Arnold, Inc.*)

11. A guitar string is tuned to play a particular note when the full length is used. Can you play a note one octave higher on the string? One octave lower?
12. Why are guitar frets (Fig. 16.45) spaced closer together as you move down the neck of the guitar and farther apart as you move up the neck?
13. What is heard when a 1000-Hz sound and a 1100-Hz sound are played together?
14. What is heard when two flutes play the same note but are not accurately tuned? How can what is heard be used to bring the flutes into tune with each other?
15. If a child continuously blows a whistle while on a merry-go-round, explain what you hear as the child comes by each time.

PROBLEMS

16.1 Wave Motion

16.1   To estimate the distance to a cliff on the opposite shore of a mountain lake, you call out, "Physics is fun!" and listen for your echo. You hear the echo 1.8 s after calling. How far away is the cliff?

16.2   A typical human ear can detect frequencies from 50 to 20 000 Hz. The speed of sound in air is about 340 m/s. To what range of wavelengths does this frequency range correspond?

16.3   A cork on the surface of a lake bobs up and down two times per second on some ripples having a wavelength of 8.50 cm. If the cork is 10.0 m from shore, how long does it take one ripple to travel from the cork to the shore?

16.4   For detecting obstacles, bats use very-high-pitched sound that has a frequency of about 50 kHz. The speed of sound in air is about 340 m/s. What is the wavelength of this sound in air?

16.5   For detecting obstacles, dolphins use very-high-pitched sound with a frequency of about 200 kHz. The speed of sound in water is about 1500 m/s. What is the wavelength of this sound in water? What is the wavelength of this sound in air (when the speed of sound in air is 350 m/s?

16.6   What is the wavelength of the sound waves produced by the note A (440 Hz) when the speed of sound is 350 m/s?

16.7   In a physics experiment, you measure sound waves that are 1.34 m long when the speed of sound in air is 350 m/s. What is the frequency of the sound?

16.8   Radio waves travel at 3.0×10^8 m/s. What is the wavelength of radio waves used by an FM station that broadcasts at 97.1 MHz?

16.9   Radio waves travel at 3.0×10^8 m/s. What is the wavelength of radio waves used by an AM station that broadcasts at 580 kHz?

16.10 During a thunderstorm, you see a bright flash of lightning and then hear the thunder caused by it 7.3 s later. How far away is the lightning?

16.11 A pulse travels down a rope at some speed. By how much must the tension be altered to cause the pulse to travel at twice this speed?

○16.12 Two students pull on opposite ends of a heavy rope. The rope is 9.0 m long and has a mass of 5.4 kg. When one end is struck, it takes 0.60 s for the pulse to travel the full length, be reflected, and travel back to the origin. With what tension are the students pulling on the rope?

○16.13 Equation 16.1 implies that the units of

$$\sqrt{\frac{\text{N}}{\text{kg/m}}}$$ reduce to meters per second. Show

that this is true.

16.2 Types of Waves

16.3 Superposition

○16.11 Figure 16.46 shows four instances of two pulses traveling on a string toward each other at 10 cm/s. All four graphs are for $t = 0$. For each case, sketch the shape of the string at 0.05-s intervals from $t = 0.05$ s to $t = 0.6$ s.

○16.15 Figure 16.47 shows two waves of equal amplitude. One has a frequency of 100 Hz; the other, 200 Hz. Carefully add these waves and graph the resulting waves.

16.4 Transverse Standing Waves

16.16 The A string on a violin is tuned to a fundamental frequency of 440 Hz. List the overtones that may be present in the audible range of frequencies.

16.17 A guitar string 0.70 m long is tuned to play A at 440 Hz when its full length vibrates. Where should a finger be placed in order to play C at 524 Hz?

16.18 A guitar string 0.70 m long is tuned to play middle C at 262 Hz when its full length vibrates. Where should a finger be placed in order to play A at 440 Hz?

16.19 A string is driven at a frequency of 550 Hz so that three loops are formed in the standing wave, as in Figure 16.48. What frequency will cause a standing wave of four loops?

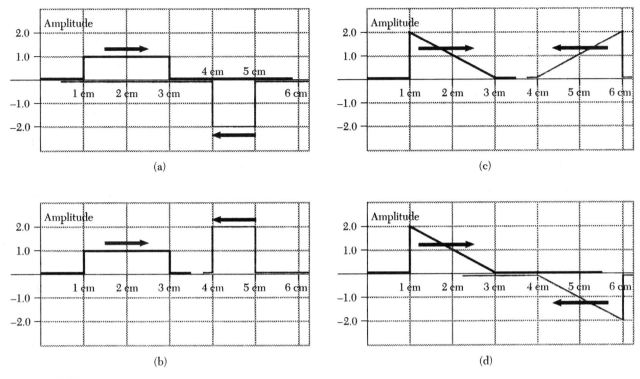

(a)

(b)

(c)

(d)

Figure 16.46
Problem 16.14. What does the superposition of two pulses look like as a function of time?

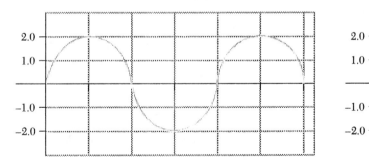

Figure 16.47
Problem 16.15. What does the superposition of these two waves look like?

Figure 16.48   Problem 16.19.

○16.20 A demonstration is carried out with the apparatus in Figure 16.49. A mass is suspended on one end of a string (so the tension in the string is the weight of the mass, mg). The string is then run up over a pulley and attached to a small-amplitude 60-Hz oscillator. As additional masses are placed on the end, standing waves appear. The distance between pulley and oscillator is 2.4 m. That amount of string has a mass of 45 g. How much mass is supported when the standing wave shown in Figure 16.49 is produced?

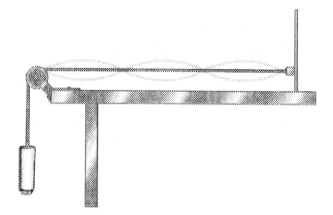

Figure 16.49   Problems 16.20 and 16.21.

○16.21 With the string and arrangement of Figure 16.49, what mass is necessary to produce a standing wave of five loops if the distance from pulley to oscillator is increased to 2.8 m?

*16.5 Longitudinal Standing Waves

16.22 An organ pipe open at one end and closed at the other is 1.0 m long. What is its fundamental frequency? Its first overtone frequency?

16.23 A whistle is made from a cylinder 12.5 cm long, one end open and the other closed. What is its fundamental frequency?

16.24 A soda bottle is 28 cm tall. What is its fundamental frequency when you blow across the top of it?

16.25 A soda straw is 20 cm long. What are the fundamental frequency and the first overtone if you blow across the top while holding your finger on the bottom?

○16.26 A soda straw is 20 cm long. What are the fundamental frequency and the second harmonic if you blow across the top while leaving the bottom open?

○16.27 When you blow across the top of a soda bottle, the sound produced by the column of air in the bottle is primarily the fundamental frequency. Consider two bottles, each 32 cm tall. One is empty. Find its fundamental frequency. What depth of water must be added to the other so that it sounds one octave higher than the first?

○16.28 A ringing tuning fork is held above a tube in a physics experiment. The tube is filled with water whose level can be altered easily, thus changing the length of the column of air below the tuning fork. As the water level is lowered, lengthening the column of air, an increase in loudness is noted when the water is 0.120, 0.360, and 0.600 m from the top of the tube. Assume the speed of sound in air is 345 m/s. What is the frequency of the tuning fork?

○16.29 A glass cylinder closed at one end and open at the other is 2.3 m long. What is its fundamental frequency? What overtones may be present in the audible range of frequencies? Take the speed of sound in air to be 340 m/s.

○16.30 A pipe open at both ends is 1.0 m long. What is its fundamental frequency? Its second harmonic frequency?

16.6 Characteristics of Sound

16.31 What is the speed of sound in air at 23°C?

16.32 What is the speed of sound in air on a very hot summer day when the temperature is 42°C?

16.33 What is the intensity level in decibels of a sound that has an intensity of 1.0×10^{-5} W/m²?

16.34 What is the intensity of a 75-dB sound?

16.35 The specifications of an amplifier list a 58-dB signal-to-noise ratio. What is the ratio of the signal intensity to the background-noise intensity?

○16.36 At some distance a stick of dynamite produces a 110-dB noise. What will the intensity level be at the same place if two sticks of dynamite are used?

○16.37 What is the combined intensity level when a 60-dB sound and a 70-dB sound are present at the same time;?

●16.38 A skyrocket explodes 100 m above the ground as shown in Figure 16.50. Three observers are spaced 100 m apart. What is the ratio of the intensity levels (in decibels) heard by observers A and B, and by A and C? The sound intensity decreases as $1/r^2$, where r is the distance between observer and source.

Figure 16.50   Problem 16.38.

*16.7 Interference

16.39 Two speakers located 1.3 m apart produce a sound of 500 Hz. Calculate two positions on a line 5.0 m in front of the speakers where you expect to find constructive interference. Calculate two positions on this line where you expect to find destructive interference.

○16.40 The ship in Figure 16.51 moves along a straight line parallel to the shore. The ship's path is 600 m from the shore. Its radio receives identical signals from the antennas at points A and B. The signal goes through destructive interference at D after passing through constructive interference at C. Determine the wavelength of the radio signals.

●16.41 Assume the ship in Figure 16.51 continues to travel along the same line, 600 m from shore, and encounters another pair of antennas transmitting with a wavelength of 500 m. If this signal is at a maximum at a point like D, directly opposite one of the antennae and there have been no other maxima since the maximum midway between the antennae at C, how far apart are the antennae?

●16.42 The two radio antennae, 1.0 km apart, shown in Figure 16.52 receive radio signals of wavelength 2.4 m from the same source in deep space. The two signals are then combined in a single receiver. The signal is maximum for $\theta = 90°$. For what other angle(s) θ will the composite signal give maximum output?

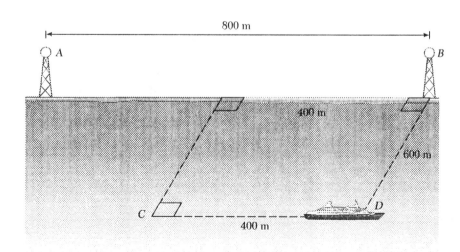

Figure 16.51
Problems 16.40 and 16.41.

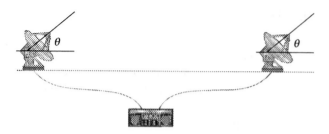

Figure 16.52   Problem 16.42.

*16.8 Beats

16.43   Two car horns sound with frequencies of 552 Hz and 556 Hz. What beat frequency is heard?

16.44   Two whistles with frequencies of 400 Hz and 403 Hz are sounded simultaneously. How many beats are heard per second?

16.45   A 524-Hz tuning fork produces a beat frequency of 4 Hz when played together with a clarinet. What is the frequency of the note played by the clarinet?

16.46   Two violins are slightly out of tune. When both players play A, beats are heard with a frequency of once per second. By how much do the notes played differ in frequency?

○16.47   Two flutes are tuned identically in a 25°C room. One flute is then taken outdoors at 0°C while the other is kept in the room. What beat frequency is heard if both play middle C (262 Hz)?

●16.48   On a day when the speed of sound in air is 350 m/s, two whistles are blown simultaneously. The whistles have frequencies of 520 Hz and 522 Hz.   (a) How many beats per second are heard?   (b) How far apart in space are adjacent regions of maximum amplitude of the sound?

*16.9 Doppler Effect

16.49   A train whistle sounds at 500 Hz. What frequency is heard by a stationary observer when the train approaches at 25 m/s? When it moves away at 25 m/s? What frequency is heard by a moving observer as she approaches the stationary train at 25 m/s? As she moves away from the stationary train at

25 m/s? Take the speed of sound to be 340 m/s.

16.50   A factory whistle beside a road emits a sound whose most intense component is at 400 Hz. What is the observed frequency heard by a passenger in a car   (a) approaching the factory at 30 m/s and   (b) moving away from the factory at 30 m/s?

16.51   A car drives at 30 m/s along a road that runs parallel to a railroad track. A train approaches traveling at 20 m/s. The train sounds a whistle that has a frequency of 400 Hz (as measured by the engineer when the train is stationary). What frequency does the driver hear as the train approaches? After the train has passed and goes away?

16.52   A police siren has a dominant frequency of 1500 Hz when measurements are made on the stationary siren by a stationary observer. What will be measured by a stationary observer as the dominant frequency when the police car approaches the observer at 30 m/s? When the observer approaches the stationary siren at 20 m/s?

○16.53   For navigational purposes, a bat emits short bursts of sound at 4.0×10^5 Hz (measured with bat and detector both stationary). The bat approaches a stationary object at 25 m/s. What is the frequency of the sound detected at the object? The object then reflects the sound with this frequency. When this reflected sound is detected by the bat, what is its frequency?

●16.54   A stationary train engineer sounds his own engine's horn, which has a frequency of 550 Hz, as another train passes. He notices a beat frequency of 2 Hz both as the train approaches and as it departs. What is the speed of the moving train? Take the speed of sound to be 340 m/s.

●16.55   An observer in a mountain town hears a train whistle and then 3.00 s later hears the start of an echo from a cliff. The echo's frequency is 0.900 that of the sound heard directly. Train, cliff, and observer all lie along a straight line.   (a) How far is the train from the cliff?   (b) How fast and in what direction is the train moving?

*16.10 Sonic Boom

16.56 When the air temperature is 0°C, an aircraft flies at Mach 0.8. What is its speed in meters per second and kilometers per hour?

16.57 When the air temperature is 30°C, an aircraft flies at Mach 1.5. What is its speed in meters per second and kilometers per hour?

16.58 Figure 16.53 shows an overhead view of a boat traveling across a still lake. The waves travel outward in a wake if the speed of the boat exceeds the wave speed. If the boat has a speed of 10.0 km/h and the wave speed is 2.0 km/h, what is the angle θ of the wake?

16.59 Show that the angle of the shock wave for a supersonic airplane is given by $\sin \theta = v_{sound}/v_{plane}$.

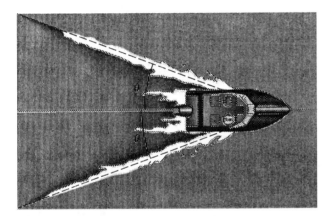

Figure 16.53   Problem 16.58.

General Problems

16.60 Magnetic recording tape moves past the recording head of a tape recorder with a speed of 3.0 cm/s. If the tape has a granularity that makes it impossible to distinguish magnetic variations closer than 1.25×10^{-5} m, what is the highest frequency it can record?

16.61 Two ropes of equal length are joined together end to end. It is observed that a pulse speeds up by a factor of 1.5 as it passes from the heavier rope to the lighter one. What is the ratio of the linear densities of the two ropes?

16.62 An organ pipe closed at one end and open at the other is tuned to 440 Hz (A) when the temperature is 0°C. What frequency is produced when the temperature is 25°C?

16.63 An organ pipe, closed at one end, is tuned to 392 Hz (G) at 20°C. What frequency is produced at 0°C?

16.64 A pipe open at both ends is 1.5 m long. What is its fundamental frequency? What overtones may be present in the audible range of frequencies? Take the speed of sound in air to be 340 m/s.

16.65 If one machine produces noise at 70 dB, how many identical machines must be operating to increase the intensity level to 80 dB?

16.66 On a day when the speed of sound in air is 335 m/s, two whistles are blown simultaneously. The wavelengths of the sound emitted are 5.50 m and 5.60 m. How many beats per second are heard?

16.67 A speaker is attached to the rim of a turntable 2.00 m in diameter. The speaker emits a sound at 520 Hz when stationary. When the turntable rotates, the observed frequency varies from 512 to 528 Hz. What is the angular velocity of the turntable?

16.68 A boat traveling at 8.6 m/s causes a bow wave that makes an angle of 20° with the boat's direction of motion. What is the speed of the water waves?

APPENDIX A

Algebra Review

An unknown quantity in an algebraic equation may be represented by any letter. It is a good idea, however, to choose a letter that is somehow related to the quantity for which the letter stands. If you are looking for the acceleration, for instance, it will be easier to remember if you label the unknown as a. You might write

$$v = v_i + at$$

or, if the initial speed v_i is zero and the speed v is 20 m/s at time $t = 5$ s, you can write

$$20 \text{ m/s} = 0 + a \,(5 \text{ s})$$

Everything in the equation is known except for the unknown a, which represents the acceleration. Throughout our general discussion here, we shall use x as the unknown.

An equation is like a double-pan balance in equilibrium (when it is "balanced"), as shown in Figure A.1. The pieces on the two sides of an equals sign are *equal* to each other. The two sides of an equation may look different from each other, but they mean the same thing; they are two ways of writing the same thing. For instance,

$$12 = 12$$

is certainly true even though it is not very interesting. A more interesting equation might be

$$3x = 12$$

which you can think of as a double-pan balance with three identical, sealed boxes, each labeled x, on the left balancing 12 kg on the right (Figure A.2). To balance the 12 kg, each box must contain 4 kg. Even though you can solve this simple equation in your head, let us look at it in detail and see what we can learn about solving equations in general.

An equation remains balanced if we *do the same thing to both sides*. To end up with only a single sealed box labeled x in the left pan, as in Figure A.3, we need to take one third of what was there originally. Doing that is the same as dividing $3x$ by 3. The balance remains in equilibrium—the equation remains true—if we do the same thing to the other side. That is, to get x by itself, we must divide *both sides* of the equation by 3:

$$\frac{3x}{3} = \frac{12}{3}$$

$$x = 4$$

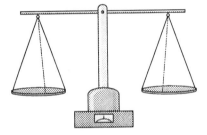

Figure A.1

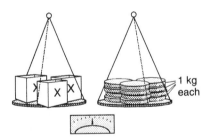

Figure A.2

1 kg each

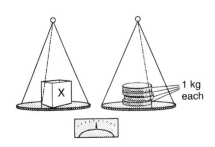

1 kg each

Figure A.3

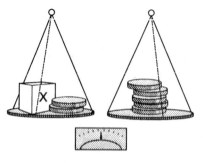

Figure A.4

Consider a second situation, where now a different sealed box, also labeled x, and 2 kg on the left are balanced by 5 kg on the right, as in Figure A.4. What is in the sealed box? To describe this situation, we can write the equation

$$x + 2 = 5$$

To get the box labeled x by itself, we need to take away the 2 kg now with it. To keep the balance in equilibrium, we must also take away 2 kg from the other side. This procedure means subtracting 2 from both sides:

$$\begin{array}{rcr} x + 2 = & & 5 \\ -\ 2 = & & -2 \\ \hline x\ \ \ \ \ = & & 3 \end{array}$$

This time, the box must have 3 kg in it, or, in terms of the equation,

$$x = 3$$

The important idea in these situations is that you must do exactly the same thing to both sides of the equation. Do whatever you need to do to get x by itself, but be sure to do the same operation to everything on that side of the equation and to everything on the other side as well.

Any mathematical operation can be used. For example, consider the equation

$$\sqrt{x} = 15$$

To get the unknown x by itself, we must square the left side so we must be sure to square the right side, too:

$$(\sqrt{x})^2 = (15)^2$$
$$x = 225$$

Sometimes several steps are needed to get the unknown by itself. For instance, consider

$$\sqrt{5x + 1} = 4$$

We cannot get x by itself in a single step. It looks like we might subtract 1, but 1 is under the square-root sign. You might think we should divide by 5, but 5 is also under the square-root sign. Before we can do either of those things, we must get rid of the square root. We do that by squaring **both** sides of the equation:

$$(\sqrt{5x + 1})^2 = (4)^2$$
$$5x + 1 = 16$$

Now we can subtract 1 from both sides to get $5x$ alone:

$$\begin{array}{rcr} 5x + 1 = & & 16 \\ -\ 1 = & & -1 \\ \hline 5x\ \ \ \ \ = & & 15 \end{array}$$

Now we can divide by 5 to get x by itself:

$$\frac{5x}{5} = \frac{15}{5}$$

$$x = 3$$

Factoring

If an unknown appears in two or more terms, it is a common factor in those terms and can be written as such. That is, the expression

$$3x + 5x$$

can be written

$$(3 + 5)x$$

which then becomes

$$8x$$

This procedure can be used in a situation such as

$$
\begin{aligned}
3x &= 48 - 5x \\
3x &= 48 - 5x \\
+ 5x &= + 5x \\
\hline
3x + 5x &= 48 \\
(3 + 5)x &= 48 \\
8x &= 48 \\
\frac{8x}{8} &= \frac{48}{8} \\
x &= 6
\end{aligned}
$$

Exponents

In an expression such as 10^3, the 10 is called the **base** and the small superscript is called the **exponent.** A base can also be an unknown, as in x^3. An exponent tells how many times the base is to be used as a factor. A positive exponent tells how many times a base is to be multiplied by itself, and a negative exponent tells how many times a base is to be used as a factor in the denominator:

$$10^3 = (10)(10)(10) = 1000$$

$$10^{-2} = \frac{1}{(10)(10)}$$

$$x^4 = xxxx$$

$$x^3 y^{-2} = \frac{xxx}{yy}$$

This rule means that in an expression like $x^3 x^2$, where the same base appears more than once, the exponents are added:

$$x^3 x^2 = (xxx)(xx) = x^5 = x^{3+2}$$

For an expression like $(x^3)^2$, where a term with an exponent is being raised to a power, the exponents are multiplied:

$$(x^3)^2 = (x^3)(x^3) = (xxx)(xxx) = x^6 = x^{3 \times 2}$$

Quadratic Equations

A quadratic equation contains the square of the unknown. The general form is

$$ax^2 + bx + c = 0$$

where a, b, and c are numbers (called **coefficients** of the equation) and x is the unknown. The solution of this general equation is

$$x = \frac{-b \pm \sqrt{b^2 - 4ac}}{2a}$$

Notice that there are two solutions for x except in the special case when $b^2 - 4ac$ vanishes. Both are perfectly good mathematical solutions to the quadratic equation. However, there are often physical restrictions that make one solution real and meaningful and the other physically meaningless.

Graphs and Linear Equations

A linear equation contains the unknown only to the first power (in other words, as x but not as x^2 or x^3). The general form of a linear equation is

$$y = mx + b$$

where m and b are numbers that relate the variables x and y. If you make a table of values that x and y can have and then construct a graph from these values, you will find that the graph is a straight line having a slope m. The slope is positive if y increases as x increases and negative if y decreases as x increases. The slope can be found from any two points (x_1, y_1) and (x_2, y_2) that lie on the line (Figure A.5):

$$m = \frac{y_2 - y_1}{x_2 - x_1} = \frac{\Delta y}{\Delta x} = \frac{\text{rise}}{\text{run}}$$

The number b, which is called the **y intercept** because it is the value at which the graph intercepts the y axis, is the value of y that corresponds to $x = 0$.

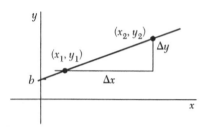

Figure A.5

Simultaneous Equations

A single equation with two unknowns, such as

$$10x + 15y = 35$$

does not have a unique solution. It is a linear equation in terms of the variables x and y and corresponds to a straight line on an x,y graph. There are an infinite number of solutions, with $(x = 0.5, y = 2)$, $(x = 2, y = 1)$, $(x = 3, y = 0.333)$, and $(x = 3.5, y = 0)$ being just a few of the possibilities.

To solve for a unique value for x and y, we need a second, independent equation, such as

$$5x - 3y = 7$$

This second linear equation corresponds to a second straight line on an x,y graph, and the point of intersection of those two lines gives the single x and y values that are consistent with both equations (Figure A.6).

In practice, it is usually easier to solve the two equations mathematically rather than graphically. There are two general approaches to solving simultaneous equations—substitution and elimination. Consider the two equations of

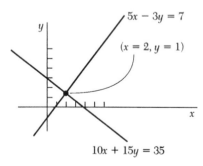

Figure A.6

our present example from both approaches. To solve by substitution, first solve one unknown in terms of the other:

$$10x + 15y = 35$$

$$10x = 35 - 15y$$

$$\frac{10x}{10} = \frac{35 - 15y}{10}$$

$$x = \frac{35 - 15y}{10}$$

Now **substitute** this value for x into the other equation:

$$5x - 3y = 7$$

$$5\left(\frac{35 - 15y}{10}\right) - 3y = 7$$

$$\left(\frac{175 - 75y}{10}\right) - \frac{30y}{10} = 7$$

$$\frac{175 - 105y}{10} = 7$$

$$10\left(\frac{175 - 105y}{10}\right) = 10(7)$$

$$175 - 105y = 70$$

$$-105y = -105$$

$$y = 1$$

Now use this value of y in either equation to solve for x:

$$5x - 3y = 7$$
$$5x - 3(1) = 7$$
$$5x - 3 = 7$$
$$\underline{+3 = +3}$$
$$5x = 10$$

$$\frac{5x}{5} = \frac{10}{5}$$

$$x = 2$$

The solution is $x = 2$ and $y = 1$.

In the elimination procedure, we multiply one or both equations by numbers until the coefficients of either x or y in both equations are the same and then add or subtract the two equations so that either the x term or the y term becomes zero. This leaves a single equation in a single unknown:

$$5x - 3y = 7 \qquad\qquad \textbf{(1)}$$
$$10x + 15y = 35 \qquad\qquad \textbf{(2)}$$

Multiply Equation 1 by the factor 2; this gives a $10x$ term in both equations. Then subtract the two equations to get rid of the x term:

$$
\begin{array}{rr}
10x - 6y = & 14 \quad\quad\text{(3)} \\
-(10x + 15y = & 35) \quad\text{(2)} \\
\hline
-21y = & -21 \\
y = & 1
\end{array}
$$

Just as before, this value for y can be put back into either equation to solve for x:

$$
\begin{array}{rcr}
10x + 15y &=& 35 \\
10x + 15(1) &=& 35 \\
10x + 15 &=& 35 \\
-15 &=& -15 \\
\hline
10x &=& 20
\end{array}
$$

$$\frac{10x}{10} = \frac{20}{10}$$

$$x = 2$$

As before, $x = 2$ and $y = 1$.

Two independent equations are necessary to solve for two unknowns. In general, n independent equations are necessary to solve for n unknowns. Multiplying an equation, as we did to get Equation 3 in the example above, only gives another statement of the same equation; the result of the multiplication is *not* an independent equation.

APPENDIX B

Trigonometry Review

Two triangles are said to be **similar** if they have the same angles even though the lengths of their sides are different. In Figure B-1, triangle ABC is similar to triangle DEF. That is, $\angle a = \angle d$, $\angle b = \angle e$, and $\angle c = \angle f$. For similar triangles, the ratios of corresponding sides are equal:

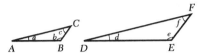

Figure B.1

$$\frac{BC}{AC} = \frac{EF}{DF}$$

$$\frac{AB}{AC} = \frac{DE}{DF}$$

$$\frac{BC}{AB} = \frac{EF}{DE}$$

We can apply this very useful property to **right** triangles. Figure B.2 shows two similar triangles, but these are similar right triangles (meaning each contains one right angle, which means a 90° angle). The angle θ is the same in both triangles. Sides AB and DE are called **adjacent sides** because they are adjacent to the angle θ. Sides BC and EF are called **opposite sides** because they are opposite θ. Sides AC and DF are called the **hypotenuse**—meaning the side opposite the right angle. Since these are similar triangles, we know the ratios of their corresponding sides are equal:

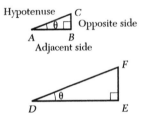

Figure B.2

$$\frac{BC}{AC} = \frac{EF}{DF} = \frac{\text{opposite}}{\text{hypotenuse}}$$

$$\frac{AB}{AC} = \frac{DE}{DF} = \frac{\text{adjacent}}{\text{hypotenuse}}$$

$$\frac{BC}{AB} = \frac{EF}{DE} = \frac{\text{opposite}}{\text{adjacent}}$$

The values of these ratios are the same for any and all right triangles that have the same angle θ, and so we could make a table of these values or program them into a calculator. These ratios have special names:

$$\frac{\text{opposite}}{\text{hypotenuse}} = \text{sine } \theta = \sin \theta$$

$$\frac{\text{adjacent}}{\text{hypotenuse}} = \text{cosine } \theta = \cos \theta$$

$$\frac{\text{opposite}}{\text{adjacent}} = \text{tangent } \theta = \tan \theta$$

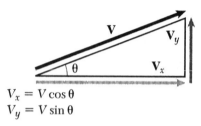

$V_x = V \cos \theta$
$V_y = V \sin \theta$

Figure B.3

These definitions can be turned around so that we can solve for various sides of the triangle. For example,

$$\text{Opposite side} = (\text{hypotenuse}) \, (\sin \theta)$$
$$\text{Adjacent side} = (\text{hypotenuse}) \, (\cos \theta)$$

This procedure becomes very useful in situations like vector addition, where the components of a vector become the sides of a right triangle and the vector becomes the hypotenuse, as in Figure B.3.

The values of sin θ, cos θ, and tan θ for all values of θ between 0 and 90° have been calculated and presented in tables of the trigonometric functions. These values are also stored in any "scientific" electronic calculator.

APPENDIX C

Spreadsheet Physics

Spreadsheets are generic scratch pads that can be used for any sort of calculations. They let us do fairly sophisticated computer programming without having to know any programming at all. We can use spreadsheets to do some interesting physics. Spreadsheets are especially useful when the same calculation needs to be done a number of times—as in multiple sets of data or to see what happens over time.

ClarisWorks and *Lotus 1-2-3* are two common spreadsheets, so this appendix presents details for both of them. All spreadsheets are quite similar, however; there may be some details of your particular one that are different from what is described here, but its basic operation is nearly the same.

A spreadsheet contains a number of **cells** in which we write numbers or ask the computer to put calculation results. Each cell is identified by a letter and a number. The letters give its horizontal position (which **column** it is in), and the numbers give its vertical position (which **row** it is in). Figure C.1 shows a small part of a spreadsheet with some of the cells labeled.

	A	B	C
1	A1	B1	C1
2	A2	B2	C2
3	A3	B3	
4	A4		
5			

Figure C.1

Click on the spreadsheet with your mouse—that is, point to someplace on the spreadsheet and make a single click with the mouse button. That cell is highlighted, the corresponding column letter and row number are highlighted, and the identity of the cell appears near the upper left. If you click in cell C4, you will see something like what is shown in Figure C.2.

C4

	A	B	C	D
1	A1	B1	C1	D1
2	A2	B2	C2	
3	A3	B3	C3	
4	A4	B4	C4	
5	A5			
6	A6			
7				
8				

Figure C.2

To type something in that cell, click the mouse near the **C4** at the top left of the screen. That line will change to look something like what appears in Figure C.3.

C4	X	✔	

	A	B	C	D
1	A1	B1	C1	D1
2	A2	B2	C2	
3	A3	B3	C3	
4	A4	B4	C4	
5	A5			
6	A6			
7				
8				
9				
10				

Figure C.3

Now, on this upper line, type in whatever you want in cell C4. It can be a number or text. When you are through, press the <return> key or use the mouse to click on the check mark [✔] on that top line. If you want to start all over, click on the [x] on that top line.

Click in cell **A1**, type 1, and press <return>. Type the numbers 1 through 10 in column **A** so that your spreadsheet looks like the one in Figure C.4.

	A	B	C	D
1	1			
2	2			
3	3			
4	4			
5	5			
6	6			
7	7			
8	8			
9	9			
10	10			

Figure C.4

In the **B** column, suppose we want to put numbers that are five times those in column **A**. We could just type in those numbers, but the computer can do the calculations for us. Click in cell **B1**. On the upper line, type an equals sign, which tells **ClarisWorks** to get ready to do a calculation. In **Lotus 1–2–3,** a plus sign has the same effect. Now type 5 and then an asterisk, which means "multiply." We want to multiply 5 by the value in cell **A1**. To tell this to the spreadsheet, we could type in 5°A1, but an easier way is to type in 5° (as we have done) and then **click** on cell **A1**. The computer then puts A1 into the formula for you. Try this and your spreadsheet should look something like Figure C.5.

B1	X	✔	=5°A1

	A	B	C	D
1	1			
2	2			
3	3			
4	4			
5	5			
6	6			
7	7			
8	8			
9	9			
10	10			

Figure C.5

After you press <return>, the spreadsheet evaluates 5°A1 and places that value in cell **B1**, so that your spreadsheet looks like Figure C.6.

	A	B	C	D
1	1	5		
2	2			
3	3			
4	4			
5	5			
6	6			
7	7			
8	8			
9	9			
10	10			

Figure C.6

Repeat this procedure for cell **B2**. We could continue with every cell in column B, but a computer is so good at repeating calculations that there is a way to let the computer repeat this procedure. Click on cell **B1** and drag down through cell **B10**. With the entire column **B** highlighted, we can tell the spreadsheet to FILL DOWN, and the same formula or procedure will be used in every highlighted cell. In *ClarisWorks* the FILL DOWN command is one of the choices under CALCULATE from the menu bar at the top of the screen. Now your spreadsheet should look like Figure C.7.

	A	B	C	D
1	1	5		
2	2	10		
3	3	15		
4	4	20		
5	5	25		
6	6	30		
7	7	35		
8	8	40		
9	9	45		
10	10	50		

Figure C.7

In column **C**, let us put a list of the squares of the numbers in column **A**. To start this, click on cell **C1**. In the upper line, type an equals sign to again tell the computer to get ready for a calculation (or a plus sign for *1–2–3*). Then click on cell **A1**, and the computer places A1 right after the equal sign. Then type asterisk to tell it to multiply. Then click on cell **A1** again. The upper line should now look like what Figure C.8 shows.

| C1 | X | ✔ | =A1°A1 |

Figure C.8

Now press <return> or click on the check mark, and the value 1 should appear in cell **C1**. You can repeat this procedure for cells **C2, C3, C4,** and so on, you can use CUT and PASTE to fill in the rest of the **C** column, or you can use the FILL DOWN command to fill in all the rest of the cells in column **C** with the same formula or the same procedure. After you do that, your spreadsheet should look like Figure C.9.

	A	**B**	**C**	**D**
1	1	5	1	
2	2	10	4	
3	3	15	9	
4	4	20	16	
5	5	25	25	
6	6	30	36	
7	7	35	49	
8	8	40	64	
9	9	45	81	
10	10	50	100	

Figure C.9

Now go back to cell **A1** and change the 1 to 10. Go on through the whole column **A** and change all the numbers. As you change a number there, the corresponding numbers in columns **B** and **C** change, too. This is the whole purpose of a spreadsheet. As you change a number, all the calculations that use that number are updated and the results are shown.

This example has been simple, but even here it is useful to label the columns so that we can remember what we have done. With good foresight, we could have left the top row blank to leave room for labels, but spreadsheets make it easy if we change our mind. We want to INSERT CELLS now to make room for these labels. Click on the 1 to the left of cell **A1** to highlight all of row 1. In *Claris Works,* choose INSERT CELLS from the CALCULATE menu. In *Lotus 1–2–3,* choose INSERT. . . from the WORKSHEET menu. Row 1 should now be blank and everything else moved down. In cell **A1**, type x. Press <return>, go to cell **B1**, type 5°x, and press <return>. In cell **C1**, type x°x and press

<return>. Now your spreadsheet should look something like what Figure C.10 shows.

	A	B	C	D
1	x	5°x	x°x	
2	10	50	100	
3	15	75	225	
4	20	100	400	
5	40	200	1600	
6	50	250	2500	
7	6	30	36	
8	7	35	49	
9	8	40	64	
10	9	45	81	

Figure C.10

Labels are important because they make it much easier to go back and understand what has been done. *Always* label entries in your spreadsheets.

As an example of spreadsheet physics, let us calculate the parabolic trajectory that an object takes if it is thrown near Earth's surface and we can ignore the effects of air resistance. If the object is thrown with an initial speed of v_{init} at an angle θ above the horizontal, we know its initial horizontal speed is

$$v_{init,x} = v_{init} \cos \theta$$

and its initial vertical speed is

$$v_{init,y} = v_{init} \sin \theta$$

Enter some labels into a spreadsheet to help keep track of things (Figure C.11).

	A	B	C	D
1	v(init) =		v(init,x) =	
2	angle =		v(init,y) =	
3	dt =			
4	time	horizontal	vertical	
5	0	0	0	
6				
7				

Figure C.11

Then type in values for v_{init} and the angle θ in cells B1 and B2. To fire a projectile at 25 m/s at an angle of 53°, you would write what is shown in Figure C.12.

	A	B	C	D
1	v(init) =	25	v(init,x) =	
2	angle =	53	v(init,y) =	
3	dt =			
4	time	horizontal	vertical	
5	0	0	0	
6				
7				

Figure C.12

Now we want to tell the spreadsheet to calculate $v_{init,x} = (25 \text{ m/s}) \cos 53°$. Most spreadsheets, like most computing languages, calculate the sine or cosine of an angle measured in radians, so you must take care of that, $53° = [(2\pi)/360°](53°)$, when you enter the formula for cells **D1** and **D2**. Click on cell **D1** and type an equals sign to tell the computer to make a calculation. Then click on cell **B1**, and it enters B1 (the value of v_{init}) into its calculation. Then type an asterisk to tell it to multiply. Then type cos (2°3.14159/360°(, where the / tells the computer to divide. Click on cell B2, and the computer puts B2 (the value of the projection angle, still in degrees) into your formula. Finish the formula by closing both parentheses. Your formula in **ClarisWorks** should now read

$$= B1°COS(2°3.14159/360°(B2))$$

Press <return>and this formula is evaluated. To get the value for $v_{init,y}$, click in cell **D2** and do the same thing—except now you want sin:

$$= B1°SIN(2°3.14159/360°(B2))$$

Lotus 1–2–3 requires an "at" symbol (@) in front of all its trigonometric functions. So the formulas in **1–2–3** should now read

$$+ B1°@COS(2°3.14159/360°(B2))$$

$$+ B1°@SIN(2°3.14159/360°(B2))$$

Now your spreadsheet should look like Figure C.13.

	A	B	C	D
1	v(init) =	25	v(init,x) =	15.05
2	angle =	53	v(init,y) =	19.97
3	dt =			
4	time	horizontal	vertical	
5	0	0	0	
6				
7				

Figure C.13

We shall make calculations for every time interval *dt*. In cell **B3**, type 0.5 for the moment. We can change the value of *dt* later if we want to. In cell **A6**, type an equals sign to tell the computer to get ready for a calculation (plus sign in *1–2–3*). Then click in cell **A5** and in cell **B3** to create the formula for A6:

$$= A5 + B3$$

Press <return> and now your spreadsheet should look like Figure C.14.

	A	B	C	D
1	v(init) =	25	v(init,x) =	15.05
2	angle =	53	v(init,y) =	19.97
3	dt =	0.5		
4	time	horizontal	vertical	
5	0	0	0	
6	0.5			
7				

Figure C.14

See what happens if you simply COPY and PASTE from cell **A6** to cell **A7**. Cell **A7** will now read something like #VALUE!. Look at the formula for cell **A7**; it is

$$= A6 + B4$$

But cell **B4** is *text!* We want the old value of the time (the value in cell **A6**) and the value of the change in time *dt* (the value in cell **B3**). To keep B3 from changing in the formula, go back to the formula for **A6** and put dollar signs in front of the B and the 3 so that the formula reads

$$= A5 + \$B\$3$$

Now COPY and PASTE from cell **A6** to cell **A7**. This time cell **A7** reads 1, as it should. Look at the formula for cell **A7**; it is

$$= A6 + \$B\$3$$

The dollar signs tell the computer to keep cell **B3** absolute—in other words, not to change it relative to the new cell (as A5 was changed to A6). Continue with COPY and PASTE (or FILL DOWN if you like) until the time reaches 4.5 s.

Now we are ready to calculate the horizontal position, given by

$$x = x_{\text{init}} + v_{\text{init},x}t$$

Click in cell **B6**, and calculate the horizontal position for time $t = 0.5$ s. As always, an equal sign tells *ClarisWorks* a calculation is beginning (likewise, a plus sign tells *1–2–3* a calculation is beginning). Type the formula

$$= B5 + D1 \degree A6$$

In this formula, B5 is the initial *x* coordinate, D1 is the initial *x* speed, and A6 is the time. When you press <return>, the computer does the calculations and puts the number 7.52 in cell **B6**. When you make the next calculation, in cell

B7, however, only the time will have changed. To keep from changing the initial x coordinate and the initial x speed, make these absolute addresses instead of relative addresses. Go back to cell **B6** and change its formula to read

$$= \$B\$5 + \$D\$1 \circ A6$$

Now use COPY and PASTE (or FILL DOWN) to fill the rest of column **B**.

The vertical position is given by

$$y = y_{init} + v_{init,y}t + \tfrac{1}{2}gt^2$$

Click on cell **C6** and write its formula as

$$= C5 + D2 \circ A6 + (0.5) \circ (-9.8) \circ A6 \circ A6$$

After you press <return>, the number 8.76 should appear in cell **C6**. Before you can COPY and PASTE (or FILL DOWN) to use this formula throughout the column, you must change C5 to C5 and D2 to D2. Rewrite the formula as

$$= \$C\$5 + \$D\$2 \circ A6 + (0.5) \circ (-9.8) \circ A6 \circ A6$$

Now use COPY and PASTE (or FILL DOWN) to fill column **C**.

Now your spreadsheet should look like Figure C.15, which is a data table of the coordinates of an object undergoing projectile motion with an initial speed of 25 m/s at 53° above the horizontal. You can also let the spreadsheet turn these data into a graph. Click and drag to highlight the data to be graphed—from cell **B5** to cell **C14** in our case. In **ClarisWorks,** click on the OPTIONS choice on the menu bar and drag down to MAKE CHART . . . . (Physicists talk about "graphs" while some spreadsheets talk about "charts.") From the choices presented there, choose X–Y line and click on OK. You should then see the graph (or chart) shown in Figure C.16.

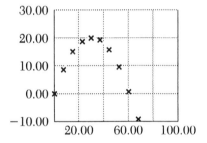

Figure C.16

	A	B	C	D
1	v(init) =	25	v(init,x) =	15.05
2	angle =	53	v(init,y) =	19.97
3	dt =	0.5		
4	time	horizontal	vertical	
5	0	0	0	
6	0.5	7.52	8.76	
7	1	15.05	15.07	
8	1.5	22.57	18.92	
9	2	30.09	20.33	
10	2.5	37.62	19.29	
11	3	45.14	15.80	
12	3.5	52.66	9.86	
13	4	60.20	1.48	
14	4.5	67.73	−9.38	

Figure C.15

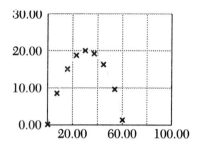

Figure C.17

We included cell **C14**, which contained the negative value of y. If you want to include only the data corresponding to $y \geq 0$, then highlight cells **B5** through **C13** and make a new graph, which should look like Figure C.17.

In *Lotus 1–2–3,* graphs like these are created by choosing GRAPH from the menu bar, dragging down to NEW GRAPH . . . , and choosing the icon showing a line graph.

With this spreadsheet set up to make the calculations for projectile motion, we can go back and change the value for v_{init}, and the coordinates will be immediately recalculated. Or we can change the value of the angle and the coordinates will be immediately recalculated. That is the power of a spreadsheet. Spreadsheets are useful in physics when repetitive calculations are required. What does the trajectory look like for $\theta = 15°$, $30°$, $45°$, $60°$, and $75°$ when the initial speed is kept constant at 25 m/s? As the value of the angle is changed in the spreadsheet, all the coordinates are recalculated and a new graph can be made and pasted into another file.

Spreadsheets are especially useful for the multiple-data-set problems that appear in most chapters in this book. Of course, computer programming languages—such as Pascal, FORTRAN, or BASIC (or TrueBASIC or QuickBASIC)—allow you to do the same calculations.

An Abbreviated Table of Isotopes

Atomic Number, Z	Element	Symbol	Mass Number, A	Atomic Mass[a]	Percent Abundance, or Decay Mode if Radioactive[b]	Half-life if Radioactive
0	(neutron)	n	1	1.008665	$\beta-$	10.6 min
1	hydrogen	H	1	1.007825	99.985	
	deuterium	D	2	2.014102	0.015	
	tritium	T	3	3.016049	$\beta-$	12.33 y
2	helium	He	3	3.016029	0.00014	
			4	4.002603	$=100$	
3	lithium	Li	6	6.015123	7.5	
			7	7.016005	92.5	
4	beryllium	Be	7	7.016930	EC, γ	53.3 days
			8	8.005305	2α	6.7×10^{-17} s
			9	9.012183	100	
5	boron	B	10	10.012938	19.8	
			11	11.009305	80.2	
6	carbon	C	11	11.011433	β^+, EC	20.4 min
			12	12.000000	98.89	
			13	13.003355	1.11	
			14	14.003242	$\beta-$	5730 y
7	nitrogen	N	13	13.005739	β^+	9.96 min
			14	14.003074	99.63	
			15	15.000109	0.37	
8	oxygen	O	15	15.003065	β^+, EC	122 s
			16	15.994915	99.76	
			18	17.999159	0.204	
9	fluorine	F	19	18.998403	100	
10	neon	Ne	20	19.992439	90.51	
			22	21.991384	9.22	
11	sodium	Na	22	21.994435	β^+, EC, γ	2.602 y
			23	22.989770	100	
			24	23.990964	β^-, γ	15.0 h
12	magnesium	Mg	24	23.985045	78.99	
13	aluminum	Al	27	26.981541	100	

Atomic Number, Z	Element	Symbol	Mass Number, A	Atomic Mass[a]	Percent Abundance, or Decay Mode if Radioactive[b]	Half-life if Radioactive
14	silicon	Si	28	27.976928	92.23	
			31	30.975364	β^-, γ	2.62 h
15	phosphorus	P	31	30.973763	100	
			32	31.973908	β^-	14.28 days
16	sulfur	S	32	31.972072	95.0	
			35	34.969033	β^-	87.4 days
17	chlorine	Cl	35	34.968853	75.77	
			37	36.965903	24.23	
18	argon	Ar	40	39.962383	99.60	
19	potassium	K	39	38.963708	93.26	
			40	39.964000	β^-, EC, γ, β^+	1.28×10^9 y
20	calcium	Ca	40	39.962591	96.94	
21	scandium	Sc	45	44.955914	100	
22	titanium	Ti	48	47.947947	73.7	
23	vanadium	V	51	50.943963	99.75	
24	chromium	Cr	52	51.940510	83.79	
25	manganese	Mn	55	54.938046	100	
26	iron	Fe	56	55.934939	91.8	
27	cobalt	Co	59	58.933198	100	
			60	59.933820	β^-, γ	5.271 y
28	nickel	Ni	58	57.935347	68.3	
			60	59.930789	26.1	
			64	63.927968	0.91	
29	copper	Cu	63	62.929599	69.2	
			64	63.929766	β^-, β^+	12.7 h
			65	64.927792	30.8	
30	zinc	Zn	64	63.929145	48.6	
			66	65.926035	27.9	
31	gallium	Ga	69	68.925581	60.1	
32	germanium	Ge	72	71.922080	27.4	
			74	73.921179	36.5	
33	arsenic	As	75	74.921596	100	
34	selenium	Se	80	79.916521	49.8	
35	bromine	Br	79	78.918336	50.69	
36	krypton	Kr	84	83.911506	57.0	
			89	88.917563	β^-	3.2 min
37	rubidium	Rb	85	84.911800	72.17	
38	strontium	Sr	86	85.909273	9.8	
			88	87.905625	82.6	
			90	89.907746	β^-	28.8 y
39	yttrium	Y	89	88.905856	100	

Atomic Number, Z	Element	Symbol	Mass Number, A	Atomic Mass[a]	Percent Abundance, or Decay Mode if Radioactive[b]	Half-life if Radioactive
40	zirconium	Zr	90	89.904708	51.5	
41	niobium	Nb	93	92.906378	100	
42	molybdenum	Mo	98	97.905405	24.1	
43	technetium	Tc	98	97.907210	β^-, γ	4.2×10^6 y
44	ruthenium	Ru	102	101.904348	31.6	
45	rhodium	Rh	103	102.90550	100	
46	palladium	Pd	106	105.90348	27.3	
47	silver	Ag	107	106.905095	51.83	
			109	108.904754	48.17	
48	cadmium	Cd	114	113.903361	28.7	
49	indium	In	115	114.90388	95.7; β^-	5.1×10^{14} y
50	tin	Sn	120	119.902199	32.4	
51	antimony	Sb	121	120.903824	57.3	
52	tellurium	Te	130	129.90623	34.5; β^-	2×10^{21} y
53	iodine	I	127	126.904477	100	
			131	130.906118	β^-, γ	8.04 days
54	xenon	Xe	132	131.90415	26.9	
			136	135.90722	8.9	
55	cesium	Cs	133	132.90543	100	
56	barium	Ba	137	136.90582	11.2	
			138	137.90524	71.7	
			144	143.922673	β^-	11.9 s
57	lanthanum	La	139	138.90636	99.911	
58	cerium	Ce	140	139.90544	88.5	
59	praesodymium	Pr	141	140.90766	100	
60	neodymium	Nd	142	141.90773	27.2	
			144	143.910096	α, 23.8	2.1×10^{15} y
61	promethium	Pm	145	144.91275	EC, α, γ	17.7 years
62	samarium	Sm	152	151.91974	26.6	
63	europium	Eu	153	152.92124	52.1	
64	gadolinium	Gd	158	157.92411	24.8	
65	terbium	Tb	159	158.92535	100	
66	dysprosium	Dy	164	163.92918	28.1	
67	holmium	Ho	165	164.93033	100	
68	erbium	Er	166	165.93031	33.4	
69	thulium	Tm	169	168.93423	100	
70	ytterbium	Yb	174	173.93887	31.6	
71	lutecium	Lu	175	174.94079	97.39	
72	hafnium	Hf	180	179.94656	35.2	
73	tantalum	Ta	181	180.94801	99.988	
74	tungsten	W	184	183.95095	30.7	

Atomic Number, Z	Element	Symbol	Mass Number, A	Atomic Mass[a]	Percent Abundance, or Decay Mode if Radioactive[b]	Half-life if Radioactive
75	rhenium	Re	187	186.95577	62.60, β^-	4×10^{10} y
76	osmium	Os	191	190.96094	β^-, γ	15.4 days
			192	191.96149	41.0	
77	iridium	Ir	191	190.96060	37.3	
			193	192.96294	62.7	
78	platinum	Pt	195	194.96479	33.8	
79	gold	Au	197	196.96656	100	
80	mercury	Hg	202	201.97063	29.8	
81	thallium	Tl	205	204.97441	70.5	
			210	209.990069	β^-	1.3 min
82	lead	Pb	204	203.973044	β^-, 1.48	1.4×10^{17} y
			206	205.97446	24.1	
			207	206.97589	22.1	
			208	207.97664	52.3	
			210	209.98418	α, β^-, γ	22.3 y
			211	210.98874	β^-, γ	36.1 min
			212	211.99188	β^-, γ	10.64 h
			214	213.99980	β^-, γ	26.8 min
83	bismuth	Bi	209	208.98039	100	
			211	210.98726	α, β^-, γ	2.15 min
			214	213.998702	β^-, α	19.7 min
84	polonium	Po	210	209.98286	α, γ	138.38 days
			214	213.99519	α, γ	164 μs
85	astatine	At	218	218.00870	α, β^-	$\approx$2 s
86	radon	Rn	222	222.017574	α, γ	3.8235 days
87	francium	Fr	223	223.019734	α, β^-, γ	21.8 min
88	radium	Ra	226	226.025406	α, γ	1.60×10^3 y
			228	228.031069	β^-	5.76 years
89	actinium	Ac	227	227.027751	α, β^-, γ	21.773 years
90	thorium	Th	228	228.02873	α, γ	1.9131 years
			232	232.038054	100, α, γ	1.41×10^{10} y
91	protactinium	Pa	231	231.035881	α, γ	3.28×10^4 y
92	uranium	U	232	232.03714	α, γ	72 years
			233	233.039629	α, γ	1.592×10^5 y
			235	235.043925	0.72; α, γ	7.038×10^8 y
			236	236.045563	α, γ	2.342×10^7 y
			238	238.050786	99.275; α, γ	4.468×10^9 y
			239	239.054291	β^-, γ	23.5 min
93	neptunium	Np	239	239.052932	β^-, γ	2.35 days
94	plutonium	Pu	239	239.052158	α, γ	2.41×10^4 y
95	americium	Am	243	243.061374	α, γ	7.37×10^3 y

Atomic Number, Z	Element	Symbol	Mass Number, A	Atomic Mass[a]	Percent Abundance, or Decay Mode if Radioactive[b]	Half-life if Radioactive
96	curium	Cm	245	245.065487	α, γ	8.5×10^3 y
97	berkelium	Bk	247	247.07003	α, γ	1.4×10^3 y
98	californium	Cf	249	249.074849	α, γ	351 years
99	einsteinium	Es	254	254.08802	α, γ, β^-	276 days
100	fermium	Fm	253	253.08518	EC, α, γ	3.0 days
101	mendelevium	Md	255	255.0911	EC, α	27 min
102	nobelium	No	255	255.0933	EC, α	3.1 min
103	lawrencium	Lr	257	257.0998	α	$\approx$35 s
104	rutherfordium	Rf	261	261.1087	α	1.1 min
105	hahnium	Ha	262	262.1138	α	0.7 min
106			263	263.1184	α	0.9 s
107	nielsbohrium	Ns	262		α	1–2 m
108	hassium	Hs	265			
109	meitnerium	Mt	266			

Note: Data taken from *Chart of the Nuclides*, 12th ed., New York, General Electric, 1977, and from C. M. Lederer and V. S. Shirley (eds.), *Table of Isotopes*, 7th ed., New York, Wiley, 1978.

[a] Masses are those for the neutral atom, including the Z electrons.

[b] The process EC stands for "electron capture."

Information for elements 104–109 courtesy of Paul J. Karol, Carnegie-Mellon University, of the American Chemical Society Nomenclature Committee.

APPENDIX E

Some Useful Tables

TABLE E.1 Mathematical Symbols Used in the Text and Their Meaning	
Symbol	**Meaning**
$=$	is equal to
$\neq$	is not equal to
$\equiv$	is defined as
$\propto$	is proportional to
$>$	is greater than
$<$	is less than
$\gg$	is much greater than
$\ll$	is much less than
$\approx$	is approximately equal to
Δx	change in x
Σx_i	sum of all quantities x_i
$\lvert x \rvert$	magnitude of x (always a positive quantity)

TABLE E.2 Standard Abbreviations of Units			
Abbreviation	**Unit**	**Abbreviation**	**Unit**
A	ampere	kcal	kilocalorie
atm	atmosphere	kg	kilogram
Btu	British thermal unit	km	kilometer
C	coulomb	kmol	kilomole
°C	degree Celsius	lb	pound
cal	calorie	m	meter
cm	centimeter	min	minute
eV	electron volt	N	newton
°F	degree Fahrenheit	rev	revolution
ft	foot	s	second
G	gauss	T	tesla
g	gram	u	atomic mass unit
H	henry	V	volt
h	hour	W	watt
hp	horsepower	Wb	weber
Hz	hertz	μm	micrometer
in.	inch	Ω	ohm
J	joule		

TABLE E.3 The Greek Alphabet

Alpha	A	α	Iota	I	ι	Rho	P	ρ
Beta	B	β	Kappa	K	κ	Sigma	Σ	σ
Gamma	Γ	γ	Lambda	Λ	λ	Tau	T	τ
Delta	Δ	δ	Mu	M	μ	Upsilon	Υ	υ
Epsilon	E	ϵ	Nu	Nu	ν	Phi	Φ	ϕ
Zeta	Z	ζ	Xi	Ξ	ξ	Chi	X	χ
Eta	H	η	Omicron	O	o	Psi	Ψ	ψ
Theta	Θ	θ	Pi	Π	π	Omega	Ω	ω

TABLE E.4 Conversion Factors

Length
1 m = 39.37 in. = 3.281 ft
1 in. = 2.54 cm
1 km = 0.621 mi
1 mi = 5280 ft = 1.609 km
1 light-year = 9.461×10^{15} m

Mass
1 kg = 10^3 g = 6.85×10^{-2} slug
1 slug = 14.59 kg
1 u = 1.66×10^{-27} kg

Time
1 min = 60 s
1 h = 3600 s
1 day = 8.64×10^4 s
1 year = 365.242 days = 3.156×10^7 s

Volume
1 liter = 1000 cm^3 = 3.531×10^{-2} ft^3
1 ft^3 = 2.832×10^{-2} m^3
1 gallon = 3.786 liter = 231 in.^{3}

Angle
$180° = \pi$ rad
1 rad = 57.30°
1° = 60 min = 1.745×10^{-2} rad

Speed
1 km/h = 0.278 m/s = 0.621 mi/h
1 m/s = 2.237 mi/h = 3.281 ft/s
1 mi/h = 1.61 km/h = 0.447 m/s = 1.467 ft/s

Force
1 N = 0.2248 lb = 10^5 dynes
1 lb = 4.448 N
1 dyne = 10^{-5} N = 2.248×10^{-6} lb

Work and energy
1 J = 10^7 erg = 0.738 ft · lb = 0.239 cal
1 cal = 4.186 J
1 ft · lb = 1.356 J
1 Btu = 1.054×10^3 J = 252 cal
1 J = 6.24×10^{18} eV
1 eV = 1.602×10^{-19} J
1 kWh = 3.60×10^6 J

Pressure
1 atm = 1.013×10^5 N/m^2 (or Pa) = 14.70 lb/in.^{2}
1 Pa = 1 N/m^2 = 1.45×10^{-4} lb/in.^{2}
1 lb/in.^{2} = 6.895×10^3 N/m^2

Power
1 hp = 550 ft · lb/s = 0.746 kW
1 W = 1 J/s = 0.738 ft · lb/s
1 Btu/h = 0.293 W

TABLE E.5 SI Base Units

Base Quantity	SI Base Unit	
	Name	**Symbol**
Length	Meter	m
Mass	Kilogram	kg
Time	Second	s
Electric current	Ampere	A
Temperature	Kelvin	K
Amount of substance	Mole	mol
Luminous intensity	Candela	cd

TABLE E.6 Derived SI Units

Quantity	Name	Symbol	Expression in Terms of Base Units	Expression in Terms of Other SI Units
Plane angle	Radian	rad	m/m	
Frequency	Hertz	Hz	s^{-1}	
Force	Newton	N	$kg \cdot m/s^2$	J/m
Pressure	Pascal	Pa	$kg/m \cdot s^2$	N/m^2
Energy and work	Joule	J	$kg \cdot m^2/s^2$	$N \cdot m$
Power	Watt	W	$kg \cdot m^2/s^3$	J/s
Electric charge	Coulomb	C	$A \cdot s$	
Electric potential (emf)	Volt	V	$kg \cdot m^2/A \cdot s^3$	W/A
Capacitance	Farad	F	$A^2 \cdot s^4/kg \cdot m^2$	C/A
Electric resistance	Ohm	Ω	$kg \cdot m^2/A^2 \cdot s^3$	V/A
Magnetic flux	Weber	Wb	$kg \cdot m^2/A \cdot s^2$	$V \cdot s$
Magnetic field intensity	Tesla	T	$kg/A \cdot s^2$	Wb/m^2
Inductance	Henry	H	$kg \cdot m^2/A^2 \cdot s^2$	Wb/A

Index

Note: Page numbers followed by n indicate footnotes.